Recent Advances on Nano-Catalysts for Biological Processes

Recent Advances on Nano-Catalysts for Biological Processes

Editors

Pritam Kumar Dikshit
Beom Soo Kim

MDPI

Basel • Beijing • Wuhan • Barcelona • Belgrade • Novi Sad • Cluj • Manchester

Editors
Pritam Kumar Dikshit
Department of Biotechnology
K. L. University
Andhra Pradesh
India

Beom Soo Kim
Dept. of Chemical Engineering
Chungbuk National University
Cheongju
Korea, South

Editorial Office
MDPI
St. Alban-Anlage 66
4052 Basel, Switzerland

This is a reprint of articles from the Special Issue published online in the open access journal *Catalysts* (ISSN 2073-4344) (available at: www.mdpi.com/journal/catalysts/special_issues/nano_cata_bio).

For citation purposes, cite each article independently as indicated on the article page online and as indicated below:

Lastname, A.A.; Lastname, B.B. Article Title. *Journal Name* **Year**, *Volume Number*, Page Range.

ISBN 978-3-7258-0214-2 (Hbk)
ISBN 978-3-7258-0213-5 (PDF)
doi.org/10.3390/books978-3-7258-0213-5

Contents

About the Editors . . . . **vii**

Preface . . . . **ix**

Pritam Kumar Dikshit and Beom Soo Kim
Recent Advances on Nano-Catalysts for Biological Processes
Reprinted from: *Catalysts* **2022**, *13*, 27, doi:10.3390/catal13010027 . . . . **1**

Pritam Kumar Dikshit, Jatin Kumar, Amit K. Das, Soumi Sadhu, Sunita Sharma and Swati Singh et al.
Green Synthesis of Metallic Nanoparticles: Applications and Limitations
Reprinted from: *Catalysts* **2021**, *11*, 902, doi:10.3390/catal11080902 . . . . **3**

Babita Tripathi, Soumya Pandit, Aparna Sharma, Sunil Chauhan, Abhilasha Singh Mathuriya and Pritam Kumar Dikshit et al.
Modification of Graphite Sheet Anode with Iron (II, III) Oxide-Carbon Dots for Enhancing the Performance of Microbial Fuel Cell
Reprinted from: *Catalysts* **2022**, *12*, 1040, doi:10.3390/catal12091040 . . . . **38**

Indrajeet Arya, Asha Poona, Pritam Kumar Dikshit, Soumya Pandit, Jatin Kumar and Himanshu Narayan Singh et al.
Current Trends and Future Prospects of Nanotechnology in Biofuel Production
Reprinted from: *Catalysts* **2021**, *11*, 1308, doi:10.3390/catal11111308 . . . . **54**

Mamata Singhvi, Minseong Kim and Beom-Soo Kim
Production of Therapeutically Significant Genistein and Daidzein Compounds from Soybean Glycosides Using Magnetic Nanocatalyst: A Novel Approach
Reprinted from: *Catalysts* **2022**, *12*, 1107, doi:10.3390/catal12101107 . . . . **73**

Lalit Goswami, Anamika Kushwaha, Anju Singh, Pathikrit Saha, Yoseok Choi and Mrutyunjay Maharana et al.
Nano-Biochar as a Sustainable Catalyst for Anaerobic Digestion: A Synergetic Closed-Loop Approach
Reprinted from: *Catalysts* **2022**, *12*, 186, doi:10.3390/catal12020186 . . . . **91**

Samie Yaseen Sharaf Zeebaree, Osama Ismail Haji, Aymn Yaseen Sharaf Zeebaree, Dunya Akram Hussein and Emad Hameed Hanna
Rapid Detection of Mercury Ions Using Sustainable Natural Gum-Based Silver Nanoparticles
Reprinted from: *Catalysts* **2022**, *12*, 1464, doi:10.3390/catal12111464 . . . . **114**

Jai Prakash, Suresh Babu Naidu Krishna, Promod Kumar, Vinod Kumar, Kalyan S. Ghosh and Hendrik C. Swart et al.
Recent Advances on Metal Oxide Based Nano-Photocatalysts as Potential Antibacterial and Antiviral Agents
Reprinted from: *Catalysts* **2022**, *12*, 1047, doi:10.3390/catal12091047 . . . . **132**

Melvin S. Samuel, Madhumita Ravikumar, Ashwini John J., Ethiraj Selvarajan, Himanshu Patel and P. Sharath Chander et al.
A Review on Green Synthesis of Nanoparticles and Their Diverse Biomedical and Environmental Applications
Reprinted from: *Catalysts* **2022**, *12*, 459, doi:10.3390/catal12050459 . . . . **161**

Arpita Roy, Amin Elzaki, Vineet Tirth, Samih Kajoak, Hamid Osman and Ali Algahtani et al.
Biological Synthesis of Nanocatalysts and Their Applications
Reprinted from: *Catalysts* **2021**, *11*, 1494, doi:10.3390/catal11121494 . . . . **185**

Ebrahim Saied, Ahmed M. Eid, Saad El-Din Hassan, Salem S. Salem, Ahmed A. Radwan and Mahmoud Halawa et al.
The Catalytic Activity of Biosynthesized Magnesium Oxide Nanoparticles (MgO-NPs) for Inhibiting the Growth of Pathogenic Microbes, Tanning Effluent Treatment, and Chromium Ion Removal
Reprinted from: *Catalysts* **2021**, *11*, 821, doi:10.3390/catal11070821 . . . . **207**

Shah Faisal, Abdullah, Hasnain Jan, Sajjad Ali Shah, Sumaira Shah and Muhammad Rizwan et al.
Bio-Catalytic Activity of Novel *Mentha arvensis* Intervened Biocompatible Magnesium Oxide Nanomaterials
Reprinted from: *Catalysts* **2021**, *11*, 780, doi:10.3390/catal11070780 . . . . **228**

About the Editors

Pritam Kumar Dikshit

Dr. Pritam Kumar Dikshit is currently working as Assistant Professor at the Department of Biotechnology, K. L. Deemed to be University, Guntur (AP). He earned his Ph.D from IIT Guwahati, India. Prior to Ph.D, he has obtained his B.Tech. and M.Tech. with specialization in Biotechnology from Biju Patnaik University of Technology, Odisha and Karunya University, Tamil Nadu, respectively.

After the completion of his PhD, he joined as Research Associate at The Energy and Resources Institute (TERI), New Delhi, and worked in the area of biological wastewater treatment coupled with membrane filtration process. In 2019, he joined as Postdoctoral researcher at Chungbuk National University, the Republic of Korea, under the Brain Korea 21 plus fellowship. Earlier, he was working as an Assistant Professor at Sharda University, Greater Noida (UP).

He has published several peer-reviewed paper in various journals and was also one of the recipients of 7th National Award from Ministry of Chemicals and Fertilizers, Department of Chemicals and Petrochemicals, Govt. of India. He is also recipient of Prof. Ashok Pandey Young Scientist Award from International Society for Energy, Environment, and Sustainability (ISEES).

His major area of interests include waste biomass valorization, biofuel and biopolymer production, biological wastewater treatment, and nanotechnology.

Beom Soo Kim

Beom Soo Kim is a professor of chemical engineering at Chungbuk National University, Cheongju, Korea. He studied chemical engineering at Seoul National University (1988), obtained a PhD in biochemical engineering at KAIST (1993), completed postdoctoral work at MIT Prof. Robert Langers lab (1998). He started his lab at Chungbuk National University (2001) and spent sabbatical research in Dr. Ching T. Hou's lab at National Center for Agricultural Utilization Research (NCAUR), United States Department of Agriculture (USDA), Peoria, Illinois (2005). He served as Editor-in-Chief for the *Korean Society for Biotechnology and Bioengineering Journal* and is an Editorial Board member of several journals such as *BioMed Research International, Biocatalysis and Agricultural Biotechnology, Korean Journal of Chemical Engineering, Biotechnology and Bioprocess Engineering, Polymers,* and *BMC Biotechnology*. He published over 150 papers, and one of his articles on the biosynthesis of silver nanoparticles has been cited more than 1200 times (Google scholar). His research interests include high-cell-density culture, biodegradable polymers, polyhydroxyalkanoates, biosynthesis and applications of nanomaterials, and biorefinery.

Preface

The use of nanomaterials has become increasingly important in recent years due to their wide range of applications in various sectors ranging from sensor technology to biomedicine and energy conversion. Nanoparticles with a size of 100 nm or less have garnered great research attention due to their high surface-to-volume ratio and unique properties. Various methods such as physical, chemical, biological, and hybrid methods are available for the synthesis of these nanoparticles. However, the use of reliable, non-toxic, and eco-friendly technologies for the synthesis of nanoparticles is of utmost importance to expand their biological applications. Several biological applications of nanoparticles can be listed, e.g., carbohydrate hydrolysis, production of biofuel, immobilization of enzymes, biotransformation, gene and drug delivery, and the detection of pathogens and proteins. Recently, various nanocarriers have also been used for the immobilization of different enzymes to produce nanobiocatalysts (NBCs) which further enhance enzyme performance.

The present Special Issue entitled "Recent Advances on Nano-Catalysts for Biological Processes" of the journal *Catalysts* features review articles as well as original research articles on the application of nanomaterials in various biological processes. The objective of the Special Issue was to highlight the recent works on this area and make it available to researchers with similar research interests. The authors' significant contributions to this Special Issue are appreciated, and we hope that the readers enjoy the content.

Pritam Kumar Dikshit and Beom Soo Kim
Editors

catalysts MDPI

Editorial

Recent Advances on Nano-Catalysts for Biological Processes

Pritam Kumar Dikshit [1,*] and Beom Soo Kim [2]

[1] Department of Biotechnology, Koneru Lakshmaiah Education Foundation, Vaddeswaram 522302, India
[2] Department of Chemical Engineering, Chungbuk National University, Cheongju 28644, Republic of Korea
* Correspondence: biotech.pritam@gmail.com

We are honored to serve as the Guest Editors of this Special Issue entitled "Recent Advances on Nano-Catalysts for Biological Processes" for the journal *Catalysts*. With the increasing demand for nanoparticles and their applications in various sectors, this Special Issue focuses primarily on the catalytic application of nanoparticles in various biological processes such as wastewater treatment, dark fermentation, biofuel production, biomass pretreatment processes, production of other value-added products, and so forth. In addition to nanoparticle applications, this Special Issue also covers the green synthesis of nanoparticles using various biological sources.

This Special Issue includes eleven articles in total, out of which five are research articles [1–5] and six are reviews [6–11]. The research article by Sharaf Zeebaree et al. [1] focuses on the colorimetric detection of mercury in water samples using natural gum-based silver nanoparticles. Natural exudate (almond gum) was used as the reducing and stabilizing agents for the production of Ag nanoparticles, which were later characterized using various analytical techniques. Singhvi et al. [2] investigated the hydrolysis of soybean-extracted glycosides using an acid-functionalized magnetic cobalt ferrite alkyl sulfonic acid ($CoFe_2O_4$-Si-ASA) nanocatalyst for the production of aglycones, i.e., daidzein and genistein. Higher conversion efficiency was achieved in the presence of nanocatalysts in comparison to the control experiment containing enzymes. Furthermore, these nanoparticles can be easily recovered from the reaction mixture using an external magnetic field and can be reused in subsequent cycle. Tripathi et al. [3] improved the performance of microbial fuel cells (MFCs) using a modified graphite sheet anode. Modification of the anode was carried out using iron (II, III) oxide (Fe_3O_4) carbon dots, which enhance the performance of MFC. The study conducted by Saied et al. [4] focused on the synthesis of MgO nanoparticles using the *Aspergillus terreus* fungal strain and its potential as an antimicrobial agent, in the treatment of tanning effluent, and in chromium ion removal was investigated. In a similar study, Faisal et al. [5] synthesized MgO using a leaf extract of *Mentha arvensis*. The synthesized nanoparticles were shown to possess good antimicrobial and antioxidant activity. In addition to this, the anti-Alzheimer, anti-cancer, and anti-*Helicobacter pylori* activities of the synthesized nanoparticles were studied.

Apart from the original research work of the aforementioned individuals, several review articles highlighting the current trends in nanoparticle synthesis and its applications in various sectors are also included in this Special Issue. Green synthesis of various nanoparticles and its application in antibacterial and antiviral agents [6], in biomedical and environmental applications [7], dye degradation and heavy metal removal [8], biofuel production [9], and its limitations [10] are thoroughly discussed in these reviews. Furthermore, Goswami et al. [11] reviewed the application of nano-biochar as a catalyst in the process of anaerobic digestion. In addition to this, the techno-economic analysis and life-cycle assessment of nano-biochar-aided anaerobic digestion were discussed in detail. Overall, this Special Issue covers the diverse biological applications of nanoparticles along with their synthesis methods.

Citation: Dikshit, P.K.; Kim, B.S. Recent Advances on Nano-Catalysts for Biological Processes. *Catalysts* **2023**, *13*, 27. https://doi.org/10.3390/catal13010027

Received: 14 December 2022
Accepted: 20 December 2022
Published: 24 December 2022

Data Availability Statement: Not applicable.

Acknowledgments: We would like to thank Keith Hohn, the Editor-in-Chief of *Catalysts*, for providing us with the opportunity to lead this Special Issue as Guest Editors. We are especially thankful to all the authors for submitting their high-quality research in this special issue and the anonymous reviewers for their time and effort in reviewing the manuscripts.

Conflicts of Interest: The authors declare no conflict of interest.

References

1. Zeebaree, S.Y.S.; Haji, O.I.; Zeebaree, A.Y.S.; Hussein, D.A.; Hanna, E.H. Rapid Detection of Mercury Ions Using Sustainable Natural Gum-Based Silver Nanoparticles. *Catalysts* **2022**, *12*, 1464. [CrossRef]
2. Singhvi, M.; Kim, M.; Kim, B.S. Production of Therapeutically Significant Genistein and Daidzein Compounds from Soybean Glycosides Using Magnetic Nanocatalyst: A Novel Approach. *Catalysts* **2022**, *12*, 1107. [CrossRef]
3. Tripathi, B.; Pandit, S.; Sharma, A.; Chauhan, S.; Mathuriya, A.S.; Dikshit, P.K.; Gupta, P.K.; Singh, R.C.; Sahni, M.; Pant, K.; et al. Modification of Graphite Sheet Anode with Iron (II, III) Oxide-Carbon Dots for Enhancing the Performance of Microbial Fuel Cell. *Catalysts* **2022**, *12*, 1040. [CrossRef]
4. Saied, E.; Eid, A.M.; Hassan, S.E.D.; Salem, S.S.; Radwan, A.A.; Halawa, M.; Saleh, F.M.; Saad, H.A.; Saied, E.M.; Fouda, A. The catalytic activity of biosynthesized magnesium oxide nanoparticles (Mgo-nps) for inhibiting the growth of pathogenic microbes, tanning effluent treatment, and chromium ion removal. *Catalysts* **2021**, *11*, 821. [CrossRef]
5. Faisal, S.; Jan, H.; Shah, S.A.; Shah, S.; Rizwan, M.; Zaman, N.; Hussain, Z.; Uddin, M.N.; Bibi, N.; Khattak, A.; et al. Bio-catalytic activity of novel Mentha arvensis intervened biocompatible magnesium oxide nanomaterials. *Catalysts* **2021**, *11*, 780. [CrossRef]
6. Prakash, J.; Krishna, S.B.N.; Kumar, P.; Kumar, V.; Ghosh, K.S.; Swart, H.C.; Bellucci, S.; Cho, J. Recent advances on metal oxide based nano-photocatalysts as potential antibacterial and antiviral agents. *Catalysts* **2022**, *12*, 1047. [CrossRef]
7. Samuel, M.S.; Ravikumar, M.; John, J.A.; Selvarajan, E.; Patel, H.; Chander, P.S.; Soundarya, J.; Vuppala, S.; Balaji, R.; Chandrasekar, N. A Review on Green Synthesis of Nanoparticles and Their Diverse Biomedical and Environmental Applications. *Catalysts* **2022**, *12*, 459. [CrossRef]
8. Roy, A.; Elzaki, A.; Tirth, V.; Kajoak, S.; Osman, H.; Algahtani, A.; Islam, S.; Faizo, N.L.; Khandaker, M.U.; Islam, M.N.; et al. Biological synthesis of nanocatalysts and their applications. *Catalysts* **2021**, *11*, 1494. [CrossRef]
9. Arya, I.; Poona, A.; Dikshit, P.K.; Pandit, S.; Kumar, J.; Singh, H.N.; Jha, N.K.; Rudayni, H.A.; Chaudhary, A.A.; Kumar, S. Current trends and future prospects of nanotechnology in biofuel production. *Catalysts* **2021**, *11*, 1308. [CrossRef]
10. Dikshit, P.K.; Kumar, J.; Das, A.K.; Sadhu, S.; Sharma, S.; Singh, S.; Gupta, P.K.; Kim, B.S. Green synthesis of metallic nanoparticles: Applications and limitations. *Catalysts* **2021**, *11*, 902. [CrossRef]
11. Goswami, L.; Kushwaha, A.; Singh, A.; Saha, P.; Choi, Y.; Maharana, M.; Patil, S.V.; Kim, B.S. Nano-biochar as a sustainable catalyst for anaerobic digestion: A synergetic closed-loop approach. *Catalysts* **2022**, *12*, 186. [CrossRef]

catalysts

MDPI

Review

Green Synthesis of Metallic Nanoparticles: Applications and Limitations

Pritam Kumar Dikshit [1,*], Jatin Kumar [1], Amit K. Das [1], Soumi Sadhu [1], Sunita Sharma [2], Swati Singh [1], Piyush Kumar Gupta [1] and Beom Soo Kim [3,*]

[1] Department of Life Sciences, School of Basic Sciences and Research, Sharda University, Greater Noida 201310, Uttar Pradesh, India; jatin.kumar1@sharda.ac.in (J.K.); amit.das@sharda.ac.in (A.K.D.); soumi.sadhu@sharda.ac.in (S.S.); swati.singh1@sharda.ac.in (S.S.); piyush.kumar1@sharda.ac.in (P.K.G.)
[2] Department of Biotechnology, School of Engineering and Technology, Sharda University, Greater Noida 201310, Uttar Pradesh, India; sunita.sharma@sharda.ac.in
[3] Department of Chemical Engineering, Chungbuk National University, Cheongju 28644, Korea
* Correspondence: pritam.kumar@sharda.ac.in (P.K.D.); bskim@chungbuk.ac.kr (B.S.K.)

Abstract: The past decade has witnessed a phenomenal rise in nanotechnology research due to its broad range of applications in diverse fields including food safety, transportation, sustainable energy, environmental science, catalysis, and medicine. The distinctive properties of nanomaterials (nano-sized particles in the range of 1 to 100 nm) make them uniquely suitable for such wide range of functions. The nanoparticles when manufactured using green synthesis methods are especially desirable being devoid of harsh operating conditions (high temperature and pressure), hazardous chemicals, or addition of external stabilizing or capping agents. Numerous plants and microorganisms are being experimented upon for an eco–friendly, cost–effective, and biologically safe process optimization. This review provides a comprehensive overview on the green synthesis of metallic NPs using plants and microorganisms, factors affecting the synthesis, and characterization of synthesized NPs. The potential applications of metal NPs in various sectors have also been highlighted along with the major challenges involved with respect to toxicity and translational research.

Keywords: green synthesis; metal nanoparticles; wastewater treatment; agriculture; food application

Citation: Dikshit, P.K.; Kumar, J.; Das, A.K.; Sadhu, S.; Sharma, S.; Singh, S.; Gupta, P.K.; Kim, B.S. Green Synthesis of Metallic Nanoparticles: Applications and Limitations. *Catalysts* **2021**, *11*, 902. https://doi.org/10.3390/catal11080902

Academic Editor: Pierluigi Barbaro

Received: 20 June 2021
Accepted: 23 July 2021
Published: 26 July 2021

1. Introduction

During the last two decades, nanotechnology has taken massive leaps to become one of the most researched and booming fields due to its applications in various fields of human welfare. Nanoparticles (NPs) are naturally occurring or engineered extremely small sized particles in the range of 1 to 100 nm. They exhibit unique and valuable physical and chemical properties. At nanoscale, particles display better catalytic, magnetic, electrical, mechanical, optical, chemical, and biological properties. Due to high surface to volume ratio, NPs show higher reactivity, mobility, dissolution properties, and strength [1]. NPs are thought to have been present on earth since its origin in the form of soil, water, volcanic dust, and minerals. Besides their natural origin, humans have also started synthesizing NPs through various methods [2]. NPs and their derived nanomaterials are finding wide application in various sectors such as food, agriculture, cosmetics, medicines, etc. Application of NPs in food sector involves food processing and preservation (nanopreservatives, toxin detection, nanoencapsulated food additives, etc.) and food packaging (nanocoatings, nanosensors, nanocomposites, edible coating NPs, etc.) In agriculture, nanotechnology is being utilized for the production of nano-fertilizers, pesticides, herbicides, and sensors. In medicine, nanotechnology involves production of various antibacterial, antifungal, antiplasmodial, anti–inflammatory, anticancer, antiviral, antidiabetic, and antioxidant agents. Nanotechnology is also useful for the early detection of life-threatening diseases such as cancer. Besides, NPs have also been used for bioremediation due to their capacity to degrade various pollutants such as organic dyes and chemicals. Given the diverse scope of nanomaterials, different countries are investing in nanotechnology with USA and China emerging at the top. In 2019, the global market of different nano products was more than 8 billion US dollars, which is expected to show annual growth rate of around 13% by 2027.

Depending on their chemical composition, four major classes of NPs are described, such as carbon-based (nanotubes and nanofibers of carbon, etc.), metal and metal oxide based (Ag, Cu, etc.), bio-organic based (liposomes, micelles, etc.), and composite based [3]. NPs can also be classified as organic and inorganic in nature [4]. Organic NPs are biodegradable in nature and include polymeric NPs, lipid based nanocarriers, liposomes, carbon-based nanomaterials, and solid lipid NPs, while inorganic NPs are based on inorganic materials comprising of metals and metal oxides such as silver oxide, zinc oxide, etc. Among all the synthesized NPs, silver NPs (Ag NPs) are the most widely employed, showing their dominance in various consumer products (more than 25%) [5]. AgNPs are majorly used as antibacterial, antifungal, and antiviral agents. With each passing year, novel varieties of NPs are being developed using state-of-art technology having diverse applications in various sectors.

The synthesis of NPs can be carried out following two different approaches, viz., (i) top-down approach, and (ii) bottom-up approach [6,7]. Furthermore, three different strategies such as physical, chemical, and biological methods are adopted for the synthesis of NPs. A schematic representation of various methods adopted for NPs synthesis and its applications is depicted in Figure 1.

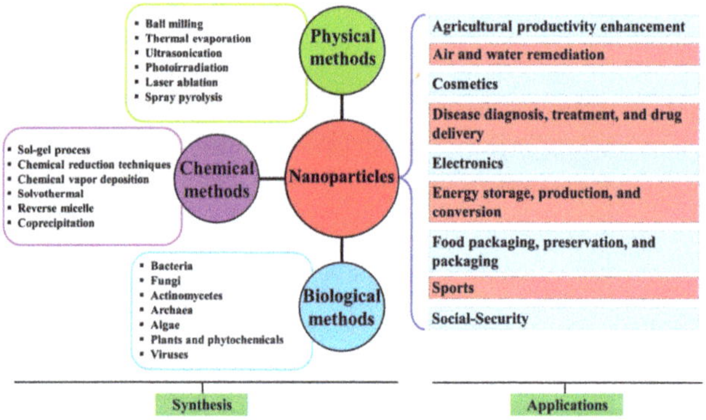

Figure 1. Schematic representation of various methods adopted for NP synthesis and its applications.

The physical methods belong to the category of top-down approach, while the chemical and biological methods follow the bottom-up approach for the synthesis NPs. Evaporation-condensation, electrolysis, diffusion, laser ablation, sputter deposition, pyrolysis, plasma arcing, and high energy ball milling are some of the most common physical methods used for the synthesis of NPs [8]. However, low production rate, expensive operations, and high energy consumption are the major limitations of these processes. Conversely, chemical synthesis methods that include chemical reduction, micro-emulsion/colloidal, electrochemical, and thermal decomposition are the conventional and most widely used methods for the synthesis of metallic NPs. The chemical reduction of NPs from their respective metal salt precursors by adding particular reducing agents is one of the most widely used methods for NPs chemical synthesis due to easy operational and equipment requirement. Several reducing agents, such as sodium borohydride ($NaBH_4$) [9], potassium bitartrate [10], formaldehyde [11], methoxypolyethylene glycol [12], hydrazine [13], etc., and stabilizing agents like dodecyl benzyl sulfate [14] and polyvinyl pyrrolidone [15] have been explored during synthesis. The chemical methods are economical for large-scale production; however, the use of toxic chemicals and production of harmful by-products cause environmental damage, thereby limiting its clinical and biomedical applications [16,17]. Hence, there is an increased demand for reliable, nontoxic, high-yielding, and eco-sustainable techniques for metallic NPs that can replace the conventional methods. The biological synthesis methods, therefore, provide an attractive alternative to the physicochemical synthesis methods.

The present article provides a critical overview on the synthesis of metallic NPs using biological methods and several factors affecting the preparation process. The applications of NPs in various sectors, such as medicine, wastewater treatment, agricultural sectors, etc., have been discussed in detail. Current challenges highlighting the toxic effects of NPs and future perspectives in each section gives us a comprehensive way forward in the near future.

2. Biological Synthesis of NPs

The biological synthesis of NPs can be carried out using a vast array of resources such as plants and plant products, algae, fungi, yeast, bacteria, and viruses. The synthesis of NPs is initiated by the mixing of noble metal salt precursors with biomaterials [18]. The presence of various compounds, such as proteins, alkaloids, flavonoids, reducing sugars, polyphenols, etc., in the biomaterials act as reducing and capping agents for the synthesis of NPs from its metal salt precursors [19]. The reduction of metal salt precursor to its successive NPs can be initially confirmed by visualizing the color change of the colloidal solution. Several studies reported the synthesis of Ag, Au, Cu, Pt, Cd, Pt, Pd, Ru, Rh, etc. using various biological agents in the recent past.

2.1. Plant-Mediated Synthesis of NPs

Figure 2 shows the Scopus search (with keywords "metal nanoparticles" and "plant extract") results of the number of research published from last 10 years on biological synthesis of NPs. An increase in the number of research publications was observed with each year and approximately 468 publications reported in the year 2020. These data further corroborate that the research interest in the area of biological NPs using plant extract is increasing significantly every year.

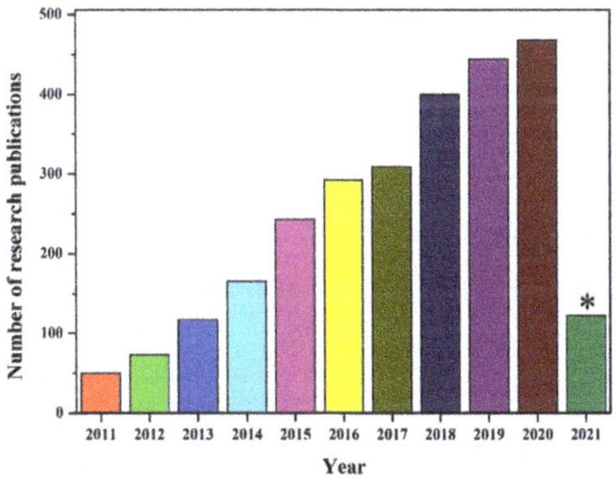

Figure 2. Number of research publications on biological synthesis of metallic NPs from last 10 years. Source: Scopus. * Number of publications reported as on 10 May 2021.

Synthesis of a wide range of metallic NPs has been reported using various plants [6,20,21]. Plant mediated synthesis of NPs can be achieved by three different methods, viz., (i) intracellularly (inside the plant), (ii) extracellularly (using plant extracts), and (iii) using individual phytochemicals. Several plants have the capability of metal accumulation and successive conversion of these accumulated metals to NPs intracellularly. The presence of several biomolecules such as amino acids, alkaloids, aldehydes, flavones, ketones, proteins, phenolics, polysaccharides, saponins, tannins, terpenoids, and vitamins in the plant plays a key role in the reduction of metals [22]. The variation in the size, shape, and properties of accumulated NPs are observed due to the variation in stabilizing and reducing potential of biomolecules present in the plant. The formation of gold NPs inside the living plant, alfalfa was reported by Gardea-Torresdey et al. [23] when the plants were grown in $AuCl_4$ rich environment. In a similar kind of study, Bali and Harris [24] observed the ability of *Medicago sativa* and *Brassica juncea* plants to accumulate Au NPs from aqueous solutions of $KAuCl_4$. The NPs were majorly located in the xylem parenchyma cells while some were also accumulated throughout the epidermis, cortex, and vascular tissue.

However, for the past several years, most of the works have been focused on using the inactive part of the plants either in powder form or as an extract for the synthesis of NPs [25]. Table 1 summarizes the green synthesis of various metal NPs using plants. Various parts of plants such as leaves, steam, flower, fruit, root, latex, seed, and seed coat are being exploited for the synthesis of metallic NPs.

Table 1. Summary on synthesis of metallic nanoparticles by various plant species.

Nanoparticles	Plant Species	Experimental Conditions	Shape and Size	References
Silver (Ag)	*Acalypha indica Linn*	Temperature: 27 °C; pH: 7.0; duration: 30 min	Spherical; 20–30 nm	[26]
	Chenopodium album leaf	Temperature: 20–100 °C; pH: 2.0–10.0; duration: 15 min	Spherical; 10–30 nm	[27]
	Hibiscus rosa sinensis leaf	pH: 7.2–8.5	Spherical; 13 nm	[28]
	Calendula officinalis seed	Temperature: 30 and 60 °C; pH: 3.0–9.0	Spherical; 7.5 nm	[29]
	Allophylus cobbe leaf	Temperature: 60 °C; pH: 8.0; duration: 6 h	Spherical; 2–10 nm	[30]
	Cissusquadrangularis leaf	Duration: 60 min	Spherical and cuboidal	[31]
	Piper nigrum, *Ziziphus Spina—Christi* and *Eucalyptus globulus* leaves	Temperature: ambient; duration: 1 h	Spherical; 8−35 nm	[32]
	Phyllanthus emblica fruit	Temperature: 65 °C; duration: 2 h	Spherical; 16.29 nm	[33]
	Blumea eriantha DC	Temperature: ambient; duration: 2–3 h	Spherical; 50 nm	[34]
	Brillantaisia patula, *Crossopteryx febrifuga* and *Senna siamea* leaf	Temperature: 70 °C; duration: 24 h	Spherical; 45–110 nm	[35]
	Ocimum tenuiflorum leaf	Temperature: ambient; duration: 10 min	Spherical and ovoid; 7–15 nm	[36]
	Annona squamosa leaf	Temperature: ambient	Spherical; 20–100 nm	[37]
	Aloe leaf	Temperature: ambient; duration: 20 min	Spherical; 20 nm	[38]
	Artocarpus heterophyllus Lam. Seed	Temperature: 121 °C; duration: 5 min	Irregular; 3–25 nm	[39]
	Trigonella foenum graecum seed	Duration: 5 min	Spherical; 17 nm	[40]
	Andrographis paniculata	Temperature: 30–95 °C	Spherical; 13–27 nm	[41]
	Podophyllum hexandrum leaf	Temperature: 20–60 °C; pH: 4.5–10.0; duration: 30–150 min	Spherical; 12–40 nm	[42]
	Syzygium cumini fruit	Temperature: ambient; pH: 7.0–9.0; duration: 2 h	Spherical; 5–20 nm	[43]
	Crassocephalum rubens leaf	Temperature: 50 °C; duration: 20 min	Spherical and hexagonal; 15–25 nm	[44]
Gold (Au)	*Cassia fistula* stem bark	Temperature: ambient	Rectangular and triangular; 55.2–98.4 nm	[45]
	Crassocephalum rubens leaf	Temperature: 50 °C; duration: 10 min	Spherical; 10–20 nm	[44]
	Simarouba glauca leaf	Duration: 15 min	Spherical and prism; <10 nm	[46]
	Hygrophila spinosa	Temperature: 30–100 °C; pH:2.0–12.0; duration: 15–60 min	Spherical, polygonal, rod and triangular; 68 nm	[47]
	Croton Caudatus Geisel leaf	Temperature: ambient	Spherical; 20–50 nm	[48]
	Moringa oleifera flower	Temperature: ambient; duration: 60 min	Triangular, hexagonal, and spherical; 5 nm	[49]
	Illicium verum	Temperature: 25–50 °C; pH: 2.0–10.00; duration: 15 min	Triangular and hexagonal; 20–50 nm	[50]
	Terminalia arjuna leaf	Temperature: ambient; duration: 15 min	Spherical; 20–50 nm	[51]
	Zingiber officinale	Temperature: 37 and 50 °C; pH: 7.4; duration: 20 min	Spherical; 5–10 nm	[52]
	Rosa hybrida petal	Temperature: ambient; duration: 5 min	Spherical, triangular, and hexagonal; 10 nm.	[53]
	Terminalia chebula seed	Temperature: ambient; duration: 20 s	Triangular, pentagonal, and spherical; 6–60 nm	[54]
	Eucommia ulmoides bark	Temperature: 30–60 °C; pH: 5.0– 13.0; duration: 30 min	Spherical	[55]
	Acorus calamus rhizome	Temperature: ambient; pH: 4.0–9.2	Spherical; 10 nm	[56]
	Curcuma pseudomontana root	Temperature: ambient; duration: 30 min	Spherical shape; 20 nm	[57]
	Citrus limon, *Citrus reticulata* and *Citrus sinensis*	Temperature: ambient; duration: 10 min	Spherical and triangular; 15–80 nm	[58]

Table 1. *Cont.*

Nanoparticles	Plant Species	Experimental Conditions	Shape and Size	References
Palladium (Pd)	*Hippophae rhamnoides* Linn leaf	Temperature: 80 °C; duration: 25 min	2.5–14 nm	[59]
	Cinnamom zeylanicum bark extract	Temperature: 30 °C; pH: 1.0–11.0; duration: 72 h	Spherical; 15–20 nm	[60]
	Banana peel extract	Temperature: 40–100 °C; pH: 2.0–5.0; duration: 3 min	50 nm	[61]
	Cinnamomum camphora leaf	Temperature: ambient; duration: 12 h	Quasi-spherical and irregular; 3.6–9.9 nm	[62]
	Catharanthus roseus leaf	Temperature: 60 °C; duration: 2 h	Spherical; 38 nm	[63]
	Terminalia chebula fruit	Temperature: ambient; duration: 40 min	-	[64]
	Rosmarinus officinalis	Temperature: ambient; duration: 24 h	Semi-spherical; 15–90 nm	[65]
	Anogeissus latifolia	Duration: 30 min	Spherical; 2.3–7.5 nm	[66]
	Daucus carota leaves	-	Rod; diameter—20 nm, length—38–48 nm	[67]
	Camellia sinensis leaves	Temperature: 100 °C; duration: 1 h	Spherical; 5–8 nm	[68]
Platinum (Pt)	*Anacardium occidentale* leaf	Temperature: ambient; pH: 6.0–8.0	Irregular rod shaped	[69]
	Cacumen platycladi	Temperature: 30–90 °C; duration: 25 h	Spherical; 2–2.9 nm	[70]
	Asparagus racemosus root	Duration: 5 min	1.0–6.0 nm	[71]
	Diopyros kaki leaf	Temperature: 25–95 °C	Spherical and plate; 2–20 nm	[72]
	Ocimum sanctum leaf	Temperature: 100 °C; duration: 1 h	Rectangular and triangular; 23 nm	[73]
Copper (Cu)	Mulberry fruit (*Morus alba* L.)	Temperature: ambient; duration: 5 h	Spherical and non-regular; 50–200 nm	[74]
	Crotalaria candicans leaf	-	Spherical; 30 nm	[75]
	Ziziphus spinachristi fruit	Temperature: 80 °C	Spherical; 5–20 nm	[76]
	Clove (*Syzygium aromaticum*) buds	Temperature: 30 °C; duration: 15 min	Spherical; 15–20 nm	[77]
Iron (Fe)	Tea leaves extract	Temperature: 80 °C; duration: 3 h	30–100 nm	[78]
	Moringa oleifera seeds	Temperature: ambient; duration: 30 min	Spherical; 2.6–6.2 nm	[79]
	Trigonella foenum-graecum seed	Temperature: 30 °C; duration: 5 min	7–14 nm	[80]
Selenium (Se)	*Ocimum tenuiflorum*	Temperature: ambient; duration: 75 h	Monodispersed and spherical; 15–20 nm	[81]
	Murraya koenigii	-	Spherical; 50–150 nm	[82]
	Zinziber officinale fruit	Temperature: ambient; pH: 9.0; duration: 75 h	Spherical; 100–150 nm	[83]
Nickel (Ni)	*Calotropis gigantea* leaves	Temperature: 80 °C; pH: 12.0; duration: 90 min	60 nm	[84]
	Desmodium gangeticum roots	Temperature: 80 °C; duration: 45 min	-	[85]

In general, the synthesis of NPs is carried out by mixing the plant biomass/extract with a metal salt solution at a desired temperature and pH. The primary confirmation of NPs synthesis can be checked by looking at the color change of the solution. The experimental procedure for the synthesis of NPs using plant biomass is depicted in Figure 3.

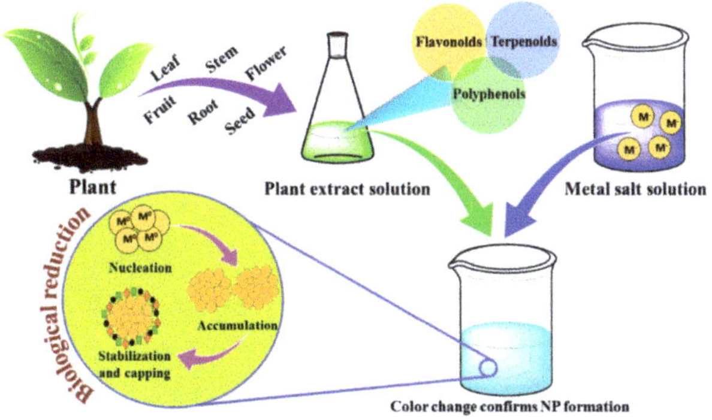

Figure 3. Schematic representation for plant-mediated biosynthesis of nanoparticles.

The plant extracts are prepared by using different methods such as hot extraction, cold extraction, and using Soxhlet apparatus, which were later used in NPs synthesis. This method of synthesis of NPs is more suitable in comparison to the intracellular method due to easy scale-up and downstream processing. Additionally, this method is renewable, non-toxic, biocompatible, and eco-friendly. Due to their biocompatible nature, these NPs are known to have various biological applications. The synthesis of metal NPs is initiated by adding the plant extract to the metal precursor solution containing the salts of respective metals. Metal precursor solutions such as $AgNO_3$, $HAuCl_4$, $PdCl_2$, H_2PtCl_6, $Cu(NO_3)_2 \cdot 3H_2O$, $FeCl_3 \cdot 6H_2O$, Na_2SeO_3, and $(NiNO_3)_2 \cdot 6H_2O$ are commonly used for the synthesis of Ag, Au, Pt, Cu, Fe, Se, and Ni NPs. The synthesis of metal NPs using plant extract mainly occurs in three stages. In the first stage, the reduction of metal ions (M^+ or M^{2+}) to metal atoms (M^0) and successive nucleation of the reduced metal atoms occurs. While in the second stage, the coalescence of small adjacent NPs into larger size particles occurs with simultaneous increase in thermodynamic stability. At the final stage, the termination of the process takes place while giving the final shape to the NPs [86,87]. The presence of various active biomolecules in the plant extract plays an important role in the reduction and stabilization of metal ions in the solution. However, due to the presence of a large number of phytochemicals in the plant extract, it is difficult to ascertain the exact reducing and stabilizing agents for NPs synthesis.

Salih et al. [32] used plant extract derived from leaves of three different plants, viz., *Piper nigrum*, *Ziziphus Spina—Christi* and *Eucalyptus globulus*, for the synthesis of AgNPs. The average particle size distribution was in the range of 8–35 nm and decreased with the increase in concentration of plant extract. In a similar kind of study, Dhar et al. [33] reported the synthesis of AgNPs using fruit extract of *Phyllanthus emblica*. The fabricated AgNPs were spherical with an average size of 60–80 nm. The extraction methods also play an important role which influences the antioxidant properties of Ag and Au NPs synthesized using leaves extract of *Crassocephalum rubens* [44]. Figure 4 depicts the size, shape, and morphological features of *C. rubens* synthesized AgNPs and AuNPs. The SEM and TEM images revealed spherical and hexagonal shapes of AgNPs with size 10–15 nm, whereas the size of AuNPs was in the range of 10–20 nm with spherical shape.

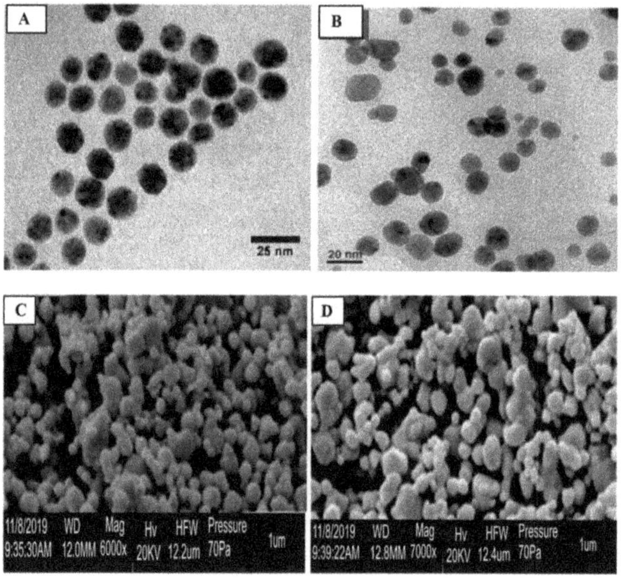

Figure 4. Characterization of AgNPs and AuNPs synthesized from *Crassocephalum rubens* leaf extract. TEM images of (**A**) AgNPs and (**B**) AuNPs; SEM images of (**C**) AgNPs and (**D**) AuNPs. Reprinted with permission from Reference [44].Copyright 2021 Elsevier.

The biosynthesized Au NPs using aqueous extract of *Hygrophila spinosa* exhibited enhanced cytotoxicity against various cancer cell lines compared to *H. spinosa* aqueous extract [47]. These green synthesized AuNPs with antioxidant and cytotoxic properties could provide a new direction for the development of nanomedicine. In addition to Au and Ag NPs, Pd NPs synthesized using *Rosmarinus officinalis* leaves extract demonstrated notable antimicrobial and antifungal activity against different bacteria and fungi [65]. Besides leaves, other plant parts such as root, fruit, flower, petal, seed, peel, bark etc. are used in the biosynthesis of different NPs [40,49,53,55,61,71,80]. Several previous studies tried to explain the mechanisms behind the antimicrobial effect of metallic NPs [88–90]. These metal ions could strongly interact with the negatively charged bacterial cell wall leading to its rupture. The other mechanism states distortion of the helical structure of bacterial DNA due to the interaction of NPs and interruption of internal and external cellular mechanisms.

2.2. Microbial Synthesis of NPs

In addition to the plant-mediated synthesis, several microorganisms such as bacteria, fungi, actinomycetes, and viruses are also reported to synthesize various metal NPs. The interaction between metals and microorganisms have been exploited in the past for various biological applications such as biomineralization, bioremediation, bioleaching, and biocorrosion [91]. However, recently microbial synthesis of NPs has emerged as a promising field of research due to certain advantages compared to other methods. The NPs are synthesized either intracellularly or extracellularly depending on the type of microorganisms [92]. Previous reports on the synthesis of metal NPs using various microorganisms have been summarized in Table 2.

Table 2. Summary on synthesis of metallic nanoparticles by microorganisms.

	Microorganisms	Nanoparticles	Shape and Size	References
Bacteria				
	Bacillus subtilis	Ag	Spherical; 3–20 nm	[93]
	Pseudomonas stutzeri	Ag	Triangular; 200 nm	[94]
	Bacillus licheniformis	Ag	40 nm	[95]
	Ochrobactrum anhtropi	Ag	Spherical; 38–85 nm	[96]
	Pantoea ananatis	Ag	Spherical; 8.06–91.31 nm	[97]
	Actinobacter	Ag	Spherical; 13.2 nm	[98]
	Pseudomonas aeruginosa	Au	15–30 nm	[99]
	Rhodopseudomonas capsulata	Au	Spherical; 10–20 nm	[100]
	Escherichia coli DH5 α	Au	Spherical, triangles, and quasi-hexagons; 25 nm	[101]
	Bacillus subtilis	Au	Spherical, 20–25 nm	[102]
	Mycobacterium sp.	Au	Spherical; 5–55 nm	[103]
	Shewanella loihica	Pt	1–10 nm	[104]
	Shewanella oneidensis MR-1	Pt	2.83–61.03 nm	[105]
	Jeotgalicoccus coquinae ZC15	Pt	Spherical; 3.74 nm	[106]
	Shewanella loihica	Pd	1–12 nm	[104]
	Shewanella oneidensis MR-1	Pd	10–100 nm	[107]
	Lysinibacillus sp. ZYM-1	Se	Cubic; 100–200 nm	[108]
	Bacillus subtilis	Se	Spherical; 50–400 nm	[109]
	Lactobacillus acidophilus	Se	Spherical; 2–15 nm	[110]
Fungi				
	Rhizopus stolonifer	Ag	Spherical; 2.86 nm	[111]
	Candida glabrata	Ag	Spherical; 2–15 nm	[112]
	Trametes trogii	Ag	Spherical and rod; 5–65 nm	[113]
	Trichoderma longibrachiatum	Ag	Spherical; 10 nm	[114]
	Fusarium oxysporum	Ag	Spherical; 21.3–37 nm	[115]
	Aspergillus terreus	Ag	Spherical; 7–23 nm	[116]
	Ganoderma sessiliforme	Ag	Spherical; 45 nm	[117]
	Candida albicans ATCC 10231	Ag	Spherical; 10–20 nm	[118]
	Cladosporium cladosporioides	Au	60 nm	[119]
	Trichoderma harzianum	Au	Spherical; 26–34 nm	[120]
	Pleurotus ostreatus	Au	Spherical; 10–30 nm	[121]
	Aspergillus sp.	Au	Spherical; 4–29 nm	[122]
	Rhizopus oryzae	Au	Spherical and flower like structure; 16–43 nm	[123]
	Penicillium chrysogenum	Pt	Spherical; 5–40 nm	[124]
	Fusarium oxysporum f. sp. *lycopersici*	Pt	Triangle, hexagons, square, and rectangles; 10–50 nm	[125]
	Fusarium oxysporum	Si	Quasi-spherical; 5–15 nm	[126]
	Fusarium oxysporum	Ti	Spherical; 6–13 nm	[126]
Yeast				
	Rhodotorula sp. ATL72	Ag	Spherical and oval; 8–21 nm	[127]
	Saccharomyces cerevisiae	Ag	Spherical; 2–20 nm	[128]
	Cryptococcus laurentii	Ag	35–400 nm	[129]
	Rhodotorula glutinis	Ag	15–220 nm	[129]
	Rhodotorula glutinis	Ag	Spherical; 15.5 nm	[130]
	Saccharomyces cerevisiae	Au	Triangle, truncated triangle, and hexagon	[131]
	Magnusiomyces ingens LHF1	Au	Spherical and pseudo-spherical; 20–28 nm	[132]
	Saccharomyces cerevisiae	Pd	Hexagonal; 32 nm	[133]
	Magnusiomyces ingens LHF1	Se	Spherical and quasi-spherical; 70–90 nm	[134]

Compared to other microorganisms, bacteria are preferred for the synthesis of NPs due to their easy maintenance, high yield, and low purification cost. In recent years, cell-free extract of endophytic bacterium, *Pantoea ananatis*, was used for the synthesis of AgNPs [97]. These synthesized spherical shaped NPs with an average size ranging from 8.06–91.32 nm exhibited significant antimicrobial activity against various pathogenic microorganisms.

In a similar kind of study, Wypij et al. [98] reported maximum antimicrobial activity of AgNPs synthesized by using acidophilic actinobacterial strain against *E. coli*, followed by *B. subtilis* and *S. aureus*. Moreover, bacteria species belonging to *Pseudomonas stutzeri, Bacillus licheniformis, Ochrobactrum anhtropi, Bacillus subtilis,* and *Actinobacter* were used for the synthesis of AgNPs (Table 2). Gold NPs were synthesized by *Pseudomonas aeruginosa, Rhodopseudomonas capsulate, Bacillus subtilis, Escherichia coli DH5α, Mycobacterium* sp., etc. A larger amount of *Actinobacter* in the medium led to the formation of smaller size with uniformly distributed spherical AuNPs [135]. Intercellular and extracellular components of microorganisms play an important role in the synthesis of AuNPs [136]. Intercellular components such as reducing sugar, fatty acids, and enzymes and extracellular such as exopolysaccharides help in the reduction of AuNPs [136].

In addition to bacteria, the synthesis of NPs using fungi has received increasing attention due to various advantages such as easy scale-up and downstream processing, economic feasibility, and increased surface region due to the presence of mycelia. Different species of fungi such as *Rhizopus stolonifera, Candida glabrata, Trametes trogii, Trichoderma longibrachiatum, Aspergillus terreus, Fusarium oxysporum, Ganoderma sessiliforme, Candida albicans ATCC 10231, Cladosporium cladosporioides, Trichoderma harzianum, Pleurotus ostreatus, Aspergillus sp., Rhizopus oryzae*, etc. are used in silver and gold NPs synthesis. Other species of fungi such as *Penicillium chrysogenum* and *Fusarium oxysporum* are used for the synthesis of Pt, Si, and Ti NPs.

Among the eukaryotic organism, yeast has been used for the synthesis of NPs like Ag, Au, Pd, Se, etc. Soliman et al. [127] synthesized AgNPs using pink yeast *Rhodotorula* sp., and the characterization revealed the NPs to be spherical and oval in shape with 8.8–21.4 nm size. These biosynthesized AgNPs exhibited significant antimicrobial activity with complete inhibition to wide range of bacteria (i.e., both Gram positive and Gram negative) as well as fungi. Similarly, the antifungal activity of biosynthesized AgNPs using two yeasts: *Rhodotorula glutinis* and *Cryptococcus laurentii* was evaluated against the phytopathogenic fungi [129]. The results of this study revealed that the antifungal activity of AgNPs from *R. glutinis* was higher than that from the ones prepared from *C. laurentii*. In another study, the morphology and size of AuNPs were controlled by varying the pH of the medium containing yeast [131]. In this method, various morphologies of gold nanoplates such as triangle, truncated triangle, and hexagonal nanoplates with uniform size were synthesized successfully (Figure 5).

(A) (B)

Figure 5. SEM images of gold nanoplates synthesized using yeast extract at different pH conditions. (**A**) Au nanoplates at low pH without NaOH, (**B**) small Au nanoplates synthesized at high pH. Reprinted with permission from Reference [131]. Copyright 2016 Springer Nature.

Similarly, the palladium (Pd) and selenium (Se) NPs are synthesized using aqueous extract of *Saccharomyces cerevisiae* and cell-free extracts of *Magnusiomyces ingens* yeast, respectively [133,134].

The NPs are formed during the microbial synthesis process is due to the oxidation/reduction of metallic ions by secreted biomolecules by microbial cells such as enzymes, sugars, carbohydrates, proteins, etc. [137]. However, a complete understanding of microbial NP synthesis is still unknown as the routes for NPs synthesis varies for each kind of microorganisms. The reduction of silver for the synthesis of extracellular and intracellular AgNPs by bacteria is mainly achieved by the action of deoxyribonucleic acid (DNA) or sulfur-containing proteins, whereas in the case of fungi, the process is carried out by nitrate-dependent reductase or carboxylic group [138]. The extracellular synthesis method is preferable due to easy and simpler purification steps. In contrast, the intracellular NPs synthesis method is challenging and expensive due to the involvement of additional separation and purification processes. Fungal-mediated synthesis of NPs holds additional advantages compared to algae or bacteria in terms of easier and simpler biomass handling and downstream processes along with the secretion of large amount protein that further increases the productivity by several folds. However, the microorganism mediated NPs synthesis process is extremely intricate and difficult due to the preparation of inoculum and growth media, isolation of strain, and maintenance of culture medium and operation conditions (pH, temperature, agitation). Conversely, the use of plant extracts or broths is simple and convenient, devoid of the complex methods of cell culture and maintenance. The time required to achieve a complete reduction of NPs using microorganisms is usually 24 to 120 h, while the reduction time is much less in using plant extract ranging from few hours to 48 h [138]. The reduction rate using the plant is much faster than microorganisms and in close agreement with the physical and chemical methods. The use of microorganisms for large-scale biosynthesis of NPs lacks feasibility compared with plants, which require less time for reaction completion. As reported in the earlier studies, the plant-mediated biosynthesis of AgNPs demonstrate better production rate, size, and morphological characteristics compared to other available biological techniques [138].

Hence, the plant-mediated synthesis of NPs proves to be a sustainable alternative not only to the other biological techniques but also to other synthesis methods such as physical and chemical. However, in-depth studies are required to understand the detailed mechanisms of action and to achieve better control over size, morphology, and production rate for making a plant-mediated synthesis method at par with chemical methods.

2.3. Factors Affecting NPs Synthesis

Adjustment of shape and size of metal NPs further enhances their functionality for various applications. The morphological parameters of NPs can be manipulated by changing various experimental parameters such as reaction time, reactant concentration, pH, temperature, aeration, salt concentration, etc. [139]. Precise control of these parameters can play a critical role during the optimization of metal NPs synthesis via the biological route. The size and shape of NPs can be controlled by varying the pH of the medium, while the acid pH leads to the formation of large-sized NPs [140,141]. During synthesis of Au NPs using oat (*Avena sativa*) biomass, Armendariz et al. [142] observed smaller sized gold NPs at pH 3.0 and 4.0 in comparison to the synthesized NPs at pH 2.0. This is due to the better accessibility of functional groups present in the extract for nucleation at higher pH compared to the presence of fewer groups at a lower pH range. In addition to pH, the concentration of biomolecules in the extract also affects the size and shape of synthesized NPs. Increase in the concentrations of *Aloe vera* leaf extract resulted in the synthesis of higher amount of spherical gold NPs instead of triangular which is due to the presence of carbonyl compounds in the extract [143]. In addition, the size of the NPs was modulated in the range of 50 to 350 nm by varying the extract concentration in the solution. The duration of reaction also plays a crucial role in the reduction of NPs and their size, which is primarily confirmed by rapid change in color of the reaction mixture. This duration can range from few minutes to few days.

A change in particle size of silver NPs was observed in the range of 10–35 nm by increasing the reaction time from 30 min to 4 h using *Azadirachta indica* leaf extract. Furthermore, the reaction temperature is one of the important parameters in the biological synthesis of NPs which also determines the shape, size, and yield of NPs. The average size of silver NPs decreased from 35 to 10 nm with the increase in reaction temperature from 25 to 60 °C using *Citrus sinensis* (sweet orange) peel extract [144].

2.4. Characterization of NPs

NPs have attracted significant attention of researchers due to their unique physical, chemical, and mechanical properties. Therefore, the physicochemical characterization of synthesized NPs is critically important before its application in various sectors. Analyzing various characteristics such as size, shape, surface morphology, surface area, structure, stability, elemental and mineral decomposition, homogeneity, intensity, etc. will provide important information about the NPs, which subsequently determined their end-use applications. Additionally, the electrical and thermal conductivity and purity of NPs can also be obtained by using these techniques. Size and shape of the synthesized NPs are mainly analyzed using X-Ray Diffraction (XRD), Scanning Electron Microscope (SEM), Field Emission Scattering Electron Microscopy (FESEM), Transmission Electron Microscopy (TEM), High-Resolution Transmission Electron Microscopy (HRTEM), Atomic Force Microscopy (AFM), Dynamic Light Scattering (DLS), Condensation Particle Counter (CPC), Photon Correlation Spectroscopy (PCS), etc. Among these techniques, XRD, SEM, and TEM are most commonly used for this purpose. Further, SEM, TEM, AFM, etc. are used for studying the surface morphology. Superconducting Quantum Interference Device (SQUID), Vibrating Sample Magnetometer (VSM), Electron paramagnetic resonance (EPR), etc. are used for the determination of magnetic properties of NPs. More details about NPs characterization techniques are reviewed by previous authors [145,146].

3. Potential Applications of Metal NPs

These synthesized metallic NPs offer a diverse platform for various applications. Some of the most important applications of these metallic NPs are summarized in this section.

3.1. Agriculture

Nanotechnology has proven its potential to benefit the agriculture sector by finding solutions to agricultural and environmental problems in order to increase food production and security [147].

3.1.1. Effect of Nano Material on Plants

Seed germination and later growth phases benefit from nano-growth stimulants [148,149]. Owing to the small size and large surface area of nanoparticles, these particles are able to seep into the seed pores eventually activating the phytohormones required for seed growth and germination [150,151]. For example, the use nano-TiO_2 and nano-SiO_2 on soybean seedlings improved nitrate reductase activity thereby enhancing seed germination. However, combining both nanomaterials (NMs) was more advantageous [152]. Seed treatment with 0.25 percent TiO_2 improved nitrogen assimilation and photosynthesis rate in spinach (*Spinacia oleracea* L.), resulting in better growth [153,154]. Watermelon (*Citrullus lanatus*) seed soaking in Fe_2O_3 NPs improved germination and initiated plant development and fruiting behavior [155]. The use of low-concentration SiO_2 NPs on tomato seeds improved germination [156]. Nanomaterials can be used in a variety of ways in in-vitro cultivation. Zn as a ZnO nanomaterial resulted in an increased calli growth and physiological parameters in tobacco (*Nicotiana tabacum* L.). Although nanoparticles have been widely utilized for promoting plant growth [149,157], NM application may also be phytotoxic [158]. The positive and negative effects, however, are dependent on the dose, size, time, exposure, and make (synthetic/biological) of the NM [158,159]. Over the past years, application of chemically or physically synthesized NM have proven to be stimulant of plant growth; however, they pose a greater threat to ecology and environment (by seeping through the soil) [160]. Utility of green/biological NM in agriculture has therefore been advantageous due to their safety and feasibility [161]. An increased usage of biologically synthesized Au, Ag, Ti, Ca, N, Fe nanoparticles either in form of nanofertilizers or nanopesticides has been employed [161–165].

3.1.2. Application of Nanomaterials in the Field of Agriculture

Over the recent years, application of nanomaterials to boost agriculture has broadly been in two major forms, either as nanofertilizers, to enhance the agricultural productivity or in the form of nanopesticides, to eradicate the pest/pathogens/weeds hampering the growth of crop plants. In this section, we will be discussing the applicability of nanofertilizers and nanopesticides, in current agricultural practices.

Nanofertilizers

Huge increase in agricultural yields, particularly grain yields, has played an important role in providing the world's food demands over the last five decades. In this context, increased usage of chemical fertilizers acts as one of the key contributors to increased crop productivity. Although, use of chemical fertilizers has increased productivity of crops, their poor use efficiency due to volatilization and leaching has led to its excessive usage [147]. On the contrary, nanofertilizers are compounds that are applied in smaller amounts and can enhance the effect of fertilizers [166]. Enhancing the effect of fertilizers on plants is usually done by governing the fertilizers in nano form which results in controlled nutrient release, eventually minimizing the risk of environmental damage [167]. With the recent advancement in the field of nanobiotechnology, nanofertilizers can be utilized as intelligent fertilizers which are able to release desired amount of nutrients just when and where they are needed by plants, thereby limiting the conversion of excess fertilizers to gaseous forms or leaking downstream [168].

To date, various NPs have been employed in developing fertilizers, some of which include hydroxyapatite, polyacrylic acid, clay minerals, chitosan, zeolite, and many more. The small size and large surface area of these NPs give them an advantage over the conventional fertilizers. For example, strong interactions of hydroxyapatite with urea lead to release of nitrogen from urea until 60 days as compared to the ammonium nitrate fertilizer (normal form of urea) which releases nitrogen only until 30 days [169].

Nanofertilizers can be broadly classified into three categories: (1) nanoparticulate nano fertilizers, (2) micronutrient nanofertilizers, and (3) macronutrient nanofertilizers. Nanoparticulate nanofertilizers include NPs, such as CNTs, TiO_2, and SiO_2, responsible for plant growth. In soybean, an amalgamation of TiO_2 and SiO_2 results in overall increase in plant growth with increased nitrogen fixation and improved seed germination [152]. As utility of TiO_2 nanoparticle in plant growth has been well established, recently several works have been done for generation of non-toxic, cheap, and environmentally safe green synthesized TiO_2 from plant extracts of *Syzgium cumini*, *Moringa oleifera*, *Cucurbita pepo*, and *Trigonella foenum* [170–173]. Micronutrients such as molybdenum (Mo), copper (Cu), iron (Fe), nickel (Ni), manganese (Mn), and Zinc (Zn) packed in NPs serve as micronutrient nano fertilizers. A mixture of three micronutrients NPs (ZnO, CuO, and B2 O3) has been successfully established to ameliorate drought stress in soybean plants [174]. Recent studies on *Zea mays* have revealed utility of biologically synthesized micronutrient nanofertilizers (iron oxide nanorods) in better plant growth as compared to the chemically synthesized NPs [164]. Similar to micronutrients, macronutrient nanofertilizers are composed of a combination of macroelements (Mg, K, N, Ca, and P) [167]. Foliar application of Mg and Fe NPs on *Vigna unguiculata*, resulted in increased seed weight and photosynthesis ability thereby resulting in an overall improvement in yield [175]. When compared to crops administered with conventional fertilizer, phosphatic nanofertilizers have been attributed a 32 percent rise in growth rate and a 20 percent rise in seed production in soybean (*Glycine max* L.) [176]. A recent study on application of green synthesized multinutrient nanofertilizer (U-NPK) made from calcium phosphate NP doped with potassium and nitrogen resulted in reduction of 40% of nitrogen requirement of plants when compared to conventional approach due to slow and gradual release of major micronutrients [165]. Overall, use of nanofertilizer results in reduction in usage of fertilizer amount by allowing slow-release products.

Nanopesticide

Plants being sessile in nature are prone to various biotic and abiotic stresses such as pests, pathogens, heat, drought, pollution, etc. These stresses directly or indirectly affect the total yield of a plant. The ill effect of pathogens on an entire crop leading to famine like situations has been extensively studied by far, and in order to have a control over these forms of biotic stress, use of pesticides had been advocated. Utility of pesticides has been established in eradication of harmful pests and pathogens from crop field resulting in crop protection [177]. However, it has been found that use of pesticides leads to deleterious effects on environment and human health. As a result, numerous pesticides have been prohibited by state or international governments. Thus, development of effective yet safe pesticides is the need of the hour. Although biopesticides have emerged as a breakthrough, their use has been limited due to significantly higher cost of production. However, nanotechnology offers a new and better approach by introduction of nanopesticides [178].

One of the widely used examples of nanopesticide is nanostructured alumina (NSA). NSA acts as negatively charged insecticide which interacts with the positively charged bodies of the insects leading to dehydration. The dehydration of insect body results in detachment of insect's cuticle eventually leading to death [167]. Potent insecticidal activity of several other NPs has also been studied over past few years. In the year 2013, Paret et al. [179] studied the antimicrobial effect of TiO_2 and ZnO NPs against *X. perforans,* casual organism of tomato spot disease. Role of Imidacloprid (IMI) as an effective systemic insecticide against several sucking insects such as *Martianus dermestoides* has been established; in addition, use of the nano-IMI being more photodegradable increases its effectiveness and environmental safety over the conventional formulation [180]. Another study found that nanoformulation of permethrin had a higher absorption than the conventional form against *Aedes aegypti*. In a recent study, Zhao et al. investigated the insecticidal activity of Cu NPs which showed potential upregulation of exogenous microbial protein within plant tissue enhancing resistance against the bollworm, under the effect of at a low dose of Cu NPs [181]. These findings are encouraging for utility of nanopesticides against various crop pests which could be an important tool in future agricultural pest management practices [147].

3.2. *Nanoparticles in Food Industry*

NPs in combination with other technologies can bring impactful innovations in the production, storage, packaging and transportation of food products. Food processing transforms raw food ingredients into a palatable format with long shelf-life and in turn, ensure efficient marketing and distribution systems for the enterprise. Fresh foods, on the other hand, require robust logistics for their transportation from source to consumer. Nanotechnology based systems play an important role to maintain the functional properties by incorporating NP based colloids, emulsions, and biopolymers solutions. Nanotechnology has provided with a new dimension and ample opportunities to develop NPs for various applications with deeper knowledge of the material. Application of nanotechnology in food industry is based on nanostructures which target food ingredients as well as sensors. Nano-food ingredients cover a wide area of applications starting from processing of food to its packaging. Nanostructure based application in food processing comprises the use as antimicrobial agents, nanoadditives, nanocarriers, anticaking agents, and nanocomposites while in food packaging, they are applied as nano-sensors for monitoring the quality of food produced [182]. The nano materials can also serve as enzyme-supports because of their large surface-to-volume ratio with respect to their conventional macro-sized counterparts. Recent nano-carriers have the potential to function as selective and exclusive delivery systems in order to carry the food additives into the food ingredients without altering the basic physicochemical properties and morphologies. For delivery of the bioactives to the target sites, particle size is the most essential factor that affects the delivery rate as the micro particles cannot be assimilated in some cell lines. A number of researchers have developed various techniques of encapsulating these bioactives using nano-sized particles or nano-emulsions resulting in enhancement of their bioavailability due to the increased surface to volume ratio. Nanoencapsulation using nano-spray-drying is another promising technique for the development of nanoparticles which can serve the food industry for the production of bioactive ingredients.

However, challenges related to performance and toxicity of nanomaterials need to be addressed to induce active development and applications of NPs. Additionally, legislation for regulating the production, application, and disposal of nanomaterials for food industry is of utmost importance. Public awareness and acceptance of the novel nano-enabled food and agriculture products are also needed to be strengthened.

3.2.1. Application of NPs in Food Preservation and Packaging

The utility and global market of NPs have developed manifold in the recent years and is expected to reach USD 125.7 billion by 2024. In the domain of food packaging, the market is expected to reach a staggering USD 44.8 billion by 2030 [183]. Nanoceuticals and Nutrition-by-nanotech are the available commercial names for food nano-supplements. Nano-sized powders and nanocochleates are used for increasing absorption and delivery of nutrients without altering the taste, flavor, and color of the food products. For better absorption of micronutrients, vitamin spray-induced nanodroplets are used. The technique of nano-encapsulation is involved when probiotics and similar targets are required to deliver into the human system with the help of Fe and Zn nanostructured capsules. NP based food supplements are more effective than their common counterparts because they are able to react more efficiently with the human cells due to their nano-size.

Food preservation systems with antimicrobial packaging provide advanced barrier properties to the food [184]. NPs or nanocomposite materials such as starch and sorbic acid-based films are being utilized in various packaging applications for their microbial growth inhibiting properties. They are effective due to their high surface-to-volume ratio as well as enhanced surface reactivity of the nano-sized antimicrobial agents which assists in inactivating microorganisms more efficaciously in comparison to micro- or macroscale agents. Metallic and semiconducting NPs are the commonly used antimicrobial NPs. Metallic NPs such as Ag with Cu, Au, and Pt demonstrate different degree of efficacies. Among the semiconducting NPs, TiO_2, ZnO, WO_3, and MgO are proven antimicrobial agents. Other antimicrobial NPs consist of natural biopolymers like chitosan (CTS) and enzymes (peroxidase, lysozyme), organically modified nanoclay (e.g., quaternary ammonium-modified MMT, Ag-zeolite), natural antimicrobial agents (e.g., nisin, thymol, carvacrol, isothiocyanate, and antibiotics) along with synthetic antimicrobial agents (quaternary ammonium salts, ethylenediaminetetraacetic acid (EDTA), propionic, benzoic, and sorbic acids) [185]. Hybrid metal-polymer matrices is a new class of materials for diverse applications due to their distinct properties such as high surface areas, orderly crystalline structures, and pores with regular size and shape. Sensor composed of graphene oxide-nickel nanoparticle biopolymer films is capable of measuring glucose concentration in the body fluids. Nevertheless, it can also be employed in food application system because of the use of biocompatible materials low toxicity of Ni and cost-effective technology [186,187]. The antibiotic resistance mechanisms are irrelevant to the development of NPs since their mode of action is only to stay in direct contact of the microbial cell walls without penetrating it. The barriers that the natural NPs create can control microbial growth and consequently spoilage of pathogens. Ag-NPs are used in biotexiles, electrical appliances, refrigerators, and other kitchen-wares as they act in bulk form and their ions have the ability to inhibit a wide range of biological processes in bacteria [188]. The incorporation of AgNPs into the gelatin-based nanocomposite film promisingly enhanced its antimicrobial activity. Further, it was observed that nanocomposites, thus developed, showed potential antibacterial properties against both Gram-negative and Gram-positive food-borne pathogens [189]. ZnO NPs have antibacterial nature which increases with decreasing particle size that can further be stimulated using visible range light to incorporate in various polymers including polypropylene [190]. The contamination of *Escherichia coli* can be inhibited by using TiO_2 as a coating in packaging material, and in combination with Ag, it improves various disinfection processes.

Recent development of smart packaging, viz., oxygen scavengers, moisture absorbers, and barrier-packaging products, account for 80% of the market share. Bakery and meat industries significantly use nano-enabled packaging technologies. The food environment is so enabled that it can continuously sense oxygen content, temperature, and microbial load. Some examples include Ag-NP-incorporated enzymes for microbial detection and gas sensing and nanofibrils of perylene-based fluorophores for detecting gaseous amines from fish and meat spoilage. Additionally, ZnO and TiO_2 nanocomposites are used for detection of volatile organic compounds. Applications of NPs in the food industry are relatively recent and have demonstrated rapid developments in this area [191]. The major developments in this area include texture alteration, components and additives encapsulation, enhancing sensory acceptance, controlled release of flavor, and enhancing bioavailability of micronutrients [192]. NPs have also altered the novelty of packaging materials enhancing their mechanical barrier and antimicrobial efficacy. Hence, the recent advancements using NPs in food preservation and packaging may be used to overcome the disadvantages of the biopolymer-based packaging technologies. Nanocomposites exhibit enhanced barrier and mechanical and thermal properties compared to their polymers and conventional counterparts.

3.2.2. Applications of NPs in Food Supplements and Value Addition

The applications of nanotechnology and the use of NPs in food science and technology appear to have emerged from various sectors viz., pharmaceuticals, cosmetics, and nutraceuticals. The advent of nanomaterials, which can interact with biological entities at a near-molecular level makes it a common technology almost for various industries including the above. Current nanotechnology applications in food industry for developing nanotextured food constituents as well as the delivery systems for nutrients require techniques of nanoemulsions, surfactant micelles, emulsion bilayers, and reverse micelles. The nanotextured food ingredients claim to offer better texture, taste, and overall acceptability [193]. Low fat nanotextured spreads, mayonnaise, ice creams, and similar products claim to be as "creamy" as their full-fat alternatives, while offering a healthier alternative to the consumers.

Nanocochleates (50 nm in size), known to protect micronutrients and antioxidants from degradation during processing and storage, are based on a phosphatidylserine carrier derived from soybean and are generally regarded as safe. The Greek term "cochleate" means a 'snail with a spiral shell'. It can be derived by adding calcium ions to small phosphatidylserine vesicles in order to influence the formation of discs which are then fused to large sheets of lipid molecules and finally rolled up into nanocrystals.

In another instance, self-assembled nanotubes were developed from a protein namely lactalbumin which is a natural alternative for nanoencapsulation of pharmaceuticals, nutrients, and supplements [194]. Nanotechnology comprises another major area, namely nanoencapsulation, which is effectively used for delivering susceptible food ingredients and additives. Microencapsulation can be employed to mask the taste and odor of tuna fish oil for enabling it to be used for supplementation for its rich omega-3 fatty acid content. Nanoencapsulated food ingredients and additives are used in a range of food products such as the delivery of live probiotic microbes for healthy metabolic function. Nanoemulsion is another use in food technology to improve the quality of sweeteners, processed foods, and beverages [195–197]. A summary on application of NPs in various aspects of food science and technology is given in Table 3.

Table 3. Applications of NPs/Nanotechnologies in various aspects of Food Science and Technology.

Application	NPs/Nanotechnology	Function	Reference
Food Production	TiO_2	Antimicrobial, coating in packaging material, detection of volatile compounds	[198]
	Nanoemulsion	Quality enhancement of beverages, sweeteners, and processed food	[195–197]
	Nanoencapsulation	Enhancement of taste, color, and odor of food materials	[194]
Food preservation and packaging	AgNPs, Ag–ZnO NPs	Packaging of meat, fruit, and dairy products by AgNPs—doped nondegradable and edible polymers and oils; antimicrobial property	[199]
	Low-density polyethylene film + Ag, ZnO NPs, TiO_2, kaolin	Orange juice, blueberry, strawberry	[200–202]
	Ethylene vinyl alcohol + AgNPs	Chicken, pork, cheese, lettuce, apples, peels, eggshells	[203]
	Polyvinylchloride + AgNPs	Minced beef	[204]
	Polyethylene + Ag, TiO_2 NPs	Fresh apples, white sliced bread, fresh carrots, soft cheese, atmosphere packaging milk powder, fresh orange juice	[205,206]
	Nanoclay-polymer nanocomposites	Meats, cheese, confectionery, cereals, boil-in-the-bag foods, extrusion-coating applications for fruit juices and dairy products, bottles for beer and carbonated drinks	[207]
	Ag-ZnO NPs	Nanostorage containers, bakeware, containers, cutting boards	[199]
	ZnNPs	Preservation and transport	[208]
Food supplement and value addition	Colloidal metal nanoparticles	Enhanced uptake	-
	Nanopowders	Increase absorption of nutrients	-
	Cellulose nanocrystal composites	Drug carrier	-
	Nanocochleates	Drug delivery, enhancement of taste and color of food materials	-

Various applications of NPs and nanotechnology provide numerous advantages for food quality and safety. From farm to fork, nanotechnology is proven impactful at every stage of food manufacturing, enhancing shelf-life, nutrition, quality control, and smart packaging. However, the unregulated applications of NPs can pose potential risks to human health and environment. Numerous studies have demonstrated the toxicological effects of NPs on biological systems. Since food contact materials are already available in the market in some countries, more data on the safety of such engineered NPs on human health are necessary to implement regulations for such products. More research on ecotoxicological effects of NPs will add on to the existing knowledge.

3.3. *Drug and Medicine*

Nanotechnology-based drugs have attracted a lot of attention in the last decade. The unique properties of NPs, viz., small size, ability to travel through fine blood capillaries, vessels, junctions, and barriers, have made them one of the most researched and studied domains [209]. They have great advantages in terms of improvement of bioavailability of drugs, solubility, toxicity safeguard, pharmacological activities, distribution, and prevention from chemical and physical degradation and increased stability of drugs inside the body [210]. Nanomedicines have shown higher capacity to bind with biomolecules as well as reduction of inflammation/oxidative stress in tissues. Thousands of different nanomedicines have been designed over the years; they have various applications in different types of diseases. Few are approved for clinical use, and many more are in the phase of clinical trials.

The use of nanomaterials as drugs and medicine implies nanotechnologies for medical application with highly advanced medical intervention at molecular levels to cure diseases.

It provides a platform for the discovery of therapeutic nanomaterials or nanomedicines. The growth in nanomedicines has introduced numerous possibilities in medical sciences, specifically in the drug delivery mechanisms. Their structural characteristics make them an excellent mode for targeting at specific sites and quick penetration inside the cell/diseased sites [211]. Depending on their application and origin, various types of NPs were discovered based on therapeutic need. Liposomal, polymeric protein, metal based, and iron oxide NPs have emerged as top-notcher. In this review, our discussion is mainly focused on the application of metal NPs.

3.3.1. Silver NPs (AgNPs)

Silver NPs are considered ideal because of their unique properties like catalytic activity and stability. They also contain anti-viral, anti-bacterial, and anti-fungal properties. One of the applications of AgNPs is used in the antibacterial nanodevices, because of its Ag^+ ion effect. They can be positively used as anti-cancer agents due to their anti-proliferative effect and ability to induce cell death [212]. AgNPs can be loaded/coated to reduce their toxicity and improve their biological retention time which allows specific targeting of cancerous cells. AgNPs from *Andrographis echioides* have been shown to inhibit the growth of MCF-2 cells and are widely used in human breast adenocarcinoma cell lines [213]. The viability of tumor cells declines with increase in AgNPs concentration. *Allium sativum* AgNPs have shown positive outcome in gastrointestinal carcinoma [214].

3.3.2. Gold NPs (AuNPs)

Gold NPs have anti-cancer properties and induce oxidative stress. They absorb photons and convert those incident photons to heat that destroys cancerous cells. Cationic gold NPs (2 nm diameter) are toxic at some dose [215]. Gold NPs exist in non-oxidized state. Smaller NPs had less protein-to-protein ratio as compared to larger ones. Reports show that gold NPs treated with Hela cervical carcinoma demonstrated increased reactive oxygen species (ROS), leading to oxidation of lipid, proteins, and other several molecules [216]. AuNPs of 10 nm size were widely distributed in organs whereas 50–250 nm (large) NPs were found to be distributed in liver, spleen, and blood when intravenously injected.

3.3.3. Iron Oxide NPs

Iron oxide NPs size lies between 1 and 100 nm in diameter. Their two main forms are magnetite (Fe_2O_3) and maghemite (γ-Fe_2O_3). Iron oxide NPs were synthesized by the process of precipitation in isobutanol (acting as surfactant) along with ammonium hydroxide and sodium hydroxide [217]. Iron oxide NPs are unstable in aqueous media without any surface coating; they aggregate and precipitate in vivo. The aggregates that are formed by unstable iron NPs inside the blood are sequestered by macrophages. Iron oxide NPs must be coated with different moieties to minimize the aggregation in certain conditions. Iron oxide magnetic NPs have many applications in anti-cancer strategy called hyperthermia; they destroy tissues nearby by generating heat. Iron oxide NPs have many properties like high solubility, stability, distribution, biocompatibility, and prolonged circulation time [218]. To increase the in vivo tumor imaging sensitivity, it is important to deliver large concentration of NPs in both tumor cells and in tumor mass. In a study, it was found that iron oxide NPs cause significant cellular morphological modifications, inducing apoptosis and necrosis in MCF-7 cell lines [219].

However, there is a need for restraint on the use and applications of these nanomaterials as drugs. A detailed understanding on possible hazards and toxicological impact of NPs on the environment as well as human health is needed prior to its application. Understanding the mechanisms of NPs access into the body, their function at cellular level, and their influences on public health is the call of the hour. Nanomaterial's characterization and understanding their surface functionality inside living systems are critical to understand their possible toxicological effects. All these parameters need more detailed studies before the approval of nanomaterial-based drugs for human usage.

3.4. Wastewater Treatment Process

The increase in population growth rate, industrialization, and excessive use of chemicals has contaminated the aquatic environment by releasing wastewater to the environment. The water from natural resources is not suitable for consumption due to the presence of organic (dyes, pesticides, surfactants, etc.), inorganic (fluoride, arsenic, copper, mercury, etc.), biological (algae, bacteria, viruses, etc.), and radiological contaminants (cesium, plutonium, uranium, etc.) [220,221]. Figure 6A depicts some of the common contaminants found in water. Several techniques such as physical, chemical, and biological have been adopted for the treatment of wastewater. However, search for new efficient technologies to improve water purification at low-cost is the current research focus. Currently, nanotechnology provides a new strategy for the removal of contaminants from wastewater with high efficiency. Several approaches have been developed in combination with various NPs for the successful removal of contaminants from wastewater as shown in Figure 6B.

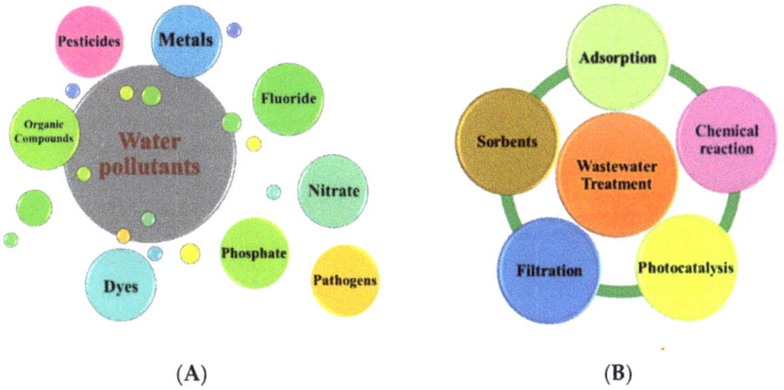

Figure 6. Common pollutants found in water and the treatment processes. (**A**) Common pollutants present in water. (**B**) Treatment approaches used for wastewater using nanoparticles.

The adsorption process is defined as the attachment of gaseous or liquid molecules over the surface of the solid and forms a layer or film of molecules. This process is mainly limited to the surface of the adsorbent where the adsorbate accumulates. The adsorption process could be physisorption or chemisorption depending upon the nature of bonding between the adsorbate and adsorbent i.e., van der Waals forces, covalent bonding, or electrostatic attraction. Adsorption is the most commonly used technique for the removal of contaminants from water due to its low-cost, easy operation, and absence of secondary pollutants formation [146]. Due to the development of nanotechnology and its wide applications in the past few decades, several nanostructured materials have been explored as adsorbent for their potential application in the treatment of industrial effluents, surface water, groundwater, and drinking water [222,223]. Nanoadsorbents exhibit higher efficiency and faster adsorption rate compared to the conventional adsorbent due to their small size, high porosity, and large active surface area [224]. Additionally, these nanoadsorbents show high reactivity and catalytic efficiency. Nanomaterials such as carbon nanotubes (CNTs), ferric oxide (Fe_3O_4), graphene, titanium oxide (TiO_2), manganese oxide (MnO_2), zinc oxide (ZnO), and magnesium oxide (MgO) are successfully used as adsorbent for the removal of contaminants such as heavy metals, azo dyes, etc. from the water [223,225]. Several nanosized metal oxide adsorbents including ferric oxide, aluminum oxides, manganese oxides, titanium oxides, magnesium oxides, and cerium oxides are proved to be promising for the removal of pollutants from water [226,227]. Furthermore, different metal oxide NPs are superparamagnetic which allows the easy separation of these adsorbents from the reaction mixture with the application of an external magnetic field. Das et al. [228] reported the removal of methylene blue dye, Cu(II), and

Co(II) from aqueous solution using green synthesized magnetite NPs from crude latex of *Jatropha curcas* and leaf extract of *Cinnamomum tamala*. The removal of Cd(II) from contaminated solution was studied by using silver NPs prepared using leaf extract of Ficus tree (*Ficus Benjamina*) [229]. In a similar type of study, the removal of cadmium ions from contaminated solution was carried out using iron oxide NPs prepared by co-precipitation method with tangerine peel extract [230]. Maximum removal efficiency 90% achieved at pH 4.0 and adsorbent dose of 0.4 g/100 mL. Zinc oxide NPs synthesized from Aloe vera and Cassava starch used as copper ion adsorbent and higher removal efficiency was observed for Aloe vera synthesized NPs with the increase in adsorbate concentration [231]. These nanoadsorbents demonstrate remarkable efficiency in the removal of pollutants from wastewater; however, the toxicity of residual NPs in the wastewater and reduced potential activity due to the use of a huge number of NPs in the treatment process to minimize the process duration are the major shortcomings of this process [232].

Filtration of contaminated water or wastewater through membranes is another way to remove the pollutants from the water. Nanofiltration is efficient and effective for the removal of different types of contaminants (organic, heavy metals, pathogens, etc.) from wastewater, and the removal efficiency is mainly dependent upon the pore size and charge characteristics of the membrane [233]. Numerous studies have focused on the development and use of a composite membrane, prepared by the introduction of NPs into the polymeric or inorganic membranes for the treatment of water. The incorporation of metal oxide NPs like silica [234], alumina [235,236], zeolite [237], and TiO_2 [161,238] into polymeric membrane improved the membrane hydrophobicity and permeability. In addition to this, the incorporation of antimicrobial NPs like silver NPs into membrane matrix hinders bacterial attachment and biofilm formation [239,240].

Metal NPs are extensively used as nano-catalysts in water treatment due to their high surface-to-volume ratio and surface catalytic activity. These nano-catalysts improve the quality of water by degrading various contaminants, viz. dyes, pesticides, herbicides, polychlorinated biphenyls, nitro aromatics, etc. [241]. Various kinds of nano-catalysts such as electrocatalysts, photocatalysts, and Fenton-based catalysts are employed in the wastewater treatment process [220]. The mechanism behind photocatalysis is the photoexcitation of electron present in the catalysts. The light irradiation causes the generation of holes (h^+) and exited electrons (e^-). Further, the generated holes (h^+) are trapped by water molecules (H_2O) in aqueous media that subsequently form the hydroxyl radicals ($^{\bullet}OH$) [242]. These hydroxyl radicals are highly reactive and powerful oxidizing agents which oxidize the organic pollutants leading to the formation of water and gaseous degradation products [242]. Numerous studies reported photocatalytic activity of green synthesized Ag, Au, Pt, and Pd NPs in degradation of different dyes [63,243–247]. Additionally, various metal oxide NPs such as ZnO, CuO, FeO, SnO_2, TiO_2, NiO, CeO_2, etc. exhibited excellent photocatalytic activity for the degradation of different organic pollutants [248].

3.5. *Antimicrobial Activity*

In the past decade, application of nanomaterials to control microbial proliferation has garnered much interest from scientists worldwide [249,250]. The increase in resistance of microorganisms to antimicrobial agents, including antibiotics, has led to a spike in health-related complications. A vast body of work has revealed that by combining three forces of material science, nanotechnology, and the inherent antimicrobial activity possessed by certain metals, innovative applications for metal NPs can be identified [251]. Previous studies have reported that metal and their counterpart metal oxide nanoparticles have displayed toxicity towards numerous microorganisms [209,250]. These NPs may be used successfully to stop the growth of various bacterial species.

The surge in development of multi-drug resistant pathogens is presenting itself as a grave problem to public health, and thus, several studies have been conducted at improving the prevailing antimicrobial treatments [251]. It has been identified that approximately 70% of bacterial infections have developed resistance to one or more of the first- and second-line drugs that have been traditionally used to treat the infection [252]. The development of resistance in bacteria to commonly used chemical antibacterial agents may occur due to the lengthy production-consumption cycle, thus leading to reduction in efficacy. Moreover, the rampant use of poor quality or over-the-counter medicines in developing countries has led to a steep rise in antimicrobial resistance [253]. The need of the hour is to speed up the research and development and the synthesis of novel antimicrobial agents which are effective as well. NPs as antibacterial agents have turned out to be an emerging technology against this challenge, which have the ability to establish an effective nanostructure, which may be used to deliver the antibacterial agents, hence targeting the bacterial growth locally and more efficiently. In addition, nanoparticles have proved to have the potency that it leaves the pathogens with little device to develop resistance against them. Most of the available metal oxide NPs have zero toxicity for mammalian cells at the concentrations that have been used to kill bacterial cells, which in turn is an advantage for using them at a larger scale [254].

Metals like gold (Au), silver (Ag), titanium (Ti), copper (Cu), and zinc (Zn) are known to have their own properties and potency and display differential activity against microorganisms. This information has been understood and utilized across various cultures for centuries [255]. Numerous kinds of nanoparticles and their derivatives have been explored for their potential antimicrobial effects against several microorganisms. Metal nanoparticles such as gold (Au), silver (Ag), silicon (Si), silver oxide (Ag_2O), titanium dioxide (TiO_2), zinc oxide (ZnO), copper oxide (CuO), calcium oxide (CaO), and magnesium oxide (MgO) have been recognized to display antimicrobial activity. In vitro studies have suggested that metal nanoparticles have the potential to inhibit several microbial species, like *Escherichia coli, Staphylococcus aureus, Bacillus subtilis, Pseudomonas aeruginosa,* etc. [256–265].

The type of materials used in formulating the nanoparticles along with their particle size are the two most significant parameters, which can have an effect on the effectiveness of antimicrobial activity. It is well established that nanoparticles tend to possess different characteristics when compared to the same material having significantly greater dimensions. This is because the surface to volume ratio of the NPs considerably increases with a decrease in the particle size [266]. Certainly, in dimensions of nanoscale, the fraction of the molecule surface noticeably increases, which in turn can lead in improvement of some of the properties of the particles. For example, it may be mass transfer, heat treatment, catalytic activity, or the dissolution rate [267]. Additionally, the morphology and physicochemical properties of NPs have also been demonstrated to wield an effect on their level of germicidal activities. Literature survey has pointed that the particle size plays a role as vital parameter that can determine the effectiveness of antimicrobial activity of the metal nanoparticles [268,269]. The use of combination therapy with metal nanoparticles has the potential to be a strategy that can help tide over the emergence of bacterial resistance to multiple antibacterial agents [270,271]. More studies need to be developed to understand if green synthesized nanoparticles have better efficacy over traditionally synthesized nanoparticles. Current studies have displayed the same level of antimicrobial effects [272,273].

The shape of nanoparticles also has major influence on their antimicrobial effects [268]. Several research studies have investigated these shape-dependent characteristics of nanoparticles. A study described antibacterial activity of Ag NPs in three dissimilar shapes, namely, spherical, rod-shaped, and truncated triangular. It came to the conclusion that the truncated triangular NPs were more inclined to be reactive owing to their high atom density surfaces and consequently displayed greater antimicrobial activity [274]. In another study, the size and shape-dependent antimicrobial activity of fluorescent Ag nanoparticles (1–5 nm) was studied against some selected Gram-positive and Gram-negative bacteria [275]. They highlighted that the size and shape of the particles generated an effect on its activity. These investigations reported that the smaller the particles size, the easier they breach the cell wall exhibiting heightened antimicrobial activity. Furthermore, the authors proposed that these AgNPs could be used for multiple diverse procedures such as wound dressing, biofilms, bio-adhesives, and coating of certain biomedical materials. It was also found that antimicrobial property of TiO_2 is related to the size, shape, and structure of its crystal [252]. It is proposed in this particular study that the generation of ROS, ultimately leading to development of oxidative stress in the cells, may be a significant mechanism for TiO_2 nanoparticles to show its germicidal activity. It is well known then that ROS has the capability to cause site specific DNA damage, ultimately leading to the death of the cell.

The exact mechanisms in which nanometals present the antibacterial effect are still an area of active investigation. However, two common options have been proposed in this aspect. Firstly, toxicity associated with free metal ions can arise due to the dissolution of the metals from the surface of NPs. Secondly, oxidative stress could be triggered through the production of reactive oxygen species (ROS) on the surface of the NPs. Based on literature review, there are some intrinsic factors that can have an influence on the ability of nanomaterials in reducing the number of cells or completely eliminating the cells [255].

NPs are a promising technology, and owing to its vast application, understanding nanotoxicity and its consequences is of utmost importance. For decades, the pharmaceutical industry has used NPs as a tool to reduce toxicity and side effects of drugs [276]; nonetheless, one needs to be careful when using NPs, as certain safety concerns still exist. Several reports have identified damage to neurological and respiratory organs issues in the circulatory system. In addition, other yet unknown toxic effects of NPs are few of the foremost apprehensions in using NPs as part of a systemic therapy. Undeniably, numerous NPs seem non-toxic, and a few of them are reduced to having non-toxic properties, which ultimately has beneficial effects on health [209,277–279]. Nevertheless, further studies need to be executed focusing on minimizing the toxicity of metal and metal oxide NPs, that will eventually be applied in therapy as a proper substitute to disinfectants and antibiotics especially in biomedical applications. Moreover, current research should make the application of antimicrobial activity of NPs in eradication of microbial infections as one of their priorities.

4. Toxicity of Metal Nanoparticles

NPs have wide applicability in different sectors such as electronics, agriculture, chemicals, pharmaceutical, food, etc. due to their unique physicochemical properties [280]. The most commonly used NPs by various sectors include metal oxide NPs such as silicon oxide (SiO_2), titanium dioxide (TiO_2), zinc oxide (ZnO), aluminum hydroxide [$Al(OH)_3$], cerium oxide (CeO), copper oxide (CuO), silver (Ag), nanoclays, carbon nanotubes, nanocellulose, etc. [281,282]. However, massive release of NPs into the environment (air, water, and soil) by various industries is resulting in production of nanowaste and proving to be dangerous for the living organisms and causing threat to ecosystem balance. Various characteristics of NPs affecting their toxicity are size, nature, reactivity, mobility, stability, surface chemistry, aggregation, storage time, etc. NPs cause adverse consequences on human health and animals. Use of NPs has intensified the risk of various diseases in humans such as diabetes, cancer, bronchial asthma, allergies, inflammation, etc. [3]. The animal reproductive system has also been shown to be affected due to toxicity of various NPs such as Au, TiO_2, etc. [283,284]. NPs enter the animal body through ingestion and inhalation and get absorbed by the cells through the processes of phagocytosis and endocytosis and induce the generation of reactive oxygen species (ROS), ultimately resulting into lipid peroxidation, mitochondrial damage, etc. Different NPs such as Ag, Cu, ZnO, Ni, etc. have also reduced the enzymatic activity of various microorganisms. In addition, excessive production of NPs is also affecting the food web of the ecosystem [285]. Toxicity effects of NPs over plants, animals, and microorganisms are shown in Figure 7.

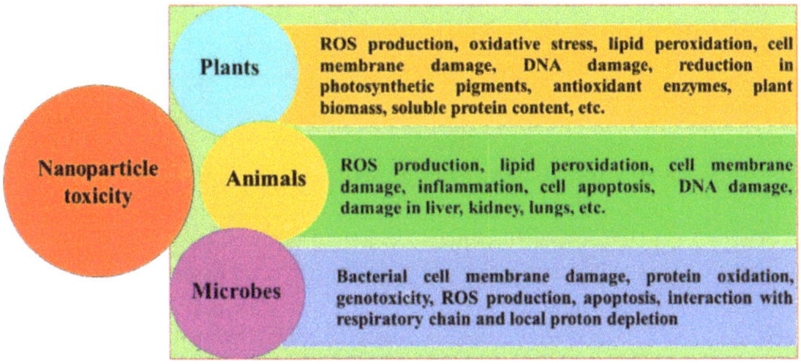

Figure 7. Toxicity effects of NPs on microbes, plants, and animals.

4.1. Impact of Nanoparticle Toxicity on Plants

Plants are of high fundamental significance as they perform photosynthesis and release oxygen in the atmosphere. As all the plant parts (roots, shoots, and leaves) are in direct contact with environmental matrices (air, water, and soil), they may get affected more by the NPs contamination as compared to the other living beings. NPs present in the atmosphere can enter into the plant body through the stomatal openings on leaves [286], while those present in soil and water can be selectively up taken by the plant roots [287].

NPs proved to be toxic to plants and hamper their growth and development. The toxicity in plants is mainly due to ROS production, causing lipid peroxidation and ultimately leading to DNA damage, reduction in photosynthetic pigments, plant biomass, soluble protein content reduction, etc. [288]. However, plants have a defensive system against oxidative stress in the form of enzymatic and non-enzymatic antioxidants, which may become inefficient under higher oxygen concentrations [289].

Assessment of NPs Phytotoxic and Genotoxic Effects on the Plants

While phytotoxic effects of NPs on the plant can be assessed by analyzing morphological and physiological changes, genotoxicity can be observed by observing DNA damage in the plant cell. By evaluating various physiological parameters such as germination, biomass production, leaf number, photosynthetic ability, root and shoot length, etc., phytotoxicity levels of NPs on the plants can be assessed. To observe genotoxic effects on the plant, assessment of cytology of plant roots for the determination of mitotic index, chromosomal abnormalities, etc. is the simplest approach [290].

SiO_2 NPs are among the top metal oxide NPs produced by various industries such as cosmetics, pharmaceuticals, food, etc. due to their worldwide demand. Positive effects of SiO_2 NPs on various plants such as *Oryza sativa*, *V. faba*, *S. lycopersicum*, *Medicago sativa*, etc. have been shown in different studies [291–294]. Toxic effects of SiO_2 NPs on plants have been investigated by Slomberg and Schoenfisch [295] in *Arabidopsis thaliana*, Karimi and Mohsenzadeh [296] in *Triticum aestivum*, and Silva and Monteiro [297] in *A. cepa*, etc. Regarding TiO_2 NPs, negative effects have been shown in *Z. mays* and *Vicia narbonensis* [298], *V. narbonensis* [299], *V. faba* [300,301], etc. ZnO NPs are majorly used in agriculture sector (as pesticides, fertilizers, etc.) [302].

In green algae, *Chlorella vulgaris* and *Dunaliella tertiolecta*, AgNPs proved to be inhibitory for their growth and displayed higher ROS production and lipid peroxidation [303]. AgNPs also proved to be toxic for the growth and development of green algae, *Pithophora oedogonia* and *Chara vulgaris* [304].

4.2. Toxicity of Nanoparticle-Based Drugs

The commercial applications of NPs as therapeutics for treatment of diseases is a double-edged sword. Even though many studies are being done worldwide to analyze the toxic effects of NM exposure, the possible mechanism of NMs interactions with biological systems and their consequences are still unknown. Research has shown that NPs can travel through the bloodstream and easily cross membrane barriers. This in turn can adversely affect tissues and organs at molecular and cellular levels [305]. NPs have demonstrated the capacity to cross the blood–brain barrier (BBB) and gain access to the brain [306]. Small size, large surface area to mass ratio (SA/MR), and surface characteristics determine nanoparticle's interaction with biological milieu and the resultant toxic effects that ensue. The unique nature of the NMs allow them to easily pass-through cell and tissue membranes and cellular compartments to cause cellular damage. The large SA/MR of NPs also remains open for active chemical interactions with cellular macromolecules. Increase in surface area of the identical chemical further enhances adsorption properties, surface reactivity, and potential toxicity [305].

NPs have the tendency to translocate across cell barriers from the entry point i.e., the respiratory tract to secondary organs, reach the cells by various mechanisms and start interacting with subcellular structures. These properties make NPs uniquely suitable for therapeutic and diagnostic uses. NPs are transported neuronally, involving retrograde and anterograde movement in axons and dendrites as well as perineural translocations. The target organs such as the central nervous system (CNS), however, bear the brunt of potential adverse effects (e.g., oxidative stress) [306]. The size of NPs has an important role in renal clearance and in avoiding immune activation, enhancing the efficacy and circulation time of the drug inside systemic circulation [209]. In case of nickel NPs, small particle sizes (less than 200 nm) are preferable for entering into epithelial cells, whereas larger NPs are phagocytosed by macrophages present. Another issue is surface charge which restraints the fate of NPs. Positively charged surface NPs with amorphous nature do not enter inside the cells, whereas, negatively charged crystalline nickel sulphide and sub-sulphide particles can enter cells by phagocytosis. Inhalation of MnO_2 NPs leads to the formation of ROS causing oxidative stress in brain [210].

Silver NPs of different sizes, i.e., 20 or 40 nm (Ag20Pep and Ag40Pep) were analyzed in THP-1-derived human macrophages through their cellular uptake. Results demonstrated a majority of the AgNPs spread throughout the cells. Formation of protein carbonyls or induction of heme oxygenase I are some of the associated responses due to oxidative stress are also observed. The charged Au NPs sized 15 nm cause cell death by apoptosis, whereas neutral Au NPs cause necrosis in HaCaT (human epidermal keratinocyte) cell lines [307]. Several NPs can penetrate inside the nuclear envelop, and they play an important role in inducing genotoxicity. Silver NPs have been found to be more toxic than gold NPs. TiO_2 NPs are considered as biological inert material in vitro and in vivo, while TiO_2 NPs larger than 15 μm are highly toxic, generating ROS. TiO_2 is toxic to PC12 cells [308]. It was reported that toxicity of NPs increases with increasing surface charges, i.e., lower positive charge NPs have less electrostatic interaction with the cells. Positive ZnO NPs have more cytotoxic effect in A549 cells as compared to same sized and shaped negatively charged ZnO [309]. Rod shaped Fe_2O_3 NPs produce high cytotoxic responses compared to spherical Fe_2O_3 NPs in RAW 264.7 (murine macrophage) cell lines [310]. Rod-shaped CeO_2 NPs cause more toxic effects and produce LDH and necrosis factor-α in RAW 264.7 cells [209]. NPs cytotoxicity depends on assay, cell line, and physical and chemical properties. Copper oxide NPs have been found to produce toxic side effects in liver and kidneys when examined on lab animals; after oral administration and interaction with gastric juice, they form reactive ionic copper. It was reported that silver NPs and iron oxide NPs can penetrate and cross the blood–brain barrier [306]. Iron oxide has the capacity to accumulate inside liver, spleen, lungs, and brain and has the capacity to cross the BBB after inhalation [305]. Iron oxide shows less cytotoxic effects at high concentration (300–500 kg/mL, 6 h) than in low concentration (25–200 μg/mL). At low concentration, it generates Reactive oxygen species (ROS), DNA damage, and causes lipid peroxidation. Silica based NPs of size 70 nm at 30 mg/kg concentration have been found to alter biochemical parameters [210]. Hence, a number of studies show evidence of NPs causing DNA and membrane damage, protein misfolding, and mitochondrial damage.

In a clinical scenario, several factors need to be addressed, viz., bioavailability, adverse reactions, cellular interactions, biodistribution, biodegradation, etc. The successful clinical translation of nanomedicines is therefore a long and onerous process of weighing the benefits against risks of toxicity involved. Scientists must proceed with caution and refrain from premature launch of nanomedicines without assessing the adverse effects involved.

Compared to physio-chemical methods, biologically synthesized NPs have been proved to be non-toxic/less toxic due to non-use of external stabilizing agents and hazardous chemicals/solvents during the synthesis process. Hasan et al. [164] compared the morphological, physiological, and biochemical responses of biologically and chemically synthesized iron oxide NPs in *Zea mays*. The biological synthesized FeO NPs promoted better plant growth as compared to the chemically derived NPs. The chemically synthesized NPs proved toxic and hampered the plant growth even at lower concentrations. In a similar kind of study, Anna et al. [311] observed better growth of green algae *P. kessleri* while using biologically synthesized NPs compared to chemical synthesized NPs.

5. Major Challenges and Future Perspective

In the recent past, research on NPs and their potential applications have progressed by leaps and bounds. Numerous studies have reported the green synthesis of metallic NPs using various biological sources such as plants, bacteria, fungi, and yeast. However, several challenges persist, which limit its large-scale production and consequent applications. Some of the major challenges observed during the synthesis are summarized below:

- Detailed optimization studies on reactants (plant extract, microorganism inoculum, fermentation medium composition, etc.) and process parameters (temperature, pH, rotational speed, etc.) are required to control the size and shape of the NPs.
- Studies also need to be focused on enhancing various physicochemical characteristics of NPs for specific applications.
- The involvement of each metabolite of plant extract and cellular components of microorganism in the synthesis of NPs should be completely analyzed.
- Scale-up of NPs production for commercial purposes using green synthesis methods needs to be prioritized.
- Improvement of NPs yield and stability with reduced reaction time is needed by optimizing various reaction parameters.

Addressing these challenges could make the green synthesis methods cost-effective and comparable to the conventional methods for the large-scale production of NPs. Additionally, the separation and purification of NPs from the reaction mixture is another important aspect that need to be explored. A detailed toxicological study of the NPs on plants and animals is necessary for expanding its application in diverse fields. In addition to wild type strains, genetically modified microorganisms with the ability to produce greater quantity of enzymes, proteins, and biomolecules could further enhance the biosynthesis as well as the stabilization of NPs. Further, enhancement of metal accumulation capacity and tolerance of genetically modified microorganisms could provide a futuristic approach for the production and application of metal NPs using the green synthesis method.

6. Conclusions

The present review focuses on the green synthesis of metal NPs derived from plants and microorganism and their applications. Green synthesis methods provide a clean, non-toxic, and eco-friendly approach for the synthesis of metal NPs compared to other conventional techniques like physical and chemical methods. A wide range of plant materials including leave extract, fruit extract, seed, fruit, bark, etc. and microorganism such as bacteria, fungi, actinomycetes, etc. have shown potential for synthesis of various metal and metal oxide NPs (viz., Au, Ag, Pt, Pd, Ni, Se, Cu, CuO, and TiO_2). The size and shape of NPs and the reaction rate strongly depend on various experimental parameters such as reaction time, reactant concentration, pH, temperature, aeration, salt concentration, etc. Different characterization techniques such as UV-VIS spectroscopy, FTIR, XRD, SEM,

TEM, EDX, and AFM have been used to determine the shape, size, and morphology of biosynthesized NPs. However, in terms of translational research, several factors, viz., bioavailability, adverse reactions, cellular interactions, biodistribution, and biodegradation, need to be addressed. The accumulation of these NPs in the environment and their uptake by biological systems can lead to disastrous consequences as a number of studies show evidence of NPs causing DNA and membrane damage, protein misfolding, and mitochondrial damage. Although numerous studies reported the biological synthesis of metal NPs, a thorough investigation is the need of the hour for widening their applications and successful commercialization.

Author Contributions: Conceptualization, P.K.D.; writing—original draft preparation, P.K.D., J.K., A.K.D., S.S. (Soumi Sadhu), S.S. (Sunita Sharma), S.S. (Swati Singh); writing—review and editing, P.K.D., J.K., A.K.D., S.S. (Soumi Sadhu), S.S. (Sunita Sharma), S.S. (Swati Singh), P.K.G., B.S.K. All authors have read and agreed to the published version of the manuscript.

Funding: This research received no external funding.

Data Availability Statement: Not applicable.

Conflicts of Interest: The authors declare no conflict of interest.

References

1. Sahoo, M.; Vishwakarma, S.; Panigrahi, C.; Kumar, J. Nanotechnology: Current applications and future scope in food. *Food Front.* **2021**, *2*, 3–22. [CrossRef]
2. Maurer-Jones, M.A.; Gunsolus, I.L.; Murphy, C.J.; Haynes, C.L. Toxicity of engineered nanoparticles in the environment. *Anal. Chem.* **2013**, *85*, 3036–3049. [CrossRef] [PubMed]
3. Pandey, R.K.; Prajapati, V.K. Molecular and immunological toxic effects of nanoparticles. *Int. J. Biol. Macromol.* **2018**, *107*, 1278–1293. [CrossRef] [PubMed]
4. Yu, H.; Park, J.Y.; Kwon, C.W.; Hong, S.C.; Park, K.M.; Chang, P.S. An overview of nanotechnology in food science: Preparative methods, practical applications, and safety. *J. Chem.* **2018**, *2018*, 5427978. [CrossRef]
5. Vance, M.E.; Kuiken, T.; Vejerano, E.P.; McGinnis, S.P.; Hochella, M.F., Jr.; Rejeski, D.; Hull, M.S. Nanotechnology in the real world: Redeveloping the nanomaterial consumer products inventory. *Beilstein J. Nanotechnol.* **2015**, *6*, 1769–1780. [CrossRef]
6. Das, R.K.; Pachapur, V.L.; Lonappan, L.; Naghdi, M.; Pulicharla, R.; Maiti, S.; Cledon, M.; Dalila, L.M.A.; Sarma, S.J.; Brar, S.K. Biological synthesis of metallic nanoparticles: Plants, animals and microbial aspects. *Nanotechnol. Environ. Eng.* **2017**, *2*, 1–21. [CrossRef]
7. Thakkar, K.N.; Mhatre, S.S.; Parikh, R.Y. Biological synthesis of metallic nanoparticles. *Nanomedicine* **2010**, *6*, 257–262. [CrossRef]
8. Iravani, S.; Korbekandi, H.; Mirmohammadi, S.V.; Zolfaghari, B. Synthesis of silver nanoparticles: Chemical, physical and biological methods. *Res. Pharm. Sci.* **2014**, *9*, 385. [PubMed]
9. Banne, S.V.; Patil, M.S.; Kulkarni, R.M.; Patil, S.J. Synthesis and characterization of silver nano particles for EDM applications. *Mater. Today* **2017**, *4*, 12054–12060. [CrossRef]
10. Tan, Y.; Dai, X.; Li, Y.; Zhu, D. Preparation of gold, platinum, palladium and silver nanoparticles by the reduction of their salts with a weak reductant-potassium bitartrate. *J. Mater. Chem.* **2003**, *13*, 1069–1075. [CrossRef]
11. Norris, C.B.; Joseph, P.R.; Mackiewicz, M.R.; Reed, S.M. Minimizing formaldehyde use in the synthesis of gold− silver core−shell nanoparticles. *Chem. Mater.* **2010**, *22*, 3637–3645. [CrossRef]
12. Mallick, K.; Witcomb, M.J.; Scurrell, M.S. Polymer stabilized silver nanoparticles: A photochemical synthesis route. *J. Mater. Sci.* **2004**, *39*, 4459–4463. [CrossRef]
13. Eluri, R.; Paul, B. Synthesis of nickel nanoparticles by hydrazine reduction: Mechanistic study and continuous flow synthesis. *J. Nanopart. Res.* **2012**, *14*, 1–14. [CrossRef]
14. Akbarzadeh, R.; Dehghani, H. Sodium-dodecyl-sulphate-assisted synthesis of Ni nanoparticles: Electrochemical properties. *Bull. Mater. Sci.* **2017**, *40*, 1361–1369. [CrossRef]
15. Pandey, G.; Singh, S.; Hitkari, G. Synthesis and characterization of polyvinyl pyrrolidone (PVP)-coated Fe_3O_4 nanoparticles by chemical co-precipitation method and removal of Congo red dye by adsorption process. *Int. Nano Lett.* **2018**, *8*, 111–121. [CrossRef]
16. Gupta, R.; Xie, H. Nanoparticles in daily life: Applications, toxicity and regulations. *J. Environ. Pathol. Toxicol. Oncol.* **2018**, *37*, 209–230. [CrossRef]
17. Hua, S.; De Matos, M.B.; Metselaar, J.M.; Storm, G. Current trends and challenges in the clinical translation of nanoparticulate nanomedicines: Pathways for translational development and commercialization. *Front. Pharmacol.* **2018**, *9*, 790. [CrossRef]
18. Sriramulu, M.; Shanmugam, S.; Ponnusamy, V.K. *Agaricus bisporus* mediated biosynthesis of copper nanoparticles and its biological effects: An in-vitro study. *Colloid Interface Sci. Commun.* **2020**, *35*, 100254. [CrossRef]

19. Kuppusamy, P.; Yusoff, M.M.; Maniam, G.P.; Govindan, N. Biosynthesis of metallic nanoparticles using plant derivatives and their new avenues in pharmacological applications—An updated report. *Saudi Pharm. J.* **2016**, *24*, 473–484. [CrossRef]
20. Mittal, A.K.; Chisti, Y.; Banerjee, U.C. Synthesis of metallic nanoparticles using plant extracts. *Biotechnol. Adv.* **2013**, *31*, 346–356. [CrossRef]
21. Narayanan, K.B.; Sakthivel, N. Green synthesis of biogenic metal nanoparticles by terrestrial and aquatic phototrophic and heterotrophic eukaryotes and biocompatible agents. *Adv. Colloid Interfac.* **2011**, *169*, 59–79. [CrossRef]
22. Nath, D.; Banerjee, P. Green nanotechnology—A new hope for medical biology. *Environ. Toxicol. Pharmacol.* **2013**, *36*, 997–1014. [CrossRef] [PubMed]
23. Gardea-Torresdey, J.L.; Parsons, J.G.; Gomez, E.; Peralta-Videa, J.; Troiani, H.E.; Santiago, P.; Yacaman, M.J. Formation and growth of Au nanoparticles inside live alfalfa plants. *Nano Lett.* **2002**, *2*, 397–401. [CrossRef]
24. Bali, R.; Harris, A.T. Biogenic synthesis of Au nanoparticles using vascular plants. *Ind. Eng. Chem. Res.* **2010**, *49*, 12762–12772. [CrossRef]
25. Mohamad, N.A.N.; Arham, N.A.; Jai, J.; Hadi, A. Plant extract as reducing agent in synthesis of metallic nanoparticles: A review. *Adv. Mat. Res.* **2014**, *832*, 350–355. [CrossRef]
26. Krishnaraj, C.; Muthukumaran, P.; Ramachandran, R.; Balakumaran, M.D.; Kalaichelvan, P.T. *Acalypha indica* Linn: Biogenic synthesis of silver and gold nanoparticles and their cytotoxic effects against MDA-MB-231, human breast cancer cells. *Biotechnol. Rep.* **2014**, *4*, 42–49. [CrossRef] [PubMed]
27. Dwivedi, A.D.; Gopal, K. Biosynthesis of silver and gold nanoparticles using *Chenopodium album* leaf extract. *Colloids Surf. A* **2010**, *369*, 27–33. [CrossRef]
28. Philip, D. Green synthesis of gold and silver nanoparticles using *Hibiscus Rosa Sinensis*. *Phys. E* **2010**, *42*, 1417–1424. [CrossRef]
29. Baghizadeh, A.; Ranjbar, S.; Gupta, V.K.; Asif, M.; Pourseyedi, S.; Karimi, M.J.; Mohammadinejad, R. Green synthesis of silver nanoparticles using seed extract of *Calendula officinalis* in liquid phase. *J. Mol. Liq.* **2015**, *207*, 159–163. [CrossRef]
30. Gurunathan, S.; Han, J.W.; Kwon, D.N.; Kim, J.H. Enhanced antibacterial and anti-biofilm activities of silver nanoparticles against Gram-negative and Gram-positive bacteria. *Nanoscale Res. Lett.* **2014**, *9*, 1–17. [CrossRef]
31. Pragathiswaran, C.; Violetmary, J.; Faritha, A.; Selvarani, K.; Nawas, P.M. Photocatalytic degradation, sensing of Cd^{2+} using silver nanoparticles synthesised from plant extract of *Cissus quadrangularis* and their microbial activity. *Mater. Today Proc.* **2021**, *45*, 3348–3356. [CrossRef]
32. Salih, T.A.; Hassan, K.T.; Majeed, S.R.; Ibraheem, I.J.; Hassan, O.M.; Obaid, A.S. In vitro scolicidal activity of synthesised silver nanoparticles from aqueous plant extract against *Echinococcus granulosus*. *Biotechnol. Rep.* **2020**, *28*, e00545. [CrossRef] [PubMed]
33. Dhar, S.A.; Chowdhury, R.A.; Das, S.; Nahian, M.K.; Islam, D.; Gafur, M.A. Plant-mediated green synthesis and characterization of silver nanoparticles using *Phyllanthus emblica* fruit extract. *Mater. Today Proc.* **2021**, *42*, 1867–1871. [CrossRef]
34. Chavan, R.R.; Bhinge, S.D.; Bhutkar, M.A.; Randive, D.S.; Wadkar, G.H.; Todkar, S.S.; Urade, M.N. Characterization, antioxidant, antimicrobial and cytotoxic activities of green synthesized silver and iron nanoparticles using alcoholic *Blumea eriantha* DC plant extract. *Mater. Today Commun.* **2020**, *24*, 101320. [CrossRef]
35. Kambale, E.K.; Nkanga, C.I.; Mutonkole, B.P.I.; Bapolisi, A.M.; Tassa, D.O.; Liesse, J.M.I.; Krause, R.W.; Memvanga, P.B. Green synthesis of antimicrobial silver nanoparticles using aqueous leaf extracts from three Congolese plant species (*Brillantaisia patula, Crossopteryx febrifuga and Senna siamea*). *Heliyon* **2020**, *6*, e04493. [CrossRef]
36. Vignesh, V.; Anbarasi, K.F.; Karthikeyeni, S.; Sathiyanarayanan, G.; Subramanian, P.; Thirumurugan, R. A superficial phyto-assisted synthesis of silver nanoparticles and their assessment on hematological and biochemical parameters in *Labeo rohita* (Hamilton, 1822). *Colloids Surf. A* **2013**, *439*, 184–192. [CrossRef]
37. Vivek, R.; Thangam, R.; Muthuchelian, K.; Gunasekaran, P.; Kaveri, K.; Kannan, S. Green biosynthesis of silver nanoparticles from *Annona squamosa* leaf extract and its in vitro cytotoxic effect on MCF-7 cells. *Process Biochem.* **2012**, *47*, 2405–2410. [CrossRef]
38. Zhang, Y.; Cheng, X.; Zhang, Y.; Xue, X.; Fu, Y. Biosynthesis of silver nanoparticles at room temperature using aqueous aloe leaf extract and antibacterial properties. *Colloids Surf. A* **2013**, *423*, 63–68. [CrossRef]
39. Jagtap, U.B.; Bapat, V.A. Green synthesis of silver nanoparticles using *Artocarpus heterophyllus* Lam. seed extract and its antibacterial activity. *Ind. Crops Prod.* **2013**, *46*, 132–137. [CrossRef]
40. Vidhu, V.K.; Philip, D. Catalytic degradation of organic dyes using biosynthesized silver nanoparticles. *Micron* **2014**, *56*, 54–62. [CrossRef]
41. Suriyakalaa, U.; Antony, J.J.; Suganya, S.; Siva, D.; Sukirtha, R.; Kamalakkannan, S.; Pichiah, P.T.; Achiraman, S. Hepatocurative activity of biosynthesized silver nanoparticles fabricated using *Andrographis paniculata*. *Colloids Surf. B* **2013**, *102*, 189–194. [CrossRef]
42. Jeyaraj, M.; Rajesh, M.; Arun, R.; MubarakAli, D.; Sathishkumar, G.; Sivanandhan, G.; Dev, G.K.; Manickavasagam, M.; Premkumar, K.; Thajuddin, N.; et al. An investigation on the cytotoxicity and caspase-mediated apoptotic effect of biologically synthesized silver nanoparticles using *Podophyllum hexandrum* on human cervical carcinoma cells. *Colloids Surf. B* **2013**, *102*, 708–717. [CrossRef] [PubMed]
43. Mittal, A.K.; Bhaumik, J.; Kumar, S.; Banerjee, U.C. Biosynthesis of silver nanoparticles: Elucidation of prospective mechanism and therapeutic potential. *J. Colloid Interface Sci.* **2014**, *415*, 39–47. [CrossRef] [PubMed]
44. Adewale, O.B.; Egbeyemi, K.A.; Onwuelu, J.O.; Potts-Johnson, S.S.; Anadozie, S.O.; Fadaka, A.O.; Osukoya, O.A.; Aluko, B.T.; Johnson, J.; Obafemi, T.O.; et al. Biological synthesis of gold and silver nanoparticles using leaf extracts of *Crassocephalum rubens* and their comparative in vitro antioxidant activities. *Heliyon* **2020**, *6*, e05501. [CrossRef]

45. Daisy, P.; Saipriya, K. Biochemical analysis of *Cassia fistula* aqueous extract and phytochemically synthesized gold nanoparticles as hypoglycemic treatment for diabetes mellitus. *Int. J. Nanomed.* **2012**, *7*, 1189. [CrossRef] [PubMed]
46. Thangamani, N.; Bhuvaneshwari, N. Green synthesis of gold nanoparticles using *Simarouba glauca* leaf extract and their biological activity of micro-organism. *Chem. Phys. Lett.* **2019**, *732*, 136587. [CrossRef]
47. Satpathy, S.; Patra, A.; Ahirwar, B.; Hussain, M.D. Process optimization for green synthesis of gold nanoparticles mediated by extract of *Hygrophila spinosa* T. Anders and their biological applications. *Phys. E* **2020**, *121*, 113830. [CrossRef]
48. Kumar, P.V.; Kala, S.M.J.; Prakash, K.S. Green synthesis of gold nanoparticles using *Croton Caudatus Geisel* leaf extract and their biological studies. *Mater. Lett.* **2019**, *236*, 19–22. [CrossRef]
49. Anand, K.; Gengan, R.M.; Phulukdaree, A.; Chuturgoon, A. Agroforestry waste *Moringa oleifera* petals mediated green synthesis of gold nanoparticles and their anti-cancer and catalytic activity. *J. Ind. Eng. Chem.* **2015**, *21*, 1105–1111. [CrossRef]
50. Sathishkumar, M.; Pavagadhi, S.; Mahadevan, A.; Balasubramanian, R. Biosynthesis of gold nanoparticles and related cytotoxicity evaluation using A549 cells. *Ecotox. Environ. Saf.* **2015**, *114*, 232–240. [CrossRef]
51. Gopinath, K.; Venkatesh, K.S.; Ilangovan, R.; Sankaranarayanan, K.; Arumugam, A. Green synthesis of gold nanoparticles from leaf extract of *Terminalia arjuna*, for the enhanced mitotic cell division and pollen germination activity. *Ind. Crops Prod.* **2013**, *50*, 737–742. [CrossRef]
52. Kumar, K.P.; Paul, W.; Sharma, C.P. Green synthesis of gold nanoparticles with *Zingiber officinale* extract: Characterization and blood compatibility. *Process Biochem.* **2011**, *46*, 2007–2013. [CrossRef]
53. Noruzi, M.; Zare, D.; Khoshnevisan, K.; Davoodi, D. Rapid green synthesis of gold nanoparticles using *Rosa hybrida* petal extract at room temperature. *Spectrochim. Acta A Mol. Biomol. Spectrosc.* **2011**, *79*, 1461–1465. [CrossRef] [PubMed]
54. Kumar, K.M.; Mandal, B.K.; Sinha, M.; Krishnakumar, V. *Terminalia chebula* mediated green and rapid synthesis of gold nanoparticles. *Spectrochim. Acta A Mol. Biomol. Spectrosc.* **2012**, *86*, 490–494. [CrossRef] [PubMed]
55. Guo, M.; Li, W.; Yang, F.; Liu, H. Controllable biosynthesis of gold nanoparticles from a *Eucommia ulmoides* bark aqueous extract. *Spectrochim. Acta A Mol. Biomol. Spectrosc.* **2015**, *142*, 73–79. [CrossRef]
56. Ganesan, R.M.; Prabu, H.G. Synthesis of gold nanoparticles using herbal *Acorus calamus* rhizome extract and coating on cotton fabric for antibacterial and UV blocking applications. *Arab. J. Chem.* **2019**, *12*, 2166–2174. [CrossRef]
57. Muniyappan, N.; Nagarajan, N.S. Green synthesis of gold nanoparticles using *Curcuma pseudomontana* essential oil, its biological activity and cytotoxicity against human ductal breast carcinoma cells T47D. *J. Environ. Chem. Eng.* **2014**, *2*, 2037–2044. [CrossRef]
58. Sujitha, M.V.; Kannan, S. Green synthesis of gold nanoparticles using Citrus fruits (*Citrus limon*, *Citrus reticulata* and *Citrus sinensis*) aqueous extract and its characterization. *Spectrochim. Acta A Mol. Biomol. Spectrosc.* **2013**, *102*, 15–23. [CrossRef]
59. Nasrollahzadeh, M.; Sajadi, S.M.; Maham, M. Green synthesis of palladium nanoparticles using *Hippophae rhamnoides* Linn leaf extract and their catalytic activity for the Suzuki-Miyaura coupling in water. *J. Mol. Catal. A Chem.* **2015**, *396*, 297–303. [CrossRef]
60. Khazaei, A.; Rahmati, S.; Hekmatian, Z.; Saeednia, S. A green approach for the synthesis of palladium nanoparticles supported on pectin: Application as a catalyst for solvent-free Mizoroki-Heck reaction. *J. Mol. Catal. A Chem.* **2013**, *372*, 160–166. [CrossRef]
61. Bankar, A.; Joshi, B.; Kumar, A.R.; Zinjarde, S. Banana peel extract mediated novel route for the synthesis of palladium nanoparticles. *Mater. Lett.* **2010**, *64*, 1951–1953. [CrossRef]
62. Yang, X.; Li, Q.; Wang, H.; Huang, J.; Lin, L.; Wang, W.; Sun, D.; Su, Y.; Opiyo, J.B.; Hong, L.; et al. Green synthesis of palladium nanoparticles using broth of *Cinnamomum camphora* leaf. *J. Nanoparticle Res.* **2010**, *12*, 1589–1598. [CrossRef]
63. Kalaiselvi, A.; Roopan, S.M.; Madhumitha, G.; Ramalingam, C.; Elango, G. Synthesis and characterization of palladium nanoparticles using *Catharanthus roseus* leaf extract and its application in the photo-catalytic degradation. *Spectrochim. Acta A Mol. Biomol. Spectrosc.* **2015**, *135*, 116–119. [CrossRef] [PubMed]
64. Kumar, K.M.; Mandal, B.K.; Kumar, K.S.; Reddy, P.S.; Sreedhar, B. Biobased green method to synthesise palladium and iron nanoparticles using *Terminalia chebula* aqueous extract. *Spectrochim. Acta A Mol. Biomol. Spectrosc.* **2013**, *102*, 128–133. [CrossRef] [PubMed]
65. Rabiee, N.; Bagherzadeh, M.; Kiani, M.; Ghadiri, A.M. *Rosmarinus officinalis* directed palladium nanoparticle synthesis: Investigation of potential anti-bacterial, anti-fungal and Mizoroki-Heck catalytic activities. *Adv. Powder Technol.* **2020**, *31*, 1402–1411. [CrossRef]
66. Kora, A.J.; Rastogi, L. Green synthesis of palladium nanoparticles using gum ghatti (*Anogeissus latifolia*) and its application as an antioxidant and catalyst. *Arab. J. Chem.* **2018**, *11*, 1097–1106. [CrossRef]
67. Joseph Kirubaharan, C.; Fang, Z.; Sha, C.; Yong, Y.C. Green synthesis of Ag and Pd nanoparticles for water pollutants treatment. *Water Sci. Technol.* **2020**, *82*, 2344–2352. [CrossRef]
68. Lebaschi, S.; Hekmati, M.; Veisi, H. Green synthesis of palladium nanoparticles mediated by black tea leaves (*Camellia sinensis*) extract: Catalytic activity in the reduction of 4-nitrophenol and Suzuki-Miyaura coupling reaction under ligand-free conditions. *J. Colloid Interface Sci.* **2017**, *485*, 223–231. [CrossRef]
69. Sheny, D.S.; Philip, D.; Mathew, J. Synthesis of platinum nanoparticles using dried *Anacardium occidentale* leaf and its catalytic and thermal applications. *Spectrochim. Acta A Mol. Biomol. Spectrosc.* **2013**, *114*, 267–271. [CrossRef] [PubMed]
70. Zheng, B.; Kong, T.; Jing, X.; Odoom-Wubah, T.; Li, X.; Sun, D.; Lu, F.; Zheng, Y.; Huang, J.; Li, Q. Plant-mediated synthesis of platinum nanoparticles and its bioreductive mechanism. *J. Colloid Interface Sci.* **2013**, *396*, 138–145. [CrossRef]
71. Raut, R.W.; Haroon, A.S.M.; Malghe, Y.S.; Nikam, B.T.; Kashid, S.B. Rapid biosynthesis of platinum and palladium metal nanoparticles using root extract of *Asparagus racemosus* Linn. *Adv. Mater. Lett.* **2013**, *4*, 650–654. [CrossRef]

72. Song, J.Y.; Kwon, E.Y.; Kim, B.S. Biological synthesis of platinum nanoparticles using *Diopyros kaki* leaf extract. *Bioproc. Biosyst. Eng.* **2010**, *33*, 159–164. [CrossRef] [PubMed]
73. Soundarrajan, C.; Sankari, A.; Dhandapani, P.; Maruthamuthu, S.; Ravichandran, S.; Sozhan, G.; Palaniswamy, N. Rapid biological synthesis of platinum nanoparticles using *Ocimum sanctum* for water electrolysis applications. *Bioproc. Biosyst. Eng.* **2012**, *35*, 827–833. [CrossRef]
74. Razavi, R.; Molaei, R.; Moradi, M.; Tajik, H.; Ezati, P.; Yordshahi, A.S. Biosynthesis of metallic nanoparticles using mulberry fruit (*Morus alba* L.) extract for the preparation of antimicrobial nanocellulose film. *Appl. Nanosci.* **2020**, *10*, 465–476. [CrossRef]
75. Lotha, R.; Shamprasad, B.R.; Sundaramoorthy, N.S.; Nagarajan, S.; Sivasubramanian, A. Biogenic phytochemicals (cassinopin and isoquercetin) capped copper nanoparticles (ISQ/CAS@ CuNPs) inhibits MRSA biofilms. *Microb. Pathog.* **2019**, *132*, 178–187. [CrossRef] [PubMed]
76. Khani, R.; Roostaei, B.; Bagherzade, G.; Moudi, M. Green synthesis of copper nanoparticles by fruit extract of *Ziziphus spina-christi* (L.) Willd: Application for adsorption of triphenylmethane dye and antibacterial assay. *J. Mol. Liq.* **2018**, *255*, 541–549. [CrossRef]
77. Rajesh, K.M.; Ajitha, B.; Reddy, Y.A.K.; Suneetha, Y.; Reddy, P.S. Assisted green synthesis of copper nanoparticles using *Syzygium aromaticum* bud extract: Physical, optical and antimicrobial properties. *Optik* **2018**, *154*, 593–600. [CrossRef]
78. Lin, Z.; Weng, X.; Owens, G.; Chen, Z. Simultaneous removal of Pb (II) and rifampicin from wastewater by iron nanoparticles synthesized by a tea extract. *J. Clean. Prod.* **2020**, *242*, 118476. [CrossRef]
79. Katata-Seru, L.; Moremedi, T.; Aremu, O.S.; Bahadur, I. Green synthesis of iron nanoparticles using *Moringa oleifera* extracts and their applications: Removal of nitrate from water and antibacterial activity against *Escherichia coli*. *J. Mol. Liq.* **2018**, *256*, 296–304. [CrossRef]
80. Radini, I.A.; Hasan, N.; Malik, M.A.; Khan, Z. Biosynthesis of iron nanoparticles using *Trigonella foenum-graecum* seed extract for photocatalytic methyl orange dye degradation and antibacterial applications. *J. Photochem. Photobiol. B* **2018**, *183*, 154–163. [CrossRef]
81. Liang, T.; Qiu, X.; Ye, X.; Liu, Y.; Li, Z.; Tian, B.; Yan, D. Biosynthesis of selenium nanoparticles and their effect on changes in urinary nanocrystallites in calcium oxalate stone formation. *3 Biotech* **2020**, *10*, 1–6. [CrossRef]
82. Yazhiniprabha, M.; Vaseeharan, B. In vitro and in vivo toxicity assessment of selenium nanoparticles with significant larvicidal and bacteriostatic properties. *Mat. Sci. Eng. C* **2019**, *103*, 109763. [CrossRef]
83. Menon, S.; KS, S.D.; Agarwal, H.; Shanmugam, V.K. Efficacy of biogenic selenium nanoparticles from an extract of ginger towards evaluation on anti-microbial and anti-oxidant activities. *Colloid Interf. Sci. Commun.* **2019**, *29*, 1–8. [CrossRef]
84. Din, M.I.; Nabi, A.G.; Rani, A.; Aihetasham, A.; Mukhtar, M. Single step green synthesis of stable nickel and nickel oxide nanoparticles from *Calotropis gigantea*: Catalytic and antimicrobial potentials. *Environ. Nanotechnol. Monit. Manag.* **2018**, *9*, 29–36. [CrossRef]
85. Sudhasree, S.; Shakila Banu, A.; Brindha, P.; Kurian, G.A. Synthesis of nickel nanoparticles by chemical and green route and their comparison in respect to biological effect and toxicity. *Toxicol. Environ. Chem.* **2014**, *96*, 743–754. [CrossRef]
86. Makarov, V.V.; Love, A.J.; Sinitsyna, O.V.; Makarova, S.S.; Yaminsky, I.V.; Taliansky, M.E.; Kalinina, N.O. "Green" nanotechnologies: Synthesis of metal nanoparticles using plants. *Acta Nat.* **2014**, *6*, 35–44. [CrossRef]
87. Si, S.; Mandal, T.K. Tryptophan-based peptides to synthesize gold and silver nanoparticles: A mechanistic and kinetic study. *Chem. Eur. J.* **2007**, *13*, 3160–3168. [CrossRef]
88. Anjana, P.M.; Bindhu, M.R.; Umadevi, M.; Rakhi, R.B. Antibacterial and electrochemical activities of silver, gold, and palladium nanoparticles dispersed amorphous carbon composites. *Appl. Surf. Sci.* **2019**, *479*, 96–104. [CrossRef]
89. Dhanavel, S.; Manivannan, N.; Mathivanan, N.; Gupta, V.K.; Narayanan, V.; Stephen, A. Preparation and characterization of cross-linked chitosan/palladium nanocomposites for catalytic and antibacterial activity. *J. Mol. Liq.* **2018**, *257*, 32–41. [CrossRef]
90. Fang, G.; Li, W.; Shen, X.; Perez-Aguilar, J.M.; Chong, Y.; Gao, X.; Chai, Z.; Chen, C.; Ge, C.; Zhou, R. Differential Pd-nanocrystal facets demonstrate distinct antibacterial activity against Gram-positive and Gram-negative bacteria. *Nat. Commun.* **2018**, *9*, 1–9. [CrossRef]
91. Klaus-Joerger, T.; Joerger, R.; Olsson, E.; Granqvist, C.G. Bacteria as workers in the living factory: Metal-accumulating bacteria and their potential for materials science. *Trends Biotechnol.* **2001**, *19*, 15–20. [CrossRef]
92. Mohamed, A.A.; Fouda, A.; Abdel-Rahman, M.A.; Hassan, S.E.D.; El-Gamal, M.S.; Salem, S.S.; Shaheen, T.I. Fungal strain impacts the shape, bioactivity and multifunctional properties of green synthesized zinc oxide nanoparticles. *Biocatal. Agric. Biotechnol.* **2019**, *19*, 101103. [CrossRef]
93. Alsamhary, K.I. Eco-friendly synthesis of silver nanoparticles by *Bacillus subtilis* and their antibacterial activity. *Saudi J. Biol. Sci.* **2020**, *27*, 2185–2191. [CrossRef] [PubMed]
94. Klaus, T.; Joerger, R.; Olsson, E.; Granqvist, C.G. Silver-based crystalline nanoparticles, microbially fabricated. *Proc. Natl. Acad. Sci. USA* **1999**, *96*, 13611–13614. [CrossRef] [PubMed]
95. Kalishwaralal, K.; Deepak, V.; Ramkumarpandian, S.; Nellaiah, H.; Sangiliyandi, G. Extracellular biosynthesis of silver nanoparticles by the culture supernatant of *Bacillus licheniformis*. *Mater. Lett.* **2008**, *62*, 4411–4413. [CrossRef]
96. Thomas, R.; Janardhanan, A.; Varghese, R.T.; Soniya, E.V.; Mathew, J.; Radhakrishnan, E.K. Antibacterial properties of silver nanoparticles synthesized by marine *Ochrobactrum* sp. *Braz. J. Microbiol.* **2014**, *45*, 1221–1227. [CrossRef] [PubMed]

97. Monowar, T.; Rahman, M.; Bhore, S.J.; Raju, G.; Sathasivam, K.V. Silver nanoparticles synthesized by using the endophytic bacterium *Pantoea ananatis* are promising antimicrobial agents against multidrug resistant bacteria. *Molecules* **2018**, *23*, 3220. [CrossRef]
98. Wypij, M.; Golinska, P.; Dahm, H.; Rai, M. Actinobacterial-mediated synthesis of silver nanoparticles and their activity against pathogenic bacteria. *IET Nanobiotechnol.* **2016**, *11*, 336–342. [CrossRef]
99. Husseiny, M.I.; Abd El-Aziz, M.; Badr, Y.; Mahmoud, M.A. Biosynthesis of gold nanoparticles using *Pseudomonas aeruginosa*. *Spectrochim. Acta A Mol. Biomol. Spectrosc.* **2007**, *67*, 1003–1006. [CrossRef]
100. He, S.; Zhang, Y.; Guo, Z.; Gu, N. Biological synthesis of gold nanowires using extract of *Rhodopseudomonas capsulata*. *Biotechnol. Progr.* **2008**, *24*, 476–480. [CrossRef]
101. Du, L.; Jiang, H.; Liu, X.; Wang, E. Biosynthesis of gold nanoparticles assisted by *Escherichia coli* DH5α and its application on direct electrochemistry of hemoglobin. *Electrochem. Commun.* **2007**, *9*, 1165–1170. [CrossRef]
102. Srinath, B.S.; Namratha, K.; Byrappa, K. Eco-friendly synthesis of gold nanoparticles by *Bacillus subtilis* and their environmental applications. *Adv. Sci. Lett.* **2018**, *24*, 5942–5946. [CrossRef]
103. Camas, M.; Camas, A.S.; Kyeremeh, K. Extracellular synthesis and characterization of gold nanoparticles using *Mycobacterium* sp. BRS2A-AR2 isolated from the aerial roots of the Ghanaian mangrove plant, *Rhizophora racemosa*. *Indian J. Microbiol.* **2018**, *58*, 214–221. [CrossRef]
104. Ahmed, E.; Kalathil, S.; Shi, L.; Alharbi, O.; Wang, P. Synthesis of ultra-small platinum, palladium and gold nanoparticles by *Shewanella loihica* PV-4 electrochemically active biofilms and their enhanced catalytic activities. *J. Saudi Chem. Soc.* **2018**, *22*, 919–929. [CrossRef]
105. Tuo, Y.; Liu, G.; Dong, B.; Yu, H.; Zhou, J.; Wang, J.; Jin, R. Microbial synthesis of bimetallic PdPt nanoparticles for catalytic reduction of 4-nitrophenol. *Environ. Sci. Pollut. Res.* **2017**, *24*, 5249–5258. [CrossRef] [PubMed]
106. Eramabadi, P.; Masoudi, M.; Makhdoumi, A.; Mashreghi, M. Microbial cell lysate supernatant (CLS) alteration impact on platinum nanoparticles fabrication, characterization, antioxidant and antibacterial activity. *Mat. Sci. Eng. C* **2020**, *117*, 111292. [CrossRef]
107. Wang, W.; Zhang, B.; He, Z. Bioelectrochemical deposition of palladium nanoparticles as catalysts by *Shewanella oneidensis* MR-1 towards enhanced hydrogen production in microbial electrolysis cells. *Electrochim. Acta* **2019**, *318*, 794–800. [CrossRef]
108. Che, L.; Dong, Y.; Wu, M.; Zhao, Y.; Liu, L.; Zhou, H. Characterization of selenite reduction by *Lysinibacillus* sp. ZYM-1 and photocatalytic performance of biogenic selenium nanospheres. *ACS Sust. Chem. Eng.* **2017**, *5*, 2535–2543. [CrossRef]
109. Wang, T.; Yang, L.; Zhang, B.; Liu, J. Extracellular biosynthesis and transformation of selenium nanoparticles and application in H_2O_2 biosensor. *Colloids Surfaces B* **2010**, *80*, 94–102. [CrossRef] [PubMed]
110. Alam, H.; Khatoon, N.; Khan, M.A.; Husain, S.A.; Saravanan, M.; Sardar, M. Synthesis of selenium nanoparticles using probiotic bacteria *Lactobacillus acidophilus* and their enhanced antimicrobial activity against resistant bacteria. *J. Clust. Sci.* **2020**, *31*, 1003–1011. [CrossRef]
111. AbdelRahim, K.; Mahmoud, S.Y.; Ali, A.M.; Almaary, K.S.; Mustafa, A.E.Z.M.; Husseiny, S.M. Extracellular biosynthesis of silver nanoparticles using *Rhizopus stolonifer*. *Saudi J. Biol. Sci.* **2017**, *24*, 208–216. [CrossRef] [PubMed]
112. Jalal, M.; Ansari, M.A.; Alzohairy, M.A.; Ali, S.G.; Khan, H.M.; Almatroudi, A.; Raees, K. Biosynthesis of silver nanoparticles from oropharyngeal *Candida glabrata* isolates and their antimicrobial activity against clinical strains of bacteria and fungi. *Nanomaterials* **2018**, *8*, 586. [CrossRef] [PubMed]
113. Kobashigawa, J.M.; Robles, C.A.; Ricci, M.L.M.; Carmarán, C.C. Influence of strong bases on the synthesis of silver nanoparticles (AgNPs) using the ligninolytic fungi *Trametes trogii*. *Saudi J. Biol. Sci.* **2019**, *26*, 1331–1337. [CrossRef] [PubMed]
114. Elamawi, R.M.; Al-Harbi, R.E.; Hendi, A.A. Biosynthesis and characterization of silver nanoparticles using *Trichoderma longibrachiatum* and their effect on phytopathogenic fungi. *Egypt. J. Biol. Pest Control* **2018**, *28*, 1–11. [CrossRef]
115. Ahmed, A.A.; Hamzah, H.; Maaroof, M. Analyzing formation of silver nanoparticles from the filamentous fungus *Fusarium oxysporum* and their antimicrobial activity. *Turk. J. Biol.* **2018**, *42*, 54–62. [CrossRef]
116. Lotfy, W.A.; Alkersh, B.M.; Sabry, S.A.; Ghozlan, H.A. Biosynthesis of silver nanoparticles by *Aspergillus terreus*: Characterization, optimization, and biological activities. *Front. Bioeng. Biotechnol.* **2021**, *9*, 633468. [CrossRef]
117. Mohanta, Y.K.; Nayak, D.; Biswas, K.; Singdevsachan, S.K.; Abd_Allah, E.F.; Hashem, A.; Alqarawi, A.A.; Yadav, D.; Mohanta, T.K. Silver nanoparticles synthesized using wild mushroom show potential antimicrobial activities against food borne pathogens. *Molecules* **2018**, *23*, 655. [CrossRef]
118. Bonilla, J.J.A.; Guerrero, D.J.P.; Sáez, R.G.T.; Ishida, K.; Fonseca, B.B.; Rozental, S.; López, C.C.O. Green synthesis of silver nanoparticles using maltose and cysteine and their effect on cell wall envelope shapes and microbial growth of *Candida* spp. *J. Nanosci. Nanotechnol.* **2017**, *17*, 1729–1739. [CrossRef]
119. Joshi, C.G.; Danagoudar, A.; Poyya, J.; Kudva, A.K.; Dhananjaya, B.L. Biogenic synthesis of gold nanoparticles by marine endophytic fungus—*Cladosporium cladosporioides* isolated from seaweed and evaluation of their antioxidant and antimicrobial properties. *Process Biochem.* **2017**, *63*, 137–144.
120. Tripathi, R.M.; Shrivastav, B.R.; Shrivastav, A. Antibacterial and catalytic activity of biogenic gold nanoparticles synthesised by *Trichoderma harzianum*. *IET Nanobiotechnol.* **2018**, *12*, 509–513. [CrossRef]
121. El Domany, E.B.; Essam, T.M.; Ahmed, A.E.; Farghali, A.A. Biosynthesis physico-chemical optimization of gold nanoparticles as anti-cancer and synergetic antimicrobial activity using *Pleurotus ostreatus* fungus. *J. Appl. Pharm. Sci.* **2018**, *8*, 119–128.

122. Shen, W.; Qu, Y.; Pei, X.; Li, S.; You, S.; Wang, J.; Zhang, Z.; Zhou, J. Catalytic reduction of 4-nitrophenol using gold nanoparticles biosynthesized by cell-free extracts of *Aspergillus* sp. WL-Au. *J. Hazard. Mater.* **2017**, *321*, 299–306. [CrossRef]
123. Kitching, M.; Choudhary, P.; Inguva, S.; Guo, Y.; Ramani, M.; Das, S.K.; Marsili, E. Fungal surface protein mediated one-pot synthesis of stable and hemocompatible gold nanoparticles. *Enzym. Microb. Technol.* **2016**, *95*, 76–84. [CrossRef]
124. Subramaniyan, S.A.; Sheet, S.; Vinothkannan, M.; Yoo, D.J.; Lee, Y.S.; Belal, S.A.; Shim, K.S. One-pot facile synthesis of Pt nanoparticles using cultural filtrate of microgravity simulated grown *P. chrysogenum* and their activity on bacteria and cancer cells. *J. Nanosci. Nanotechnol.* **2018**, *18*, 3110–3125. [CrossRef] [PubMed]
125. Riddin, T.L.; Gericke, M.; Whiteley, C.G. Analysis of the inter-and extracellular formation of platinum nanoparticles by *Fusarium oxysporum* f. sp. *lycopersici* using response surface methodology. *Nanotechnology* **2006**, *17*, 3482.
126. Bansal, V.; Rautaray, D.; Bharde, A.; Ahire, K.; Sanyal, A.; Ahmad, A.; Sastry, M. Fungus-mediated biosynthesis of silica and titania particles. *J. Mater. Chem.* **2005**, *15*, 2583–2589. [CrossRef]
127. Soliman, H.; Elsayed, A.; Dyaa, A. Antimicrobial activity of silver nanoparticles biosynthesised by *Rhodotorula* sp. strain ATL72. *Egypt. J. Basic Appl. Sci.* **2018**, *5*, 228–233. [CrossRef]
128. Korbekandi, H.; Mohseni, S.; Mardani Jouneghani, R.; Pourhossein, M.; Iravani, S. Biosynthesis of silver nanoparticles using *Saccharomyces cerevisiae. Artif. Cell Nanomed. Biotechnol.* **2016**, *44*, 235–239. [CrossRef]
129. Fernández, J.G.; Fernández-Baldo, M.A.; Berni, E.; Camí, G.; Durán, N.; Raba, J.; Sanz, M.I. Production of silver nanoparticles using yeasts and evaluation of their antifungal activity against phytopathogenic fungi. *Process Biochem.* **2016**, *51*, 1306–1313. [CrossRef]
130. Cunha, F.A.; da CSO Cunha, M.; da Frota, S.M.; Mallmann, E.J.; Freire, T.M.; Costa, L.S.; Paula, A.J.; Menezes, E.A.; Fechine, P.B. Biogenic synthesis of multifunctional silver nanoparticles from *Rhodotorula glutinis* and *Rhodotorula mucilaginosa*: Antifungal, catalytic and cytotoxicity activities. *World J. Microbiol. Biotechnol.* **2018**, *34*, 1–15. [CrossRef]
131. Yang, Z.; Li, Z.; Lu, X.; He, F.; Zhu, X.; Ma, Y.; He, R.; Gao, F.; Ni, W.; Yi, Y. Controllable biosynthesis and properties of gold nanoplates using yeast extract. *Nano Micro Lett.* **2017**, *9*, 1–13. [CrossRef]
132. Qu, Y.; You, S.; Zhang, X.; Pei, X.; Shen, W.; Li, Z.; Li, S.; Zhang, Z. Biosynthesis of gold nanoparticles using cell-free extracts of *Magnusiomyces ingens* LH-F1 for nitrophenols reduction. *Bioprocess Biosyst. Eng.* **2018**, *41*, 359–367. [CrossRef]
133. Sriramulu, M.; Sumathi, S. Biosynthesis of palladium nanoparticles using *Saccharomyces cerevisiae* extract and its photocatalytic degradation behaviour. *Adv. Nat. Sci. Nanosci. Nanotechnol.* **2018**, *9*, 025018. [CrossRef]
134. Lian, S.; Diko, C.S.; Yan, Y.; Li, Z.; Zhang, H.; Ma, Q.; Qu, Y. Characterization of biogenic selenium nanoparticles derived from cell-free extracts of a novel yeast *Magnusiomyces ingens*. *3 Biotech* **2019**, *9*, 1–8. [CrossRef]
135. Wadhwani, S.A.; Shedbalkar, U.U.; Singh, R.; Vashisth, P.; Pruthi, V.; Chopade, B.A. Kinetics of synthesis of gold nanoparticles by *Acinetobacter* sp. SW30 isolated from Environment. *Indian J. Microbiol.* **2016**, *56*, 439–444. [CrossRef]
136. Tan, K.B.; Sun, D.; Huang, J.; Odoom-Wubah, T.; Li, Q. State of arts on the bio-synthesis of noble metal nanoparticles and their biological application. *Chin. J. Chem. Eng.* **2020**, *30*, 272–290. [CrossRef]
137. Prabhu, S.; Poulose, E.K. Silver nanoparticles: Mechanism of antimicrobial action, synthesis, medical applications, and toxicity effects. *Int. Nano Lett.* **2012**, *2*, 32. [CrossRef]
138. Sabri, M.A.; Umer, A.; Awan, G.H.; Hassan, M.F.; Hasnain, A. Selection of suitable biological method for the synthesis of silver nanoparticles. *Nanomater. Nanotechnol.* **2016**, *6*, 29. [CrossRef]
139. Gurunathan, S.; Han, J.; Park, J.H.; Kim, J.H. A green chemistry approach for synthesizing biocompatible gold nanoparticles. *Nanoscale Res. Lett.* **2014**, *9*, 1–11. [CrossRef]
140. Dubey, S.P.; Lahtinen, M.; Sillanpää, M. Tansy fruit mediated greener synthesis of silver and gold nanoparticles. *Process Biochem.* **2010**, *45*, 1065–1071. [CrossRef]
141. Sathishkumar, M.; Sneha, K.; Yun, Y.S. Immobilization of silver nanoparticles synthesized using *Curcuma longa* tuber powder and extract on cotton cloth for bactericidal activity. *Bioresour. Technol.* **2010**, *101*, 7958–7965. [CrossRef]
142. Armendariz, V.; Herrera, I.; Jose-yacaman, M.; Troiani, H.; Santiago, P.; Gardea-Torresdey, J.L. Size controlled gold nanoparticle formation by *Avena sativa* biomass: Use of plants in nanobiotechnology. *J. Nanoparticle Res.* **2004**, *6*, 377–382. [CrossRef]
143. Chandran, S.P.; Chaudhary, M.; Pasricha, R.; Ahmad, A.; Sastry, M. Synthesis of gold nanotriangles and silver nanoparticles using Aloevera plant extract. *Biotechnol. Progr.* **2006**, *22*, 577–583. [CrossRef]
144. Kaviya, S.; Santhanalakshmi, J.; Viswanathan, B.; Muthumary, J.; Srinivasan, K. Biosynthesis of silver nanoparticles using *Citrus sinensis* peel extract and its antibacterial activity. *Spectrochim. Acta A Mol. Biomol. Spectrosc.* **2011**, *79*, 594–598. [CrossRef]
145. Mourdikoudis, S.; Pallares, R.M.; Thanh, N.T. Characterization techniques for nanoparticles: Comparison and complementarity upon studying nanoparticle properties. *Nanoscale* **2018**, *10*, 12871–12934. [CrossRef]
146. Punia, P.; Bharti, M.K.; Chalia, S.; Dhar, R.; Ravelo, B.; Thakur, P.; Thakur, A. Recent advances in synthesis, characterization, and applications of nanoparticles for contaminated water treatment—A review. *Ceram. Int.* **2021**, *47*, 1526–1550. [CrossRef]
147. Usman, M.; Farooq, M.; Wakeel, A.; Nawaz, A.; Cheema, S.A.; ur Rehman, H.; Ashraf, I.; Sanaullah, M. Nanotechnology in agriculture: Current status, challenges and future opportunities. *Sci. Total Environ.* **2020**, *721*, 137778. [CrossRef] [PubMed]
148. Nadiminti, P.P.; Dong, Y.D.; Sayer, C.; Hay, P.; Rookes, J.E.; Boyd, B.J.; Cahill, D.M. Nanostructured liquid crystalline particles as an alternative delivery vehicle for plant agrochemicals. *ACS Appl. Mater. Interfaces* **2013**, *5*, 1818–1826. [CrossRef]
149. Aslani, F.; Bagheri, S.; Muhd Julkapli, N.; Juraimi, A.S.; Hashemi, F.S.G.; Baghdadi, A. Effects of engineered nanomaterials on plants growth: An overview. *Sci. World J.* **2014**, *2014*, 641759. [CrossRef] [PubMed]

150. Khodakovskaya, M.; Dervishi, E.; Mahmood, M.; Xu, Y.; Li, Z.; Watanabe, F.; Biris, A.S. Carbon nanotubes are able to penetrate plant seed coat and dramatically affect seed germination and plant growth. *ACS Nano* **2009**, *3*, 3221–3227. [CrossRef] [PubMed]
151. Kasote, D.M.; Lee, J.; Jayaprakasha, G.K.; Patil, B.S. Seed priming with iron oxide nanoparticles modulate antioxidant potential and defense linked hormones in watermelon seedlings. *ACS Sust. Chem. Eng.* **2019**, *7*, 5142. [CrossRef]
152. Lu, C.; Zhang, C.; Wen, J.; Wu, G.; Tao, M. Research of the effect of nanometer materials on germination and growth enhancement of *Glycine max* and its mechanism. *Soybean Sci.* **2002**, *21*, 168–171.
153. Zheng, L.; Hong, F.; Lu, S.; Liu, C. Effect of nano-TiO_2 on strength of naturally aged seeds and growth of spinach. *Biol. Trace Elem. Res.* **2005**, *104*, 83–91. [CrossRef]
154. Yang, F.; Hong, F.; You, W.; Liu, C.; Gao, F.; Wu, C.; Yang, P. Influence of nano-anatase TiO_2 on the nitrogen metabolism of growing spinach. *Biol. Trace Elem. Res.* **2006**, *110*, 179–190. [CrossRef]
155. Li, J.; Chang, P.R.; Huang, J.; Wang, Y.; Yuan, H.; Ren, H. Physiological effects of magnetic iron oxide nanoparticles towards watermelon. *J. Nanosci. Nanotechnol.* **2013**, *13*, 5561–5567. [CrossRef]
156. Siddiqui, M.H.; Al-Whaibi, M.H. Role of nano-SiO_2 in germination of tomato (*Lycopersicum esculentum*) seeds Mill. *Saudi J. Biol. Sci.* **2014**, *21*, 13–17. [CrossRef] [PubMed]
157. Raliya, R.; Tarafdar, J.C. ZnO nanoparticle biosynthesis and its effect on phosphorous-mobilizing enzyme secretion and gum contents in Clusterbean (*Cyamopsis tetragonoloba L.*). *Agric. Res.* **2013**, *2*, 48–57. [CrossRef]
158. Mazaheri-Tirani, M.; Dayani, S. In vitro effect of zinc oxide nanoparticles on *Nicotiana tabacum* callus compared to ZnO micro particles and zinc sulfate ($ZnSO_4$). *Plant Cell Tissue Organ Cult.* **2020**, *140*, 279–289. [CrossRef]
159. Noori, A.; Donnelly, T.; Colbert, J.; Cai, W.; Newman, L.A.; White, J.C. Exposure of tomato (*Lycopersicon esculentum*) to silver nanoparticles and silver nitrate: Physiological and molecular response. *Int. J. Phytoremediation* **2020**, *22*, 40–51. [CrossRef] [PubMed]
160. Anjum, N.A.; Rodrigo, M.A.M.; Moulick, A.; Heger, Z.; Kopel, P.; Zitka, O.; Adam, V.; Lukatkin, A.S.; Duarte, A.C.; Pereira, E. Transport phenomena of nanoparticles in plants and animals/humans. *Environ. Res.* **2016**, *151*, 233–243. [CrossRef]
161. Irshad, M.A.; Nawaz, R.; ur Rehman, M.Z.; Adrees, M.; Rizwan, M.; Ali, S.; Ahmad, S.; Tasleem, S. Synthesis, characterization and advanced sustainable applications of titanium dioxide nanoparticles: A review. *Ecotoxicol Environ. Saf.* **2021**, *212*, 111978. [CrossRef] [PubMed]
162. Acharya, P.; Jayaprakasha, G.K.; Crosby, K.M.; Jifon, J.L.; Patil, B.S. Green-synthesized nanoparticles enhanced seedling growth, yield, and quality of onion (*Allium cepa* L.). *ACS Sust. Chem. Eng.* **2019**, *7*, 14580–14590. [CrossRef]
163. Sehnal, K.; Hosnedlova, B.; Docekalova, M.; Stankova, M.; Uhlirova, D.; Tothova, Z.; Kepinska, M.; Milnerowicz, H.; Fernandez, C.; Ruttkay-Nedecky, B.; et al. An assessment of the effect of green synthesized silver nanoparticles using sage leaves (*Salvia officinalis* L.) on germinated plants of maize (*Zea mays* L.). *Nanomaterials* **2019**, *9*, 1550. [CrossRef]
164. Hasan, M.; Rafique, S.; Zafar, A.; Loomba, S.; Khan, R.; Hassan, S.G.; Khan, M.W.; Zahra, S.; Zia, M.; Mustafa, G.; et al. Physiological and anti-oxidative response of biologically and chemically synthesized iron oxide: *Zea mays* a case study. *Heliyon* **2020**, *6*, 1–7. [CrossRef]
165. Ramírez-Rodríguez, G.B.; Dal Sasso, G.; Carmona, F.J.; Miguel-Rojas, C.; Pérez-De-Luque, A.; Masciocchi, N.; Guagliardi, A.; Delgado-López, J.M. Engineering biomimetic calcium phosphate nanoparticles: A green synthesis of slow-release multinutrient (NPK) nanofertilizers. *ACS Appl. Bio Mater.* **2020**, *3*, 1344–1353. [CrossRef]
166. Rameshaiah, G.N.; Pallavi, J.; Shabnam, S. Nano fertilizers and nano sensors—An attempt for developing smart agriculture. *Int. J. Eng. Res. Gen. Sci.* **2015**, *3*, 314–320.
167. Mittal, D.; Kaur, G.; Singh, P.; Yadav, K.; Ali, S.A. Nanoparticle-Based Sustainable Agriculture and Food Science: Recent Advances and Future Outlook. *Front. Nanotechnol.* **2020**, *2*, 579954. [CrossRef]
168. Mastronardi, E.; Tsae, P.; Zhang, X.; Monreal, C.; DeRosa, M.C. Strategic role of nano-technology in fertilizers: Potential and limitations. In *Nanotechnologies in Food and Agriculture*; Springer: Berlin/Heidelberg, Germany, 2015; pp. 25–67.
169. Kottegoda, N.; Munaweera, I.; Madusanka, N.; Karunaratne, V. A green slow-release fertilizer composition based on urea-modified hydroxyapatite nanoparticles encapsulated wood. *Curr. Sci.* **2011**, *101*, 73–78.
170. Sethy, N.K.; Arif, Z.; Mishra, P.K.; Kumar, P. Green synthesis of TiO_2 nanoparticles from *Syzygium cumini* extract for photo-catalytic removal of lead (Pb) in explosive industrial wastewater. *Green Process. Synth.* **2020**, *9*, 171–181. [CrossRef]
171. Patidar, V.; Jain, P. Green synthesis of TiO_2 nanoparticle using *Moringa oleifera* leaf extract. *Int. Res. J. Eng. Technol.* **2017**, *4*, 470–473.
172. Abisharani, J.M.; Devikala, S.; Kumar, R.D.; Arthanareeswari, M.; Kamaraj, P. Green synthesis of TiO_2 nanoparticles using *Cucurbita pepo* seeds extract. *Mater. Today Proc.* **2019**, *14*, 302–307. [CrossRef]
173. Subhapriya, S.; Gomathipriya, P. Green synthesis of titanium dioxide (TiO_2) nanoparticles by *Trigonella foenum-graecum* extract and its antimicrobial properties. *Microb. Pathog.* **2018**, *116*, 215–220. [CrossRef]
174. Dimkpa, C.O.; Bindraban, P.S.; Fugice, J.; Agyin-Birikorang, S.; Singh, U.; Hellums, D. Composite micronutrient nanoparticles and salts decrease drought stress in soybean. *Agro. Sustain. Dev.* **2017**, *37*, 5. [CrossRef]
175. Delfani, M.; Baradarn Firouzabadi, M.; Farrokhi, N.; Makarian, H. Some physiological responses of black-eyed pea to iron and magnesium nanofertilizers. *Commun. Soil Sci. Plant Anal.* **2014**, *45*, 530–540. [CrossRef]
176. Liu, R.; Lal, R. Synthetic apatite nanoparticles as a phosphorus fertilizer for soybean (*Glycine max*). *Sci. Rep.* **2014**, *4*, 5686. [CrossRef] [PubMed]

177. Jampílek, J.; Králová, K. Nanopesticides: Preparation, targeting, and controlled release. In *New Pesticides and Soil Sensors*; Grumezescu, A.M., Ed.; Academic Press: Cambridge, MA, USA, 2017; pp. 81–127.
178. Sasson, Y.; Levy-Ruso, G.; Toledano, O.; Ishaaya, I. Nanosuspensions: Emerging novel agrochemical formulations. In *Insecticides Design Using Advanced Technologies*; Springer: Berlin/Heidelberg, Germany, 2007; pp. 1–39.
179. Paret, M.L.; Vallad, G.E.; Averett, D.R.; Jones, J.B.; Olson, S.M. Photocatalysis: Effect of light-activated nanoscale formulations of TiO_2 on *Xanthomonas perforans* and control of bacterial spot of tomato. *Phytopathology* **2013**, *103*, 228–236. [CrossRef]
180. Guan, Y.; Pearce, R.C.; Melechko, A.V.; Hensley, D.K.; Simpson, M.L.; Rack, P.D. Pulsed laser dewetting of nickel catalyst for carbon nanofiber growth. *Nanotechnology* **2008**, *19*, 235604. [CrossRef]
181. Zhao, L.; Lu, L.; Wang, A.; Zhang, H.; Huang, M.; Wu, H. Nano-biotechnology in agriculture: Use of nanomaterials to promote plant growth and stress tolerance. *J. Agricul. Food. Chem.* **2020**, *68*, 1935–1947. [CrossRef] [PubMed]
182. Ezhilarasi, P.N.; Karthik, P.; Chhanwal, N.; Anandharamakrishnan, C. Nanoencapsulation techniques for food bioactive components: A review. *Food Bioprocess Tech.* **2013**, *6*, 628–647. [CrossRef]
183. China Research and Intelligence. *Global Nanotechnology for Food Packaging Market to Grow at 13.6% CAGR, Which Is Anticipated to Reach USD 44.8 billion by 2030*; China Research and Intelligence Co. Ltd.: Shanghai, China, 2021.
184. Assefa, Z.; Admassu, S. Development and characterization of antimicrobial packaging films. *J. Food Process Technol.* **2013**, *4*, 1–9. [CrossRef]
185. Kalita, D.; Baruah, S. The impact of nanotechnology on food. In *Nanomaterials Applications for Environmental Matrices*; Do Nascimento, R., Ferreira, O.P., De Paula, A.J., Neto, V.D.O.S., Eds.; Elsevier: Amsterdam, The Netherlands, 2019; pp. 369–379.
186. Krishna, R.; Campina, J.M.; Fernandes, P.M.V.; Ventura, J.; Titus, E.; Silva, A.F. Reduced graphene oxide-nickel nanoparticles/biopolymer composite films for sub-millimolar detection of glucose. *Analyst* **2016**, *141*, 4151–4161. [CrossRef]
187. Yang, L.; Chen, Y.; Shen, Y.; Yang, M.; Li, X.; Han, X.; Jiang, X.; Zhao, B. SERS strategy based on the modified Au nanoparticles for highly sensitive detection of bisphenol A residues in milk. *Talanta* **2018**, *179*, 37–42. [CrossRef]
188. Ahmed, S.; Ahmad, M.; Swami, B.; Ikram, S. A review on plants extract mediated synthesis of silver nanoparticles for antimicrobial applications: A green expertise. *J. Adv. Res.* **2016**, *7*, 17–28. [CrossRef]
189. Kanmani, P.; Rhim, J.W. Physical, mechanical and antimicrobial properties of gelatin based active nanocomposite films containing AgNPs and nanoclay. *Food Hydrocoll.* **2014**, *35*, 644–652. [CrossRef]
190. Shankar, S.; Teng, X.; Li, G.; Rhim, J.W. Preparation, characterization, and antimicrobial activity of gelatin/ZnO nanocomposite films. *Food Hydrocoll.* **2015**, *45*, 264–271. [CrossRef]
191. He, X.; Hwang, H.M. Nanotechnology in food science: Functionality, applicability, and safety assessment. *J. Food Drug Anal.* **2016**, *24*, 671–681. [CrossRef] [PubMed]
192. Aguilera, J.M. *Nanotechnology in Food Products: Workshop Summary*; Institute of Medicine (US) Food Forum; National Academies Press: Washington, DC, USA, 2009.
193. Singh, T.; Shukla, S.; Kumar, P.; Wahla, V.; Bajpai, V.K.; Rather, I.A. Application of nanotechnology in food science: Perception and overview. *Front. Microbiol.* **2017**, *8*, 1501. [CrossRef]
194. Graveland-Bikker, J.F.; De Kruif, C.G. Unique milk protein based nanotubes: Food and nanotechnology meet. *Trends Food Sci. Technol.* **2006**, *17*, 196–203. [CrossRef]
195. Oberdörster, G.; Stone, V.; Donaldson, K. Toxicology of nanoparticles: A historical perspective. *Nanotoxicology* **2007**, *1*, 2–25. [CrossRef]
196. Fernandez, A.; Cava, D.; Ocio, M.J.; Lagaron, M. Perspectives for biocatalysts in food packaging. *Trends Food Sci. Technol.* **2008**, *19*, 198–206. [CrossRef]
197. Pradhan, N.; Singh, S.; Ojha, N.; Shrivastava, A.; Barla, A.; Rai, V.; Bose, S. Facets of nanotechnology as seen in food processing, packaging, and preservation industry. *BioMed Res. Int.* **2015**, *2015*, 365672. [CrossRef]
198. Weir, A.; Westerhoff, P.; Fabricius, L.; Hristovski, K.; Goetz, N.V. Titanium dioxide nanoparticles in food and personal care products. *Environ. Sci. Technol.* **2012**, *46*, 2242–2250. [CrossRef]
199. Carbone, M.; Tommas, D.; Gianfranco, D.D.; Sabbatella, G.; Antiochia, R. Silver nanoparticles in polymeric matrices for fresh food packaging. *J. King Saud Univ. Sci.* **2016**, *28*, 273–279. [CrossRef]
200. Emamifar, A.; Kadivar, M.; Shahedi, M.; Soleimanian-Zad, S. Evaluation of nanocomposite packaging containing Ag and ZnO on shelf life of fresh orange juice. *Innov. Food Sci. Emerg. Technol.* **2010**, *11*, 742–748. [CrossRef]
201. Motlagh, N.V.; Mosavian, M.T.H.; Mortazavi, S.A. Effect of polyethylene packaging modified with silver particles on the microbial, sensory and appearance of dried barberry. *Packag. Technol. Sci.* **2012**, *26*, 39–49. [CrossRef]
202. Yang, F.M.; Li, H.M.; Li, F.; Xin, Z.H.; Zhao, L.Y.; Zheng, Y.H.; Hu, Q.H. Effect of nano-packing on preservation quality of fresh strawberry (*Fragaria ananassa* Duch. Cv Fengxiang) during storage at 4 °C. *J. Food Chem.* **2010**, *75*, C236–C240. [CrossRef] [PubMed]
203. Martinez-Abad, A.; Lagaron, J.M.; Ocio, M.J. Development and characterization of silver-based antimicrobial ethylene-vinyl alcohol copolymer (EVOH) films for food-packaging applications. *J. Agric. Food Chem.* **2012**, *60*, 5350–5359. [CrossRef] [PubMed]
204. Mahdi, S.S.; Vadood, R.; Nourdahr, R. Study on the antimicrobial effect of nanosilver tray packaging of minced beef at refrigerator temperature. *Glob. Vet.* **2012**, *9*, 284–289.
205. Metak, A.M.; Ajaal, T.T. Investigation on polymer based nanosilver as food packaging materials. *Int. J. Biol. Food Vet. Agric. Eng.* **2013**, *7*, 772–777.

206. Metak, A.M. Effects of nanocomposites based nano-silver and nano-titanium dioxideon food packaging materials. *Int. J. Appl. Sci. Technol.* **2015**, *5*, 26–40.
207. Akbari, Z.; Ghomashchi, T.; Aroujalian, A. Potential of nanotechnology for food packaging industry. In Proceedings of the "Nano Micro Technologies in the Food Health Food Industries" Conference organised by Institute of Nanotechnology, Amsterdam, The Netherland, 25–26 October 2006.
208. Khalaf, H.H.; Sharoba, A.M.; El-Tanahi, H.H.; Morsy, M.K. Stability of antimicrobial activity of pullulan edible films incorporated with nanoparticles and essential oils and their impact on turkey deli meat quality. *J. Food Dairy Sci.* **2013**, *4*, 557–573. [CrossRef]
209. Elsaesser, A.; Howard, C. Toxicology of nanoparticles. *Adv. Drug Deliver. Rev.* **2012**, *64*, 129–137. [CrossRef] [PubMed]
210. Zoroddu, M.; Medici, S.; Ledda, A.; Nurchi, V.M.; Lachowicz, J.I.; Peana, M. Toxicity of nanoparticles. *Curr. Med. Chem.* **2014**, *21*, 3837–3853. [CrossRef] [PubMed]
211. Sebastian, R. Nanomedicine—The future of cancer treatment: A Review. *J. Cancer Prev. Curr. Res.* **2017**, *8*, 00265. [CrossRef]
212. Venugopal, K.; Rather, H.A.; Rajagopal, K.; Shanthi, M.P.; Sheriff, K.; Illiyas, M.; Rather, R.A.; Manikandan, E.; Uvarajan, S.; Bhaskar, M.; et al. Synthesis of silver nanoparticles (Ag NPs) for anticancer activities (MCF 7 breast and A549 lung cell lines) of the crude extract of *Syzygium aromaticum*. *J. Photochem. Photobiol. B* **2017**, *167*, 282–289. [CrossRef]
213. Elangovan, K.; Elumalai, D.; Anupriya, S.; Shenbhagaraman, R.; Kaleena, P.; Murugesan, K. Phyto mediated biogenic synthesis of silver nanoparticles using leaf extract of *Andrographis echioides* and its bio-efficacy on anticancer and antibacterial activities. *J. Photochem. Photobiol. B* **2015**, *151*, 118–124. [CrossRef] [PubMed]
214. Arivazhagan, S.; Velmurugan, B.; Bhuvaneswari, V.; Nagini, S. Effects of aqueous extracts of garlic (*Allium sativum*) and neem (*Azadirachta indica*) leaf on hepatic and blood oxidant-antioxidant status during experimental gastric carcinogenesis. *J. Med. Food* **2004**, *7*, 334–339. [CrossRef]
215. Adewale, O.; Davids, H.; Cairncross, L.; Roux, S. Toxicological behavior of gold nanoparticles on various models: Influence of physicochemical properties and other factors. *Int. J. Toxicol.* **2019**, *38*, 357–384. [CrossRef]
216. Rao, P.; Nallappan, D.; Madhavi, K.; Rahman, S.; Jun Wei, L.; Gan, S. Phytochemicals and Biogenic Metallic Nanoparticles as Anticancer Agents. *Oxid. Med. Cell. Longev.* **2016**, 1–15. [CrossRef]
217. Farshchi, H.; Azizi, M.; Jaafari, M.; Nemati, S.; Fotovat, A. Green synthesis of iron nanoparticles by Rosemary extract and cytotoxicity effect evaluation on cancer cell lines. *Biocatal. Agric. Biotechnol.* **2018**, *16*, 54–62. [CrossRef]
218. Kostyukova, D.; Chung, Y. Synthesis of iron oxide nanoparticles using isobutanol. *J. Nanomater.* **2016**, *2016*, 1–9. [CrossRef]
219. Peng, X.H.; Qian, X.; Mao, H.; Wang, A.Y.; Chen, Z.G.; Nie, S.; Shin, D.M. Targeted magnetic iron oxide nanoparticles for tumor imaging and therapy. *Int. J. Nanomed.* **2008**, *3*, 311–321.
220. Anjum, M.; Miandad, R.; Waqas, M.; Gehany, F.; Barakat, M.A. Remediation of wastewater using various nano-materials. *Arab. J. Chem.* **2019**, *12*, 4897–4919. [CrossRef]
221. Sharma, S.; Bhattacharya, A. Drinking water contamination and treatment techniques. *Appl. Water Sci.* **2017**, *7*, 1043–1067. [CrossRef]
222. Gautam, R.K.; Chattopadhyaya, M.C. *Nanomaterials for Wastewater Remediation*; Elsevier: Amsterdam, The Netherlands, 2016.
223. Sadegh, H.; Ali, G.A.; Gupta, V.K.; Makhlouf, A.S.H.; Shahryari-ghoshekandi, R.; Nadagouda, M.N.; Sillanpää, M.; Megiel, E. The role of nanomaterials as effective adsorbents and their applications in wastewater treatment. *J. Nanostruct. Chem.* **2017**, *7*, 1–14. [CrossRef]
224. Dil, E.A.; Ghaedi, M.; Asfaram, A. The performance of nanorods material as adsorbent for removal of azo dyes and heavy metal ions: Application of ultrasound wave, optimization and modeling. *Ultrason. Sonochem.* **2017**, *34*, 792–802. [CrossRef] [PubMed]
225. Santhosh, C.; Velmurugan, V.; Jacob, G.; Jeong, S.K.; Grace, A.N.; Bhatnagar, A. Role of nanomaterials in water treatment applications: A review. *Chem. Eng. J.* **2016**, *306*, 1116–1137. [CrossRef]
226. El-Sayed, M.E. Nanoadsorbents for water and wastewater remediation. *Sci. Total Environ.* **2020**, *739*, 139903. [CrossRef] [PubMed]
227. Zhao, C.; Wang, B.; Theng, B.K.; Wu, P.; Liu, F.; Wang, S.; Lee, X.; Chen, M.; Li, L.; Zhang, X. Formation and mechanisms of nano-metal oxide-biochar composites for pollutant removal: A review. *Sci. Total Environ.* **2021**, *767*, 145305. [CrossRef]
228. Das, C.; Sen, S.; Singh, T.; Ghosh, T.; Paul, S.S.; Kim, T.W.; Jeon, S.; Maiti, D.K.; Im, J.; Biswas, G. Green synthesis, characterization and application of natural product coated magnetite nanoparticles for wastewater treatment. *Nanomaterials* **2020**, *10*, 1615. [CrossRef]
229. Al-Qahtani, K.M. Cadmium removal from aqueous solution by green synthesis zero valent silver nanoparticles with Benjamina leaves extract. *Egypt. J. Aquat. Res.* **2017**, *43*, 269–274. [CrossRef]
230. Ehrampoush, M.H.; Miria, M.; Salmani, M.H.; Mahvi, A.H. Cadmium removal from aqueous solution by green synthesis iron oxide nanoparticles with tangerine peel extract. *J. Environ. Health Sci.* **2015**, *13*, 1–7. [CrossRef] [PubMed]
231. Primo, J.D.O.; Bittencourt, C.; Acosta, S.; Sierra-Castillo, A.; Colomer, J.F.; Jaerger, S.; Teixeira, V.C.; Anaissi, F.J. Synthesis of Zinc Oxide nanoparticles by ecofriendly routes: Adsorbent for copper removal from wastewater. *Front. Chem.* **2020**, *8*, 571790. [CrossRef]
232. Zhu, Y.; Liu, X.; Hu, Y.; Wang, R.; Chen, M.; Wu, J.; Wang, Y.; Kang, S.; Sun, Y.; Zhu, M. Behavior, remediation effect and toxicity of nanomaterials in water environments. *Environ. Res.* **2019**, *174*, 54–60. [CrossRef]
233. Wang, P.; Wang, F.; Jiang, H.; Zhang, Y.; Zhao, M.; Xiong, R.; Ma, J. Strong improvement of nanofiltration performance on micropollutant removal and reduction of membrane fouling by hydrolyzed-aluminum nanoparticles. *Water Res.* **2020**, *175*, 115649. [CrossRef]

234. Bottino, A.; Capannelli, G.; D'asti, V.; Piaggio, P. Preparation and properties of novel organic-inorganic porous membranes. *Sep. Purif. Technol.* **2001**, *22*, 269–275. [CrossRef]
235. Fujiwara, M.; Imura, T. Photo induced membrane separation for water purification and desalination using azobenzene modified anodized alumina membranes. *ACS Nano* **2015**, *9*, 5705–5712. [CrossRef]
236. Maximous, N.; Nakhla, G.; Wong, K.; Wan, W. Optimization of Al_2O_3/PES membranes for wastewater filtration. *Sep. Purif. Technol.* **2010**, *73*, 294–301. [CrossRef]
237. Yurekli, Y. Removal of heavy metals in wastewater by using zeolite nano-particles impregnated polysulfone membranes. *J. Hazard. Mater.* **2016**, *309*, 53–64. [CrossRef]
238. Bet-Moushoul, E.; Mansourpanah, Y.; Farhadi, K.; Tabatabaei, M. TiO_2 nanocomposite based polymeric membranes: A review on performance improvement for various applications in chemical engineering processes. *Chem. Eng. J.* **2016**, *283*, 29–46. [CrossRef]
239. Mauter, M.S.; Wang, Y.; Okemgbo, K.C.; Osuji, C.O.; Giannelis, E.P.; Elimelech, M. Antifouling ultrafiltration membranes via post-fabrication grafting of biocidal nanomaterials. *ACS Appl. Mater. Inter.* **2011**, *3*, 2861–2868. [CrossRef]
240. Zodrow, K.; Brunet, L.; Mahendra, S.; Li, D.; Zhang, A.; Li, Q.; Alvarez, P.J. Polysulfone ultrafiltration membranes impregnated with silver nanoparticles show improved biofouling resistance and virus removal. *Water Res.* **2009**, *43*, 715–723. [CrossRef]
241. Patanjali, P.; Singh, R.; Kumar, A.; Chaudhary, P. Nanotechnology for water treatment: A green approach. In *Green Synthesis, Characterization and Applications of Nanoparticles*; Shukla, A.K., Iravani, S., Eds.; Elsevier: Amsterdam, The Netherlands, 2019; pp. 485–512.
242. Anjum, M.; Al-Makishah, N.H.; Barakat, M.A. Wastewater sludge stabilization using pre-treatment methods. *Process Saf. Environ.* **2016**, *102*, 615–632. [CrossRef]
243. Hemmati, S.; Mehrazin, L.; Ghorban, H.; Garakani, S.H.; Mobaraki, T.H.; Mohammadi, P.; Veisi, H. Green synthesis of Pd nanoparticles supported on reduced graphene oxide, using the extract of *Rosa canina* fruit, and their use as recyclable and heterogeneous nanocatalysts for the degradation of dye pollutants in water. *RSC Adv.* **2018**, *8*, 21020–21028. [CrossRef]
244. Kumar, B.; Smita, K.; Cumbal, L. Phytosynthesis of gold nanoparticles using Andean Ajı′ (*Capsicum baccatum* L.). *Cogent Chem.* **2015**, *1*, 1120982. [CrossRef]
245. Kumar, B.; Smita, K.; Cumbal, L.; Debut, A. One pot synthesis and characterization of gold nanocatalyst using Sacha inchi (*Plukenetia volubilis*) oil: Green approach. *J. Photochem. Photobiol. B* **2016**, *158*, 55–60. [CrossRef]
246. Sumi, M.B.; Devadiga, A.; Shetty, K.V.; Saidutta, M.B. Solar photocatalytically active, engineered silver nanoparticle synthesis using aqueous extract of mesocarp of *Cocos nucifera* (Red Spicata Dwarf). *J. Exp. Nanosci.* **2017**, *12*, 14–32. [CrossRef]
247. Tahir, K.; Nazir, S.; Li, B.; Khan, A.U.; Khan, Z.U.H.; Ahmad, A.; Khan, F.U. An efficient photo catalytic activity of green synthesized silver nanoparticles using *Salvadora persica* stem extract. *Sep. Purif. Technol.* **2015**, *150*, 316–324. [CrossRef]
248. Nagajyothi, P.C.; Prabhakar Vattikuti, S.V.; Devarayapalli, K.C.; Yoo, K.; Shim, J.; Sreekanth, T.V.M. Green synthesis: Photocatalytic degradation of textile dyes using metal and metal oxide nanoparticles-latest trends and advancements. *Crit. Rev. Environ. Sci. Technol.* **2020**, *50*, 2617–2723. [CrossRef]
249. Huh, A.J.; Kwon, Y.J. Nanoantibiotics: A new paradigm for treating infectious diseases using nanomaterials in the antibiotics resistant era. *J. Control. Release* **2011**, *156*, 128–145. [CrossRef]
250. Seil, J.T.; Webster, T.J. Antimicrobial applications of nanotechnology: Methods and literature. *Int. J. Nanomed.* **2012**, *7*, 2767–2781.
251. Pelgrift, R.Y.; Friedman, A.J. Nanotechnology as a therapeutic tool to combat microbial resistance. *Adv. Drug Deliv. Rev.* **2013**, *65*, 1803–1815. [CrossRef] [PubMed]
252. Allahverdiyev, A.M.; Abamor, E.S.; Bagirova, M.; Rafailovich, M. Antimicrobial effects of TiO_2 and Ag_2O nanoparticles against drug-resistant bacteria and leishmania parasites. *Future Microbiol.* **2011**, *6*, 933–940. [CrossRef] [PubMed]
253. Tenover, F.C.; Hughes, J.M. The challenges of emerging infectious diseases: Development and spread of multiply-resistant bacterial pathogens. *JAMA* **1996**, *275*, 300–304. [CrossRef] [PubMed]
254. Dizaj, S.M.; Lotfipour, F.; Barzegar-Jalali, M.; Zarrintan, M.H.; Adibkia, K. Antimicrobial activity of the metals and metal oxide nanoparticles. *Mater. Sci. Eng. C* **2014**, *44*, 278–284. [CrossRef]
255. Besinis, A.; De Peralta, T.; Handy, R.D. The antibacterial effects of silver, titanium dioxide and silica dioxide nanoparticles compared to the dental disinfectant chlorhexidine on *Streptococcus* mutans using a suite of bioassays. *Nanotoxicology* **2014**, *8*, 1–16. [CrossRef]
256. Lok, C.-N.; Ho, C.-M.; Chen, R.; He, Q.-Y.; Yu, W.-Y.; Sun, H.; Tam, P.K.-H.; Chiu, J.-F.; Che, C.-M. Silver nanoparticles: Partial oxidation and antibacterial activities. *J. Biol. Inorg. Chem.* **2007**, *12*, 527–534. [CrossRef]
257. Jiang, W.; Mashayekhi, H.; Xing, B. Bacterial toxicity comparison between nano- and micro-scaled oxide particles. *Environ. Pollut.* **2009**, *157*, 1619–1625. [CrossRef] [PubMed]
258. Chen, S.F.; Li, J.P.; Qian, K.; Xu, W.P.; Lu, Y.; Huang, W.X.; Yu, S.H. Large scale photochemical synthesis of M@TiO_2 nanocomposites (M = Ag, Pd, Au, Pt) and their optical properties, CO oxidation performance, and antibacterial effect. *Nano Res.* **2010**, *3*, 244–255. [CrossRef]
259. Rajendra, R.; Balakumar, C.; Ahammed, H.A.M.; Jayakumar, S.; Vaideki, K.; Rajesh, E. Use of zinc oxide nano particles for production of antimicrobial textiles. *Int. J. Eng. Sci. Technol.* **2010**, *2*, 202–208. [CrossRef]
260. Kuang, Y.; He, X.; Zhang, Z.; Li, Y.; Zhang, H.; Ma, Y.; Wu, Z.; Chai, Z. Comparison study on the antibacterial activity of nano- or bulk-cerium oxide. *J. Nanosci. Nanotechnol.* **2011**, *11*, 4103–4108. [CrossRef]

261. Li, S.-T.; Qiao, X.-L.; Chen, J.-G.; Wu, C.-L.; Mei, B. The investigation of antibacterial characteristics of magnesium oxide and it's nano-composite materials. *J. Funct. Mater.* **2005**, *11*, 1651–1654.
262. Srivastava, M.; Singh, S.; Self, W.T. Exposure to silver nanoparticles inhibits selenoprotein synthesis and the activity of thioredoxin reductase. *Environ. Health Perspect.* **2012**, *120*, 56–61. [CrossRef] [PubMed]
263. Singh, S.; Patel, P.; Jaiswal, S.; Prabhune, A.A.; Ramana, C.V.; Prasad, B.L.V. A direct method for the preparation of glycolipid-metal nanoparticle conjugates: Sophorolipids as reducing and capping agents for the synthesis of water re-dispersible silver nanoparticles and their antibacterial activity. *New J. Chem.* **2009**, *33*, 646–652. [CrossRef]
264. Masoumbaigi, H.; Rezaee, A.; Hosseini, H.; Hashemi, S. Water disinfection by zinc oxide nanoparticle prepared with solution combustion method. *Desalin. Water Treat.* **2015**, *56*, 2376–2381. [CrossRef]
265. Makhluf, S.; Dror, R.; Nitzan, Y.; Abramovich, Y.; Jelinek, R.; Gedanken, A. Microwave assisted synthesis of nanocrystalline MgO and its use as a bacteriocide. *Adv. Funct. Mater.* **2005**, *15*, 1708–1715. [CrossRef]
266. Abbaszadegan, A.; Ghahramani, Y.; Gholami, A.; Hemmateenejad, B.; Dorostkar, S.; Nabavizadeh, M.; Sharghi, H. The effect of charge at the surface of silver nanoparticles on antimicrobial activity against gram-positive and gram-negative bacteria: A preliminary study. *J. Nanomater.* **2015**, *16*, 1–8. [CrossRef]
267. Adibkia, K.; Alaei-Beirami, M.; Barzegar-Jalali, M.; Mohammadi, G.; Ardestani, M.S. Evaluation and optimization of factors affecting novel diclofenac sodium-eudragit RS100 nanoparticles. *Afr. J. Pharm. Pharmacol.* **2012**, *6*, 941–947.
268. Buzea, C.; Pacheco, I.I.; Robbie, K. Nanomaterials and nanoparticles: Sources and toxicity. *Biointerphases* **2007**, *2*, MR17–MR71. [CrossRef]
269. Adibkia, K.; Barzegar-Jalali, M.; Nokhodchi, A.; Siahi Shadbad, M.; Omidi, Y.; Javadzadeh, Y.; Mohammadi, G. A review on the methods of preparation of pharmaceutical nanoparticles. *J. Pharm. Sci.* **2010**, *15*, 303–314.
270. Hoseinzadeh, E.; Alikhani, M.-Y.; Samarghandi, M.-R.; Shirzad-Siboni, M. Antimicrobial potential of synthesized zinc oxide nanoparticles against gram positive and gram negative bacteria. *Desalin. Water Treat.* **2014**, *52*, 4969–4976. [CrossRef]
271. Mirhosseini, M.; Firouzabadi, F.B. Reduction of listeria monocytogenes and *Bacillus cereus* in milk by zinc oxide nanoparticles. *Iran J. Pathol.* **2015**, *10*, 97–104.
272. Wongyai, K.; Wintachai, P.; Maungchang, R.; Rattanakit, P. Exploration of the antimicrobial and catalytic properties of gold nanoparticles greenly synthesized by *Cryptolepis buchanani Roem.* and Schult extract. *J. Nanomater.* **2020**, *2020*, 1320274. [CrossRef]
273. Loo, Y.Y.; Rukayadi, Y.; Nor-Khaizura, M.A.R.; Kuan, C.H.; Chieng, B.W.; Nishibuchi, M.; Radu, S. In vitro antimicrobial activity of green synthesized silver nanoparticles against selected Gram-negative food borne pathogens, *Front. Microbiol.* **2018**, *9*, 1555.
274. Pal, S.; Tak, Y.K.; Song, J.M. Does the antibacterial activity of silver nanoparticles depend on the shape of the nanoparticle? A study of the gram-negative bacterium *Escherichia coli*. *Appl. Environ. Microbiol.* **2007**, *73*, 1712–1720. [CrossRef] [PubMed]
275. Bera, R.K.; Mandal, S.M.; Raj, C.R. Antimicrobial activity of fluorescent Ag nanoparticles. *Lett. Appl. Microbiol.* **2014**, *58*, 520–526. [CrossRef]
276. Song, Y.; Chen, L. Effect of net surface charge on physical properties of the cellulose nanoparticles and their efficacy for oral protein delivery. *Carbohyd. Polym.* **2015**, *121*, 10–17. [CrossRef] [PubMed]
277. Tsuji, J.S.; Maynard, A.D.; Howard, P.C.; James, J.T.; Lam, C.-W.; Warheit, D.B.; Santamaria, A.B. Research strategies for safety evaluation of nanomaterials, part IV: Risk assessment of nanoparticles. *Toxicol. Sci.* **2006**, *89*, 42–50. [CrossRef] [PubMed]
278. De Jong, W.H.; Borm, P.J. Drug delivery and nanoparticles: Applications and hazards. *Int. J. Nanomed.* **2008**, *3*, 133. [CrossRef]
279. Zinjarde, S.S. Bio-inspired nanomaterials and their applications as antimicrobial agents. *Chron. Young Sci.* **2012**, *3*, 1–74. [CrossRef]
280. Khan, I.; Saeed, K.; Khan, I. Nanoparticles: Properties, applications and toxicities. *Arab. J. Chem.* **2019**, *12*, 908–931. [CrossRef]
281. Pulit-Prociak, J.; Banach, M. Silver nanoparticles—A material of the future? *Open Chem.* **2016**, *14*, 76–91. [CrossRef]
282. Giorgetti, L. Effects of nanoparticles in plants: Phytotoxicity and genotoxicity assessment. *Nanomater. Plants Algae Microorg.* **2019**, *2*, 65–87.
283. Semmler-Behnke, M.; Lipka, J.; Wenk, A.; Hirn, S.; Schäffler, M.; Tian, F.; Schmid, G.; Oberdörster, G.; Kreyling, W.G. Size dependent translocation and fetal accumulation of gold nanoparticles from maternal blood in the rat. *Part. Fibre Toxicol.* **2014**, *11*, 1–12. [CrossRef]
284. Gao, G.; Ze, Y.; Li, B.; Zhao, X.; Zhang, T.; Sheng, L.; Hu, R.; Gui, S.; Sang, X.; Sun, Q.; et al. Ovarian dysfunction and gene–expressed characteristics of female mice caused by long–term exposure to titanium dioxide nanoparticles. *J. Hazard. Mater.* **2012**, *243*, 19–27. [CrossRef] [PubMed]
285. Dash, S.R.; Kundu, C.N. Promising opportunities and potential risk of nanoparticle on the society. *IET Nanobiotechnol.* **2020**, *14*, 253–260. [CrossRef]
286. Wang, W.N.; Tarafdar, J.C.; Biswas, P. Nanoparticle synthesis and delivery by an aerosol route for watermelon plant foliar uptake. *J. Nanoparticle. Res.* **2013**, *15*, 1417. [CrossRef]
287. Tripathi, A.; Liu, S.; Singh, P.K.; Kumar, N.; Pandey, A.C.; Tripathi, D.K.; Chauhan, D.K.; Sahi, S. Differential phytotoxic responses of silver nitrate ($AgNO_3$) and silver nanoparticle (AgNps) in *Cucumis sativus* L. *Plant Gene* **2017**, *11*, 255–264. [CrossRef]
288. Zhu, Y.; Wu, J.; Chen, M.; Liu, X.; Xiong, Y.; Wang, Y.; Feng, T.; Kang, S.; Wang, X. Recent advances in the biotoxicity of metal oxide nanoparticles: Impacts on plants, animals and microorganisms. *Chemosphere* **2019**, *237*, 124403. [CrossRef]
289. Verma, S.K.; Das, A.K.; Patel, M.K.; Shah, A.; Kumar, V.; Gantait, S. Engineered nanomaterials for plant growth and development: A perspective analysis. *Sci. Total Environ.* **2018**, *630*, 1413–1435. [CrossRef] [PubMed]

290. Barbafieri, M.; Giorgetti, L. Contaminant bioavailability in soil and phytotoxicity/genotoxicity tests in *Vicia faba* L.: A case study of boron contamination. *Environ. Sci. Pollut. Res. Int.* **2016**, *23*, 24327–24336. [CrossRef]
291. Adhikari, T.; Kundu, S.; Subba Rao, A. Impact of SiO_2 and Mo nanoparticles on seed germination of rice (*Oryza sativa* L.). *Int. J. Agric. Food Sci. Technol.* **2013**, *4*, 809–816.
292. Qados, A.M.S.A. Mechanism of nanosilicon–mediated alleviation of salinity stress in faba bean (*Vicia faba* L.). *Plants Am. J. Exp. Agric.* **2015**, *7*, 78–95. [CrossRef]
293. Almutairi, Z.M. Effect of nano-silicon application on the expression of salt tolerance genes in germinating tomato ("*Solanum lycopersicum*" L.) seedlings under salt stress. *Plant Omics* **2016**, *9*, 106–114.
294. Zmeeva, O.; Daibova, E.; Proskurina, L.; Petrova, L.V.; Kolomiets, N.E.; Svetlichny, V.A.; Lapin, I.N.; Karakchieva, N.I. Effects of silicon dioxide nanoparticles on biological and physiological characteristics of *Medicago sativa* L. nothosubsp. varia (Martyn) in natural agroclimaticc of the subtaiga zone in western Siberia. *Bionanoscience* **2017**, *7*, 672–679. [CrossRef]
295. Slomberg, D.L.; Schoenfisch, M.H. Silica nanoparticle phytotoxicity to *Arabidopsis thaliana*. *Environ. Sci. Technol.* **2012**, *46*, 10247–10254. [PubMed]
296. Karimi, J.; Mohsenzadeh, S. Effects of silicon oxide nanoparticles on growth and physiology of wheat seedlings. *Russ. J. Plant Physiol.* **2016**, *63*, 119–123. [CrossRef]
297. Silva, G.H.; Monteiro, R.T.R. Toxicity assessment of silica nanoparticles on *Allium cepa*. *Ecotoxicol. Environ. Contam.* **2017**, *12*, 25–31. [CrossRef]
298. Ruffini Castiglione, M.; Giorgetti, L.; Geri, C.; Cremonini, R. The effects of nano-TiO_2 on seed germination, development and mitosis of root tip cells of *Vicia narbonensis* L. and *Zea mays* L. *J. Nanoparticle Res.* **2011**, *13*, 2443–2449. [CrossRef]
299. Ruffini Castiglione, M.; Giorgetti, L.; Cremonini, R.; Bottega, S.; Spano, C. Impact of TiO_2 nanoparticles on *Vicia narbonensis* L.: Potential toxicity effects. *Protoplasma* **2014**, *251*, 1471–1479. [CrossRef]
300. Ruffini Castiglione, M.; Giorgetti, L.; Becarelli, S.; Siracusa, G.; Lorenzi, R.; Di Gregorio, S. Polycyclic aromatic hydrocarbon-contaminated soils: Bioaugmentation of autochthonous bacteria and toxicological assessment of the bioremediation process by means of *Vicia faba* L. *Environ. Sci. Pollut. Res. Int.* **2016**, *23*, 7930–7941. [CrossRef]
301. Ruffini Castiglione, M.; Giorgetti, L.; Bellani, L.; Muccifora, S.; Bottega, S.; Spano, C. Root responses to different types of TiO_2 nanoparticles and bulk counterpart in plant model system *Vicia faba* L. *Environ. Exp. Bot.* **2016**, *130*, 11–21. [CrossRef]
302. Kah, M.; Hofmann, T. Nanopesticide research: Current trends and future priorities. *Environ. Int.* **2014**, *63*, 224–235. [CrossRef]
303. Oukarroum, A.; Bras, S.; Perreault, F.; Popovic, R. Inhibitory effects of silver nanoparticles in two green algae, *Chlorella vulgaris* and *Dunaliella tertiolecta*. *Ecotoxicol. Environ. Saf.* **2012**, *78*, 80–85. [CrossRef]
304. Dash, A.; Singh, A.P.; Chaudhary, B.R.; Singh, S.K.; Dash, D. Effect of silver nanoparticles on growth of eukaryotic green algae. *Nanomicro Lett.* **2012**, *4*, 158–165. [CrossRef]
305. Bakand, S.; Hayes, A. Toxicological considerations, toxicity assessment, and risk management of inhaled nanoparticles. *Int. J. Mol. Sci.* **2016**, *17*, 929. [CrossRef]
306. Oberdörster, G.; Elder, A.; Rinderknecht, A. Nanoparticles and the brain: Cause for concern? *J. Nanosci. Nanotechnol.* **2009**, *9*, 4996–5007. [CrossRef] [PubMed]
307. Haase, A.; Tentschert, J.; Jungnickel, H. Toxicity of silver nanoparticles in human macrophages: Uptake, intracellular distribution and cellular responses. *J. Phys. Conf. Ser.* **2011**, *304*, 012030. [CrossRef]
308. Baranowska-Wójcik, E.; Szwajgier, D.; Oleszczuk, P.; Winiarska-Mieczan, A. Effects of titanium dioxide nanoparticles exposure on human health—A review. *Biol. Trace Elem. Res.* **2020**, *193*, 118–129. [CrossRef] [PubMed]
309. Baek, M.; Kim, M.; Cho, H.; Lee, J.; Yu, J.; Chung, H.; Choi, S. Factors influencing the cytotoxicity of zinc oxide nanoparticles: Particle size and surface charge. *J. Phys. Conf. Ser.* **2011**, *304*, 012044. [CrossRef]
310. Zhang, L.; Tan, S.; Liu, Y.; Xie, H.; Luo, B.; Wang, J. In vitro inhibition of tumor growth by low-dose iron oxide nanoparticles activating macrophages. *J. Biomater. Appl.* **2019**, *33*, 935–945. [CrossRef] [PubMed]
311. Anna, M.; Oksana, V.; Jana, K. Effect of chemically and biologically synthesized Ag nanoparticles on the algae growth inhibition. *AIP Conf. Proc.* **2017**, *1918*, 020008.

catalysts MDPI

Article

Modification of Graphite Sheet Anode with Iron (II, III) Oxide-Carbon Dots for Enhancing the Performance of Microbial Fuel Cell

Babita Tripathi [1,†], Soumya Pandit [2,*,†], Aparna Sharma [2,†], Sunil Chauhan [1,*,†], Abhilasha Singh Mathuriya [3], Pritam Kumar Dikshit [4], Piyush Kumar Gupta [2,5], Ram Chandra Singh [1], Mohit Sahni [1], Kumud Pant [5] and Satyendra Singh [6]

[1] Department of Physics, Sharda School of Basic Sciences and Research, Sharda University, Greater Noida 201310, India
[2] Biopositive Lab, Department of Life Sciences, Sharda School of Basic Sciences and Research, Sharda University, Greater Noida 201310, India
[3] Ministry of Environment, Forest and Climate Change, Indira Paryavaran Bhawan, Jor Bagh, New Delhi 110003, India
[4] Department of Bio-Technology, Koneru Lakshmaiah Education Foundation, Vaddeswaram 522302, India
[5] Department of Biotechnology, Graphic Era Deemed to be University, Dehradun, Uttarakhand 248002, India
[6] Special Centre for Nanoscience, Jawaharlal Nehru University, New Delhi 110067, India
* Correspondence: sounip@gmail.com (S.P.); sunil.chauhan@sharda.ac.in (S.C.)
† These authors contributed equally to this work.

Citation: Tripathi, B.; Pandit, S.; Sharma, A.; Chauhan, S.; Mathuriya, A.S.; Dikshit, P.K.; Gupta, P.K.; Singh, R.C.; Sahni, M.; Pant, K.; et al. Modification of Graphite Sheet Anode with Iron (II, III) Oxide-Carbon Dots for Enhancing the Performance of Microbial Fuel Cell. *Catalysts* **2022**, *12*, 1040. https://doi.org/10.3390/catal12091040

Academic Editor: Ioannis V. Yentekakis

Received: 12 August 2022
Accepted: 6 September 2022
Published: 13 September 2022

Abstract: The present study explores the use of carbon dots coated with Iron (II, III) oxide (Fe_3O_4) for its application as an anode in microbial fuel cells (MFC). Fe_3O_4@PSA-C was synthesized using a hydrothermal-assisted probe sonication method. Nanoparticles were characterized with XRD, SEM, FTIR, and RAMAN Spectroscopy. Different concentrations of Fe_3O_4- carbon dots (0.25, 0.5, 0.75, and 1 mg/cm^2) were coated onto the graphite sheets (Fe_3O_4@PSA-C), and their performance in MFC was evaluated. Cyclic voltammetry (CV) of Fe_3O_4@PSA-C (1 mg/cm^2) modified anode indicated oxidation peaks at −0.26 mV and +0.16 mV, respectively, with peak currents of 7.7 mA and 8.1 mA. The fluxes of these anodes were much higher than those of other low-concentration Fe_3O_4@PSA-C modified anodes and the bare graphite sheet anode. The maximum power density (Pmax) was observed in MFC with a 1 mg/cm^2 concentration of Fe_3O_4@PSA-C was 440.01 mW/m^2, 1.54 times higher than MFCs using bare graphite sheet anode (285.01 mW/m^2). The elevated interaction area of carbon dots permits pervasive Fe_3O_4 crystallization providing enhanced cell attachment capability of the anode, boosting the biocompatibility of Fe_3O_4@PSA-C. This significantly improved the performance of the MFC, making Fe_3O_4@PSA-C modified graphite sheets a good choice as an anode for its application in MFC.

Keywords: bioanode; carbon dots; iron (ii, iii) oxide; power density; microbial fuel cell; internal resistance; hydrothermal assisted probe sonication; biocompatibility

1. Introduction

Bioelectricity generated by electrochemically active bacteria (EAB) has gained much attention, with promising significance for renewable energy generation, wastewater remediation, and metallic nanoparticle fabrication [1,2]. In such fuel cells, waste disposal can be converted to electricity with the aid of EAB that colonizes on the anode surface. On the surface of the anode, biodegradable waste is oxidized and generates electrons, protons, and other by-products, protons pass through the cation exchange membrane from anode to cathode, and electrons via an outer circuit, electrons, and protons combined with oxygen at the cathode to complete the reaction [3]. Anode plays a major role in determining the power output of the MFC [4]. However, sluggish EAB biofilm growth impedes the development of the anode half-cell potential and causes voltage loss; consequently, the MFC productivity has not achieved its full potential, resulting in decreased power output [5]. MFC power interruptions occur due to undesirable physicochemical conditions for bacterial consortiums and system architecture [6,7]. The anode composition has the largest impact on energy output in MFCs because it combines microbiology and electrochemistry. The anode's conductance and biocatalytic efficiency must be improved to reduce potential losses. In MFCs, graphite disc anodes are treated with Au and Pd nanoparticles, for example, outperformed untreated anodes in biocompatibility and power generation [8]. As a result, changing the anode makeup might be a good technique for dealing with overpotential problems [9]. Anode remodeling alters the physical and chemical characteristics of the anode, allowing for improved microbial adhesion by increasing the electrochemically specific surface area and enhancing electron transport due to the increased permeability of the anode [10].

Interesting and informative research has been conducted in order to bring about positive changes in MFC anodes [11]. For example, a carbon cloth anode treated with formic acid boosted the ideal power density of a single-chambered MFC by 38.1 percent (from 611.56 mW/m^2 to 877.935 mW/m^2) [12]. Likewise, electrolysis of the carbon fiber anode in nitric acid followed by rinsing in aqueous ammonia gave a noteworthy power density of 3.20 $\pm$ 0.05 W/m^2 for two-chambered MFCs [13]. Chaetoceros phytoplankton and lauric acid were used as methanotrophic inhibitors in MFCs implanted with mixed fermented sludge in the anodic compartment [14]. Rajesh et al. found 45.18% and 11.6% optimum coulombic efficiency, respectively [15]. Spontaneous elimination of AQS diazonium cations generated in situ immobilized an electron transfer accelerator, anthraquinone-2-sulfonic acid (AQS), onto the surface of a graphite sheet anode [16]. They used AQS in MFC, which showed a maximum power density increasing from 967 $\pm$ 33 mW/m^2 to 1872 $\pm$ 42 mW/m^2.

Entities including neutral red [17] and lignin [18] have also been used to improve the bio-catalytic effectiveness of the anode in MFCs. Anode characteristics like transmittance and the percentage surface area available for bacterial attachment are crucial, and both can be improved using graphene to improve electronic conductivity. A family of adaptable nanomaterials called carbon quantum dots (C-dots) has not yet been described in MFCs. Prior attempts at creating C-dots from coconut shell were said to include surface functional groups for hydroxyl and carboxyl. The C-dots' capacity to absorb and transport electrons was made possible by the existence of these ligands on a composite. C-dots were added as a suspension to the anode compartment of an MFC, which increased maximum power density by 22.5 percent. Compared to methylene blue, a common electron shuttle used in MFCs, C-dots demonstrated superior efficiency as electron shuttles [19]. Most of the EABs are metal reducing in natural condition. One of the primary limiting phases of MFC performance that requires attention is the sluggish exocellular electron transfer (EET) process in the anode. Most of the time, EET is improved by employing a modified anode. Anodes in MFCs are often made of biocompatible conductive carbon paper/cloth, graphite, and stainless steel. These materials, however, lack electrochemical activity for anode microbial processes. As a result, it is critical to create appropriate anode materials for efficient bacterial adhesion and electron collecting capabilities for better EET. The electron transfer efficiency of the outermost cytochromes (Omcs) [Omc C and Omc A] may be increased in the presence of redox-active biocompatible metal oxides and/or hydroxides. However, there have been relatively few reports on the C-dots metal oxides modified anode in MFC to date. As a result, the incorporation of C-dots and Iron oxide might address the shortcomings of the electrode materials, i.e., non-capacitive anode.

In the present research work, Fe_3O_4@PSA-C (0.25 mg/cm^2, 0.5 mg/cm^2, 0.75 mg/cm^2, and 1 mg/cm^2) were used to change graphite sheet anodes. Fe_3O_4@PSA-C assigned for Iron oxide @ sodium poly-acrylate-Carbon dot where Iron oxide (Fe_3O_4) is synthesized with sono- chemical method in presence of sodium poly-acrylate and after chemical sonication process with C-dots, finally we got Fe_3O_4@PSA-C-dots nanocomposite. The performance of MFCs using these anodes was compared to that of MFCs using a raw graphite sheet as an anode. Higher amount of Fe_3O_4@PSA-C like 1.25 mg/cm^2 was showing minute increase/ no increase in power density as compared to 1 mg/cm^2 which can be neglected hence the amount of Fe_3O_4@PSA-C was taken till the highest power density observed. This study aimed to examine if changing the anode to Fe_3O_4@PSA-C may enhance MFC power density (PD) at a fair cost to create a replacement anode composition that could improve MFC performance.

2. Results and Discussion

2.1. Structural Properties

2.1.1. X-ray Diffraction

The phase composition and crystallite size were determined using an X-ray Diffraction (XRD) technique to find the structure of nanoparticles. Figure 1 shows the XRD pattern of the as-synthesized nanoparticles, which shows the typical diffraction peaks for the synthesized nanocomposites as a pure phase with no secondary phases. Figure 1a depicts the XRD diffraction peak for C-dots at 2θ value 18° consist of (002) plane. Figure 1b reveals the XRD peaks for Fe_3O_4@PSA-C and Fe_3O_4@PSA samples. The XRD pattern of the nanoparticles synthesized by the sonication process was identical to the XRD pattern of the functionalized Fe_3O_4@PSA nanoparticles. The diffraction peaks may be assigned to magnetite Fe_3O_4 (JCPDS 19-0629). The intense diffracted peaks were recorded at 2θ = 18.21; 30.19; 35.58; 43.16; 53.45; 57.00; 62.75; 71.19; 74.24, 75.96 and 78.77, indicating the presence of magnetite as a phase pure with a spherical structure for (111), (220), (311), (400), (422), (511), (440), (620), (622) and (444) planes, which validate the presence of Fe_3O_4 in Fe_3O_4@PSA synthesized sample [20].

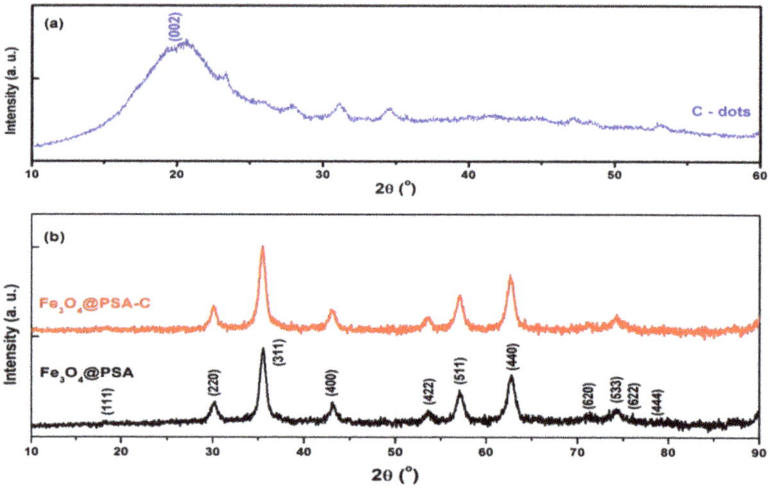

Figure 1. XRD patterns for (**a**) C-dots (**b**) Fe_3O_4@PSA-C and Fe_3O_4@PSA nanoparticles.

The assigned XRD peaks for Fe_3O_4@PSA-C functionalized hetero-structure nanoparticles were confirmed to have the presence of all validated positions of Fe_3O_4@PSA and C-dots as given in Figure 2b [20,21]. The calculated dimension of the magnetic nanoparticle crystallites at room temperature using the Debye Scherrer equation, based on the broadening of the most intense peak (311–35.58), was recorded in the XRD graph (Figure 1). The average crystallite size of the produced nanoparticles is 10 nm computed using the Debye–Scherrer equation [22,23].

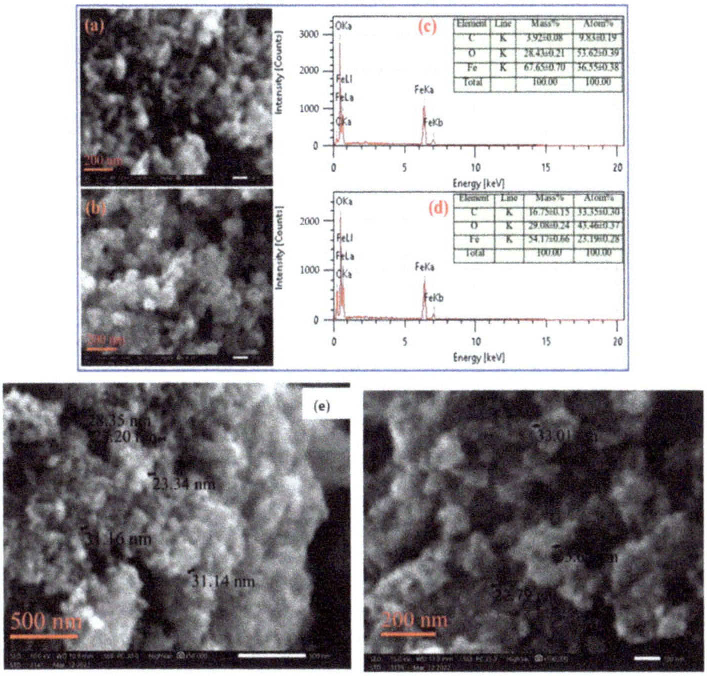

Figure 2. SEM graph for (**a**) Fe_3O_4@PSA, (**b**) Fe_3O_4@PSA-C nanoparticles, (**c**) EDX mapping of Fe_3O_4@PSA and (**d**) EDX mapping of Fe_3O_4@PSA-C, (**e**) SEM images of Fe_3O_4@PSA-C nanoparticles with grain size nanoparticles.

2.1.2. Scanning Electron Microscopy (SEM)

Figure 2a–d shows the SEM images of Fe_3O_4@PSA and Fe_3O_4@PSA-C nanoparticles with their energy dispersive X-ray (EDX) elemental mapping. The SEM micrograph results show that synthesized samples are spherical agglomerated structures due to the composition of numerous small nanoparticles and their particle size in the range of 10 nm to 30 nm. The particle size decreased due to sodium polyacrylate ($C_3H_3NaO_2$) n as CA, which chelates the Fe ions during the sonochemical process. The compositions of Fe_3O_4@PSA and Fe_3O_4@PSA-C nanocomposite was determined by X-ray energy-dispersive spectroscopy (EDS), as shown in Figure 2c,d. The EDS analysis of the Fe_3O_4@PSA samples showed peaks with weight percentages for Fe (67.65%), O (28.43%) and C (3.92%). The EDS analysis of the Fe_3O_4@PSA-C composite showed peaks for Fe (54.17%), O (29.08%) and C (16.75%), which reveal the implantation of the C-dots in the Fe_3O_4@PSA nanoparticles. The carbon- dots agglomerated over the Fe_3O_4@PSA surface.

2.1.3. Fourier Transform Infrared Spectroscopy (FTIR)

The chemical bond and function of Fe_3O_4@PSA, C-dots, and Fe_3O_4@PSA-Cdots nanoparticles have been identified by FTIR analysis in Figure 3. The presence of the Fe-O bond in tetrahedral and octahedral positions can be seen in the bands around 450–700 cm^{-1}. The peaks at 1440–1550 cm^{-1} represent the coordination between COO-group vibration and Fe-cations. The absorption bands around 2900–3600 cm^{-1} have been supported by the presence of O-H and C-H bonds [24]. Cow milk-based carbon dots contain various chemical compositions with polarities [25].

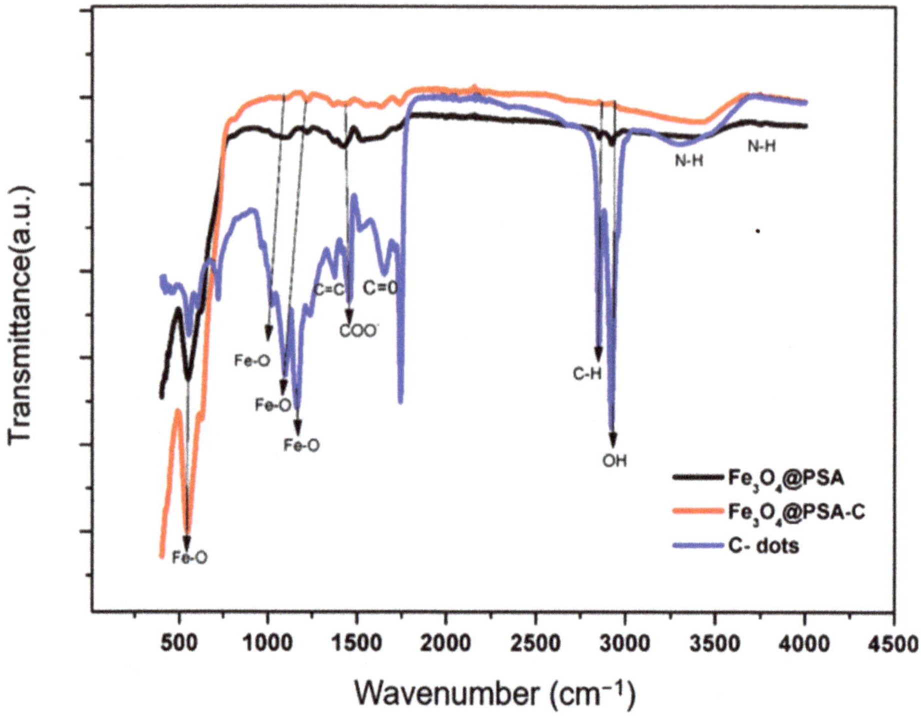

Figure 3. FTIR spectra of C-dots; Fe_3O_4@PSA-C and Fe_3O_4@PSA nanoparticles.

The band between 3100–3600 cm^{-1} shows the presence of O-H and N-H bonds, whereas the bands around 2800 and 2900 cm^{-1} are attributed to the C-H bond. The peaks at 1400 cm^{-1},1550 cm^{-1}, and 1630 cm^{-1} were assigned to C=C, N-H, and C=O starch bonds with π-π* and n-π* transitions. The FTIR spectra peaks of Fe_3O_4@PSA-C nanocomposite represent the slightly shifted band at 1400 cm^{-1},1550 cm^{-1}, and 1630 cm^{-1} ascribed to C=C, N-H, and C=O bonds due to the surface interaction between the functional moieties of nanocomposites. The peaks at 1100 cm^{-1} and 1300 cm^{-1} show the interaction between Fe-O bonds in Fe_3O_4@PSA and the carbonyl group present in C-dots. All such results confirm the hetero-structure nanocomposite of Fe_3O_4@PSA-C [26].

2.1.4. RAMAN Spectroscopy

In Figure 4, the Raman spectroscopy technique was used to identify the modes of C-dots, Fe_3O_4@PSA-C and Fe_3O_4@PSA nanoparticles. The cation distribution, non-stoichiometry, and defect conditions influence the results of Raman modes. The estimated values of polarized intensities also help to identify the Fe_3O_4@PSA Raman modes. The weak anti-stokes spectrum is present in the Fe_3O_4@PSA Raman band around 195 cm^{-1}. The ascribed vibrational spectrum of the observed modes was based on polarization induced due to the orientation of Fe_3O_4 nanoparticles like T2g at 193 cm^{-1}; T1u at 201 cm^{-1}; 390 cm^{-1} and 450 cm^{-1}; Eg at 296 cm^{-1}, T2g at 500 cm^{-1} and 590 cm^{-1} whereas, A1g at 660 cm^{-1} as shown in Figure 5. The bands around 1340 cm^{-1} and 1600 cm^{-1} in Raman spectra were assigned to D-band and G-bands for C-dots. The presence of sp3 and sp2 hybridized carbon atoms shows C-dots nanoparticles' graphitic nature. In Fe_3O_4@PSA-C spectra, the presence of T2g, T1u, Eg, T2g, A1g, D, and G vibrational modes with a modest frequency shift, shows that the Fe_3O_4 and C-dots were well coupled in the hetero-structure nanomaterial [27].

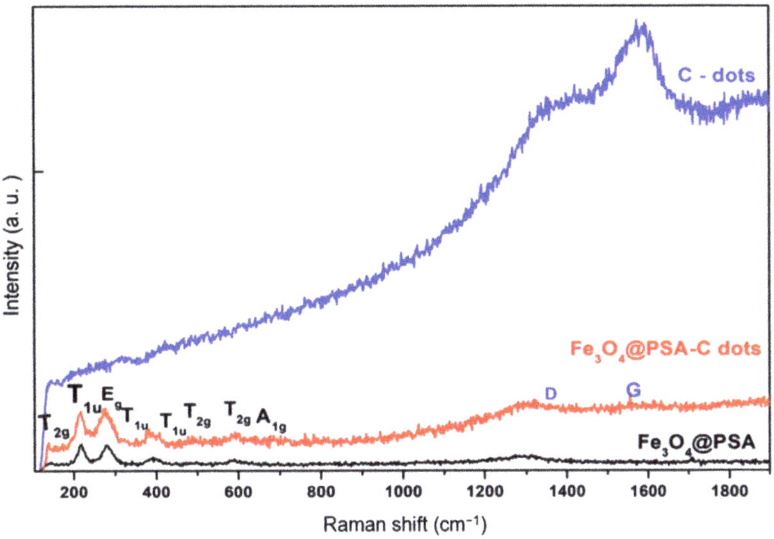

Figure 4. Raman spectra of C-dots; Fe_3O_4@PSA-Cdots and Fe_3O_4@PSA nanoparticles.

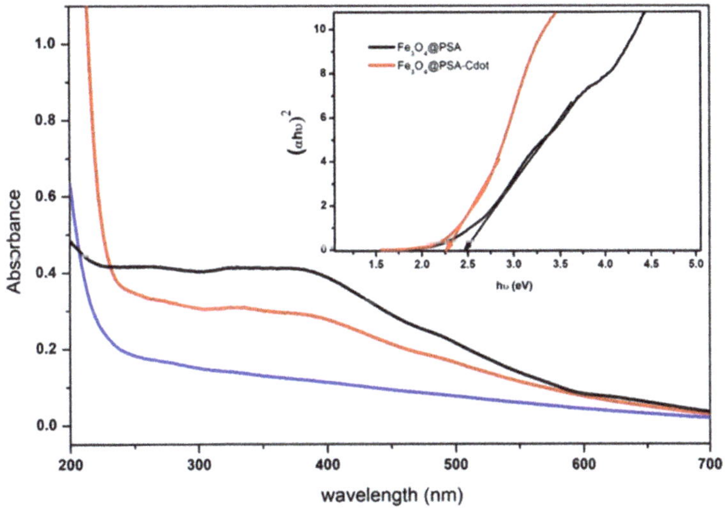

Figure 5. The plot $(\alpha h\nu)^2$ vs. hν represents the energy bandgap (Eg) of the Fe_3O_4 and Fe_3O_4@PSA-C nanoparticles.

2.2. Optical Properties

The optical properties of functionalized nanoparticles depend on their electronic characteristics and energy bandgaps (Eg). These optical properties of pristine samples were analyzed by the UV-Visible absorption spectra shown in Figure 6. The energy bandgap (Eg) was calculated using 'Tauc's relation.'

$$\alpha = \left(A\left[\left(h\nu - E_g \right) \right]^{\frac{1}{2}} \right) / h\nu$$

where A refers to constant, hν for photon energy, Eg for band gap, and α for absorption coefficient [22].

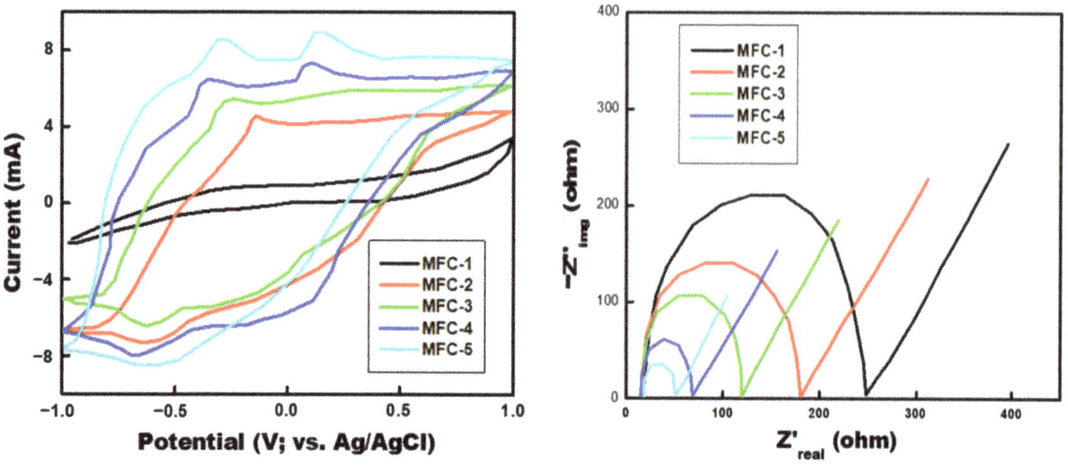

(a) (b)

Figure 6. (**a**) Cyclic Voltammetry analysis of all five MFCs; (**b**) Nyquist plot of operational MFCs.

The bandgap was determined from the $(\alpha h\nu)^2$ vs. $h\nu$ curve by drawing an extrapolation of the data point to the photon energy axis where $(\alpha h\nu)^2 = 0$, giving the optical band gap Eg. Figure 3 shows the energy bandgap (Eg) values for Fe_3O_4@PSA and Fe_3O_4@PSA-C synthesized materials. The Eg values for Fe_3O_4@PSA were estimated to be around 2.4 eV and 2.25 eV for Fe_3O_4@PSA-C nanocomposite [24]. The bandgap was reduced after the composition of C-dots due to the agglomerated particle size and functionalization of the surface of the composite nanomaterial. The composition of C-dots results in a charge imbalance, which triggers the presence of oxygen vacancies in prepared samples. The variation in the angle of Fe-O-Fe towards 180° also affects their electronic structure and creates an impurity band. This impurity band enhances charge transfer between the carriers by alteration of the energy bandgap. The substitution of C-dots in pristine Fe_3O_4@PSA shows lower energy band gap values, which resulted in improved optical properties of the Fe_3O_4@PSA-C nanocomposite.

2.3. Microbial Electrochemical Study

MFCs having anode with different concentrations of Fe_3O_4@PSA were operated in a fed-batch mode. Following the biofilm development throughout the ninth batch rotation, the anode's electrochemical kinetics were studied in operational MFCs. The CV approach has been widely utilized to evaluate the biocompatibility of anode components for electro-catalyzing substrates using microbes as biocatalysts [28]. In this study, the functionalization of anodes enhanced with various components like raw graphite sheet and different amounts of Fe_3O_4@PSA-C mixture was assessed only based on the MFCs' electrochemical properties. Five MFC system were built and named; for instance, MFC having raw graphite as anode was named as MFC-1, followed by MFC with increasing amount of Fe_3O_4@PSA-C 0.25 mg/cm^2, 0.50 mg/cm^2, 0.75 mg/cm^2 and 1 mg/cm^2 as MFC-2, MFC-3, MFC-4 and MFC-5, respectively (Table 1). Figure 6a summarizes the output of the cyclic voltammetry (CV) evaluation of the various anode components employed in these tests. By using a linearly cycled potential sweep among two or more preset values, cyclic voltammetry is an electrochemical method for determining how much current flows through a redox-active solution. The rates of electronic-transfer events, the energy levels of the analyte, and the thermodynamics of redox reactions can all be easily ascertained using this method. The Fe_3O_4@PSA-C (1 mg/cm^2) composite exhibited separate redox maxima, which were determined to be greater than the Fe_3O_4@PSA-C (0.25 mg/cm^2, 0.5 mg/cm^2, and 0.75 mg/cm^2) and reference anode, as per CV. During the oxidative examination of CV, two oxidation maxima at −0.26 mV and +0.16 mV at current values of 7.7 mA and 8.1 mA were recorded in MFC utilizing Fe_3O_4@PSA-C (1 mg/mL).

Table 1. Comparison of different quantities of Fe_3O_4@PSA-C (0.25, 0.50, 0.75, and 1 mg/cm^2) in the anode with the standard graphite sheet.

MFC Serial No.	Anode	OCV (mV)	PD (mW/m^2)	Charge Transfer Resistance, R_{ct} (Ohm)
MFC-1	Raw graphite sheet	701	285.01	230
MFC-2	Fe_3O_4@PSA-C modified anode (0.25 mg/cm^2)	726	334.41	164.46
MFC-3	Fe_3O_4@PSA-C modified anode (0.50 mg/cm^2)	746	379.56	103.14
MFC-4	Fe_3O_4@PSA-C modified anode (0.75 mg/cm^2)	765	420.33	52.96
MFC-5	Fe_3O_4@PSA-C modified anode (1 mg/cm^2)	771	440.01	33.45

Similarly, MFC having Fe_3O_4@PSA-C (0.75 mg/cm^2) also reported two oxidation peaks at −0.40 mV and +0.02 mV at current values of 5.7 mA and 6.8 mA. A single oxidation spike was observed at −0.12 mV with a current of 3.7 mA in the case of MFC having Fe_3O_4@PSA-C 0.5 mg/cm^2 and −0.2 mV with a current of 4.8 mA in the case of MFC having Fe- 0.25 mg/cm^2, respectively. Still, the reference anode showed no enhanced oxidation spike, perhaps because of poor anode kinematics.

The oxidation of essential metabolites may have caused the initial oxidation spike at −0.26 mV. Furthermore, electroactive enhancers such as Fe_3O_4@PSA-C-treated electrodes have been discovered to boost the electronic conductivity of the electroactive species, resulting in greater signal responsiveness [29]. FCDMA, for instance, had several oxidation spikes that were not hazy, while other anodes had crushed maxima. Moreover, the oxidation potential in the Fe_3O_4@PSA-C-modified anode was substantially greater than in the control anode, which might be attributable to decreased oxidation overpotential failure due to the FCDMA anode's super-hydrophilicity. In contrast, as opposed to the control, the Fe_3O_4@PSA-C (0.25 mg/cm^2 and 0.5 mg/cm^2) had greater efficacy and generated more oxidative potential.

Solution resistance (R_s), charge transfer resistance of the electrode (R_{ct}), and diffusion resistance (R_d) have all been demonstrated to be effective in assessing anodic impedance in MFCs using EIS [30]. The dielectric characteristics of materials, and the complex resistance are determined through electrochemical impedance spectroscopy (EIS). This is determined by the interaction of the external field with the specific sample's dipole moment, which is often expressed by the permittivity. Additionally, it is thought of as an experimental method for describing electrochemical systems. With this technique, system impedance is measured across a range of frequencies. As a result, a frequency response involving energy storage and dissipation characteristics is provided. The majority of the time, Nyquist plots are used to visually represent the data obtained by electrochemical impedance spectroscopy. The EC-Lab V11.36 tool was used to create the EIS evaluation findings. A calculated Nyquist plot of hypothetical impedance vs. real impedance is shown in Figure 6b. The width of the semi-largest circle may be used to compute R_{ct} readings. Any electrode's R_{ct} score is a statistic that represents the electrode's oxidation or reduction activation energy. A higher R_{ct} value suggests a larger potential barrier for redox processes and vice versa. As a result, a change in R_{ct}'s value might impact anode dynamics. Improved electrochemical rates resulted in a lower R_{ct} of 52.96 and 33.45 for electron transport in MFCs using Fe_3O_4@PSA-C (0.75 mg/cm^2 and 1 mg/cm^2), approximately 6.9-fold lower than the control anode (230 Ω). The R_{ct} of the Fe_3O_4@PSA-C (0.25 mg/cm^2 and 0.5 mg/cm^2) was 164.46 and 103.14, respectively, 2.22 times lower than the control anode.

In the investigation, Fe_3O_4@PSA-C (0.75mg/cm^2 and 1 mg/cm^2) has been transformed into highly conductive oxide in the anode during MFC operation, enhancing extracellular electron transport [31]. Furthermore, the magnitudes of the Fe_3O_4@PSA-C—0.25 mg/cm^2, 0.5 mg/cm^2, 0.75 mg/cm^2, and 1 mg/mL R_s were found to be almost equivalent (~18 ohm), showing that the mixture should be conceptually similar.

2.4. Power Generation

The power output reached a stable state after two fed-batch periods. At infinite impedance, the maximum OCV improved from 701 ± 23 mV in MFC-1 to 726 ± 20 mV in MFC-2, 746 ± 22 mV in MFC-3, 765 ± 25 mV in MFC-4, and 771 ± 28 mV in MFC-5 (reported as open-circuit voltage, OCV). Stronger electrogenic biofilm formation on the anode in MFCs using Fe_3O_4@PSA-C and other assessed MFCs resulted in enhanced OV and OCV, leading to enhanced electrogenic yield.

When the entire anodic chamber volume and actual anode surface area were normalized, polarisation experiments were performed to establish the appropriate power density. Figure 7a shows that MFC-1 with an unmodified anode had a lower power density of 285.01 ± 10 mW/m^2, compared to MFC-2 (0.25 mg/ cm^2 Fe_3O_4@PSA-C)which had a power density of 334.41 ± 6 mW/m^2 (4.86 ± 0.8 W/m^3), MFC-3 (0.50 mg/cm^2 Fe_3O_4@PSA-C) with 379.56 ± 8 (5.52 ± 0.5 W/m^3), MFC-4 (0.75 mg/cm^2 Fe_3O_4@PSA-C with 420.33 ± 12 mW/m^2 (6.11 ± 0.3 W/m^3) and MFC-5 (1 mg/cm^2 Fe_3O_4@PSA-C) with 440 ± 7 mW/m^2 (6.40 ± 0.4 W/m^3).

These results were superior to those previously reported for single-chambered MFCs with various anode promoters. Despite this, the MFC-2 and MFC-3 with lower Fe_3O_4@PSA-C altered anode had a lower inbuilt impedance than the MFC-1, which might be due to the reduced carbon dots' lower charge transmission obstacle. Because Fe_3O_4@PSA-C on the anode side worked as a final electron acceptor for extracellular electron transport, reduced the internal impedance, and decreased the anode's conductance (Table 1).

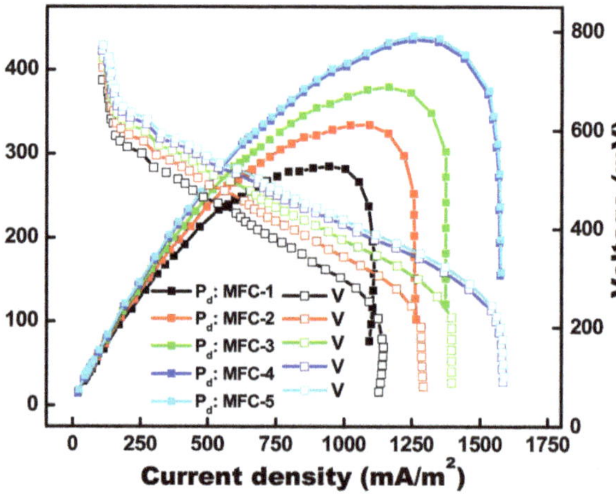

(a)

Figure 7. *Cont.*

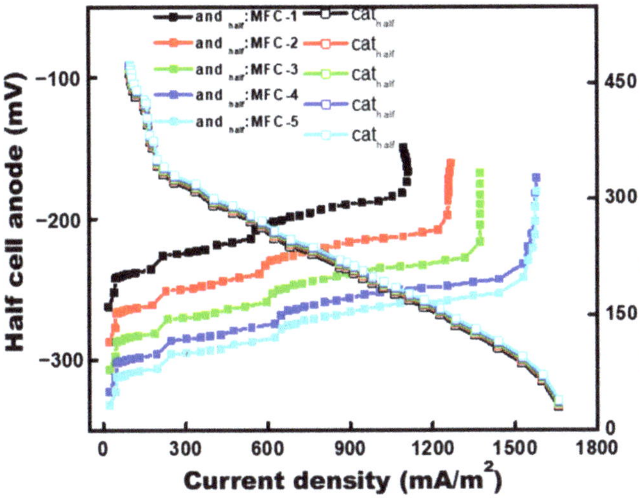

(**b**)

Figure 7. (**a**) Polarization study of unaltered and Fe_3O_4@PSA-C altered anode; the power density and voltage data points are presented as solid and open symbols, respectively. (**b**) Anode and cathode half-cell analysis to understand the variation in anode after modification with Fe_3O_4@PSA-C. Open and solid symbols represent cathode and anode half-cell voltages, respectively.

The existence of rapid energy losses in MFC-1, MFC-2, and MFC-3 instead of a gradual current decline in MFC-4 and MFC-5 indicates that MFC-1, MFC-2, and MFC-3 have a larger activation rate (Figure 7a). The anode potential in MFC-4 (−322 mV) and MFC-5 (−332 mV) was found to be the lowest when compared to MFC-1 (−262 mV), MFC-2 (−287 mV) and MFC-3 (−307 mV), as shown in Figure 7b. Using carbon dots in conjunction with Fe^{3+} can increase the anode's surface area on a micro/nanoscale, allowing additional cell adhesion and catalytic regions. The Fe_3O_4-carbon dot modified anode-based MFC's increased cell adhesion and catalytic regions were combined with increased energy output. Moreover, because most electricigens employ hydrophilic redox messengers, the interface between the anode contacts region and the biofilm impacts the MFC's energy output [32,33]. Increased microbe adhesion to the anode elevated anode voltage, resulting in improved electrogenic productivity in MFCs using Fe_3O_4@PSA-C modified anode (FCDMA).

3. Materials and Methods

3.1. Synthesis of Fe-Carbon Dots

Materials Required: Iron Chloride ($FeCl_3{\cdot}6H_2O$); Iron Sulphate ($FeSO_4{\cdot}7H_2O$); Ammonium Hydroxide (NH_4OH); Tri-Sodium Citrate ($Na_3C_6H_5O_7$) as capping agent (CA) and Cow milk. All chemicals were purchased from Sigma Aldrich. The Fe_3O_4@PSA-C hetero-structure was synthesized by the hydrothermal-assisted probe sonication method. The C-dots synthesis process was performed using the hydrothermal method, whereas further Fe_3O_4@PSA-C nanoparticle synthesis was done using the sonochemical method (Probe Sonicator), as shown in Figure 8.

Synthesis of C-dots: For the synthesis of C-dots, a fixed amount of cow milk was taken for the hydrothermal treatment at 120 °C for 4 h in a 100 mL stainless steel autoclave [25]. After cooling at room temperature, the sample was filtered and washed continuously using double distilled water until it reached pH-7. The C-dots filtered particles were dried at 60 °C, and the prepared C-dots nanoparticles became ready for characterization and participated in the Fe-C-dot synthesis process. For the synthesis of C-dots, we have used the cow milk. The Amul cow milk was purchased from the local market. The specifications and content of the Amul cow milk are given below (Table 2).

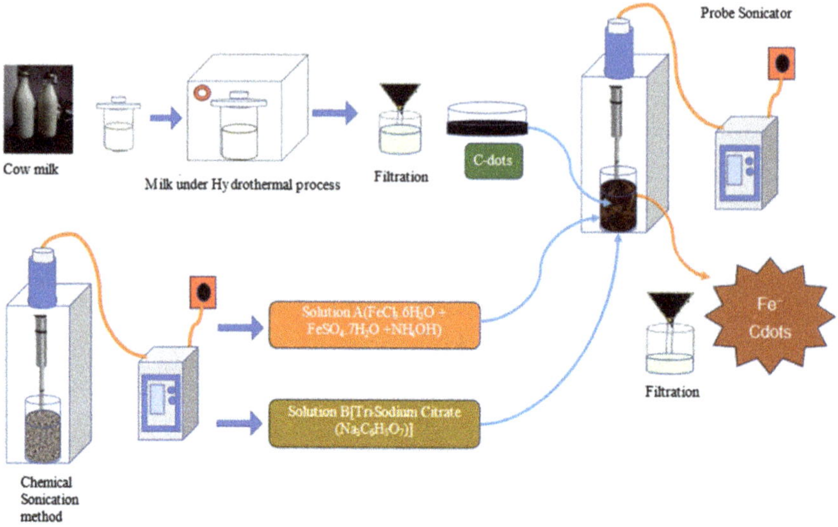

Figure 8. The synthesis process of C-dots; Fe_3O_4@PSA-C and Fe_3O_4@PSA samples using the Hydrothermal assisted chemical sonication method.

Table 2. Specification and content of Amul cow milk.

Packaging Type	Poly Packet
Brand	Amul
Serving Size	200 mL/100 mL
Amount per	100 mL/100 mL
Energy	62 Kcal/100 mL
Energy from FAT	32 Kcal/100 mL
Total FAT	3.5 g/100 mL
Saturated FAT	2.3 g/100 mL
Cholesterol	8 mg/100 mL
Total Carbohydrate	4.7 g/100 mL
Pack Size	500 mL

The C-dots were synthesized using the Amul cow milk many times in our lab; further the XRD, FTIR, and Raman measurements were performed. The results are reproducible.

Synthesis of Fe_3O_4@PSA-C: The whole sonication synthesis process was divided into three steps, and the prepared solutions in different steps were marked as solution A, Solution B, and F-Cdots. A certain amount of $FeCl_3{\cdot}6H_2O$ and $FeSO_4{\cdot}7H_2O$ (1:1 ratio) were dissolved in 15 mL of DD water to prepare solution A. The prepared solution was kept in the probe sonicator system (Labman Pro650, Labman Scientific Instruments Pvt. Ltd., Chennai, India) with frequency 20–25 kHz and 6 mm probe. The settings for the probe sonicator are pulse rate (3 sec on/off), power 585 watt and the cut off temperature 65 °C. Subsequently, a 7 mL NH_4OH solution was added to the prepared solution and sonicated for the next 5 min under the same conditions. The prepared solution was marked as Solution A. Parallelly; solution B was prepared in 10 mL of DD water using a certain amount of sodium polyacrylate $(C_3H_3NaO_2)_n$ as CA. Finally, solution B and 0.2 gm C-dots (pre-synthesized using cow milk) were mixed with solution A under vigorous sonication for 10–15 min. The resultant solution was filtered and continuously washed with DD water and acetone till pH-7 was achieved. The prepared sample was dried in an oven at 70 °C for 12 h to obtain Fe_3O_4@PSA-C hetero-structure nanoparticles in pure blackish powder form [34].

3.2. Characterization of Fe-Carbon Dots

The prepared samples were structurally characterized by an X-Ray diffractometer (XRD, MiniFlex 600, Rigaku, Japan). The measurements were carried out at room temperature using a CuK_α radiation source (λ = 1.5406 Å, operated at 40 KV and 40 mA). The data were collected in the range 2θ = 7° to 90° with a step size of 0.02° and 1.2 s count time at each step. Scanning electron microscopic images were recorded by using a Zeiss Ultra scanning electron microscope (SEM, JSMIT200, JEOL, New Delhi, India) operated at 10–15 kV. Raman spectroscopy was carried out in the backscattering configuration (LabRAM HR, HORIBA India Private Limited, Kolkata, India) with a charge-coupled device detector and 532 nm Laser excitation sources. The Laser power was kept below 10 mW in order to avoid any sample heating. The vibrational properties were characterized by Fourier Transformed Infrared (FTIR) spectroscopy performed on a spectrum BX-II spectrophotometer (PerkinElmer, New Delhi, India) with a spectral resolution of 1 cm^{-1}. The FTIR measurements of all the samples were done in the reflectance method in the wavenumber range of 400–4000 cm^{-1} at room temperature. The optical properties of the synthesized samples were analyzed by using a LABMAN-900S UV–visible spectrometer (Labman Scientific Instruments Pvt. Ltd., Chennai, India).

3.3. Construction of Electrodes

A titanium horn edge sonicator was used to disperse the Fe_3O_4-carbon dots in 50 mL of deionized water (Piezo-U-Sonic, Kolkata, India). After 3 h of sonication, the composite mixture was transferred to 0.5 mL of 5% polyvinyl alcohol (PVA) and sonicated for another 30 min. The Fe_3O_4@PSA-C concentrations in the composite mixture were kept at 0.25 mg/cm^2, 0.5 mg/cm^2, 0.75 mg/cm^2 and 1 mg/cm^2, respectively. Clumps of graphite sheet with an approximate contact area of 9 cm^2 (3.0 cm × 3.0 cm) were washed in 1 M HCl. These pieces were sonicated for 30 min with distilled water to remove adhering particles, then rinsed multiple times using distilled water and 35% ethanol, followed by thermal treatment in a muffle furnace for over 30 min at 400 °C. A fabricated graphite sheet was submerged in the combination above of Fe-Carbon dots in a new Petri plate overnight at 60 °C to generate Fe_3O_4@PSA-C-modified anodes (FCDMA). A raw graphite sheet anode was made using the same procedure but without including Fe_3O_4@PSA-C in the solution. The redesigned anodes were heated in a hot air oven at 60 °C for 24 h. Measurement of dried anodes provided the correct surface mass deposition of Fe_3O_4@PSA-C mixture of raw graphite sheet.

The MFCs' air cathodes were constructed using stainless steel (SS) wire mesh potential collectors (6 cm × 6 cm). Pt-C catalysts (0.5 mg/cm^2, Sigma-Aldrich, St. Louis, MO, USA) and 33.33 L/cm^2 PDMS (Polydimethylsiloxane) adhesive were combined in acetone and coated on SS wire mesh. Fabricated air cathodes were heated for 6 h at 80 °C in a muffle furnace before being stored in a desiccator for MFCs [35].

3.4. MFC Fabrication and Operation

Acrylic sheets were used to make five comparable single-chambered MFCs with a 30 mL anodic chamber size. In such MFCs, four FCDMA (with different amounts of Fe_3O_4@PSA-C) and a raw graphite sheet anode (reference) were utilized, with the rest of the constituents remaining the same. MFC-1, MFC-2, MFC-3, MFC-4, and MFC-5 were assigned to MFCs that used raw graphite sheet, Fe_3O_4@PSA-C (with 0.25 mg/cm^2, 0.5 mg/cm^2, 0.75 mg/cm^2, and 1 mg/cm^2) covered anodes, respectively. The anodic chamber and cathode of MFCs were separated by a 16 cm^2 cation exchange membrane (Ultrex, Ventura, CA, USA), which allowed H^+ ions to move from the anodic chamber to the air-cathode interaction junction. High-quality concealed copper wires were employed to connect the anode and cathode over a 100 Ω applied load. *Shewanella putrefaciens* MTCC 8104, an EAB, was used as an inoculant in the anodic chamber [36]. Twenty milliliters of inoculum was used as an inoculant in the anodic chamber of each MFC [37].Synthetic effluent containing 3 g/L sodium acetate and an adequate feeding media were supplied to the MFCs [36,38]. At the top of the anode chamber, there were two ports: one for the electrode terminal and the other for the reference electrode (Ag/AgCl, saturated KCl; +197 mV, Equiptronics, Mumbai, India) for sampling [39]. In the testing, five single-chambered MFCs with varied customized anodes were employed. However, the performance results may be trusted because the tests were done in triplicate for many feed gaps. These MFCs were kept in a batch mode with a three-day fresh eating interval at ambient temperatures varying from 33 to 37 °C.

3.5. Computation and Analysis

A digital multi-meter was utilized to record current output daily (HTC 830L). The polarisation experiment was conducted with the help of a data logger and a variable external resistance stage. The impedance ranged from 20,000 Ω to 5 Ω, and the related power was measured for at least 30 min (Agilent 34970A, Selangor Darul Ehsan, Malaysia). The power density in proportion to the expected surface area of the anode was calculated using Pd (mW/m^2) = EI/A, in which E and I are voltage and current, respectively, connected to specifically applied loads, and A is the anode surface area. The inherent resistance of the MFC was calculated using the current interruption method [39].

3.6. Analyses of Improved Anodes

The physical characterization of several anodes after remodeling and during biofilm growth on the relevant anodes while performing was carried out using a scanning electron microscope (SEM, ZEISS, Oberkochen, Germany).

3.7. Electrochemical Analysis

Electrochemical methods like electrochemical impedance spectroscopy (EIS) and cyclic voltammetry were used to investigate the anodic degradation kinetics of the different anodes used. Each MFC's platinum rod and anode were used as counter and operating electrodes in EClab electrochemical interrogations to analyze MFCs with a 3-electrode set-up. The voltage and current related to the Ag/AgCl (+197 mV vs. SHE) counter electrode w measured. The CV of the anodes was achieved by scanning the administered voltage from −1 V to +1 V at a frequency of 10 mV/s. The EIS was performed using alternating current (AC) at a rate of 100 kHz to 100 MHz and a voltage intensity of 5 mV. The EIS spectra were modeled in an analogous network to determine charge transfer resistance (R_{ct}) and solution resistance (R_s).

4. Conclusions

The present study used carbon dots coated Iron (II, III) oxide (Fe_3O_4@PSA-C) as a biocompatible anode to enhance power generation in a microbial fuel cell (MFC).

The Fe_3O_4@PSA-C hetero-structure was synthesized by the hydrothermal-assisted probe sonication method. The C-dots synthesis process was performed using the hydrothermal method, whereas further Fe_3O_4@PSA-C nanoparticle synthesis was completed using the sonochemical method. Nanoparticles were characterized with XRD, SEM, FTIR, and RAMAN Spectroscopy. The structural characterizations confirm the desired nanoparticles' pure phase formation with crystallite sizes ranging between 10–30 nm. Different concentrations of Fe_3O_4@PSA-C were coated onto graphite sheets and used as anodes in a single-chambered air cathode MFC. A graphite sheet anode enhanced with a Fe_3O_4@PSA-C combination was used to examine the building of a more bio-compatible anode for boosting

MFC efficiency. Power density in MFCs using Fe_3O_4@PSA-C was 1.54 times greater than in MFCs using raw graphite sheet anodes in a *Shewanella* -grown MFC.

The physical characterization of Fe_3O_4@PSA-C showed that the high surface area and several active sites were responsible for increased adhesion and low activation energy for the electroactive bacteria, thereby increasing the extracellular electron transfer. The coupling of Fe_3O_4 with carbon dots enhanced hydrophilicity with only a few sheets of flat Fe_3O_4@PSA-C, generating stronger stickiness and a broad contact surface for microorganisms to cling on the anode, resulting in better anode dynamics and efficiency in MFC. As a result, an anode made of carbon-based materials coated with Fe_3O_4@PSA-C may be employed in MFCs to improve coulombic efficiency and collect more energy while eliminating sewage in the anode compartment. The Fe_3O_4@PSA-C-modified bio-anodes were shown to boost the redox peak current during CV (Figure 6a). The im-proved redox profile with modified bioanodes is thought to be owing to EAB's intrin-sic capacity to transport electrons from its OmcC to insoluble Fe(III) metal centers of Fe_3O_4@PSA-C. A capacitive anode modifier may provide a capacitive bridge between the EAB and the anode, allowing the regulated passage of electrons from the EAB to the anode.

As a consequence, the larger the capacitance of the anode modifier, the better the efficiency. The various valence metal centers of Fe_3O_4@PSA-C may act as a redox cou-ple to link the EAB and anode interfaces. Furthermore, it is hypothesized that the con-stant interconversion of Fe(II) to Fe(III) contributes to transient electron storage through microbial or electrochemical reduction/oxidation, resulting in an increase in transient charge storage on Fe_3O_4@PSA-C loaded anodes. EIS analysis was used at OCP to investigate the interfacial charge transfer activity of several bioanodes [40]. Nyquist charts depict the impedance characteristics of all electrodes (Figure 5). CPE was uti-lized to analyze the varied resistive and capacitive properties of bioanodes using an equivalent circuit consisting of R_s, R_{ct}, and Z_W. The presence of Fe^{+3} and C-dot in re-dox active Fe_3O_4@PSA-C might be responsible for improved charge transfer and consequently decreased R_{ct} (Figure 6b). The half-cell potential measurements show that at the same current, all MFCs with bare graphite sheet anodes have equivalent half-cell potentials (Figure 7b, curve with solid symbol). Thus, the discrepancies in power generation between these five MFC reactors were caused by variances in cathode potentials (Figure 6b, curve with open symbol).

Author Contributions: Conceptualization, S.P., S.C. and A.S.M.; experiment, B.T., A.S., S.P. and S.C.; writing—original draft preparation, B.T., A.S., S.S., S.P. and S.C.; writing—review, P.K.D., A.S.M., P.K.G., M.S. and R.C.S.; editing, S.C., S.P., R.C.S. and K.P.; supervision, S.C. and S.P. All authors have read and agreed to the published version of the manuscript.

Funding: Authors duly acknowledge the grant received from Sharda University seed grant project (SUSF2001/01). Authors duly acknowledge the grant received from the Life Sciences Research Board, Defense Research and Development Organization (DRDO) (File No. LSRB/81/48222/LSRB-368/BTB/2020) and Department of Science and Technology (File No. CRD/2018/000022) for the accomplishment of this work. Gratitude to Science and Engineering Research Board (SERB), File Number TAR/2019/000210, Govt. of India to accomplish the current research work.

Data Availability Statement: Not applicable.

Conflicts of Interest: The authors declare no conflict of interest.

References

1. Logan, B.E.; Rabaey, K. Conversion of Wastes into Bioelectricity and Chemicals by Using Microbial Electrochemical Technologies. *Science* **2012**, *337*, 686–690. [CrossRef] [PubMed]
2. Wu, R.; Tian, X.; Xiao, Y.; Ulstrup, J.; Christensen, H.E.M.; Zhao, F.; Zhang, J. Selective Electrocatalysis of Biofuel Molecular Oxidation Using Palladium Nanoparticles Generated on Shewanella Oneidensis MR-1. *J. Mater. Chem. A* **2018**, *6*, 10655–10662. [CrossRef]
3. Rojas Flores, S.; Nazario-Naveda, R.; Betines, S.; Cruz–Noriega, M.; Cabanillas-Chirinos, L.; Valdiviezo Dominguez, F. Sugar Industry Waste for Bioelectricity Generation. *Environ. Res. Eng. Manag.* **2021**, *77*, 15–22. [CrossRef]

4. Chou, H.-T.; Lee, H.-J.; Lee, C.-Y.; Tai, N.-H.; Chang, H.-Y. Highly Durable Anodes of Microbial Fuel Cells Using a Reduced Graphene Oxide/Carbon Nanotube-Coated Scaffold. *Bioresour. Technol.* **2014**, *169*, 532–536. [CrossRef] [PubMed]
5. Mehdinia, A.; Ziaei, E.; Jabbari, A. Multi-Walled Carbon Nanotube/SnO2 Nanocomposite: A Novel Anode Material for Microbial Fuel Cells. *Electrochim. Acta* **2014**, *130*, 512–518. [CrossRef]
6. Min, B.; Román, O.B.; Angelidaki, I. Importance of Temperature and Anodic Medium Composition on Microbial Fuel Cell (MFC) Performance. *Biotechnol. Lett.* **2008**, *30*, 1213–1218. [CrossRef]
7. Torres, C.I.; Krajmalnik-Brown, R.; Parameswaran, P.; Marcus, A.K.; Wanger, G.; Gorby, Y.A.; Rittmann, B.E. Selecting Anode-Respiring Bacteria Based on Anode Potential: Phylogenetic, Electrochemical, and Microscopic Characterization. *Environ. Sci. Technol.* **2009**, *43*, 9519–9524. [CrossRef]
8. Fan, Y.; Xu, S.; Schaller, R.; Jiao, J.; Chaplen, F.; Liu, H. Nanoparticle Decorated Anodes for Enhanced Current Generation in Microbial Electrochemical Cells. *Biosens. Bioelectron.* **2011**, *26*, 1908–1912. [CrossRef]
9. Neethu, B.; Bhowmick, G.D.; Ghangrekar, M.M. Improving Performance of Microbial Fuel Cell by Enhanced Bacterial-Anode Interaction Using Sludge Immobilized Beads with Activated Carbon. *Process. Saf. Environ. Prot.* **2020**, *143*, 285–292. [CrossRef]
10. Bhowmick, G.D.; Noori, M.D.T.; Das, I.; Neethu, B.; Ghangrekar, M.M.; Mitra, A. Bismuth Doped TiO_2 as an Excellent Photocathode Catalyst to Enhance the Performance of Microbial Fuel Cell. *Int. J. Hydrogen Energy* **2018**, *43*, 7501–7510. [CrossRef]
11. Ahilan, V.; Bhowmick, G.D.; Ghangrekar, M.M.; Wilhelm, M.; Rezwan, K. Tailoring Hydrophilic and Porous Nature of Polysiloxane Derived Ceramer and Ceramic Membranes for Enhanced Bioelectricity Generation in Microbial Fuel Cell. *Ionics* **2019**, *25*, 5907–5918. [CrossRef]
12. Liu, W.; Cheng, S.; Guo, J. Anode Modification with Formic Acid: A Simple and Effective Method to Improve the Power Generation of Microbial Fuel Cells. *Appl. Surf. Sci.* **2014**, *320*, 281–286. [CrossRef]
13. Zhang, J.; Li, J.; Ye, D.; Zhu, X.; Liao, Q.; Zhang, B. Enhanced Performances of Microbial Fuel Cells Using Surface-Modified Carbon Cloth Anodes: A Comparative Study. *Int. J. Hydrogen Energy* **2014**, *39*, 19148–19155. [CrossRef]
14. Rajesh, P.P.; Noori, M.D.T.; Ghangrekar, M.M. Controlling Methanogenesis and Improving Power Production of Microbial Fuel Cell by Lauric Acid Dosing. *Water Sci. Technol.* **2014**, *70*, 1363–1369. [CrossRef] [PubMed]
15. Rajesh, P.P.; Jadhav, D.A.; Ghangrekar, M.M. Improving Performance of Microbial Fuel Cell While Controlling Methanogenesis by Chaetoceros Pretreatment of Anodic Inoculum. *Bioresour. Technol.* **2015**, *180*, 66–71. [CrossRef] [PubMed]
16. Tang, X.; Li, H.; Du, Z.; Ng, H.Y. Spontaneous Modification of Graphite Anode by Anthraquinone-2-Sulfonic Acid for Microbial Fuel Cells. *Bioresour. Technol.* **2014**, *164*, 184–188. [CrossRef]
17. Park, D.H.; Laivenieks, M.; Guettler, M.V.; Jain, M.K.; Zeikus, J.G. Microbial Utilization of Electrically Reduced Neutral Red as the Sole Electron Donor for Growth and Metabolite Production. *Appl. Env. Microbiol.* **1999**, *65*, 2912–2917. [CrossRef]
18. Sakdaronnarong, C.; Ittitanakam, A.; Tanubumrungsuk, W.; Chaithong, S.; Thanosawan, S.; Sinbuathong, N.; Jeraputra, C. Potential of Lignin as a Mediator in Combined Systems for Biomethane and Electricity Production from Ethanol Stillage Wastewater. *Renew. Energy* **2015**, *76*, 242–248. [CrossRef]
19. Vishwanathan, A.S.; Aiyer, K.S.; Chunduri, L.A.A.; Venkataramaniah, K.; Siva Sankara Sai, S.; Rao, G. Carbon Quantum Dots Shuttle Electrons to the Anode of a Microbial Fuel Cell. *3 Biotech* **2016**, *6*, 228. [CrossRef] [PubMed]
20. Abbas, M.W.; Soomro, R.A.; Kalwar, N.H.; Zahoor, M.; Avci, A.; Pehlivan, E.; Hallam, K.R.; Willander, M. Carbon Quantum Dot Coated Fe_3O_4 Hybrid Composites for Sensitive Electrochemical Detection of Uric Acid. *Microchem. J.* **2019**, *146*, 517–524. [CrossRef]
21. Wang, H.; Wei, Z.; Matsui, H.; Zhou, S. Fe_3O_4/Carbon Quantum Dots Hybrid Nanoflowers for Highly Active and Recyclable Visible-Light Driven Photocatalyst. *J. Mater. Chem. A* **2014**, *2*, 15740–15745. [CrossRef]
22. Chauhan, S.; Kumar, M.; Yousuf, A.; Rathi, P.; Sahni, M.; Singh, S. Effect of Na/Co Co-Substituted on Structural, Magnetic, Optical and Photocatalytic Properties of BiFeO3 Nanoparticles. *Mater. Chem. Phys.* **2021**, *263*, 124402. [CrossRef]
23. Lemine, O.M. Transformation of Goethite to Hematite Nanocrystallines by High Energy Ball Milling. *Adv. Mater. Sci. Eng.* **2014**, *2014*, e589146. [CrossRef]
24. Periyasamy, M.; Sain, S.; Sengupta, U.; Mandal, M.; Mukhopadhyay, S.; Kar, A. Bandgap Tuning of Photo Fenton-like Fe_3O_4/C Catalyst through Oxygen Vacancies for Advanced Visible Light Photocatalysis. *Mater. Adv.* **2021**, *2*, 4843–4858. [CrossRef]
25. Wang, L.; Zhou, H.S. Green Synthesis of Luminescent Nitrogen-Doped Carbon Dots from Milk and Its Imaging Application. *Anal. Chem.* **2014**, *86*, 8902–8905. [CrossRef]
26. Ahmadian-Fard-Fini, S.; Salavati-Niasari, M.; Ghanbari, D. Hydrothermal Green Synthesis of Magnetic Fe_3O_4-Carbon Dots by Lemon and Grape Fruit Extracts and as a Photoluminescence Sensor for Detecting of E. Coli Bacteria. *Spectrochim. Acta A Mol. Biomol. Spectrosc.* **2018**, *203*, 481–493. [CrossRef]
27. Kumar, G.G.; Sarathi, V.G.S.; Nahm, K.S. Recent Advances and Challenges in the Anode Architecture and Their Modifications for the Applications of Microbial Fuel Cells. *Biosens. Bioelectron.* **2013**, *43*, 461–475. [CrossRef]
28. Fricke, K.; Harnisch, F.; Schröder, U. On the Use of Cyclic Voltammetry for the Study of Anodic Electron Transfer in Microbial Fuel Cells. *Energy Environ. Sci.* **2008**, *1*, 144–147. [CrossRef]
29. Xu, Y.; Bai, H.; Lu, G.; Li, C.; Shi, G. Flexible Graphene Films via the Filtration of Water-Soluble Noncovalent Functionalized Graphene Sheets. *J. Am. Chem. Soc.* **2008**, *130*, 5856–5857. [CrossRef]
30. Baranitharan, E.; Khan, M.R.; Prasad, D.M.R.; Teo, W.F.A.; Tan, G.Y.A.; Jose, R. Effect of Biofilm Formation on the Performance of Microbial Fuel Cell for the Treatment of Palm Oil Mill Effluent. *Bioprocess. Biosyst. Eng.* **2015**, *38*, 15–24. [CrossRef]

31. Yuan, Y.; Zhou, S.; Zhao, B.; Zhuang, L.; Wang, Y. Microbially-Reduced Graphene Scaffolds to Facilitate Extracellular Electron Transfer in Microbial Fuel Cells. *Bioresour. Technol.* **2012**, *116*, 453–458. [CrossRef] [PubMed]
32. Pandit, S.; Khilari, S.; Roy, S.; Ghangrekar, M.M.; Pradhan, D.; Das, D. Reduction of Start-up Time through Bioaugmentation Process in Microbial Fuel Cells Using an Isolate from Dark Fermentative Spent Media Fed Anode. *Water Sci. Technol.* **2015**, *72*, 106–115. [CrossRef] [PubMed]
33. Zhou, M.; Chi, M.; Luo, J.; He, H.; Jin, T. An Overview of Electrode Materials in Microbial Fuel Cells. *J. Power Sources* **2011**, *196*, 4427–4435. [CrossRef]
34. Neto, D.M.A.; Freire, R.M.; Gallo, J.; Freire, T.M.; Queiroz, D.C.; Ricardo, N.M.P.S.; Vasconcelos, I.F.; Mele, G.; Carbone, L.; Mazzetto, S.E.; et al. Rapid Sonochemical Approach Produces Functionalized Fe_3O_4 Nanoparticles with Excellent Magnetic, Colloidal, and Relaxivity Properties for MRI Application. *J. Phys. Chem. C* **2017**, *121*, 24206–24222. [CrossRef]
35. Khilari, S.; Pandit, S.; Ghangrekar, M.M.; Das, D.; Pradhan, D. Graphene Supported α-MnO 2 Nanotubes as a Cathode Catalyst for Improved Power Generation and Wastewater Treatment in Single-Chambered Microbial Fuel Cells. *RSC Adv.* **2013**, *3*, 7902–7911. [CrossRef]
36. Panda, J.; Chowdhury, R. Growth Kinetic Study of Electrochemically Active Bacterium Shewanella Putrefaciens MTCC 8104 on Acidic Effluent of Jute Stick Pyrolysis. *Indian Chem. Eng.* **2021**, *63*, 193–205. [CrossRef]
37. More, T.T.; Ghangrekar, M.M. Improving Performance of Microbial Fuel Cell with Ultrasonication Pre-Treatment of Mixed Anaerobic Inoculum Sludge. *Bioresour. Technol.* **2010**, *101*, 562–567. [CrossRef]
38. Behera, M.; Ghangrekar, M.M. Performance of Microbial Fuel Cell in Response to Change in Sludge Loading Rate at Different Anodic Feed PH. *Bioresour. Technol.* **2009**, *100*, 5114–5121. [CrossRef]
39. Pandit, S.; Khilari, S.; Roy, S.; Pradhan, D.; Das, D. Improvement of Power Generation Using Shewanella Putrefaciens Mediated Bioanode in a Single Chambered Microbial Fuel Cell: Effect of Different Anodic Operating Conditions. *Bioresour. Technol.* **2014**, *166*, 451–457. [CrossRef]
40. Khilari, S.; Pandit, S.; Varanasi, J.L.; Das, D.; Pradhan, D. Bifunctional Manganese Ferrite/Polyaniline Hybrid as Electrode Material for Enhanced Energy Recovery in Microbial Fuel Cell. *ACS Appl. Mater. Interfaces* **2015**, *7*, 20657–20666. [CrossRef]

catalysts

MDPI

Review

Current Trends and Future Prospects of Nanotechnology in Biofuel Production

Indrajeet Arya [1,†], Asha Poona [1,†], Pritam Kumar Dikshit [2,†], Soumya Pandit [1], Jatin Kumar [1], Himanshu Narayan Singh [3], Niraj Kumar Jha [4], Hassan Ahmed Rudayni [5], Anis Ahmad Chaudhary [5,*] and Sanjay Kumar [1,*]

1 Department of Life Sciences, School of Basic Sciences and Research, Sharda University, Greater Noida 201310, India; 2019003610.indrajeet@pg.sharda.ac.in (I.A.); 2019006742.asha@pg.sharda.ac.in (A.P.); soumya.pandit@sharda.ac.in (S.P.); jatin.kumar1@sharda.ac.in (J.K.)
2 Department of Biotechnology, KoneruLakshmaiah Education Foundation, Vaddeswaram, Guntur 522502, India; pritamkumar@kluniversity.in
3 Department of System Biology, Columbia University Irving Medical Center, New York, NY 10032, USA; hs3290@columbia.edu
4 Department of Biotechnology, School of Engineering and Technology, Sharda University, Greater Noida 201310, India; niraj.jha@sharda.ac.in
5 Department of Biology, College of Science, Imam Mohammad Ibn Saud Islamic University (IMSIU), Riyadh 11623, Saudi Arabia; harudayni@imamu.edu.sa
* Correspondence: aachaudhary@imamu.edu.sa (A.A.C.); sanjay.kumar7@sharda.ac.in (S.K.)
† These authors share equal contribution.

Citation: Arya, I.; Poona, A.; Dikshit, P.K.; Pandit, S.; Kumar, J.; Singh, H.N.; Jha, N.K.; Rudayni, H.A.; Chaudhary, A.A.; Kumar, S. Current Trends and Future Prospects of Nanotechnology in Biofuel Production. *Catalysts* **2021**, *11*, 1308. https://doi.org/10.3390/catal11111308

Academic Editor: Rafael Luque

Received: 5 October 2021
Accepted: 26 October 2021
Published: 28 October 2021

Abstract: Biofuel is one of the best alternatives to petroleum-derived fuels globally especially in the current scenario, where fossil fuels are continuously depleting. Fossil-based fuels cause severe threats to the environment and human health by releasing greenhouse gases on their burning. With the several limitations in currently available technologies and associated higher expenses, producing biofuels on an industrial scale is a time-consuming operation. Moreover, processes adopted for the conversion of various feedstock to the desired product are different depending upon the various techniques and materials utilized. Nanoparticles (NPs) are one of the best solutions to the current challenges on utilization of biomass in terms of their selectivity, energy efficiency, and time management, with reduced cost involvement. Many of these methods have recently been adopted, and several NPs such as metal, magnetic, and metal oxide are now being used in enhancement of biofuel production. The unique properties of NPs, such as their design, stability, greater surface area to volume ratio, catalytic activity, and reusability, make them effective biofuel additives. In addition, nanomaterials such as carbon nanotubes, carbon nanofibers, and nanosheets have been found to be cost effective as well as stable catalysts for enzyme immobilization, thus improving biofuel synthesis. The current study gives a comprehensive overview of the use of various nanomaterials in biofuel production, as well as the major challenges and future opportunities.

Keywords: nanoparticles; biofuel; transesterification; catalyst; immobilization

1. Introduction

It is very well known that the consumption of fossil fuels is increasing rapidly with an increase in population growth rate and urbanization, leading towards the exhaustion of petroleum-derived fuels in the near future. The limited availability of fossil fuels is a major problem around the globe. Moreover, high dependency on petroleum-derived fuels has raised many questions about its adverse effects on the environment, economy, and energy saving. Therefore, significant research has been focused on the search for an alternative source that can reduce the consumption of fossil-derived fuels [1].

Biofuels are considered as an alternative to fossil-based fuels and have gained worldwide attention in recent years due to their distinct properties [2]. The production of biofuel is being carried out using many plant sources such as vegetables, corn, soybean, sugarcane, palm oil, and Jatropha (used in Africa) as feedstock in almost every continent [3,4]. In countries such as the USA and Brazil, bioethanol is successfully being applied as biodiesel for otto-cycle engines in combination with gasoline [1]. Biodiesel on the other hand is an important type of biofuel, having the capability to either substitute or replace fossil-based diesel. The production of biodiesel is carried out through a trans-esterification process using renewable bio lipids [5]. Some of the potential feedstocks used to produce biodiesel are oil extracts of seeds or kernels of non-edible crops. An important non-edible oil plant is Jatropha which is native to Central and South America and is being considered as a reliable source for the production of biodiesel due to its high oil content [6]. In addition, edible oils obtained from plants like sunflower, soybean, palm, etc. are also being used as substrates for biodiesel production [7–9].

The use of nanotechnology and nanomaterials in biofuel research has risen as a promising tool in providing cost-effective techniques to improve production quality. There are multiple advantages to using nanoparticles (NPs) over other sources for biofuel synthesis due to their size and unique properties such as the high surface area to volume ratio and special attributes such as a significant extent of crystallinity, catalytic activity, adsorption capacity, and stability [10–12]. Carbon nanotubes and metal oxide nanoparticles are generally used as nano-catalysts for biofuel production because of their additional properties which aid in high potential recovery [13]. Nanotechnology in combination with other processes such as gasification, pyrolysis, hydrogenation, and anaerobic digestion has proven to be useful for the synthesis of biofuels [14,15]. The present review addresses the advancement of NPs over biofuel production in terms of their applications and challenges, with future perspectives.

2. Biofuel Types

Biofuels are generated from renewable sources, thus protecting the environment and solving the problem of depletion of fossil fuels and are considered as an alternative to fossil fuels. Biofuels are mainly divided into three main generations, namely first generation, second generation, and third-generation [16,17]. First-generation biofuel requires edible sources such as vegetable oils, starch and sugar as raw material for conversion. Microorganisms and enzymes are mainly utilized to act as a catalyst to convert saccharides into alcohol during the fermentation process. Production of biofuel is low due to limited feedstock supplies and the biofuel produced is costlier than that of petrol-based fuel. The production of biofuels in the second generation is expensive and requires a non-edible source for its production [18]. Third-generation biofuel requires advanced instruments (Figure 1). Biofuel production in this type of generation is mainly from algal and lignocellulosic biomass [19]. Advancements in each generation have led to the utilization of non-usable biomass, making the production cost-effective and increasing the biofuel production in lesser time. Therefore, to overcome these issues nanotechnology has been employed for biofuel-based processes. Several biofuels such as biohydrogen, biodiesel, bioethanol, biogas, etc.,have been produced with the application of nanotechnology. Two main reactions, trans-esterification and esterification, are adopted for the conversion of triglyceride to biofuels. Nano-catalysts such as nanotubes, nanosheets, and nanoparticles are largely available from microbial fuel cells for biofuel generation.

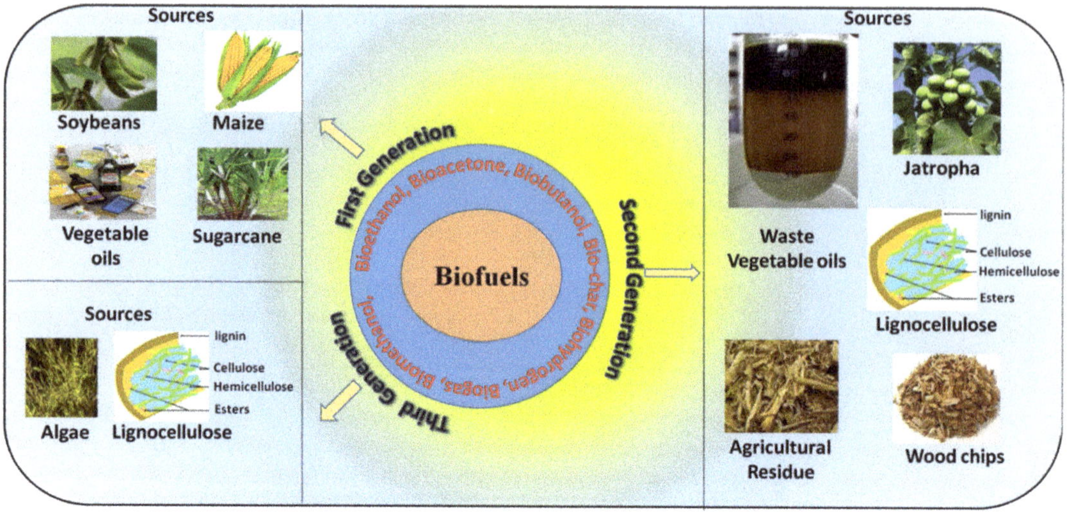

Figure 1. Representation of biofuel types and their sources.

3. Different Nanoparticles in Biofuel Production

Nanoparticles possess large surface area and super magnetic properties under the applied field which make them easier to separate from a biofuel cell and help in the recycling of enzymes. Several nanoparticles are used for biofuel production and form a support system for the catalyst to form a nano-catalyst. Some of these are magnetic nanoparticles and carbon nanotubes (CNTs), which act as a support system for enzymes. Other than these, metal, metal oxide, heterogeneous catalysts, acid-functionalized, etc. are used.

3.1. Carbon Nanotubes (CNTs)

CNTs are allotropes of carbon formed by rolling up sheets of graphene to a cylindrical shape. Due to their potential in carrying redox reactions and electron transfer kinetics, these nanotubes are primarily used in the fabrication of biosensors and microbial fuel cells [20]. The CNTs are of two types, Multi-Walled Carbon Nanotubes (MWCNTs), having multiple layers of graphene, and Single-Walled Carbon Nanotubes (SWCNTs), consisting of a single atomic layer of carbon atoms [21–23]. CNTs have characteristic features such as stability, high surface area, and less toxicity, and are used as a catalyst for biofuel production. Various studies have been performed on CNTs for biofuel synthesis. As their precursors are renewable, CNTs emerged as a promising nanomaterial because of their cost-effectiveness [24]. Liu et al. reported that the addition of 100 mg/L CNTs into anaerobic sludge blanket (UASB) reactors enhanced biohydrogen production with a production rate of 5.55 L/L/d and hydrogen yield of 2.45 mol/mol glucose [20]. The addition of CNTs during the anaerobic digestion process resulted in a reduction of start-up period and enhanced performance as compared to other activated carbon (AC) particles. In a similar kind of study, the immobilization of *Enterobacter aerogenes* over functionalized multi-walled carbon nanotube (MWCNT-COOH) enhanced the hydrogen production rate (2.72 L/L/h), hydrogen yield (2.2 mol/mol glucose), and glucose degradation efficiency (96.20%) in comparison to the free cells [25]. The immobilization process also reduced the lag phase duration of the anaerobic digestion process as compared to the free cell. There are different ways to synthesize CNTs, such as chemical vapor deposition, laser removal, arc discharge, etc. These particles are made up of graphite sheets rolled up into round and hollow shapes and are used for enzyme immobilization [26]. CNTs have a high capacity for loading enzymes due to their large surface area [27]. Enzymes can be immobilized on CNTs, thereby increasing the reusability and maintaining the catalytic activity of the immobilized enzyme [28–30]. It has been shown that MWNTs functionalized with amino groups improve the thermal stability of CNT [31]. Furthermore, usage of CNTs in biofuel generation increases the overall enzyme concentration and some properties of CNTs, such as porosity and conductivity, make it an important candidate for enzyme immobilization [1]. It was reported that functionalization of carbon nanotubes with ferrocene (Fc) and 2, 2′-azino-bis (3ethylbenzothiazoline-6-sulfonate) diammonium salt (ABTS)as mediator improves the catalytic activity in comparison to a glass carbon electrode, and maximum power output was also found to be 100 times greater than that of a carbon electrode without the incorporation of nanotubes. Here, ferrocene was utilized as an anodic mediator and ABTS as a cathodic mediator on the developed CNTs. 100 μWcm^{-2} power was generated when both anode and cathode were paired with nanostructures and their suitable mediators [32].

In another study conducted on Multi-Walled Carbon Nanotubes (s-MWCNTs), these were sulfonated, turned to s-MWCNT and tested for different parameters such as possess high catalytic activity [29]. In just half an hour from 1.5–2.0 h, 95.12% methanol was converted to oleic acid at temperature 210°C when increased from 180°C using s-MWCNT as a catalyst. The stability of carbon nanostructures was demonstrated after treatment with H_2SO_4, where no effects on the structure of carbon nanotubes were found. Additionally, a coupled reaction was performed to produce oleic acid, first when only the reaction was carried out, and later when the equilibrium was shifted by removing water. The first reaction was stopped after 4.5 h, while the coupling reaction stopped after 1.5 h and methanol recycled from each process was again reacted for a further 3 h to give 95.46 wt.% yield from the first process and 98.28 wt.% from the latter. Again, the reaction continued, for 1.5 h for the first process, but the yield remained unchanged, and for the coupling reaction the yield increased to 99.10 wt.% in just 1 h [33].

In several investigations, it was observed that MWCNTs functioned better than SWCNTs due to enzyme immobilization being consistent with their structural configuration, which increased the catalytic activity of the immobilized enzymes. MWCNTs surpassed the cellulose hydrolysis from *Aspergillus niger* with an efficiency of 85–97% and maintained their recyclable potential at 52–75% following 6 cycles of hydrolysis [34,35].

3.2. Magnetic Nanoparticles

Enzymes like cellulases and lipases are frequently used in the biofuel industries [36,37]. Many studies on magnetic nanoparticles suggest that they play an important role in the immobilization of enzymes for biofuel generation. Enzymes can be reused after immobilizing them to a support matrix coated with certain nanomaterial and this process is suitable for hydrolysis of lignocellulosic biomass [38]. The immobilization of enzymes used for hydrolysis of lignocellulosic biomass can be improved by altering various properties of enzymes [39]. The super magnetic property of magnetic nanoparticles is useful in the separation of immobilized enzymes, thus increasing reusability [40]. Many such attempts have been made to immobilize cellulose on magnetic nanoparticles for hydrolysis of biomass [41].

$CaSO_4/Fe_2O_3$-SiO_2 NPs are being used in a study to demonstrate biodiesel production from *Jatropha curcus* [42]. The pore size of nanoparticles is measured at 90 nm and a volume of 0.55 cm^3/g with a high surface area of 391 m^2/g. In optimum conditions, biodiesel production from crude Jatropha is measured at 94%, but after four cycles, it decreased to 85% and then gradually decreased further due to the inactivation of the nanoparticles. Further investigation was done to find the reason behind the inactivation of NPs. The results showed that the deposition of components of the reaction medium was blocking the pores after the fourth, seventh and ninth cycles. The surface area was also reduced to 252 m^2/g, which was less than earlier.

In another study, Fe_3O_4-NH_2 and reduced graphene oxide were incorporated into aniline for the formation of a nanocomposite, a polyaniline (PANI) matrix. The nanocomposite is shown to have improved the function in bio-electro catalysis of glucose oxidase. Investigation of the performance of rGO/PANI/f-Fe_3O_4/Frt/GOx, a bio anode, was carried out.

Glucose oxidase immobilized on rGO/PANI/f-Fe_3O_4 showed a very high catalytic current. Furthermore, reduced graphene oxide coated with PANI has a high surface area and a high electrical conductivity. The results have demonstrated that the nanocomposite is efficient in electron transfer. When applied to an enzymatic biofuel cell (EBFC), the maximum current produced was 32.9 mA cm^{-2}at a glucose concentration of 50 mM [43].

In biodiesel production, magnetic nano ferrites doped with calcium have been observed to have a significant effect, enhancing production yield by almost 85% from soybean cooking oils [44]. It was demonstrated that employing sugarcane leaves and MnO_2 nanoparticles increased bioethanol production. At various stages, it catalyzed this process. Sugarcane leaves are transformed to bioethanol in this technique and due to their large surface area, MnO_2 nanoparticles are accountable for the binding of enzymes to their active sites, improving ethanol synthesis [45]. It was discovered that the immobilization of yeast cells on the magnetic nanoparticles resulted in the higher production of ethanol [46,47]. Previous research has demonstrated the potential of implementing MNPs to hydrolyze the microalgae cell wall by immobilizing cellulase on MNPs accompanied by lipid extraction (Figure 2) [48]. Mahmood et al. studied the effect of iron nanoparticles addition over anaerobic digestion and hydrogen production using an aquatic weed, water hyacinth (*Eichhornia crassipes*) as the substrates [49]. Results of this study revealed that a specific concentration of iron nanoparticles enhanced the hydrogen yield reaching 57 mL/g of the dry weight of plant biomass. The enhancement of hydrogen production while using glucose as the substrate has also been reported in some studies [50,51]. In addition to zero-valent nanoparticles, iron oxides nanoparticles, such as Fe_2O_3 and Fe_3O_4 have been explored for the production of biohydrogen using glucose, wastewater, and sugarcane bagasse [52–54]. Nano zero-valent iron (nZVI) and Fe_2O_3 have also been explored for the enhancement of biogas production using waste-activated sludge [55]. The addition of 10 mg/g of total suspended solids (TSS)nZVI and 100 mg/g TSS Fe_2O_3NPs increased the methane production by 120% and 117% of control. These results confirmed that the addition of a low concentration of NPs promoted microbial growth as well as activities of key enzymes, leading to higher biogas production.

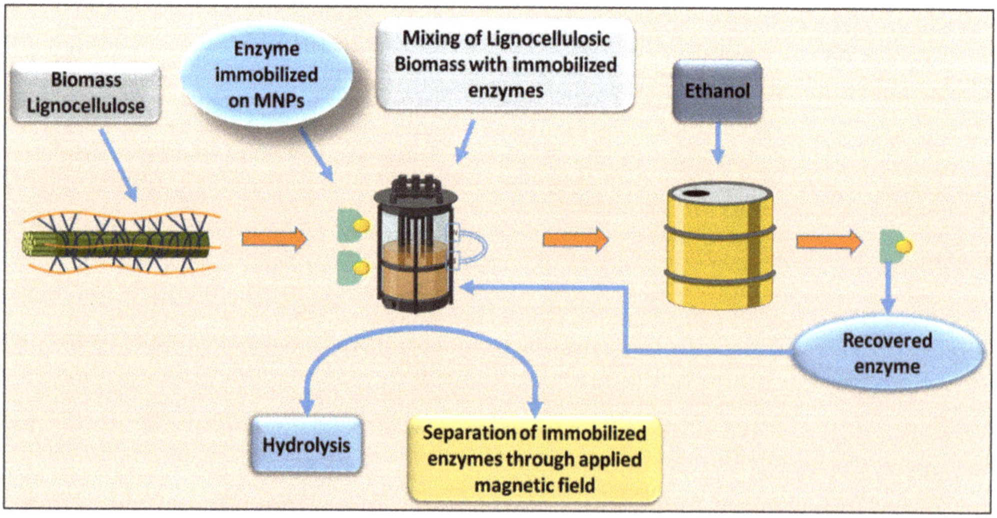

Figure 2. Biofuel production with the use of cellulase incorporated in MNPs (magnetic nanoparticles) to break down cellulose.

3.3. Acid Functionalized Nanoparticles

The potential pre-treatment strategies for lignocellulosic biomass include different methodologies based on acid and base. In this context, acid-functionalized nanoparticles are believed to play a key part in the hydrolysis of different biomasses, which are further used for bio-fuel generation. Transesterification and esterification methods are generally employed for triglycerides and fatty acids, respectively, and by making use of acid and base catalysts to improve reaction for FAME (fatty acid methyl ester)biodiesel production (Figure 3). The base-catalyzed process is somewhat easier than the acid-catalyzed process. On the other hand, acid-catalyzed reaction processes are cheaper in terms of the biomass they utilize [56].

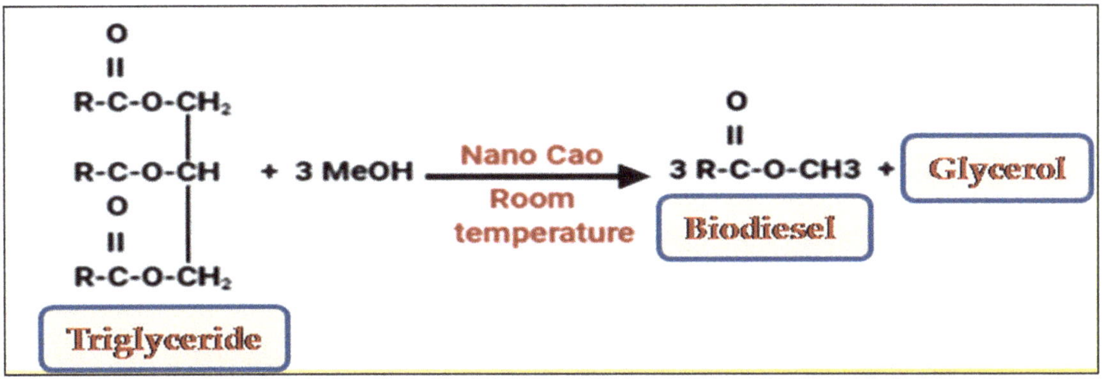

Figure 3. Preparation of FAME (fatty acid methyl ester)Biodiesel from Triglyceride.

According to Wang et al., silica-coated Fe/Fe_3O_4 MNPs assisted by sulfamic acid and sulfonic acids have been used for biodiesel production [57]. Transesterification of glyceryl trioleate and esterification of oleic acid have been carried out using sulfonic acid-functionalized/sulfamic acid-functionalized magnetic nanoparticles. It has been shown that 88% conversion of glyceryl trioleate in sulfonic acid-functionalized and 100% in sulfamic acid-functionalized in trans-esterification process was achieved at 100 °C in 20 h and it was 100% for oleic acid in just 4 h through an esterification process. Moreover, only 62% conversion was recorded when the sulfonic acid-functionalized catalyst was used, and 95% conversion was achieved for the sulfamic acid-functionalized process in the fifth cycle consecutively.

A recent study demonstrated the capability of nanotechnology by using acid-functionalized magnetic nanoparticles (MNPs) as catalysts in the hydrolysis of cellobiose from lignocellulose biomass. It was observed in the study that acid-functionalized MNPs with a 6% sulfur concentration resulted in 96% conversion of cellobiose, higher than the traditional conversion of 32.8%, in the absence of the catalyst [58]. Due to their nano-catalyst characteristics for the immobilization of various enzymes, these acid functionalized MNPs could accelerate the hydrolysis reaction. Apart from this, the high surface-to-volume ratio of such MNPs promotes the hydrolysis rate in comparison to the chemical pre-treatment. It was revealed that sulfonate-supported silica MNPs may be used to hydrolyze lignocellulose biomass, making them viable hydrolysis catalysts. Furthermore, these nanoparticles are thermally stable and can be easily separated from reaction mixture [59]. Enzymes associated with the production of biodiesel or bioethanol can be immobilized using MNPs as a medium. MNPs' strong coercivity and paramagnetic properties during the methanogenesis process also make them suitable for biogas production [60,61].

3.4. Metallic Nanoparticles

Although metallic nanoparticles have not been explored widely, various studies have been performed to check their efficiency in biofuel production. Metallic NPs are known for their higher surface area and nano-size that enable many enzymes such as oxidoreductase to bind with magnetic nanoparticles, consequently improving electron transfer [62]. Many catalytic nanoparticles have been constructed for a higher rate of ion transfer and oxygen reduction rate activity. It has been hypothesized that metallic NPs may be incorporated in a structured way to enhance electrocatalytic activity and creating a biofuel cell with high loading capacity and good electron transfer rate when employed in a layer-by-layer assembly with suitable polymers and enzymes [63].

Hybrid nano-catalysts have been designed using metallic NPS of gold, platinum, and Pt0.75-Tin0.25 by installation in acid-functionalized Multi-Walled NCTs, whereas gold NPs were encapsulated in poly (amidoamine) PAMAM dendrimer structure in another method. HR-TEM analyses have shown that dendrimer encapsulated NPs are highly arranged and very efficient. Biofuel cells have been configured with gold, platinum, and Pt0.75-Tin0.25 supported by MWCNTs, whereas gold NPs demonstrated great electrical conductivity and biocompatibility, and better catalytic activity than platinum NPs. The combination of platinum and tin NPs showed high oxidation activity for ethanol [64].

In another study, gold NPs were synthesized via laser ablation in an aqueous solution, which demonstrated good catalytic activity even on the 10th cycle, with great electrocatalytic efficiency. The LA-Au NPs outperformed, even with a lower surface area. This makes LA-Au a suitable candidate for biofuel cell development [65].

Various forms of nanomaterial have been used for the synthesis of biohydrogen. Gold nanoparticles (5 nm) improve substrate utilization capacity by 56% and boost biohydrogen generation rate by 46% [66]. Because of their smaller size and larger surface area, Au nanoparticles facilitate biohydrogen generation by adhering microbial cells to active sites. These nanoparticles also increase the enzymatic activity in the biohydrogen synthesis machinery, which is essential for biohydrogen production. Silver nanoparticles have also been observed to optimize substrate utilization which in turn promotes biohydrogen production. These nanoparticles shorten the lag period of bacterial and algal development while also activating the acetic reaction, which is the primary biohydrogen generation pathway. In photosynthetic microbes, nanoparticles promote the synthesis of biohydrogen. Nanoparticles added to the growth medium improve microbial growth, physiological processes and photosynthetic efficiency, protein synthesis, and nitrogen metabolism and. as observed in *Chlorella Vulgaris,* the optimal concentrations of Ag and Au nanoparticles increased its photosynthetic activity [67]. It has been demonstrated that the addition of zerovalent iron nanoparticles enhances biogas generation from waste matter [68,69]. Nickel nanoparticles have also been widely utilized in the hydrogenation process for converting glucose into sorbitol [70].

3.5. *Metal-Oxide Nanoparticles*

The synthesis of metal oxide NPs is fundamental for successful application and solution phase methods that give a great deal of control over the synthesis product. Metal oxide NPs are frequently arranged using the sol-gel technique, where the reaction is stopped before gelation occurs, like precipitation strategies. The properties of NPs are ascertained by the development, nucleation, and aging mechanisms.

Metal oxide NPs are known for their uses in sensors, catalysis, natural remediation, and electronic materials. Metal oxides have been used for the conversion of vegetable oil to biofuel. The metal oxides used as a catalyst are KOH, MoO_3, ZnO, V_2O_5, Co_3O_4, and NiO, and have the capacity to catalyze the transformation of oil into organic liquid products [71].

Metal oxides have been used as a support system with high catalytic activity but lower selectivity. Production of biodiesel has been carried out with the use of nano-catalysts CaO and Al_2O_3. Jatropha oil has been a good source of feedstock and biodiesels are synthesized by the process of transesterification with 82.3% yield using methanol and oil [72].

The metal oxide catalyst of ZrO_2 has been shown to employ both esterification and transesterification contemporaneously using a mixed feedstock of free fatty acids and soybean oil. ZrO_2 has been reported as highly stable, hard, and having both acidic and basic properties. The yield of fatty acid methyl ester (FAME)of 89–86% has been shown in both contemporaneous processes. The higher temperature condition for the metal oxide catalyst of ZrO_2 has been reported to achieve higher FAME yields [73]. Moreover, other NP catalysts such as MeO-SBA-15, ZnO-SBA-15, La_2O_3-SBA-15, etc., have been used to increase the biofuel production from waste cooking oils [74].

In biohydrogen production via the dark fermentation process, silica (SiO_2) nanoparticles have been employed. The nanocomposite produced by the combination of SiO2 nanoparticles with iron oxide (Fe_3O_4) has higher catalytic activity and stability, hence making them increasingly significant in biohydrogen production. Moreover, these nanocomposites provide additional advantages such as stability at high temperatures and low toxicity [75–77]. It has been reported that, with the addition of Fe_3O_4/ZnMg(Al)O nanoparticles, biodiesel productivity increased [78]. Nanoparticle functionalization is another process for increased biodiesel production. For example, Fe_3O_4/SiO_2 nanoconjugates can be used in biodiesel synthesis. Using nano-conjugates, biodiesel production can be increased by up to 97.1%. Various types of cooking and algal oils have been utilized in this technique and, with the availability of these ion-silica nanocomposites, algal oils have a high productivity rate [79]. Si-NPs are often deposited on the surface of nanoparticles for the immobilization of lignocellulolytic enzymes such as cellulase. Si-NPs have been shown to increase catalytic activity in the synchronous saccharification step for bioethanol synthesis using *Trichodermaviride* cellulose [80].

4. Nanoparticles in Heterogeneous Catalysis

Heterogeneous catalysts have emerged as an advancement on homogenous catalysts as they do not need too much water and are easy to separate from the reaction mixture [81]. The heterogeneous catalyst has been used for biofuel production [82,83]. Their separation is easy, and one can obtain contaminant-free products, which are normally non-corrosive, eco-friendly, and with high selectivity and long lifetimes. In some studies, the conversion of lignocellulosic biomass to biofuel has been demonstrated using NPs as heterogeneous catalysts [84]. The catalytic activity and selectivity of dispersed metal nanoparticle catalysts in heterogeneous catalysis were improved by using hybrid support made up of metal-organic framework (MOF) crystals and partially reduced graphene oxide (PRGO) nanosheets. Palladium nano-catalysts incorporated into a 3D hierarchical nanocomposite, Pd/PRGO/Ce-MOF, consisting of a Ce-based MOF wrapped in thin PRGO nanosheets, providing a heterogeneous tandem catalyst for the hydrodeoxygenation of vanillin, a common component in lignin-derived bio-oil, under mild reaction conditions. The developed heterogeneous catalyst Pd/PRGO/Ce-MOF has been shown to maintain its optimal catalytic activity and selectivity over four runs [85].

5. Applications

Biofuel as an alternative fuel source has many applications in various sectors globally. It may alleviate the problem of the constant degradation of petroleum-derived fuel. Nano-catalysts can increase the catalytic activity of biofuel-based reactions. These nano-catalyst/particles are of various types and have been developed continuously for their incorporation into biofuel cells as discussed in the above sections.

5.1. Biohydrogen Production

Generally, two different fermentation methods, i.e., (i) photo fermentation and (ii) dark fermentation are utilized for biological hydrogen production. Photo fermentation is carried out by microorganisms such as cyanobacteria and green algae in the presence of sunlight and water during the oxygen photosynthesis process. In the case of dark fermentation, anaerobic bacteria play a major role in the degradation of substrate or biomass for the production of biohydrogen [86,87]. Although this method is the most commonly adopted for the production of biohydrogen, the formation of by-products during the fermentation process inhibits the hydrogen production. Low hydrogen yield, the major limitations of this process can be solved by the application of nanoparticles. The unique physical and chemical properties of the nanoparticles have diversified their application in dark fermentation process leading to enhancement in hydrogen production. Several metal (Ag, Au, Cu, Fe, Ni) and metal oxide (Fe_2O_3, Fe_3O_4,TiO_2) nanoparticles have been successfully explored over the last few years. A summary of the application of various NPs in biohydrogen production is given in Table 1.

Table 1. Summary of application of nanoparticles in biohydrogen production process.

Nanoparticles	Substrate/ Feedstock	Reaction Conditions	Summary	Reference
Ag	Glucose	Mixed culture; pH–8.5; temperature–35 °C; rotation–120 rpm;	Higher hydrogen yield (2.48 mol/mol glucose) observed compared to blank. Reduction in lag phase observed with addition of Ag NPs. Reduction in ethanol production observed in presence of Ag NPs.	[86]
Au	Acetate	Anaerobic sludge; pH–7.2; temperature–35 °C	The hydrogen production rate reached 105 2 mL/L per day with the addition of Ag NPs.	[87]
Au	Artificial wastewater	Anaerobic culture; pH–7.2; temperature–35 °C	Maximum cumulative hydrogen production 4.48 mol per mol sucrose achieved with 5 nm Au NPs. The conversion efficiency of sucrose to hydrogen reached 56%.	[66]
Cu	Glucose	*Enterobacter cloacae* 811101 and *Clostridium acetobutylicum* NCIM 2337; pH–7.0 (*E. cloacae*), 6.0 (*C.acetobutylicum*);temperature–37 °C;duration–24 h	The Cu-NPs were found to have a more inhibitory effect on biohydrogen production. Addition of Cu NPs in fermentative process showed higher inhibitory effect than the $CuSO_4$ supplementation. Cu NPs with concentration less than 2.5 mg/L enhanced hydrogen production.	[88]
Fe	Glucose	*Anaerobic sludge*, pH–5.5; temperature–37 °C	The hydrogen and biogas yield of the control test were 247 and 391 mL/g VS, respectively. Addition Ni^{2+} ions improved hydrogen production by 55%.	[89]
Fe	Water hyacinth	*Mixed culture and Clostridium butyricum TISTR*, temperature–35 °C; duration–4 days	A maximum hydrogen yield 57mL/g of the plant biomass equal to 85.50% of the theoretical maximum is obtained.	[35]
Fe	Glucose	*Enterobacter cloacae DH-89*, pH–7.0; temperature–37 °C	Supplementation of Fe NPs significantly improved the hydrogen yield. A maximum H_2 yield 1.9 mol $mol^{-}1$ glucose utilized was observed with addition of 100 mg/L FeNPs, which increases the glucose conversion by two-fold.	[51]
Ni	Industrial wastewater	Anaerobic sludge; pH–7.0; temperature–55 °C; rotation–180 rpm	Ni-Gr NC dose of 60 mg/L exhibited the highest improvement (105%) in H_2 production. H_2 production was improved by 67% compared with supplementation of Ni nanoparticles.	[90]
Iron oxide	Glucose	*E. cloacae 811101*; pH–7.0; temperature–37 °C; duration–24 h;	Maximum hydrogen yields 2.07 mol H_2/mol glucose and 5.44 mol H_2/mol sucrose were achieved with addition of 125 mg/L and 200 mg/L iron oxide NPs. Enhancement of hydrogen production was higher with addition of iron oxide NPs compared to ferrous iron supplementation.	[91]

Table 1. *Cont.*

Nanoparticles	Substrate/ Feedstock	Reaction Conditions	Summary	Reference
Fe_2O_3	Glucose	Anaerobic sludge; pH–5.5; temperature–60 °C; rotation–150 rpm	Maximum hydrogen yield reached 1.92 mol H_2/mol glucose with a hydrogen content of 51%. Metal NPs are not consumed by the microbes and only act as hydrogen production enhancer.	[52]
Fe_3O_4	Wastewater	Mixed culture; pH–6.0; temperature–37 °C; rotation–200 rpm	The maximum hydrogen production rate and specific hydrogen yield reached 80.7 mL/h and 44.28 mL H_2/g COD with supplementation of NPs. Highest cumulative volume of hydrogen (380 mL), hydrogen content (62.14%) and % COD reduction (72.5) was obtained under the optimal conditions.	[53]
Fe_3O_4	Sugarcane bagasse	Anaerobic sludge; pH–5.0; temperature–30 °C	Addition of 200 mg/LFe^{2+} and magnetite NPs enhanced the HY by 62.1% and 69.6%, respectively. Highest hydrogenase gene activity was confirmed by immobilized cultures on magnetite nanoparticles.	[54]
TiO_2	Malate	*R. sphaeroides* NMBL–02; pH–8.0; temperature–32 °C	Hydrogen production rate enhanced by 1.54 fold and duration by 1.88 fold in the presence of 60 mg/mL of TiO_2 NPs in comparison to the control. Maximum hydrogen production 1900 mL/L with 63.27% malate conversion achieved.	[92]

5.2. Effectiveness of Nanoparticles in Biogas Generation for Industrial Benefits

Biogas generation has four main phases: (a) Hydrolysis, that converts organic waste into simple monomeric or dimeric units, (b) Acidogenesis, where the hydrolysis product is utilized for the fermentation, (c) Acetogenesis, which leads to the formation of acetate with H_2 and CO_2, and (d) Methanogenesis, which is the final stage where methane is produced from the early generated acetate, H_2 and CO_2 [93]. Nanotechnology plays an important role in biogas and methane production as it has a bio-stimulating effect on the methanogenic phase. Some studies have suggested that trace element-based NPs (Co, Fe, Fe_3O_4 and Ni) at various levels of concentration with significant particle size decrease the duration of lag phase as well as the time taken to attain the peak conversion rate [94]. Nano zero valence iron (NZVI) has been shown to affect the anaerobic digestion by increasing the production of biogas and methane. Moreover, NZVI stimulates the methanogenesis in the process of AD while inhibiting dichlorination [95]. Different types of NP have demonstrated their use for the synthesis of biogas (Table 2).

Table 2. Summary of application of nanoparticles in biogas production process.

Nanoparticles	Substrate/Feedstock	Reaction Conditions	Summary	Reference
Ni	Manure slurry	Temperature–37 °C; rotation–20 rpm (in 1 min interval)	Addition of 2 mg/L Ni NPs enhanced the biogas production by 1.74 times in comparison to control. The methane volume increased by 2.01 times. Highest specific biogas (614.5 mL per g VS) and methane (361.6 mL per g VS) production were attained with 2 mg/L Ni NPs.	[94]

Table 2. *Cont.*

Nanoparticles	Substrate/Feedstock	Reaction Conditions	Summary	Reference
Nano zero–valent iron (nZVI)	Waste activated sludge	Temperature–35 °C; rotation–120 rpm; duration–30 days	Addition of 10 mg/g total suspended solids (TSS)nZVI increased methane production to 120% of the control. Low concentrations of nZVI promoted a number of microbes (Bacteria and Archaea) and activities of key enzymes.	[55]
nZVI	Sewage sludge	pH–7.0; temperature–37 °C; duration–30 days	Methane yield enhanced by 25.2% in the presence of nZVI. COD removal efficiency was 54.4% in presence of nZVI, higher compared to control (44.6%). The addition of nZVI showed positive impact on the removal of chlorinated pharmaceutical and personal care products.	[96]
nZVI	Domestic sludge	Temperature–37 °C; duration–14 days	Methane content was stimulated up to 88% with addition of nZVI.	[97]
Co	Manure slurry	Temperature–37 °C; rotation–20 rpm (in 1 min interval)	Addition of 1 mg/L Ni NPs enhanced the biogas production by 1.64 times in comparison to control. The methane volume increased by 1.86 times.	[94]
Cu	Granular sludge	pH–7.2; temperature–30 °C; rotation–120 rpm	Cu NPs caused severe methanogenic inhibition. The 50% inhibiting concentrations determined towards aceto-clastic and hydrogenotrophic methanogens were 62 and 68 mg/L.	[98]
ZnO	Waste activated Sludge	Temperature–37 °C; duration–14 days	100 mg/L Zn^{2+} exhibited 53.7% reduction in methane production compared to control. Less VFA consumed during methanogenesis when more ZnO ENMs were present.	[99]
ZnO	Granular sludge	pH–7.2; temperature–30 °C; rotation–120 rpm	The 50% inhibiting concentrations determined towards aceto-clastic and hydrogenotrophic methanogens were 87 and 250 mg/L. Methanogenic inhibition is due to the release of toxic divalent Zn ions caused by corrosion and dissolution of the NPs.	[98]
CuO	Municipal waste activated sludge	Temperature–35 °C	Increase in CuO NP concentration from 5 to 1000 mg per gTS, and an increase in the inhibition of AD from 5.8 to 84.0% was observed. EC_{50} values of short- and long-term inhibitions were calculated as 224.2 mg CuO per g TS and 215.1 mg CuO per g TS, respectively.	[100]

5.3. Bioethanol

In contrast to petroleum derivatives, NPs have been utilized to improve gas-liquid mass transfer, which in turn improves cell mass concentration for the generation of bioethanol by syngas fermentation [101]. Bioethanol is considered a reasonable and eco-accommodating biofuel. It has been reported that bioethanol is has favorable chemical properties such as high dissipation enthalpy and a high-octane number. Currently, bioethanol is delivered from edible and non-edible vegetable oils, squander materials, algal, and bacterial biomass. Initially, microalgae have been a good source of bioethanol in terms of their quantity [102–104]. Genetically engineered microorganisms have been shown to produce a higher quantity of bioethanol than normal microorganisms [105]. Different types of NP have growing applications in the generation of bioethanol. Practically, it has been shown that MnO_2 nanoparticles increase the production of bioethanol utilizing agricultural waste and sugarcane leaves [46]. Various NPs utilized in ethanol production are shown in Table 3.

Table 3. Summary of application of nanoparticles in bioethanol production process.

Nanoparticles	Substrate/ Feedstock	Reaction Conditions	Summary	Reference
NiO	Potato peel waste	*S. cerevisiae* BY4743; Instantaneous saccharificationfermentation (NIISF); temperature–37 °C, rotation–120 rpm, duration–24 h	• 59.96% enhancement in bioethanol production. • Addition of nanoparticle improved bioethanol productivity by 145% and acetic acid concentration by 110%.	[106]
NiOand Fe_3O_4	Potato peel waste	*Saccharomyces cerevisiae* BY4743; temperature–30 °C; rotation–120 rpm; duration–72 h	• Maximum ethanol yield of 0.26 g/g, 0.22 g/L/h ethanol productivity and 51% fermentation efficiency at 0.01 wt%. • 1.60-fold and 1.13-fold using NiO and Fe_3O_4 NPs, respectively	[107]
ZnO	Rice straw	*Fusariumoxysporum*;temperature–20 to 25 °C; pH–6.0 to 8.0; rotation –100 to 200 rpm; duration –72 h	• Maximum ethanol yield of 0.0359 g/g of dry weight-based plant biomass was obtained at 200 mg/L concentration of ZnO nanoparticles. • Characterization of nanoparticles was carried out using UV–Vis spectroscopy, FTIR, XRD, SEM, TGA and DTA analysis.	[108]
Magnetic nanoparticles	Corn starch	Immobilized Saccharomyces cerevisiae; pH–4.0; temperature –60 °C	• Ethanol productivity reached 264 g/L.h. • The prepared immobilized cells were stable at 4°C in saline for more than 1 month.	[47]

5.4. Biodiesel

Biodiesel has many promising future applications due to the emission of fewer pollutants, is eco-friendly, and is produced from edible as well as non-edible oils. Oils are converted to biodiesel through the process of transesterification. The process utilizes homogeneous and heterogeneous catalysts [109]. Nanomaterials have promising results in biodiesel production. NPs can enhance the catalytic reaction during transesterification, thereby improving the production of biodiesel [110]. It is reported that the biodiesel production yield was enhanced in the presence of CaO based nano-catalysts as heterogeneous catalysts [111]. Microalgae biomass was also reported as a potential source to produce biodiesel [112]. Vegetable oils containing triglycerides have been utilized to produce biodiesel, which acts as a substitute for diesel. The process of transesterification is carried out to lower the viscosity of the vegetable oil [113].

Nanostructure provides emerging immobilization support due to the nanoscale size and large surface area. Microbial enzymes such as lipase from *Pseudomonas cepacia* are immobilized on the surface of nanoparticles and enhance the production of biofuel due to an enhanced transesterification reaction. Fictionalization of the nanoparticle process also increases the production of biodiesel. Nanoconjugates have also been shown to increase the production of biodiesel. Iron-silica nanoconjugates such as Fe_3O_4/SiO_2 have emerging applications in biodiesel production [114]. In this process, various types of cooking and algal oils have been used. Algal oils have a high yield production in the presence of these iron-silica nanocomposites [79]. The use of different NPs in biodiesel production is explained in Table 4.

Table 4. Summary of application of nanoparticles in biodiesel production process.

Nanoparticles	Substrate/Feedstock	Reaction Conditions	Summary	Reference
Fe_3O_4/ZnMg(Al)O	Microalgal oil	Temperature–65 °C;duration–3 h;methanol to oil ratio: 12:1	Biodiesel yield reached 94% under the optimal conditions. 82% biodiesel yield was observed after 7 times regeneration. Increase of the molar ratio of methanol to oil increased biodiesel yield.	[78]
SiO_2 and SiO_2–CH_3	*Chlorella vulgaris*	Methanol/sulfuric acid–85:15 v/v;temperature–70 °C;duration–40 min	Dry cell weight increased by 177% and 210% by adding SiO_2 and SiO_2–CH_3 NPs. Addition of NPs increased CO_2 mass transfer rate.	[115]
CaO and MgO	Waste cooking oil	For CaO: weight–1.5%; methanol to oil ratio–1:7; duration–6 h. For MgO: weight–3% (0.7 g of Nano CaO and 0.5 g of Nano MgO); alcohol to oil ratio–1:7; duration–6 h.	Nano MgO alone is not capable of catalysing the transesterification reaction due to weaker affinity. Nano MgO in combination with CaO increased the transesterification yield. The biodiesel yield reached 98.95% of weight.	[116]
Ni doped ZnOnanocatalyst	Castor oil	Methanol to oil ratio–1:8; catalyst loading –11% (w/w); temperature–55 °C, duration–60 min	95.20% higher biodiesel yield was observed under optimum conditions. The reusability study of nano-catalysts showed efficient for 3 cycles.	[117]
$Ni_{0.5}Zn_{0.5}Fe_2O_4$ doped with Cu	Soybean oil	Methanol to oil ratio–1:20; catalyst loading–4% (wt); temperature–180 °C, duration– 1 h,	Presence of Cu ions facilitated an increase of 5.5–85% in the conversion values in methyl esters. Cu^{2+} ions doping influenced in the structure, morphology and magnetic properties of nano-ferrites.	[44]
CaO	Bombaxceiba oil	Methanol to oil ratio–30.37:1; catalystloading–1.5% (wt); temperature–65 °C;duration–70.52 min	96.2% yield of methyl ester was achieved under optimum conditions. CaO-NPs reused for five consecutive cycles with minimum loss of activity.	[118]
Calcite/Au	Sunflower oil	Methanol to oil ratio–9:1; catalyst loading: 0.3% (wt); temperature–65 °C;duration– 6 h	The oil conversion was in the range of 90–97% under optimum conditions. The nano-catalysts were stable up to 10 cycles without loss of activity.	[119]
MgO/$MgAl_2O_4$	Sunflower oil	Methanol to oil ratio–12:1; catalyst loading– 3% (wt); temperature–110 °C; time–3 h	95.7% conversion of sunflower oil achieved. The prepared catalyst was stable for 6 cycles. Size, shape and crystallinity of catalysts are important parameters affecting biodiesel production.	[120]
Hydrotalcite particles with Mg/Al	Jatropha oil	Methanol to oil ratio–0.4:1 (v/v); catalyst loading– 1% (wt); temperature–44.85 °C;duration–1.5 h; anhydrous methanol–40 mL; sulfuric acid–4 mL	95.2% biodiesel yield was achieved under optimal conditions. The catalyst showed reliable performance for 8 consecutive cycles.	[121]
TiO_2–ZnO	Palm oil	Methanol to oil ratio –6:1; temperature–50–80 °C; duration– 5 h	92.2% FAME conversion and 92% yield was attained within 5 h at 60 °C. The synthesized catalysts were characterized by XRD, FT–IR, and FE–SEM.	[122]
CaO	Rice bran oil	Methanol to oil ratio–30:1; temperature–65 °C; duration–120 min; catalyst loading = 0.4%(wt)	93% FAME yield observed after 120 min under optimum conditions. The reusability of catalyst revealed that the FAME yield decreased significantly after fifth cycle.	[123]

Table 4. *Cont.*

Nanoparticles	Substrate/Feedstock	Reaction Conditions	Summary	Reference
CaO	Microalgae oil	Methanol to oil ratio–10:1; temperature–70 °C, duration– 3.6 h, methanol/oil; catalyst loading–1.7% (wt)	The nanoparticles are of spherical shape with average particle size of 75 nm. 86.41% microalgal biodiesel yield reported under optimal conditions. Reusability study of catalyst revealed 86.41% to 67.87% loss in biodiesel production after the sixth cycle.	[124]
ZnO	Waste cooking oil	Methanol to oil ratio–6:1; temperature–60 °C; duration– 15 min; catalyst loading–1.5% (wt)	FAME conversions yield up to 96% achieved under ultrasonic irradiation. Synthesized biodiesel properties such as density and viscosity were at par with standard biodiesel.	[125]

6. Current Challenges and Future Perspectives for Biofuel Production with the Implementation of Nanotechnology

Biofuel is the future of petroleum-based industries, as it is more environmentally friendly, cleaner, renewable and safe to use. Furthermore, the limited availability and increasing demand have led to price hikes for petroleum-derived fuels, prompting researchers to think about biofuels as a suitable alternative [113]. Even though it is safer and cleaner to use, the production of biofuels is still a complex process.

The main factor in the production of biofuel is the availability of biomass which can be easily obtained from woods, plants, organic waste, agricultural waste, municipal solid waste, etc. Still, there are many challenges and opportunities available for improvement in order to replace commercially available petroleum-based oils. Pre-treatment strategies for lignocellulosic biomass require high operation costs [126]. Algal biomass is also being used for biodiesel production as it is oil-rich, carbon-neutral, and can grow rapidly. It is considered that this may replace fossil fuels for biodiesel production. On the other hand, the cultivation of algal biomass is costly, and the lipid extraction is energy intensive [127]. Implementing nanotechnology to produce biofuels at an industrial scale is challenging as nano-catalyst based biofuel production has not fully emerged. In addition, studies are still improving biofuel production using available resources. Up to now, usually edible crops such as maize, sugarcane, etc., have been utilized for the large-scale production of biofuels. Biofuel production from non-edible sources is lower in comparison to edible sources. Nanotechnology is accelerating biofuel production and increasing the amount of biofuel produced from non-edible sources. It will still be problematic to replace petroleum-derived fuel with commercially available biofuel because it must be mixed with other fuels for usage and it is not cost-effective. The possibility of using biofuel as an alternative and green energy source will be significantly higher in the near future.

7. Conclusions

It is clear from the current review that the incorporation of nanoparticles during biofuel production enhanced this significantly. This enhancement is mainly due to the unique physico–chemical properties of nanoparticles such as large surface-area-to-volume ratio, high reactivity, good dispersibility, high specificity, etc. Several nanoparticles such as metal, metal oxide, magnetic, and carbonous materials are successfully used for enhancement of biofuel production from various substrates. Apart from the production process, nanoparticles are also used in the pretreatment process to enhance the digestibility of substrate leading to enhanced biofuel production. However, successful commercialization of this process requires the addressing of several technical barriers. These barriers include synthesis and application nanoparticles that are non-toxic to microorganisms, use of less expensive and environment friendly nanoparticles, and adaptation of biological nanoparticle synthesis methods in place of chemical methods, which requires stringent operational conditions.

Author Contributions: Conceptualization and supervision: S.K., P.K.D., and S.P.; writing—original draft preparation: I.A. and A.P.; review and editing, artwork and schemes: S.K., H.N.S., A.A.C., H.A.R., S.P., J.K., P.K.D., N.K.J. All authors have read and agreed to the published version of the manuscript.

Funding: This work required no external funding.

Data Availability Statement: This study did not report any data.

Acknowledgments: All the authors are grateful to the Department of Life Sciences, School of Basic Sciences and Research, Sharda University, Greater Noida, for providing the infrastructure and facilities for this research. Biorender software is also highly acknowledged for artwork and schemes.

Conflicts of Interest: All authors declare no competing interests with the work presented in the manuscript.

References

1. Rai, M.; Dos Santos, J.C.; Soler, M.F.; Franco Marcelino, P.R.; Brumano, L.P.; Ingle, A.P.; Gaikwad, S.; Gade, A.; Da Silva, S.S. Strategic Role of Nanotechnology for Production of Bioethanol and Biodiesel. *Nanotechnol. Rev.* **2016**, *5*, 231–250. [CrossRef]
2. Bhattarai, K.; Stalick, W.M.; Mckay, S.; Geme, G.; Bhattarai, N. Biofuel: An Alternative to Fossil Fuel for Alleviating World Energy and Economic Crises. *J. Environ. Sci. Health Part A Toxic* **2011**, *46*, 1424–1442. [CrossRef]
3. Shalaby, E.A. Biofuel: Sources, Extraction and Determination. In *Liquid, Gaseous and Solid Biofuels*; Fang, Z., Ed.; IntechOpen: Rijeka, Croatia, 2013.
4. Folaranmi, J. Production of Biodiesel (B100) from Jatropha Oil Using Sodium Hydroxide as Catalyst. *J. Pet. Eng.* **2013**, *2013*, 1–6. [CrossRef]
5. Forde, C.J.; Meaney, M.; Carrigan, J.B.; Mills, C.; Boland, S.; Hernon, A. Biobased Fats (Lipids) and Oils from Biomass as a Source of Bioenergy. *Bioenergy Res. Adv. Appl.* **2014**, 185–201. [CrossRef]
6. Ahmad, M.; Ajab, M.; Zafar, M.; Sult, S. Biodiesel from Non Edible Oil Seeds: A Renewable Source of Bioenergy. *Econ. Eff. Biofuel Prod.* **2011**, *2005*. [CrossRef]
7. Mohadi, R.; Harahap, A.H.; Hidayati, N.; Lesbani, A. Transesterification of Tropical Edible Oils to Biodiesel Using Catalyst From Scylla Serrata. *Sriwij. J. Environ.* **2016**, *1*, 24–27. [CrossRef]
8. Thirumarimurugan, M.; Sivakumar, V.M.; Xavier, A.M.; Prabhakaran, D.; Kannadasan, T. Preparation of Biodiesel from Sunflower Oil by Transesterification. *IJBBB* **2012**, 441–444. [CrossRef]
9. Ahmmed, B.; Samaddar, O.U.; Kibria, K.Q. Production of Biodiesel from Used Vegetable Oils Production of Biodiesel from Used Vegetable Oils. *Int. J. Sci. Res. Sci. Technol.* **2019**, *6*. [CrossRef]
10. Do Nascimento, R.O.; Rebelo, L.M.; Sacher, E. Physicochemical Characterizations of Nanoparticles Used for Bioenergy and Biofuel Production. In *Nanotechnology for Bioenergy and Biofuel Production*; Rai, M., da Silva, S.S., Eds.; Springer: Cham, Switzerland, 2017; pp. 173–191. ISBN 978-3-319-45459-7.
11. Dikshit, P.K.; Kumar, J.; Das, A.K.; Sadhu, S.; Sharma, S.; Singh, S.; Gupta, P.K.; Kim, B.S. Green Synthesis of Metallic Nanoparticles: Applications and Limitations. *Catalysts* **2021**, *11*, 902. [CrossRef]
12. Saoud, K. Nanocatalyst for Biofuel Production: A Review. In *Green Nanotechnology for Biofuel Production*; Srivastava, N., Srivastava, M., Pandey, H., Mishra, P.K., Ramteke, P.W., Eds.; Springer: Cham, Switzerland, 2018; pp. 39–62. ISBN 978-3-319-75052-1.
13. Singh, N.; Dhanya, B.S.; Verma, M.L. Nano-Immobilized Biocatalysts and Their Potential Biotechnological Applications in Bioenergy Production. *Mater. Sci. Energy Technol.* **2020**, *3*, 808–824. [CrossRef]
14. Ali, S.; Shafique, O.; Mahmood, S.; Mahmood, T.; Khan, B.A.; Ahmad, I. Biofuels Production from Weed Biomass Using Nanocatalyst Technology. *Biomass Bioenergy* **2020**, *139*, 105595. [CrossRef]
15. Hussain, S.T.; Ali, S.A.; Bano, A.; Mahmood, T. Use of Nanotechnology for the Production of Biofuels from Butchery Waste. *Int. J. Phys. Sci.* **2011**, *6*, 7271–7279. [CrossRef]
16. Kumar, Y.; Yogeshwar, P.; Bajpai, S.; Jaiswal, P.; Yadav, S.; Pathak, D.P.; Sonker, M.; Tiwary, S.K. Nanomaterials: Stimulants for Biofuels and Renewables, Yield and Energy Optimization. *Mater. Adv.* **2021**, *2*, 5318–5343. [CrossRef]
17. Malode, S.J.; Prabhu, K.K.; Mascarenhas, R.J.; Shetti, N.P.; Aminabhavi, T.M. Recent Advances and Viability in Biofuel Production. *Energy Convers. Manag. X* **2021**, *10*, 100070. [CrossRef]
18. Hirani, A.H.; Javed, N.; Asif, M.; Basu, S.K.; Kumar, A. A Review on First- and Second-Generation Biofuel Productions. In *Biofuels: Greenhouse Gas Mitigation and Global Warming: Next Generation Biofuels and Role of Biotechnology*; Kumar, A., Ogita, S., Yau, Y.-Y., Eds.; Springer: New Delhi, India, 2018; pp. 141–154. ISBN 978-81-322-3763-1.
19. Dahman, Y.; Syed, K.; Begum, S.; Roy, P.; Mohtasebi, B. 14-Biofuels: Their characteristics and analysis. In *Biomass, Biopolymer-Based Materials, and Bioenergy*; Verma, D., Fortunati, E., Jain, S., Zhang, X., Eds.; Woodhead Publishing Series in Composites Science and Engineering; Woodhead Publishing: Cambridge, UK, 2019; pp. 277–325. ISBN 978-0-08-102426-3.

20. Liu, Z.; Lv, F.; Zheng, H.; Zhang, C.; Wei, F.; Xing, X.-H. Enhanced Hydrogen Production in a UASB Reactor by Retaining Microbial Consortium onto Carbon Nanotubes (CNTs). *Int. J. Hydrog. Energy* **2012**, *37*, 10619–10626. [CrossRef]
21. Saifuddin, N.; Raziah, A.Z.; Junizah, A.R. Carbon Nanotubes: A Review on Structure and Their Interaction with Proteins. *J. Chem.* **2013**, *2013*, 1–18. [CrossRef]
22. Ando, Y.; Zhao, X.; Shimoyama, H.; Sakai, G.; Kaneto, K. Physical Properties of Multiwalled Carbon Nanotubes. *Int. J. Inorg. Mater.* **1999**, *1*, 77–82. [CrossRef]
23. Dresselhaus, M.S.; Dresselhaus, G.; Eklund, P.C.; Rao, A.M. Carbon Nanotubes. In *The Physics of Fullerene-Based and Fullerene-Related Materials*; Andreoni, W., Ed.; Springer: Dordrecht, The Netherlands, 2000; pp. 331–379. ISBN 978-94-011-4038-6.
24. Peng, F.; Zhang, L.; Wang, H.; Lv, P.; Yu, H. Sulfonated Carbon Nanotubes as a Strong Protonic Acid Catalyst. *Carbon* **2005**, *43*, 2405–2408. [CrossRef]
25. Boshagh, F.; Rostami, K.; Moazami, N. Biohydrogen Production by Immobilized Enterobacter Aerogenes on Functionalized Multi-Walled Carbon Nanotube. *Int. J. Hydrog. Energy* **2019**, *44*, 14395–14405. [CrossRef]
26. Feng, W.; Ji, P. Enzymes Immobilized on Carbon Nanotubes. *Biotechnol. Adv.* **2011**, *29*, 889–895. [CrossRef]
27. Lee, D.G.; Ponvel, K.M.; Kim, M.; Hwang, S.; Ahn, I.S.; Lee, C.H. Immobilization of Lipase on Hydrophobic Nano-Sized Magnetite Particles. *J. Mol. Catal. B Enzym.* **2009**, *57*, 62–66. [CrossRef]
28. Pavlidis, I.V.; Tsoufis, T.; Enotiadis, A.; Gournis, D.; Stamatis, H. Functionalized Multi-Wall Carbon Nanotubes for Lipase Immobilization. *Adv. Eng. Mater.* **2010**, *12*, B179–B183. [CrossRef]
29. Khan, M.; Anwer, T.; Mohammad, F. Sensing Properties of Sulfonated Multi-Walled Carbon Nanotube and Graphene Nanocomposites with Polyaniline. *J. Sci. Adv. Mater. Devices* **2019**, *4*, 132–142. [CrossRef]
30. Deep, A.; Sharma, A.L.; Kumar, P. Lipase Immobilized Carbon Nanotubes for Conversion of Jatropha Oil to Fatty Acid Methyl Esters. *Biomass Bioenergy* **2015**, *81*, 83–87. [CrossRef]
31. Verma, M.; Naebe, M.; Barrow, C.; Puri, M. Enzyme Immobilisation on Amino-Functionalised Multi-Walled Carbon Nanotubes: Structural and Biocatalytic Characterisation. *PLoS ONE* **2013**, *8*, e73642. [CrossRef] [PubMed]
32. Nazaruk, E.; Sadowska, K.; Biernat, J.F.; Rogalski, J.; Ginalska, G.; Bilewicz, R. Enzymatic Electrodes Nanostructured with Functionalized Carbon Nanotubes for Biofuel Cell Applications. *Anal. Bioanal. Chem.* **2010**, *398*, 1651–1660. [CrossRef] [PubMed]
33. Shu, Q.; Zhang, Q.; Xu, G.; Wang, J. Preparation of Biodiesel Using S-MWCNT Catalysts and the Coupling of Reaction and Separation. *Food Bioprod. Process.* **2009**, *87*, 164–170. [CrossRef]
34. Ahmad, R.; Khare, S.K. Immobilization of Aspergillus Niger Cellulase on Multiwall Carbon Nanotubes for Cellulose Hydrolysis. *Bioresour. Technol.* **2018**, *252*, 72–75. [CrossRef]
35. Mubarak, N.M.; Wong, J.R.; Tan, K.W.; Sahu, J.N.; Abdullah, E.C.; Jayakumar, N.S.; Ganesan, P. Immobilization of Cellulase Enzyme on Functionalized Multiwall Carbon Nanotubes. *J. Mol. Catal. B Enzym.* **2014**, *107*, 124–131. [CrossRef]
36. Tran, D.T.; Chen, C.L.; Chang, J.S. Immobilization of Burkholderia Sp. Lipase on a Ferric Silica Nanocomposite for Biodiesel Production. *J. Biotechnol.* **2012**, *158*, 112–119. [CrossRef]
37. Verma, M.L.; Chaudhary, R.; Tsuzuki, T.; Barrow, C.J.; Puri, M. Immobilization of β-Glucosidase on a Magnetic Nanoparticle Improves Thermostability: Application in Cellobiose Hydrolysis. *Bioresour. Technol.* **2013**, *135*, 2–6. [CrossRef]
38. Puri, M.; Barrow, C.J.; Verma, M.L. Enzyme Immobilization on Nanomaterials for Biofuel Production. *Trends Biotechnol.* **2013**, *31*, 215–216. [CrossRef]
39. Singh, O.V.; Chandel, A.K. *Sustainable Biotechnology-Enzymatic Resources of Renewable Energy*; Springer: Berlin/Heidelberg, Germany, 2018; ISBN 9783319954806.
40. Alftrén, J.; Hobley, T.J. Covalent Immobilization of β-Glucosidase on Magnetic Particles for Lignocellulose Hydrolysis. *Appl. Biochem. Biotechnol.* **2013**, *169*, 2076–2087. [CrossRef] [PubMed]
41. Huang, P.J.; Chang, K.L.; Hsieh, J.F.; Chen, S.T. Catalysis of Rice Straw Hydrolysis by the Combination of Immobilized Cellulase from Aspergillus Niger on β -Cyclodextrin-Fenanoparticles and Ionic Liquid. *BioMed Res. Int.* **2015**, *2015*, 1–9. [CrossRef]
42. Teo, S.H.; Islam, A.; Chan, E.S.; Thomas Choong, S.Y.; Alharthi, N.H.; Taufiq-Yap, Y.H.; Awual, M.R. Efficient Biodiesel Production from Jatropha Curcus Using CaSO4/Fe2O3-SiO2 Core-Shell Magnetic Nanoparticles. *J. Clean. Prod.* **2019**, *208*, 816–826. [CrossRef]
43. Shakeel, N.; Ahamed, M.I.; Ahmed, A.; Rahman, M.M; Asiri, A.M. Functionalized Magnetic Nanoparticle-Reduced Graphene Oxide Nanocomposite for Enzymatic Biofuel Cell Applications. *Int. J. Hydrog. Energy* **2019**, *44*, 28294–28304. [CrossRef]
44. Dantas, J.; Leal, E.; Mapossa, A.B.; Cornejo, D.R.; Costa, A.C.F.M. Magnetic nanocatalysts of Ni0.5Zn0.5Fe2O4 doped with Cu and performance evaluation in transesterification reaction for biodiesel production. *Fuel* **2017**, *191*, 463–471. [CrossRef]
45. Cherian, E.; Dharmendirakumar, M.; Baskar, G. Immobilization of Cellulase onto MnO2 Nanoparticles for Bioethanol Production by Enhanced Hydrolysis of Agricultural Waste. *Chin. J. Catal.* **2015**, *36*, 1223–1229. [CrossRef]
46. Ivanova, V.; Petrova, P.; Hristov, J. Application in the Ethanol Fermentation of Immobilized Yeast Cells in Matrix of Alginate/Magnetic Nanoparticles, on Chitosan-Magnetite Microparticles and Cellulose-Coated Magnetic Nanoparticles. *arXiv preprint* **2011**, arXiv:1105.0619.
47. Lee, K.H.; Choi, I.S.; Kim, Y.-G.; Yang, D.-J.; Bae, H.-J. Enhanced Production of Bioethanol and Ultrastructural Characteristics of Reused Saccharomyces Cerevisiae Immobilized Calcium Alginate Beads. *Bioresour. Technol.* **2011**, *102*, 8191–8198. [CrossRef]
48. Duraiarasan, S.; Razack, S.A.; Manickam, A.; Munusamy, A.; Syed, M.B.; Ali, M.Y.; Ahmed, G.M.; Mohiuddin, M.S. Direct Conversion of Lipids from Marine Microalga C. Salina to Biodiesel with Immobilised Enzymes Using Magnetic Nanoparticle. *J. Environ. Chem. Eng.* **2016**, *4*, 1393–1398. [CrossRef]

49. Mahmood, T.; Zada, B.; Malik, S.A. Effect of Iron Nanoparticles on HyacinthÃ¢â‚¬â„¢s Fermentation. *Int. J. Sci.* **2013**, *2*, 106–121.
50. Taherdanak, M.; Zilouei, H.; Karimi, K. Investigating the Effects of Iron and Nickel Nanoparticles on Dark Hydrogen Fermentation from Starch Using Central Composite Design. *Int. J. Hydrog. Energy* **2015**, *40*, 12956–12963. [CrossRef]
51. Nath, D.; Manhar, A.K.; Gupta, K.; Saikia, D.; Das, S.K.; Mandal, M. Phytosynthesized Iron Nanoparticles: Effects on Fermentative Hydrogen Production by Enterobacter Cloacae DH-89. *Bull. Mater. Sci.* **2015**, *38*, 1533–1538. [CrossRef]
52. Engliman, N.S.; Abdul, P.M.; Wu, S.-Y.; Jahim, J.M. Influence of Iron (II) Oxide Nanoparticle on Biohydrogen Production in Thermophilic Mixed Fermentation. *Int. J. Hydrog. Energy* **2017**, *42*, 27482–27493. [CrossRef]
53. Malik, S.N.; Pugalenthi, V.; Vaidya, A.N.; Ghosh, P.C.; Mudliar, S.N. Kinetics of Nano-Catalysed Dark Fermentative Hydrogen Production from Distillery Wastewater. *Energy Procedia* **2014**, *54*, 417–430. [CrossRef]
54. Reddy, K.; Nasr, M.; Kumari, S.; Kumar, S.; Gupta, S.K.; Enitan, A.M.; Bux, F. Biohydrogen Production from Sugarcane Bagasse Hydrolysate: Effects of PH, S/X, Fe^{2+}, and Magnetite Nanoparticles. *Environ. Sci.Pollut. Res.* **2017**, *24*, 8790–8804. [CrossRef]
55. Wang, T.; Zhang, D.; Dai, L.; Chen, Y.; Dai, X. Effects of Metal Nanoparticles on Methane Production from Waste-Activated Sludge and Microorganism Community Shift in Anaerobic Granular Sludge. *Sci. Rep.* **2016**, *6*, 25857. [CrossRef]
56. Wang, A.; Wang, J.; Lu, C.; Xu, M.; Lv, J.; Wu, X. Esterification for Biofuel Synthesis over an Eco-Friendly and Efficient Kao- 610 linite-Supported $SO_4{}^{2-}/ZnAl_2O_4$ Macroporous Solid Acid Catalyst. *Fuel* **2018**, *234*, 430–440. [CrossRef]
57. Wang, H.; Covarrubias, J.; Prock, H.; Wu, X.; Wang, D.; Bossmann, S.H. Acid-Functionalized Magnetic Nanoparticle as Heterogeneous Catalysts for Biodiesel Synthesis. *J. Phys. Chem. C* **2015**, *119*, 26020–26028. [CrossRef]
58. Peña, L.; Hohn, K.L.; Li, J.; Sun, X.S.; Wang, D. Synthesis of Propyl-Sulfonic Acid-Functionalized Nanoparticles as Catalysts for Cellobiose Hydrolysis. *J. Biomater. Nanobiotechnol.* **2014**, *5*, 241–253. [CrossRef]
59. Erdem, S.; Erdem, B.; Öksüzoğlu, R.M. Magnetic Nano-Sized Solid Acid Catalyst Bearing Sulfonic Acid Groups for Biodiesel Synthesis. *Open Chem.* **2018**, *16*, 923–929. [CrossRef]
60. Lai, D.; Deng, L.; Guo, Q.; Fu, Y. Hydrolysis of Biomass by Magnetic Solid Acid. *Energy Environ. Sci.* **2011**, *4*, 3552–3557. [CrossRef]
61. Antunes, F.A.F.; Gaikwad, S.; Ingle, A.P.; Pandit, R.; dos Santos, J.C.; Rai, M.; da Silva, S.S. Bioenergy and Biofuels: Nanotechnological Solutions for Sustainable Production. In *Nanotechnology for Bioenergy and Biofuel Production*; Rai, M., da Silva, S.S., Eds.; Springer: Cham, Switzerland, 2017; pp. 3–18. ISBN 978-3-319-45459-7.
62. Vincent, K.A.; Li, X.; Blanford, C.F.; Belsey, N.A.; Weiner, J.H.; Armstrong, F.A. Enzymatic Catalysis on Conducting Graphite Particles. *Nat. Chem. Biol.* **2007**, *3*, 761–762. [CrossRef]
63. Kwon, C.H.; Ko, Y.; Shin, D.; Kwon, M.; Park, J.; Bae, W.K.; Lee, S.W.; Cho, J. High-Power Hybrid Biofuel Cells Using Layer-by-Layer Assembled Glucose Oxidase-Coated Metallic Cotton Fibers. *Nat. Commun.* **2018**, *9*, 1–11. [CrossRef] [PubMed]
64. Aquino Neto, S.; Almeida, T.S.; Palma, L.M.; Minteer, S.D.; De Andrade, A.R. Hybrid Nanocatalysts Containing Enzymes and Metallic Nanoparticles for Ethanol/O2 Biofuel Cell. *J. Power Sources* **2014**, *259*, 25–32. [CrossRef]
65. Hebié, S.; Holade, Y.; Maximova, K.; Sentis, M.; Delaporte, P.; Kokoh, K.B.; Napporn, T.W.; Kabashin, A.V. Advanced Electrocatalysts on the Basis of Bare Au Nanomaterials for Biofuel Cell Applications. *ACS Catal.* **2015**, *5*, 6489–6496. [CrossRef]
66. Zhang, Y.; Shen, J. Enhancement Effect of Gold Nanoparticles on Biohydrogen Production from Artificial Wastewater. *Int. J. Hydrog. Energy* **2007**, *32*, 17–23. [CrossRef]
67. Eroglu, E.; Eggers, P.K.; Winslade, M.; Smith, S.M.; Raston, C.L. Enhanced Accumulation of Microalgal Pigments Using Metal Nanoparticle Solutions as Light Filtering Devices. *Green Chem.* **2013**, *15*, 3155–3159. [CrossRef]
68. Su, L.; Shi, X.; Guo, G.; Zhao, A.; Zhao, Y. Stabilization of Sewage Sludge in the Presence of Nanoscale Zero-Valent Iron (NZVI): Abatement of Odor and Improvement of Biogas Production. *J Mater Cycles Waste Manag.* **2013**, *15*, 461–468. [CrossRef]
69. Karri, S.; Sierra-Alvarez, R.; Field, J.A. Zero Valent Iron as an Electron-Donor for Methanogenesis and Sulfate Reduction in Anaerobic Sludge. *Biotechnol. Bioeng.* **2005**, *92*, 810–819. [CrossRef] [PubMed]
70. Kobayashi, H.; Hosaka, Y.; Hara, K.; Feng, B.; Hirosaki, Y.; Fukuoka, A. Control of Selectivity, Activity and Durability of Simple Supported Nickel Catalysts for Hydrolytic Hydrogenation of Cellulose. *Green Chem.* **2014**, *16*, 637–644. [CrossRef]
71. Yigezu, Z.D.; Muthukumar, K. Catalytic Cracking of Vegetable Oil with Metal Oxides for Biofuel Production. *Energy Convers. Manag.* **2014**, *84*, 326–333. [CrossRef]
72. Hashmi, S.; Gohar, S.; Mahmood, T.; Nawaz, U.; Farooqi, H. Biodiesel Production by Using CaO-Al2O3 Nano Catalyst. *Int. J. Eng. Res. Sci.* **2016**, *2*, 2395–6992.
73. Kim, M.; DiMaggio, C.; Salley, S.O.; Ng, K.S. A New Generation of Zirconia Supported Metal Oxide Catalysts for Converting Low Grade Renewable Feedstocks to Biodiesel. *Bioresour. Technol.* **2012**, *118*, 37–42. [CrossRef] [PubMed]
74. Cao, X.; Li, L.; Shitao, Y.; Liu, S.; Hailong, Y.; Qiong, W.; Ragauskas, A.J. Catalytic Conversion of Waste Cooking Oils for the Production of Liquid Hydrocarbon Biofuels Using In-Situ Coating Metal Oxide on SBA-15 as Heterogeneous Catalyst. *J. Anal. Appl. Pyrolysis* **2019**, *138*, 137–144. [CrossRef]
75. Abbas, M.; Parvatheeswara Rao, B.; Nazrul Islam, M.; Naga, S.M.; Takahashi, M.; Kim, C. Highly Stable- Silica Encapsulating Magnetite Nanoparticles (Fe3O4/SiO2) Synthesized Using Single Surfactantless- Polyol Process. *Ceram. Int.* **2014**, *40*, 1379–1385. [CrossRef]
76. Kunzmann, A.; Andersson, B.; Vogt, C.; Feliu, N.; Ye, F.; Gabrielsson, S.; Toprak, M.S.; Buerki-Thurnherr, T.; Laurent, S.; Vahter, M.; et al. Efficient Internalization of Silica-Coated Iron Oxide Nanoparticles of Different Sizes by Primary Human Macrophages and Dendritic Cells. *Toxicol. Appl. Pharmacol.* **2011**, *253*, 81–93. [CrossRef]

77. Mohan, S.V.; Mohanakrishna, G.; Reddy, S.S.; Raju, B.D.; Rao, K.S.R.; Sarma, P.N. Self-Immobilization of Acidogenic Mixed Consortia on Mesoporous Material (SBA-15) and Activated Carbon to Enhance Fermentative Hydrogen Production. *Int. J. Hydrog. Energy* **2008**, *33*, 6133–6142. [CrossRef]
78. Chen, Y.; Liu, T.; He, H.; Liang, H. Fe_3O_4/ZnMg(Al)O Magnetic Nanoparticles for Efficient Biodiesel Production. *Appl. Organomet. Chem.* **2018**, *32*, e4330. [CrossRef]
79. Chiang, Y.-D.; Dutta, S.; Chen, C.-T.; Huang, Y.-T.; Lin, K.-S.; Wu, J.C.S.; Suzuki, N.; Yamauchi, Y.; Wu, K.C.-W. Functionalized Fe3O4@Silica Core–Shell Nanoparticles as Microalgae Harvester and Catalyst for Biodiesel Production. *ChemSusChem* **2015**, *8*, 789–794. [CrossRef]
80. Papadopoulou, A.; Zarafeta, D.; Galanopoulou, A.P.; Stamatis, H. Enhanced Catalytic Performance of Trichoderma Reesei Cellulase Immobilized on Magnetic Hierarchical Porous Carbon Nanoparticles. *Protein J.* **2019**, *38*, 640–648. [CrossRef] [PubMed]
81. Ferreira, R.; Menezes, R.; Sampaio, K.; Batista, E. Heterogeneous Catalysts for Biodiesel Production: A Review. *Food Public Health* **2019**, *9*, 125–137. [CrossRef]
82. Semwal, S.; Arora, A.K.; Badoni, R.P.; Tuli, D.K. Biodiesel Production Using Heterogeneous Catalysts. *Bioresour. Technol.* **2011**, *102*, 2151–2161. [CrossRef]
83. Narasimhan, M.; Chandrasekaran, M.; Govindasamy, S.; Aravamudhan, A. Heterogeneous Nanocatalysts for Sustainable Biodiesel Production: A Review. *J. Environ. Chem. Eng.* **2021**, *9*, 104876. [CrossRef]
84. Akia, M.; Yazdani, F.; Motaee, E.; Han, D.; Arandiyan, H. A Review on Conversion of Biomass to Biofuel by Nanocatalysts. *Biofuel Res. J.* **2014**, *1*, 16–25. [CrossRef]
85. Ibrahim, A.A.; Lin, A.; Zhang, F.; AbouZeid, K.M.; El-Shall, M.S. Palladium Nanoparticles Supported on Hybrid MOF-PRGO for Catalytic Hydrodeoxygenation of Vanillin as a Model for Biofuel Upgrade Reactions. *ChemCatChem* **2016**, *9*, 469–480. [CrossRef]
86. Zhao, W.; Zhang, Y.; Du, B.; Wei, D.; Wei, Q.; Zhao, Y. Enhancement Effect of Silver Nanoparticles on Fermentative Biohydrogen Production Using Mixed Bacteria. *Bioresour. Technol.* **2013**, *142*, 240–245. [CrossRef]
87. Khan, M.M.; Lee, J.; Cho, M.H. Electrochemically Active Biofilm Mediated Bio-Hydrogen Production Catalyzed by Positively Charged Gold Nanoparticles. *Int. J. Hydrog. Energy* **2013**, *38*, 5243–5250. [CrossRef]
88. Mohanraj, S.; Anbalagan, K.; Rajaguru, P.; Pugalenthi, V. Effects of Phytogenic Copper Nanoparticles on Fermentative Hydrogen Production by Enterobacter Cloacae and Clostridium Acetobutylicum. *Int. J. Hydrog. Energy* **2016**, *41*, 10639–10645. [CrossRef]
89. Taherdanak, M.; Zilouei, H.; Karimi, K. The Effects of Fe0 and Ni0 Nanoparticles versus Fe2+ and Ni2+ Ions on Dark Hydrogen Fermentation. *Int. J. Hydrog. Energy* **2016**, *41*, 167–173. [CrossRef]
90. Elreedy, A.; Ibrahim, E.; Hassan, N.; El-Dissouky, A.; Fujii, M.; Yoshimura, C.; Tawfik, A. Nickel-Graphene Nanocomposite as a Novel Supplement for Enhancement of Biohydrogen Production from Industrial Wastewater Containing Mono-Ethylene Glycol. *Energy Convers. Manag.* **2017**, *140*, 133–144. [CrossRef]
91. Mohanraj, S.; Kodhaiyolii, S.; Rengasamy, M.; Pugalenthi, V. Phytosynthesized Iron Oxide Nanoparticles and Ferrous Iron on Fermentative Hydrogen Production Using Enterobacter Cloacae: Evaluation and Comparison of the Effects. *Int. J. Hydrog. Energy* **2014**, *39*, 11920–11929. [CrossRef]
92. Pandey, A.; Gupta, K.; Pandey, A. Effect of Nanosized TiO2 on Photofermentation by Rhodobacter Sphaeroides NMBL-02. *Biomass Bioenergy* **2015**, *72*, 273–279. [CrossRef]
93. Nzila, A. Mini Review: Update on Bioaugmentation in Anaerobic Processes for Biogas Production. *Anaerobe* **2017**, *46*, 3–12. [CrossRef]
94. Abdelsalam, E.; Samer, M.; Attia, Y.A.; Abdel-Hadi, M.A.; Hassan, H.E.; Badr, Y. Influence of Zero Valent Iron Nanoparticles and Magnetic Iron Oxide Nanoparticles on Biogas and Methane Production from Anaerobic Digestion of Manure. *Energy* **2017**, *120*, 842–853. [CrossRef]
95. Ganzoury, M.A.; Allam, N.K. Impact of Nanotechnology on Biogas Production: A Mini-Review. *Renew. Sustain. Energy Rev.* **2015**, *50*, 1392–1404. [CrossRef]
96. Suanon, F.; Sun, Q.; Li, M.; Cai, X.; Zhang, Y.; Yan, Y.; Yu, C.-P. Application of Nanoscale Zero Valent Iron and Iron Powder during Sludge Anaerobic Digestion: Impact on Methane Yield and Pharmaceutical and Personal Care Products Degradation. *J. Hazard. Mater.* **2017**, *321*, 47–53. [CrossRef]
97. Amen, T.W.M.; Eljamal, O.; Khalil, A.M.E.; Matsunaga, N. Biochemical Methane Potential Enhancement of Domestic Sludge Digestion by Adding Pristine Iron Nanoparticles and Iron Nanoparticles Coated Zeolite Compositions. *J. Environ. Chem. Eng.* **2017**, *5*, 5002–5013. [CrossRef]
98. Gonzalez-Estrella, J.; Sierra-Alvarez, R.; Field, J.A. Toxicity Assessment of Inorganic Nanoparticles to Acetoclastic and Hydrogenotrophic Methanogenic Activity in Anaerobic Granular Sludge. *J. Hazard. Mater.* **2013**, *260*, 278–285. [CrossRef]
99. Zhang, L.; He, X.; Zhang, Z.; Cang, D.; Nwe, K.A.; Zheng, L.; Li, Z.; Cheng, S. Evaluating the Influences of ZnO Engineering Nanomaterials on VFA Accumulation in Sludge Anaerobic Digestion. *Biochem. Eng. J.* **2017**, *125*, 206–211. [CrossRef]
100. Ünşar, E.K.; Çığgın, A.S.; Erdem, A.; Perendeci, N.A. Long and Short Term Impacts of CuO, Ag and CeO_2 Nanoparticles on Anaerobic Digestion of Municipal Waste Activated Sludge. *Environ. Sci. Process. Impacts* **2016**, *18*, 277–288. [CrossRef]
101. Kim, Y.K.; Lee, H. Use of Magnetic Nanoparticles to Enhance Bioethanol Production in Syngas Fermentation. *Bioresour. Technol.* **2016**, *204*, 139–144. [CrossRef]
102. Lam, M.K.; Lee, K.T. *Chapter 12—Bioethanol Production from Microalgae*; Kim, S.-K., Ed.; Academic Press: Boston, MD, USA, 2015; pp. 197–208. ISBN 978-0-12-800776-1.

103. De Farias Silva, C.E.; Bertucco, A. Bioethanol from Microalgae and Cyanobacteria: A Review and Technological Outlook. *Process Biochem.* **2016**, *51*, 1833–1842. [CrossRef]
104. Velazquez, J.; Rodriguez-Jasso, R.; Colla, L.; Galindo, A.; Cervantes, D.; Aguilar, C.; Fernandes, B.; Ruiz, H. Microalgal Biomass Pretreatment for Bioethanol Production: A Review. *Biofuel Res. J.* **2018**, *5*, 780–791. [CrossRef]
105. Parambil, L.K.; Sarkar, D. In Silico Analysis of Bioethanol Overproduction by Genetically Modified Microorganisms in Coculture Fermentation. *Biotechnol. Res. Int.* **2015**, *2015*, 238082. [CrossRef]
106. Sanusi, I.A.; Suinyuy, T.N.; Kana, G.E.B. Impact of Nanoparticle Inclusion on Bioethanol Production Process Kinetic and Inhibitor Profile. *Biotechnol. Rep.* **2021**, *29*, e00585. [CrossRef] [PubMed]
107. Sanusi, I.A.; Faloye, F.D.; Gueguim Kana, E.B. Impact of Various Metallic Oxide Nanoparticles on Ethanol Production by Saccharomyces Cerevisiae BY4743: Screening, Kinetic Study and Validation on Potato Waste. *Catal. Lett.* **2019**, *149*, 2015–2031. [CrossRef]
108. Gupta, K.; Chundawat, T.S. Zinc Oxide Nanoparticles Synthesized Using Fusarium Oxysporum to Enhance Bioethanol Production from Rice-Straw. *Biomass Bioenergy* **2020**, *143*, 105840. [CrossRef]
109. Bohlouli, A.; Mahdavian, L. Catalysts Used in Biodiesel Production: A Review. *Biofuels* **2018**, *12*, 885–898. [CrossRef]
110. Kumar, L.R.; Ram, S.K.; Tyagi, R.D. Application of Nanotechnology in Biodiesel Production. In *Biodiesel Production: Technologies, Challenges, and Future Prospects*; American Society of Civil Engineers: Reston, VA, USA, 2021; pp. 397–419.
111. Banković-Ilić, I.B.; Miladinović, M.R.; Stamenković, O.S.; Veljković, V.B. Application of Nano CaO–Based Catalysts in Biodiesel Synthesis. *Renew. Sust. Energ. Rev.* **2017**, *72*, 746–760. [CrossRef]
112. Zhang, X.; Yan, S.; Tyagi, R.D.; Surampalli, R.Y. Biodiesel Production from Heterotrophic Microalgae through Transesterification and Nanotechnology Application in the Production. *Renew. Sustain. Energy Rev.* **2013**, *26*, 216–223. [CrossRef]
113. Demirbas, A. Biofuels Sources, Biofuel Policy, Biofuel Economy and Global Biofuel Projections. *Energy Convers. Manag.* **2008**, *49*, 2106–2116. [CrossRef]
114. Feyzi, M.; Norouzi, L. Preparation and Kinetic Study of Magnetic Ca/Fe_3O_4@SiO_2 Nanocatalysts for Biodiesel Production. *Renew. Energy* **2016**, *94*, 579–586. [CrossRef]
115. Jeon, H.-S.; Park, S.E.; Ahn, B.; Kim, Y.-K. Enhancement of Biodiesel Production in Chlorella Vulgaris Cultivation Using Silica Nanoparticles. *Biotechnol.Bioproc. E* **2017**, *22*, 136–141. [CrossRef]
116. Tahvildari, K.; Anaraki, Y.N.; Fazaeli, R.; Mirpanji, S.; Delrish, E. The Study of CaO and MgO Heterogenic Nano-Catalyst Coupling on Transesterification Reaction Efficacy in the Production of Biodiesel from Recycled Cooking Oil. *J. Environ. Health Sci.Eng.* **2015**, *13*, 73. [CrossRef]
117. Baskar, G.; Aberna Ebenezer Selvakumari, I.; Aiswarya, R. Biodiesel Production from Castor Oil Using Heterogeneous Ni Doped ZnO Nanocatalyst. *Bioresour. Technol.* **2018**, *250*, 793–798. [CrossRef]
118. Harsha Hebbar, H.R.; Math, M.C.; Yatish, K.V. Optimization and Kinetic Study of CaO Nano-Particles Catalyzed Biodiesel Production from Bombax Ceiba Oil. *Energy* **2018**, *143*, 25–34. [CrossRef]
119. Bet-Moushoul, E.; Farhadi, K.; Mansourpanah, Y.; Nikbakht, A.M.; Molaei, R.; Forough, M. Application of CaO-Based/Au Nanoparticles as Heterogeneous Nanocatalysts in Biodiesel Production. *Fuel* **2016**, *164*, 119–127. [CrossRef]
120. Rahmani Vahid, B.; Haghighi, M.; Toghiani, J.; Alaei, S. Hybrid-Coprecipitation vs. Combustion Synthesis of Mg-Al Spinel Based Nanocatalyst for Efficient Biodiesel Production. *Energy Convers. Manag.* **2018**, *160*, 220–229. [CrossRef]
121. Deng, X.; Fang, Z.; Liu, Y.; Yu, C.-L. Production of Biodiesel from Jatropha Oil Catalyzed by Nanosized Solid Basic Catalyst. *Energy* **2011**, *36*, 777–784. [CrossRef]
122. Madhuvilakku, R.; Piraman, S. Biodiesel Synthesis by TiO_2–ZnO Mixed Oxide Nanocatalyst Catalyzed Palm Oil Transesterification Process. *Bioresour. Technol.* **2013**, *150*, 55–59. [CrossRef]
123. Mazaheri, H.; Ong, H.C.; Masjuki, H.H.; Amini, Z.; Harrison, M.D.; Wang, C.-T.; Kusumo, F.; Alwi, A. Rice Bran Oil Based Biodiesel Production Using Calcium Oxide Catalyst Derived from Chicoreus Brunneus Shell. *Energy* **2018**, *144*, 10–19. [CrossRef]
124. Pandit, P.R.; Fulekar, M.H. Egg Shell Waste as Heterogeneous Nanocatalyst for Biodiesel Production: Optimized by Response Surface Methodology. *J. Environ. Manag.* **2017**, *198*, 319–329. [CrossRef] [PubMed]
125. Varghese, R.; Henry, J.P.; Irudayaraj, J. Ultrasonication-Assisted Transesterification for Biodiesel Production by Using Heterogeneous ZnO Nanocatalyst. *Environ. Prog. Sustain. Energy* **2018**, *37*, 1176–1182. [CrossRef]
126. Cheah, W.Y.; Sankaran, R.; Show, P.L.; Ibrahim, T.N.B.T.; Chew, K.W.; Culaba, A.; Chang, J.S. Pretreatment Methods for Lig-781 nocellulosic Biofuels Production: Current Advances, Challenges and Future Prospects. *Biofuel Res. J.* **2020**, *7*, 1115–1127. [CrossRef]
127. Pattakrine, M.V.; Pattakrine, V.M. Nanotechnology for Algal Biofuels. In *The Science of Algal Fuels*; Springer: Dordrecht, The Netherlands, 2012; Volume 25, pp. 147–163.

catalysts

Article

Production of Therapeutically Significant Genistein and Daidzein Compounds from Soybean Glycosides Using Magnetic Nanocatalyst: A Novel Approach

Mamata Singhvi, Minseong Kim and Beom-Soo Kim *

Department of Chemical Engineering, Chungbuk National University, Cheongju 28644, Korea
* Correspondence: bskim@chungbuk.ac.kr; Tel.: +82-43-261-2372; Fax: +82-43-269-2370

Highlights:

- Production of therapeutical aglycone compounds was attempted using a nanocatalyst.
- The maximum amount of diadzein (8.91 g/L) and genistein (12.0 g/L) was generated at 80 °C in 3 h with yields of 0.590 and 0.621 g/g substrate, respectively..
- Reused nanocatalyst demonstrated ~35% catalytic efficiency even after third recycle.
- Nanocatalyst can be a better substitute for costly enzymes in aglycone production.

Abstract: Genistein and daidzein are well-known biologically active pharmaceutical compounds that play significant roles in the treatment of various diseases such as cardiovascular problems, cancer, etc. In some plants, the glycosides daidzin and genistin are present in ample amounts that can be converted into aglycones, daidzein and genistein, through hydrolysis. Here, magnetic cobalt ferrite alkyl sulfonic acid ($CoFe_2O_4$-Si-ASA) nanocatalyst was used for the hydrolysis of glycosides into aglycones. The application of $CoFe_2O_4$-Si-ASA nanocatalyst generated a maximum 8.91 g/L diadzein and 12.0 g/L genistein from 15.1 g/L daidzin and 19.3 g/L genistin with conversion efficiencies of 59.0% and 62.2%, respectively, from soybean glycosides at 80 °C in 3 h. The use of a modern nanocatalyst is preferred over enzymes because of its lower production cost, higher rate of reaction, higher stability, etc. To our knowledge, this is the first report on using nanocatalyst for the production of genistein and daidzein in a sustainable manner.

Keywords: cobalt ferrite alkyl sulfonic acid nanocatalyst; soybean; genistin; daidzin; genistein; daidzein; β-glucosidase

Citation: Singhvi, M.; Kim, M.; Kim, B.-S. Production of Therapeutically Significant Genistein and Daidzein Compounds from Soybean Glycosides Using Magnetic Nanocatalyst: A Novel Approach. *Catalysts* **2022**, *12*, 1107. https://doi.org/10.3390/catal12101107

Academic Editor: Anwar Sunna

Received: 19 July 2022
Accepted: 23 September 2022
Published: 25 September 2022

1. Introduction

Phytoestrogens are plant-derived compounds possessing health-promoting properties, which are classified into groups such as isoflavones, phytosterols, and lignans [1]. Isoflavones belong to the flavonoid class, which has drawn considerable attention in the medical field chiefly because of their antioxidant and estrogenic activities [2]. Most of the reported studies have designated that particularly the unconjugated forms of isoflavones, i.e., aglycones, exhibit greater beneficial effects than the isolated forms [3]. In general, legumes are the most essential source of isoflavones. Among them, soybean (*Glycine max*) is considered the ultimate source, which principally occurs in the form of β-glucosides [4]. These glycosides can be further hydrolyzed to be converted into their bioactive forms, i.e., aglycone forms, which have a substantial effect on human health [5].

Daidzein (7-hydroxy-3-(4-hydroxyphenyl)-4-benzopyrone) and genistein (5,7-dihydroxy-3-(4-hydroxyphenyl)-4H-1-benzopyran-4-one) are significant isoflavones that are usually present in their respective glycoside forms, daidzin and genistin, in legumes [6]. These aglycone compounds (daidzein and genistein) possess a 7-hydroxyisoflavone frame, which is structurally equivalent to the utmost effective estrogen hormone, estradiol-17β. Owing to the structural homologies between estrogen and isoflavone compounds, aglycone compounds are capable of binding to estrogen receptors (such as ERα and ERβ), which can mimic the estrogenic actions of the antagonistic or agonistic types [7]. Aglycone compounds have been widely used in the treatment of various hormone-associated diseases such as breast cancer, menopausal problems, cardiovascular diseases, osteoporosis, anticarcinogenic, antioxidant, anti-inflammatory, and antimicrobial agents [8,9].

Numerous studies related to the conversion of glycosides into their respective aglycone moieties using different routes have been reported. Aglycone compounds are significant, considering their effective therapeutic activities. Being lower-molecular-weight compounds, aglycones are easily absorbed through the gastrointestinal tract [10]. After removing sugar moieties from isoflavone glycosides (daidzin, genistin), highly significant isoflavone aglycones such as daidzein and genistein can be generated. Usually, chemical (acid, alkali, etc.) and biological methods (enzymes and microbes) are employed to extract aglycone compounds from conjugated glycosides, albeit with several advantages and disadvantages. Although a chemical process offers advantages with respect to cost and a faster rate of reaction, production of undesired inhibitor compounds because of a lack of specificity toward the substrate makes it unsuitable for application [11]. The hydrolysis of soybean substrates using acids results in the formation of genotoxic compounds, hydroxymethylfurfural and ethoxymethylfurfural, utilizing constituent oligosaccharides [12]. On the other hand, biological methods employing β-glucosidase enzymes and microbial cells [13–16] for the production of aglycones have issues of stability, production cost, and the generation of mixtures of several end products. Therefore, searching for a sustainable and proficient option is crucial for the conversion of glycoside into aglycone compounds.

Regarding this concern, rather than using traditional methods, attention has been paid toward the development of a magnetic acid-functionalized nanocatalyst that possesses catalytic properties and can be easily recovered from the reaction mixture because of its magnetic characteristic [17,18]. Several studies have been reported on the use of various nanobiocatalysts in the pretreatment of different biomass materials [19,20]. Peña et al. [21] used different acid-functionalized magnetic nanocatalysts for cellobiose hydrolysis and among the used nanomaterials, alkylsulfonic acid-functionalized nanomaterial exhibited higher efficiency with 78% cellobiose conversion. In addition, carbon-based nanomaterials functionalized with sulfonic acids have been studied for cellobiose hydrolysis, which generated monomeric glucose with 84% conversion efficiency [22]. Fe-based nanomaterials have also been known to possess exceptional enzyme-mimicking properties, along with a magnetic property. Considering all these aspects, we attempted to mimic the β-glucosidase activity of an acid-functionalized cobalt ferrite alkyl sulfonic acid ($CoFe_2O_4$-Si-ASA) nanocatalyst in an effective manner to hydrolyze the glycosidic linkages in soybean-derived glycoside moieties to obtain aglycones. In this study, a $CoFe_2O_4$-Si-ASA nanocatalyst was prepared and the hydrolytic efficiency was assessed for hydrolysis of glycosides with the aim of developing a simple, sustainable, and ecofriendly approach for the generation of aglycone compounds from glycoside conjugates. An unexplored acid-functionalized nanocatalyst was used for the conversion of soybean-derived glycosides into aglycones, as demonstrated in Figure 1.

The hydrolysis of soybean-extracted glycosides was conducted using a magnetic $CoFe_2O_4$-Si-ASA nanocatalyst and compared with enzymatic hydrolysis. For enzymatic hydrolysis, β-glucosidase produced by *Fusarium verticillioides* was used in this study. *F. verticillioides* has been reported earlier to produce cellulases under submerged fermentation conditions for application in cellulose hydrolysis [23]. In the current investigation, we optimized the solid-state fermentation conditions for cellulase production to obtain enzymes in concentrated form with higher activities using *F. verticillioides*. To accomplish our objectives, optimization studies for hydrolysis of soybean-derived isoflavone glycoside using an acid-functionalized $CoFe_2O_4$-Si-ASA magnetic nanocatalyst were conducted to obtain higher titers of aglycones. To date, microbial enzymes have been used for the generation of aglycone compounds, which is a costlier process than using a nanocatalyst. To our knowledge, this is the first study on the application of nanocatalyst to generate aglycones from glycosides. The application of this nanomaterial-assisted approach of generating aglycone moieties from plant glycosides will have abundant prospectives in the therapeutic industry.

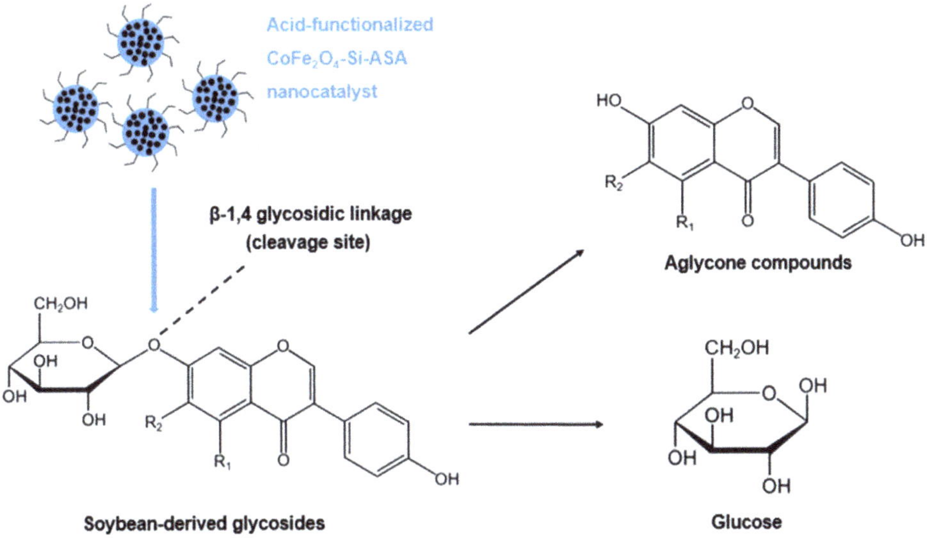

Figure 1. Illustrative representation of $CoFe_2O_4$-Si-ASA nanocatalyst-mediated conversion of soybean-derived glycosides into aglycones, i.e., daidzein and genistein.

2. Materials and Methods

2.1. Chemicals

Cellulose, p-nitrophenyl-β-D-glucopyranoside (p-NPG), isoflavone glycoside standards (genistin and daidzin), and their aglycone counterparts (genistein and daidzein) were purchased from Sigma-Aldrich (St. Louis, MO, USA). Soybean flour and wheat bran were procured from the local market. Ammonium hydroxide, isopropanol, and toluene were purchased from Fisher Scientific (Pittsburgh, PA, USA). 3-mercaptopropyltrimethoxysilane (MPTMS), 4-(triethoxysilyl)-butyronitrile (98%), diethylamine (99%), methylamine (40%), sodium dodecyl sulfate, and tetraethylorthosilicate (TEOS) (99.999%) were purchased from Sigma-Aldrich. Iron (II) chloride tetrahydrate ($FeCl_2 \cdot 4H_2O$) (97%), and cobalt (II) chloride hexahydrate ($COCl_2 \cdot 6H_2O$) (97%) were purchased from Junsei Chemical Co. Ltd., Tokyo, Japan. All other chemicals and reagents used in this study were purchased locally.

2.2. Synthesis and Functionalization of Nanoparticles

The synthesis of cobalt ferrite ($CoFe_2O_4$) nanoparticles (NPs) was carried out using the method described by Wang et al. [24] with slight modifications. Furthermore, silica coating and acid functionalization of the synthesized magnetic $CoFe_2O_4$ NPs were performed as per the method described by Ingle et al. [25].

2.2.1. Synthesis of Magnetic $CoFe_2O_4$ Nanoparticles ($CoFe_2O_4$ MNPs)

For the synthesis of $CoFe_2O_4$ NPs, 10 mM of $COCl_2 \cdot 6H_2O$ was mixed with 20 mM of $FeCl_2 \cdot 4H_2O$ solution to make the volume 250 mL of aqueous solution, and then 50 mM sodium dodecyl sulfate was added to the mixture. This solution mixture was kept for stirring initially at 30 °C for half an hour before elevating the temperature to 70 °C. In the meantime, another solution (500 mL) was prepared with 75 mL of methylamine (40% *w/w*) and distilled water, which was further heated at 70 °C. Finally, both solutions were mixed and stirred for 3 h. The synthesized NPs were separated magnetically and washed thrice with distilled water followed by ethanol. The isolated MNPs were dried in a hot-air oven at 60 °C for characterization studies.

2.2.2. Synthesis of Silica Coated $CoFe_2O_4$ Nanoparticles ($CoFe_2O_4$–Si MNPs)

The silica coating of $CoFe_2O_4$ MNPs was conducted by combining the process described by Rajkumari et al. [26] and Peña et al. [21]. The synthesized $CoFe_2O_4$ MNPs were dispersed in ethanol and sonicated for 15–20 min. Then 25 mL of the ethanol-dispersed $CoFe_2O_4$ MNPs solution was added to 430 mL of isopropanol and 20 mL of distilled water. This solution was further sonicated for 15 min and then 50 mL of concentrated NH4OH was added to the mixture. The solution of TEOS/isopropanol (1:40) was added dropwise to the earlier solution and sonicated with stirring for 4 h. The synthesized NPs were separated magnetically, washed with distilled water, and finally dried in a hot-air oven at 80 °C for 24 h.

2.2.3. Acid Functionalization of $CoFe_2O_4$-Si MNPs Using Alkylsulfonic Acid ($CoFe_2O_4$-Si-ASA)

An acid-functionalized catalyst, $CoFe_2O_4$-Si-ASA, was synthesized by functionalizing silica-coated $CoFe_2O_4$ MNPs with ASA by following the method reported by Wang et al. [24] with slight variations. For the preparation of acid functionalized $CoFe_2O_4$-Si nanocatalyst, a solution mixture containing MPTMS (5.0 mL), ethanol (50 mL), and distilled water (45 mL) was prepared and then 0.5 g of $CoFe_2O_4$-Si MNPs was added. This solution mixture was sonicated for 30 min and further kept overnight for stirring at 70 °C. $CoFe_2O_4$–Si MNPs with attached thiol groups were collected magnetically and washed thrice with distilled water. A solution of 50% H_2O_2 (30 mL), distilled water (30 mL), and methanol (30 mL) was added to the recovered thiol-attached $CoFe_2O_4$-Si MNPs and the mixture was kept at room temperature for oxidation of thiol to sulfonic acid groups. After oxidation, the formed acid-functionalized $CoFe_2O_4$-Si-ASA nanocatalyst was recovered magnetically and washed with distilled water thrice. The synthesized acid-functionalized $CoFe_2O_4$-Si-ASA nanocatalyst was further reacidified with 3 M H2SO4 (25 mL), washed two or three times with distilled water, and finally dried in a hot-air oven at 80 °C for 24 h. Furthermore, the synthesis of all these nanomaterials was validated by characterization studies.

2.3. *Characterization of Synthesized Nanoparticles*

To study the surface morphology of the synthesized NPs, scanning electron microscopy (SEM, LEO-1530) analysis was conducted. The samples were prepared by platinum coating and images were recorded at an accelerating voltage of 3 kV. Furthermore, the sizes of the synthesized NPs were estimated using transmission electron microscopy (TEM, Carl Zeiss, Libra 120, Jena, Germany) at a voltage of 120 kV. For TEM analysis, samples were prepared by dispersing the synthesized NPs in water, which was then sonicated for about 5–10 min. These dispersed samples were further adsorbed onto 200-mesh copper grids for 30–60 s at room temperature and then viewed by TEM. To determine the elemental composition, the synthesized NPs were further analyzed using energy dispersive X-ray spectroscopy (EDX, Thermo Fisher Scientific, Waltham, MA, USA). Fourier transform–infrared spectroscopy (FTIR) analysis was used to examine the absorption peaks and functional groups present on the synthesized nanomaterials. FTIR analysis of the synthesized NPs was carried out using an IR200 spectrometer (Thermo Fisher Scientific) in a wavenumber range of 400–4000 cm^{-1} by following the KBr pellet method as reported earlier [23]. X-ray diffraction (XRD) was carried out to determine the degree of cellulose crystallinity for raw and pretreated corncob biomass using a high-power X-ray diffractometer (model: JP/SmartLab, 9 kW). Samples were analyzed from $2\theta = 5$ to 70° at a scanning speed of 3°/min [23].

2.4. *Microbial Strain, Medium, and β-Glucosidase Enzyme Production*

In this study, the previously isolated fungal strain, *Fusarium verticillioides* (Accession no. PRJNA664836), was used for cellulase enzyme, particularly β-glucosidase, production.

The strain was regularly maintained and sub-cultured every three months on potato dextrose agar (PDA) slants under optimized growth conditions as reported earlier [23]. The method for the medium preparation and its composition used in this study was similar as reported earlier [16]. The response surface methodology-optimized basal medium (RSM-BM) consisted of (per L): KH_2PO_4 (1.0 g), $CaCl_2$ (0.15 g), urea (0.60 g), $(NH_4)_2SO_4$ (0.50 g), yeast extract (0.15 g), peptone (0.45 g), $MgSO_4{\cdot}7H_2O$ (0.15 g), $FeSO_4{\cdot}7H_2O$ (0.0075 g), $MnSO_4{\cdot}7H_2O$ (0.0024 g), $ZnSO_4{\cdot}7H_2O$ (0.0021 g), $CoCl_2$ (0.003 g), and Tween 80 (1.0 mL). The cellulase production experiments were conducted under solid-state fermentation conditions using the *F. verticillioides* fungal strain.

Solid-state fermentation was carried out in 250 mL Erlenmeyer flasks containing 4 g of wheat bran, 1 g of cellulose, and 8 mL of production medium (RSM-BM) to moisten the substrate. The spore suspension from the sporulated fungal slant was prepared using saline water containing triton X-100 (0.1% *v*/*v*). The flasks were inoculated with spores (about 10^7–10^8) derived from fungal strain grown on PDA slants and incubated at 28 °C under static conditions. The inoculated flasks were extracted by adding 50 mL of sodium citrate buffer (50 mM, pH 4.5) after every 2-day interval. For enzyme extraction, flasks were incubated at 37 °C with shaking at 200 rpm for 2 h. The extracted broth was centrifuged at 8000 rpm for 10 min and the supernatant was collected. All supernatant samples were analyzed to determine extracellular β-glucosidase activity and protein concentration. The extracted enzyme supernatant was further used for the hydrolysis of soybean-derived glycosides.

2.5. Hydrolysis of Soybean-Derived Glycosides Using Acid-Functionalized $CoFe_2O_4$-Si-ASA Nanocatalyst and β-Glucosidase Enzyme

The soybean flour was extracted using methanol, as reported previously [16] with slight modifications. For the glycoside extraction experiment, soybean flour was defatted with three volumes of *n*-hexane by stirring for 30 min at 30 °C, centrifuging at 10,000 rpm for 20 min, and air drying. The defatted soybean flour (10 g) was further treated with 50 mL of methanol (80%), stirring for 2 h at 80 °C and centrifuging at 10,000 rpm for 10 min to obtain the extract containing glycosides.

To evaluate and compare the catalytic efficiency of the acid-functionalized $CoFe_2O_4$-Si-ASA nanocatalyst with β-glucosidase enzymes, this soybean extract containing glycosides was hydrolyzed using a $CoFe_2O_4$-Si-ASA nanocatalyst and β-glucosidase enzymes derived from *F. verticillioides* in two different flasks. In the case of hydrolysis using acid-functionalized $CoFe_2O_4$-Si-ASA nanocatalyst, different concentrations of NPs (50, 150, 250, 350, and 450 mg per g of substrate) were used for hydrolysis of soy-flour extract. The initial optimization studies for soybean-derived glycoside hydrolysis were performed using different concentrations of $CoFe_2O_4$-Si-ASA nanocatalyst at 80 °C. After optimizing the concentration of the nanocatalyst, the hydrolysis performance was checked at higher temperature and pressure conditions, i.e., at 120 °C and 15 psi pressure, respectively, for 1 h. After hydrolysis, the samples were analyzed to determine the content of aglycones using high-performance liquid chromatography (HPLC) analysis.

Furthermore, to assess the catalytic performance of the $CoFe_2O_4$-Si-ASA nanocatalyst, in-house-generated β-glucosidase enzymes were used for comparative glycoside hydrolysis. For enzymatic hydrolysis, 10 mL of soybean-flour extract was hydrolyzed with 5.0 IU of β-glucosidase at 60 °C for different time intervals as optimized previously [16]. The obtained samples after hydrolysis were dried under vacuum and resuspended in methanol (1 mL) prior to HPLC analysis. The reaction mixture without $CoFe_2O_4$-Si-ASA nanocatalyst/enzyme was used as control.

2.6. Recovery and Reuse of $CoFe_2O_4$-Si-ASA Nanocatalyst for Hydrolysis Experiments

After the first hydrolysis reaction, the acid-functionalized $CoFe_2O_4$-Si-ASA MNPs were easily separated by applying a magnetic field and the remaining liquid hydrolysate was collected to analyze the concentration of aglycones (daidzein and genistein) formed after hydrolysis. Hence, the recovered NPs were further washed two to three times with distilled water and dried at 70 °C in a hot-air oven overnight and reused in the subsequent hydrolysis reaction. The recovered $CoFe_2O_4$-Si-ASA MNPs after the first cycle of hydrolysis were reused in the next cycle of hydrolysis. All these hydrolysis experiments using reused MNPs were performed at similar conditions to those used for optimization studies (at 80 °C for 3 h), as mentioned in the above Section 2.5.

2.7. Analytical Methods

To evaluate the acidity of the synthesized nanocatalyst, an ion-exchange titration method was used, as reported earlier [27]. To remove the hydronium ions from the nanocatalyst, 50 mg of catalyst was added to 20 mL of a sodium chloride/tetramethylammonium chloride (2 M). This test solution was sonicated, kept static at room temperature overnight, and then titrated to neutrality with a 10 mM NaOH solution. The extracted solid-state enzymes and $CoFe_2O_4$-Si-ASA nanocatalyst were tested for β-glucosidase-mimicking activity using a *p*-NPG substrate, as reported earlier [16]. The enzyme supernatant was analyzed to determine protein concentration using the bicinchoninic acid assay method [28]. The concentration of glycosides (daidzin and genistin) in soybean flour and the obtained aglycones (daidzein and genistein) in the hydrolysate were determined using reverse phase HPLC (YL 9100, Younglin Inc., Anyang, Korea), as reported earlier [16]. HPLC analysis was conducted with a YL 9100 system and an Eclipse plus C18 (4.6 mm × 250 mm, 5 μm) column at 260 nm for 60 min and the column temperature was set at 35 °C [29]. An isocratic elution was performed using an acetonitrile (10%, *v*/*v*) mobile phase at a flow rate of 0.3 mL/min. The unknown amounts of glycosides and aglycones were determined by correlating with standards. For further authentication of the glycosides and aglycones present in the control and hydrolysate samples, the mass spectrum was determined by LC/MS (maXis 4G, Bruker Biosciences, Billerica, MA, USA) using an ESI detector.

3. Results and Discussion

3.1. Synthesis and Characterization of All Synthesized Nanoparticles

All three MNPs, i.e., $CoFe_2O_4$, $CoFe_2O_4$-Si, and acid-functionalized $CoFe_2O_4$-Si-ASA, used in this study were prepared as mentioned above (Materials and Methods Section 2.2). To confirm the synthesis of all three MNPs, characterization studies were conducted using SEM, TEM, EDX, FTIR, and XRD analyses.

To determine the surface morphology and shape of all three synthesized nanomaterials, the $CoFe_2O_4$, $CoFe_2O_4$-Si, and acid-functionalized $CoFe_2O_4$-Si-ASA were subjected to SEM analysis. All three MNPs were observed to be spherical and uniform in shape (Figures S1A,B and 2A). Furthermore, TEM analysis confirmed that the synthesis of all MNPs was successful from their nano-sizes ranging from 5–25 nm. The diameters of $CoFe_2O_4$, $CoFe_2O_4$-Si, and acid-functionalized $CoFe_2O_4$-Si-ASA were recorded to be in the range of 7.74–25.2, 5.07–4.11, and 5.2–12.2 nm (as presented in Figure S1C,D and Figure 2B), respectively. Among all three MNPs, acid–functionalized $CoFe_2O_4$-Si-ASA (Figure 2B) and $CoFe_2O_4$ NPs (Figure S1C) were well dispersed, whereas the silica-coated $CoFe_2O_4$-Si NPs were observed to be agglomerated (Figure S1D).

EDX analysis is usually used to determine the elemental composition of any material. Furthermore, the elemental composition of all three MNPs characterized using EDX confirmed the occurrence of metal constituents, oxygen, carbon, and other components in each nanomaterial. As shown in Figure S2, strong signals for specific metals (Fe: 0.64 eV; Co: 0.70 eV; Si: 0.18 eV; S: 0.24 eV; C: 0.27 eV, and O: 0.51 eV) confirmed the formation of the desired nanomaterials. The details of the elemental composition of all nanomaterials are given in Table 1. The EDX analysis exhibited 15.13 wt% of Fe and 23.47 wt% of Co in case of the $CoFe_2O_4$ MNPs. The presence of elemental Fe and Co in major amounts assured us that the synthesized NPs are $CoFe_2O_4$ MNPs. Besides this, the peaks corresponding to the element Si were observed in the cases of both silica-coated $CoFe_2O_4$-Si and acid-functionalized $CoFe_2O_4$-Si-ASA MNPs designating the effective surface amendment by silica on bare

$CoFe_2O_4$ MNPs [30]. Furthermore, a substantial occurrence of sulfur (S) in the case of $CoFe_2O_4$-Si-ASA (5.05 wt%) indicated the strong functionalization of acid on $CoFe_2O_4$-Si MNPs. In the case of acid-functionalized $CoFe_2O_4$-Si-ASA MNPs, a large amount of C was detected with respect to the S element, specifying the probable loss of the sulfonic acid groups during the MNP synthesis process. The XRD spectra of the $CoFe_2O_4$-Si-ASA MNPs are shown in Figure S3. The obtained XRD spectrum for $CoFe_2O_4$-Si-ASA MNPs agrees well with reported studies [21]. The XRD pattern for the $CoFe_2O_4$-Si-ASA MNPs shows all the respective signals for the $CoFe_2O_4$ crystals; this spectrum has an additional peak between 20 and 30° (2θ), which is associated with amorphous silica. The size of the $CoFe_2O_4$-Si-ASA MNPs was calculated as 21.6 nm, which seems to be larger than those obtained with TEM. The larger sizes calculated from the XRD patterns could be due to the possible aggregation of MNPs. In addition, the magnetic behavior of $CoFe_2O_4$ MNPs was confirmed by providing a magnetic field, as shown in Figure S4.

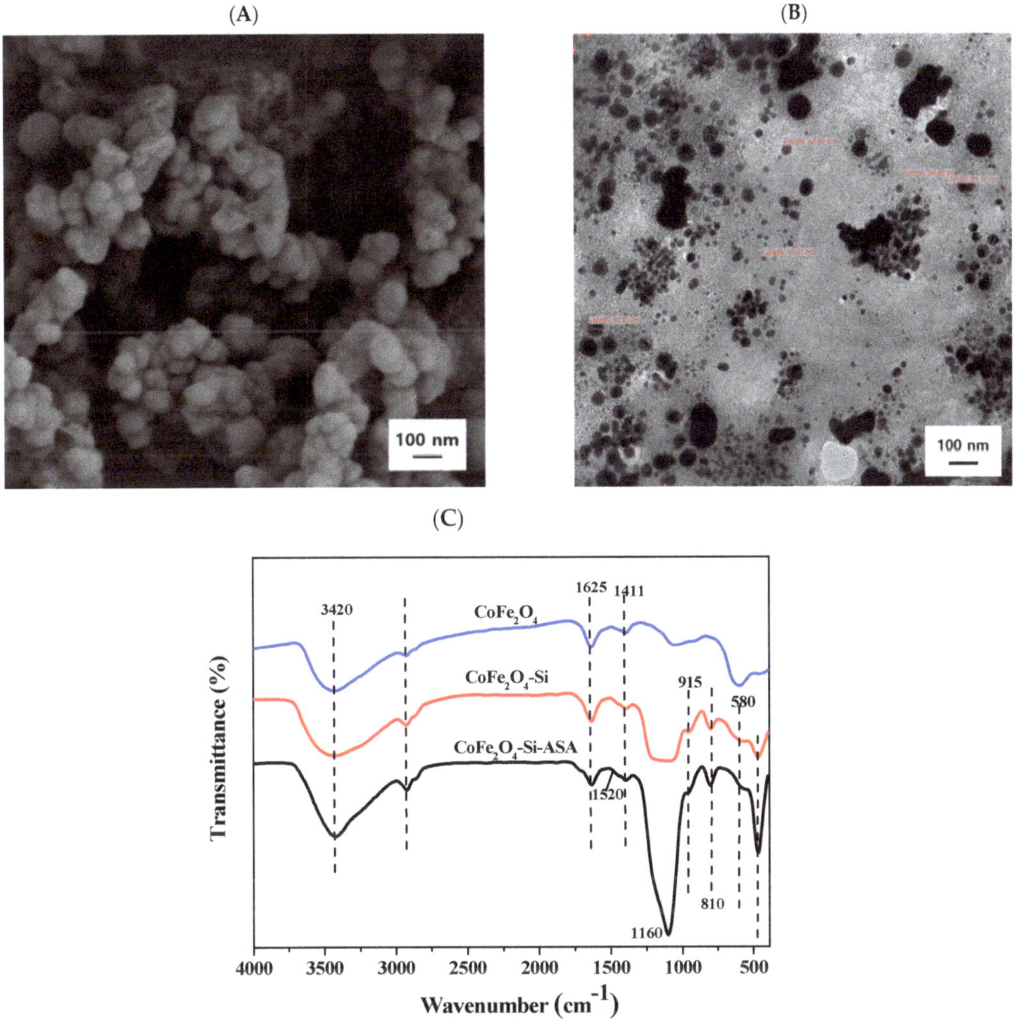

Figure 2. Characterization of the synthesized $CoFe_2O_4$-Si-ASA nanocatalyst using (**A**) SEM, (**B**) TEM, and (**C**) FTIR analyses.

Table 1. Elemental composition of nanoparticles used in this study.

Synthesized Nanoparticles	Atomic Weight Percentage (%) [a]						Acid Capacity (mM H^+/g) [b]
	C	O	Fe	Co	Si	S	
$CoFe_2O_4$	34.19	27.20	15.13	23.47	-	-	-
$CoFe_2O_4$-Si	37.16	53.01	1.13	1.12	7.59	-	-
$CoFe_2O_4$-Si-ASA	51.23	38.69	0.18	0.21	4.64	5.05	0.92

[a] data obtained from EDX analysis. [b] calculated using titration method

FTIR analysis is generally used to identify the functional groups present in the experimental samples. Figure 2C depicts the FTIR spectra of all three synthesized MNPs, $CoFe_2O_4$, $CoFe_2O_4$-Si, and acid-functionalized $CoFe_2O_4$-Si-ASA. As shown in Figure 2C, the significant peak was detected in the wavelength range of 520–560 cm^{-1} in all three MNPs assigning to the stretching of the Fe-O bond [31], which is one of the confirmatory features of $CoFe_2O_4$-MNPs. The common absorption peaks at wavelengths of 800 and 910 cm^{-1}, ascribed to the stretching vibrations of the Si-O-Si and Si-O-H groups, appeared only in $CoFe_2O_4$-Si and acid-functionalized $CoFe_2O_4$-Si-ASA [24]. These peaks were not observed in $CoFe_2O_4$ MNPs, specifying the surface coating of silica in the case of $CoFe_2O_4$-Si and acid-functionalized $CoFe_2O_4$-Si-ASA MNPs only. Similarly, stretching vibrations at 1160 and 1380 cm^{-1}, assigned to Si-O and O=S=O stretching vibrations, were observed in the spectra of both $CoFe_2O_4$-Si and acid-functionalized $CoFe_2O_4$-Si-ASA samples [32]. The presence of peaks at a wavelength of about 1420 cm^{-1}, attributed to undissociated SO_3H groups, was found only in acid-functionalized $CoFe_2O_4$-Si-ASA MNPs [33], further confirming the effective functionalization of acids on silica-coated MNPs. The obtained FTIR spectra for all three MNPs indicated common peaks at the wavelengths of 3420, 2810, and 1625 cm^{-1}, which correspond to the stretching vibrations of the O-H and C-H groups, respectively [21]. The results obtained from FTIR analysis revealed the presence of all of the expected functional groups in the case of all three synthesized MNPs.

The overall results obtained after characterizing all these materials revealed that strong magnetic $CoFe_2O_4$ NPs were synthesized and appropriately coated with silica, which was confirmed by the FTIR and EDX analyses. Similarly, TEM analysis confirmed that the synthesis of all NPs was successful from their nano-sizes in the range of 5–25 nm. Thus, after confirming the synthesis of all three MNPs through characterization studies, acid-functionalized $CoFe_2O_4$-Si-ASA nanocatalyst was further evaluated for its hydrolytic activity to generate therapeutic aglycone compounds from soybean-derived glycosides.

3.2. β-Glucosidase Enzyme Production by Solid-State Fermentation

The isolated fungal strain, *F. verticillioides* was assessed for production of extracellular β-glucosidase enzyme in flasks using RSM-BM [16] containing wheat bran (4.0 g) and cellulose (1.0 g) substrates under solid-state fermentation conditions. As presented in Table 2, the optimized RSM-BM containing wheat bran (2.5%) and cellulose (1%) exhibited maximum β-glucosidase activity (1.959 ± 0.098 IU/g) under solid-state fermentation conditions. Previously, we have reported the production of cellulases under submerged fermentation conditions, which exhibited the highest activity of β-glucosidase (2.91 ± 0.12 IU/mL) on the 8th day of fermentation [23]. As enzymes produced under submerged fermentation conditions are usually in the diluted form, we attempted to produce enzymes in the concentrated form using solid-state fermentation in this study. As shown in Table 2, the higher enzyme activities were obtained on the 6th day of fermentation, which reduced the fermentation time and increased β-glucosidase activities as compared to previously reported enzyme activities using submerged fermentation [23].

Table 2. Determination of β-glucosidase activities produced by *Fusarium verticillioides* under solid-state fermentation conditions.

Fermentation Time (Days)	β-Glucosidase (IU/g)	Protein (mg/g)
2nd day	0.453 ± 0.021	0.153 ± 0.0082
4th day	1.858 ± 0.101	0.858 ± 0.049
6th day	1.959 ± 0.098	0.895 ± 0.052
8th day	2.052 ± 0.111	0.782 ± 0.043

A fermentation experiment was carried out at 28 °C under solid-state conditions in the fermentation medium containing wheat bran (4 g) and cellulose (1 g) substrates. The standard deviation values represented in the table are derived from the experiment performed in triplicate.

There are very few studies conducted on the production of cellulases with such higher activities under solid-state fermentation using *Fusarium* sp. Ramanathan et al. [34] reported cellulase activities produced by the *F. oxysporum* strain under submerged fermentation conditions, which exhibited β-glucosidase activities of 1.784 IU/mL, along with endoglucanase (1.921 IU/mL) and exoglucanase (1.342 IU/mL) activities. The *F. verticillioides* strain demonstrated both β-glucosidase (0.39 IU/mL) and endoglucanase (6.5 IU/mL) enzyme activities using gamba grass as a substrate [35], which are lower than those obtained in the present study. Hence, β-glucosidase produced by *F. verticillioides* was further used as control in the conversion of soybean-derived glycosides.

3.3. *Comparative Studies on Hydrolysis of Soybean-Derived Glycosides Using $CoFe_2O_4$-Si-ASA Nanocatalyst and β-Glucosidase Enzyme*

Glycosides such as daidzin and genistin reveal medicinal properties only when processed into their aglycone forms, i.e., daidzein and genistein, respectively [36]. Hence, it is practical to generate daidzein and genistein in bulk amounts from glycosides through a hydrolysis reaction by breaking down β-glycosidic linkages. Traditionally, β-glucosidase enzymes have been used for the generation of aglycones, which is not a viable option considering the high cost of enzyme production. Nowadays, enzyme-mimicking nanomaterials have been employed for several applications such as biomass hydrolysis. In the present study, we have applied β-glucosidase-mimicking $CoFe_2O_4$-Si-ASA nanocatalyst to generate daidzein and genistein from soybean-derived glycosides and further compared their catalytic efficiencies with β-glucosidase enzyme during hydrolysis reaction.

3.3.1. Hydrolysis of Soybean-Derived Glycosides Using $CoFe_2O_4$-Si-ASA Nanocatalyst

First, the synthesized $CoFe_2O_4$-Si-ASA nanocatalyst was tested for its catalytic ability in converting soybean glycosides into aglycones. As mentioned above in the Materials and Methods Section 2.5, optimization studies for hydrolysis of soybean-extracted glycosides were carried out at 80 °C using various concentrations of nanocatalyst (i.e., 50, 150, 250, 350, and 450 mg/g of substrate used) and samples were taken after every hour. The maximum conversions of glycosides were obtained after 3 h of hydrolysis reaction and there was no further increase in glycoside conversion observed. Hence, the hydrolysis reaction was stopped after 3 h of hydrolysis (data not shown). As the amount of nanocatalyst used in the hydrolysis reaction majorly affects the activity, different concentrations of $CoFe_2O_4$-Si-ASA were verified as depicted in Figure 3. At lower concentration (i.e., 50 mg/g), the nanocatalyst showed very slight conversion. With increasing quantities of nanocatalyst, the hydrolysis caused by the acidic sulfonic groups seemed to be activated and ultimately resulted in a higher conversion of glycosides. As illustrated in Figure 3, the hydrolysis profile exhibited that the rate of hydrolysis and the generation of daidzein and genistein compounds was influenced by the concentrations of the $CoFe_2O_4$-Si-ASA nanocatalyst used. An increase in the concentration of nanocatalyst also upturns the generation of aglycones after hydrolyzing glycosides, which was examined by HPLC analysis. In the case of using a 350 mg/g concentration of $CoFe_2O_4$-Si-ASA nanocatalyst, the maximum 8.91 g/L of diadzein and 12 g/L of genistein was achieved from 15.1 g/L daidzin and

19.3 g/L genistin with the conversion efficiency of 59% and 62.2%, respectively, which were relatively higher than the other concentrations of nanocatalyst used. Thus, for further hydrolysis experiments, the optimized concentration of nanocatalyst (350 mg/g) was used.

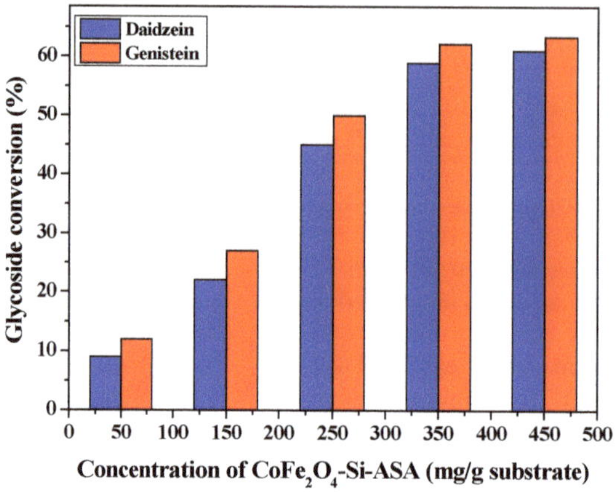

Figure 3. Hydrolysis of soybean-derived glycosides into aglycones using different concentrations of $CoFe_2O_4$-Si-ASA nanocatalyst at 80 °C for 3 h.

Moreover, to check the effect of higher temperature and pressure conditions on the catalytic performance of the $CoFe_2O_4$-Si-ASA nanocatalyst, hydrolysis experiments were carried out at 120 °C and 15 psi pressure for 1 h. As presented in Table 3, the hydrolysis reaction performed at 120 °C and 15 psi pressure generated the maximum 10.7 g/L of daidzein and 13.7 g/L genistein with conversion efficiencies of 69.2% and 73.2%, respectively. This reaction condition exhibited 10–11% enhancement in the generation of aglycone compounds as compared to the aglycones produced during the hydrolysis reaction at 80 °C (as compared in Table 3). Although a slight improvement was observed in the hydrolytic performance of the nanocatalyst at higher temperature and pressure conditions, the hydrolysis reaction at 80 °C is preferred, concerning the lower energy requirement, no formation of inhibitor compounds, and better recyclability of the nanocatalyst. Thus, further hydrolysis experiments employing the nanocatalyst were performed at 80 °C for 3 h.

Table 3. Comparative analysis of hydrolytic conversion of soybean-derived glycosides into aglycone compounds using acid-functionalized cobalt ferrite alkyl sulfonic acid ($CoFe_2O_4$-Si-ASA) nanocatalyst and enzyme β-glucosidase.

Catalyst Used	Glycosides (g/L)		Aglycones (g/L)	
	Daidzin	Genistin	Daidzein	Genistein
$CoFe_2O_4$-Si-ASA (80 °C for 3 h)	15.1 ± 0.51	19.3 ± 0.68	8.91 ± 0.45 (59.0%)	12.0 ± 0.48 (62.2%)
$CoFe_2O_4$-Si-ASA (120 °C at 15 psi for 1 h)	15.5 ± 0.68	18.7 ± 0.99	10.7 ± 0.41 (69.2%)	13.7 ± 0.51 (73.3%)
β-glucosidase enzyme (60 °C for 2 h)	14.8 ± 0.81	17.9 ± 0.91	12.1 ± 0.72 (81.8%)	15.6 ± 0.62 (87.2%)

The standard deviation values are from three independent experiments.

3.3.2. Hydrolysis of Soybean-Derived Glycosides Using β-Glucosidase Enzymes

For comparative evaluation, β-glucosidase enzymes produced by *F. verticilliodes* were used as biocatalyst for the hydrolysis of soybean-derived glycosides. As demonstrated in Table 3, the enzymatic hydrolysis of soybean-derived glycosides converted daidzin (14.8 g/L) and genistin (17.9 g/L) into daidzein (12.1 g/L) and genistein (15.6 g/L) with conversion efficiencies of 81.7% and 87.5%, respectively. The application of enzyme exhibited better catalytic ability than the nanocatalyst toward the hydrolysis of soy flour glycosides. There are many studies reported on the production of various aglycone compounds by hydrolyzing soybean-derived isoflavones by β-glucosidases [37–39]. Our present studies exhibited slightly lower conversion efficiencies using nanocatalyst (69–70%) than enzymes (81–87%). However, the higher cost of enzymes, slower rate of reaction, and lower stability, as compared to modern nanocatalyst, constrain their application in the conversion of glycosides into aglycones on a commercial scale. To address these issues, we attempted to employ a magnetic nanocatalyst for the generation of aglycone compounds, which is more sustainable option than biocatalysts.

All the hydrolyzed samples were analyzed using HPLC, demonstrating that the nanocatalyst was active in the conversion of soybean-derived glycosides. The standards of both glycosides (daidzin and genistin) and aglycones (daidzein and genistein) were analyzed using HPLC to determine their retention times and to quantify compounds in control and nanocatalyst-hydrolyzed samples under the conditions mentioned in the above Section 2.7. As shown in Figure 4A, the HPLC profile of the control sample (i.e., unhydrolyzed soybean-derived glycosides) exhibited the peaks for diadzin and geinstin accompanied by few other unknown components. In addition, the HPLC profile of the nanocatalyst-hydrolyzed samples designated the presence of diadzein and genistein peaks by matching the retention times with the standards (Figure 4B). Moreover, the peaks observed in the HPLC analysis for both samples were confirmed by LC-MS analysis through the mass spectrum $[M + H]^+$ peak. The LC-MS analysis authenticated that the obtained compounds were glycosides (daidzin and genistin) in the control samples (Figure 5A) and aglycones (daidzein and genistein) in the nanocatalyst-hydrolyzed samples (Figure 5B) from their molecular weights.

In this study, the conversion efficiencies of the generated diadzein (69.2%) and genistein (73.2%) from soybean-derived glycosides using $CoFe_2O_4$-Si-ASA nanocatalyst were somewhat lower than enzymatic hydrolysis (as shown in Table 3). Hitherto, there have been no studies reported on the conversion of plant-based glycosides into aglycones using any kind of nanocatalyst. Hence, very limited information is available related to the application of nanomaterials in the conversion of glycoside into aglycones. The overall results obtained in this study were corroborated with some reported studies related to cellobiose hydrolysis into simple glucose sugar molecules using various nanomaterials. Recently, Carlier and Hermans [22] reported higher cellobiose hydrolysis (84%) caused by the catalytic action of carbon nanomaterials functionalized with sulfonic acids, i.e., SO_3H/reduced graphene oxide catalyst at a temperature of 130 °C in only 2 h. Previously, Peña et al. [21] tested the catalytic efficiency of two different nanomaterials, i.e., acid-functionalized perfluoroalkylsulfonic and alkylsulfonic acid, in the cellobiose hydrolysis at 175 °C, which exhibited 75% and 78% conversion into glucose, respectively. These MNPs were further evaluated for pretreating wheat straw [32] and corn stover [19] at different temperatures, which gave higher hydrolysis as compared to the control. Wang et al. [40] proved the exceptional catalytic abilities of acid-functionalized silica-coated Fe_3O_4 MNPs in different catalytic processes. Hence, such nanocatalyst has been used for the pretreatment of various lignocellulosic substrates.

Some of the reported studies, as discussed above, exhibit slightly better conversion efficiencies as compared to our process, but most studies have been conducted under extreme conditions, e.g., higher temperature and pressure. Conversely, our current strategy can be considered as most promising since hydrolysis reaction was conducted under milder conditions. To our knowledge, no studies have been reported on the employment of acid-functionalized nanocatalyst for the production of pharmaceutically significant aglycone compounds. Thus, in this study, we demonstrated a prospective process for the production of aglycones from soybean-derive glycosides using a nanocatalyst.

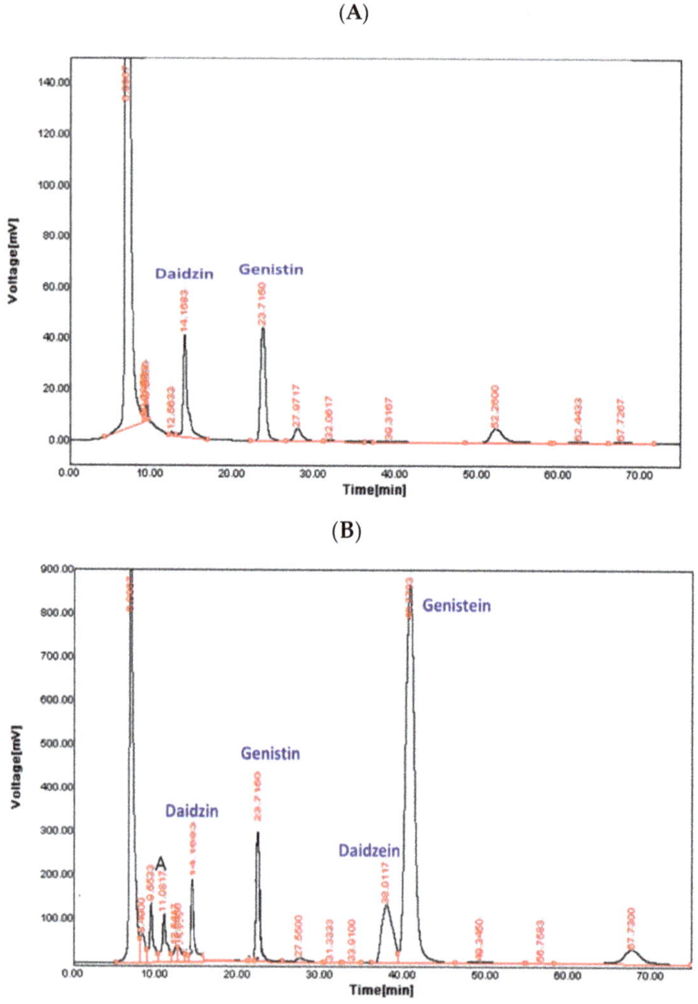

Figure 4. HPLC analysis of soybean-derived glycoside components (daidzin and genistin) and their hydrolyzed products containing aglycone (daidzein and genistein) compounds: (**A**) control (0 h), (**B**) soy-flour extract after hydrolysis with acid nanocatalyst $CoFe_2O_4$-Si-ASA at 80 °C for 3 h.

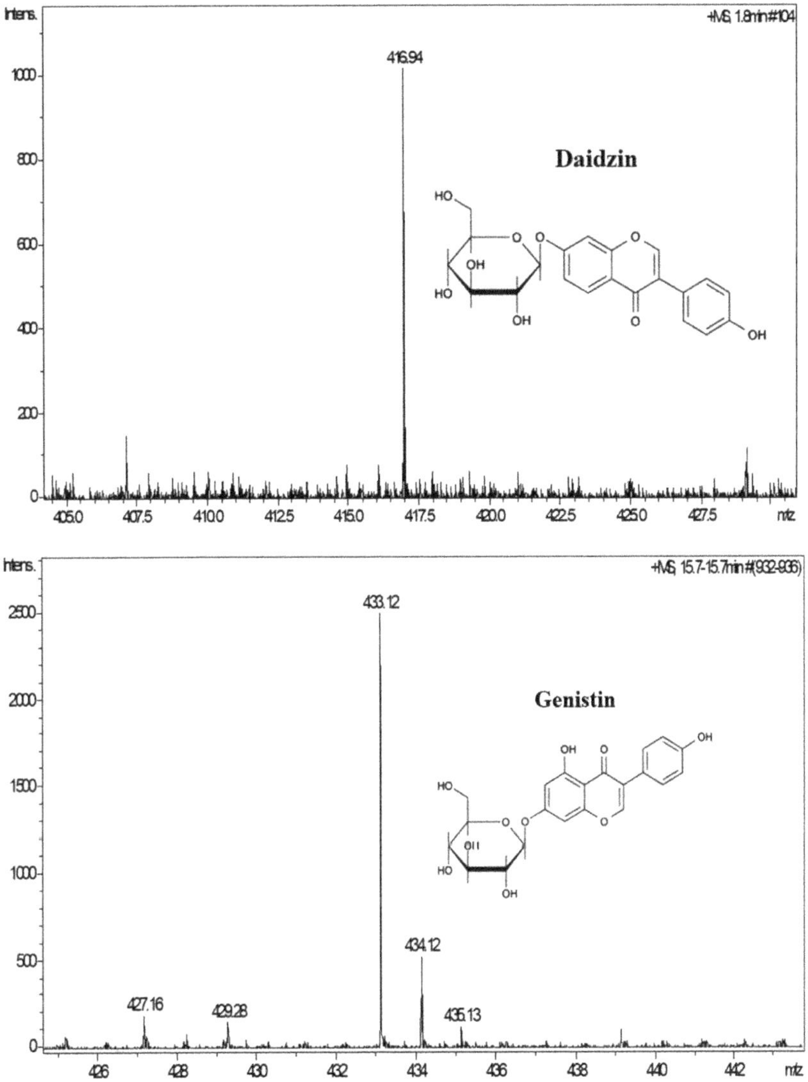

(A)

Figure 5. *Cont.*

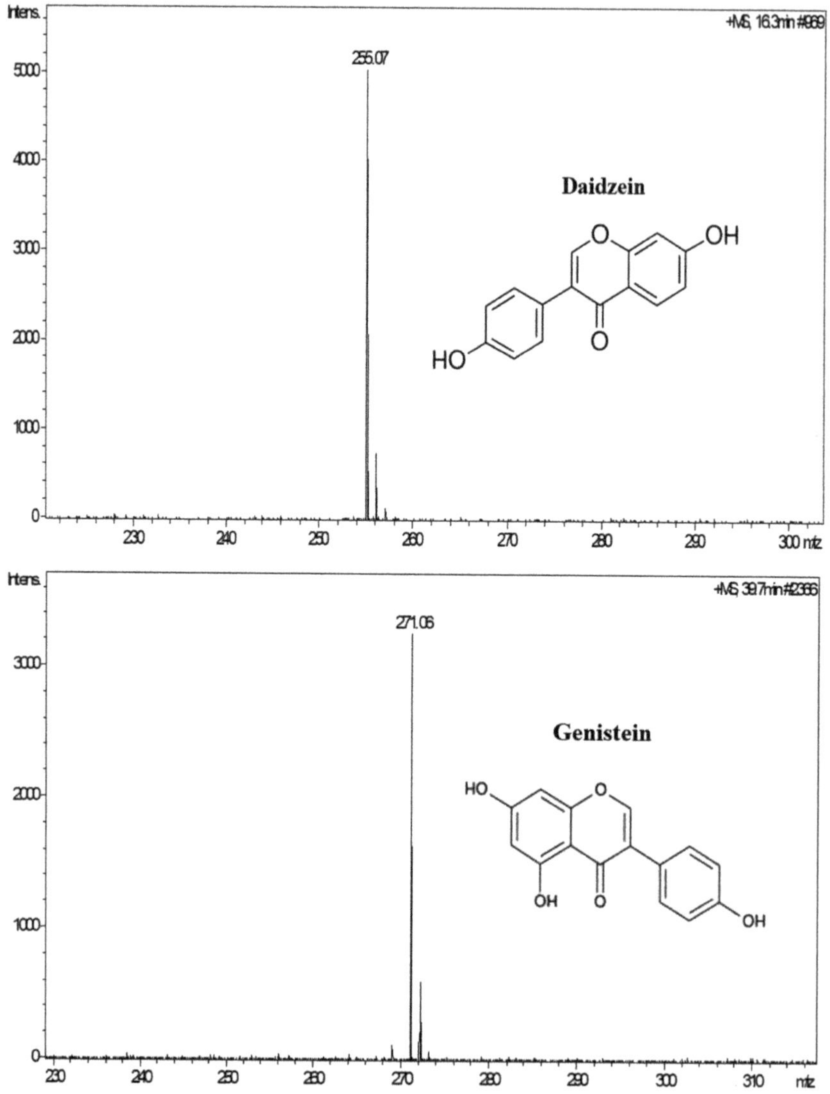

(B)

Figure 5. LC-MS analysis of soybean-derived glycoside components (daidzin and genistin) and their hydrolyzed products containing aglycones (daidzein and genistein) compounds: (**A**) control (0 h), (**B**) soy-flour extract after hydrolysis with acid nanocatalyst $CoFe_2O_4$-Si-ASA at 80 °C for 3 h.

3.4. Recycling of Used $CoFe_2O_4$-Si-ASA Nanocatalyst

Subsequently, the magnetic $CoFe_2O_4$-Si-ASA nanocatalyst used during the hydrolysis reaction was easily recovered by providing a magnetic field, which was further reused for the next cycles of the hydrolysis experiment. For recycling experiments, similar hydrolysis conditions were used as for the optimization studies (350 mg catalyst per g of substrate), hydrolyzed for 3 h at 80 °C.

Reusability of the nanocatalyst was tested for three cycles of the hydrolysis experiment, as demonstrated in Figure 6. A minor decline in the hydrolytic performance was observed during the recycling of the nanocatalyst for the second and third cycle. A trivial decrease in the production of daidzein and genistein, i.e., from 59 and 62.2% to 43.2 and 35%, respectively, was noted after the second recycling experiment. For the third recycling of the nanocatalyst, a further ~35% decrease in conversion efficiency was observed. These results indicated that the bond between sulfonic acid groups and silica-coated $CoFe_2O_4$ catalyst remained strong even after reusing the nanocatalyst for hydrolysis reaction. However, the gradual decline in glycoside conversion was detected during subsequent cycles, which can be implicated to the loss of reacted/unreacted sulfonic acid groups on the silica surface [24,25]. The acidity of the reused nanocatalyst after the third cycle was calculated to be 0.73 mM H^+/g, which was less than that of the unused catalyst (0.92 mM H^+/g). Hence, the possible improvement in the process of functionalizing acid groups on MNPs to achieve stronger bonding can improve the catalytic efficiency of the nanocatalyst. However, the selectivity in terms of the generation of daidzein and genistein components was still maintained even after the last third recycle because of the occurrence of robust sulfonic acid functional groups, which are supposed to be catalytic sites for hydrolysis reaction. In view of these observations, it is clear that acid-functionalized nanocatalyst can prove to be one of the promising options for the generation of aglycone compounds by breaking down β-glycosidic linkages present in the glycosides.

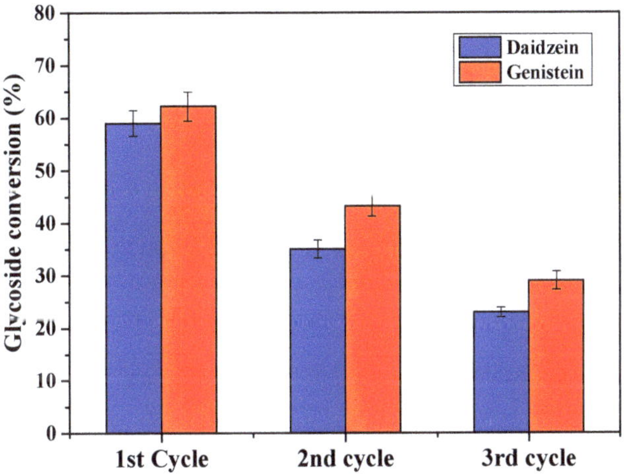

Figure 6. Recycling of used acid-functionalized cobalt ferrite alkyl sulfonic acid ($CoFe_2O_4$-Si-ASA) nanocatalyst for a hydrolysis experiment. Hydrolysis reaction was performed at 80 °C for 3 h.

4. Conclusions

The synthesis of an acid-functionalized $CoFe_2O_4$-Si-ASA nanocatalyst possessing strong sulfonic acid groups was confirmed by characterization studies and the acid titration method. Hydrolysis of glycosides into aglycones was achieved with conversion efficiencies of 59.0–62.2% and 69.2–73.2% using $CoFe_2O_4$-Si-ASA nanocatalyst at 80 °C for 3 h and 120 °C with 15 psi pressure for 1 h, respectively. Most importantly, hydrolysis experiments were performed at a lower temperature (80 °C) for only 3 h, which is promising for hydrolysis of glycosidic linkages in various substrates. In addition, $CoFe_2O_4$-Si-ASA nanocatalyst can be easily recovered and recycled because of its magnetic properties. In view of the expedient hydrolytic efficiency of these $CoFe_2O_4$-Si-ASA MNPs, it is assumed that such a nanocatalyst can be a better alternate candidate for conventional enzymes used for the breakdown of glycosidic linkages. Hitherto, no studies have been reported on the exploitation of such an acid-functionalized nanocatalyst, $CoFe_2O_4$-Si-ASA, for the generation of aglycones by converting glycoside moieties. Although few studies reported the use of solid acid catalysts for cellobiose conversion, reactions were performed under extreme conditions (e.g., high temperature, high pressure, etc.) and for a longer period. On the contrary, the process standardized in the current study was conducted at considerably lower temperature and shorter reaction time, which makes the strategy more expedient. Moreover, the easy recovery of the magnetic nanocatalyst because of its magnetic property makes it possible to reuse the $CoFe_2O_4$-Si-ASA nanocatalyst for succeeding cycles of hydrolysis. All these parameters and accomplished outcomes from this study would certainly aid in making the process commercially viable by reducing the extensive cost involved in the process. Ultimately, this process can substantiate to be an effective unconventional approach for the synthesis of therapeutically relevant aglycone compounds on a commercial scale.

Supplementary Materials: The following supporting information can be downloaded at https://www.mdpi.com/article/10.3390/catal12101107/s1: Figure S1: SEM (A and B) and TEM (C and D) analysis of cobalt ferrite ($CoFe_2O_4$) and silica coated of cobalt ferrite ($CoFe_2O_4$-Si) nanocatalyst, respectively; Figure S2: EDX analysis of the synthesized nanoparticles: (A) cobalt ferrite ($CoFe_2O_4$), (B) silica coated of cobalt ferrite ($CoFe_2O_4$-Si), and (C) acid-functionalized cobalt ferrite alkyl sulfonic acid ($CoFe_2O_4$-Si-ASA); Figure S3: Cobalt ferrite ($CoFe_2O_4$) nanoparticles showing magnetic property; Figure S4: $CoFe_2O_4$ MNPs showing magnetic property.

Author Contributions: Conceptualization, M.S.; methodology, M.S. and M.K.; formal analysis, M.S.; investigation, M.S.; data curation, M.S.; writing—original draft preparation, M.S.; writing—review and editing, B.-S.K.; supervision, B.-S.K. All authors have read and agreed to the published version of the manuscript.

Funding: This research was funded by the National Research Foundation of Korea (NRF- 2021R1I1A1A01043785).

Institutional Review Board Statement: Not applicable.

Informed Consent Statement: Not applicable.

Data Availability Statement: Data is available upon request.

Conflicts of Interest: The authors declare that they have no competing interest to the reported studies in this paper.

References

1. Albertazzi, P.; Purdie, D.W. The nature and utility of the phytoestrogens: A review of the evidence. *Maturitas* **2002**, *42*, 173–185. [CrossRef]
2. Nemitz, M.C.; Picada, J.N.; da Silva, J.; Garcia, A.L.H.; Papke, D.K.; Grivicich, I.; Teixeira, H.F. Determination of the main impurities formed after acid hydrolysis of soybean extracts and the in vitro mutagenicity and genotoxicity studies of 5-ethoxymethyl-2-furfural. *J. Pharm. Biomed. Anal.* **2016**, *129*, 427–432. [CrossRef]
3. Křížová, L.; Dadáková, K.; Kašparovská, J.; Kašparovský, T. Isoflavones. *Molecules* **2019**, *24*, 1076. [CrossRef]
4. Barnes, S. The biochemistry, chemistry and physiology of the isoflavones in soybeans and their food products. *Lymphat. Res. Biol.* **2010**, *8*, 89–98. [CrossRef]
5. He, F.J.; Chen, J.Q. Consumption of soybean, soy foods, soy isoflavones and breast cancer incidence: Differences between Chinese women and women in Western countries and possible mechanisms. *Food Sci. Hum. Wellness* **2013**, *2*, 146–161. [CrossRef]
6. Halabalaki, M.; Alexi, X.; Aligiannis, N.; Lambrinidis, G.; Pratsinis, H.; Florentin, I.; Mitakou, S.; Mikros, E.; Skaltsounis, A.-L.; Alexis, M.N.; et al. Estrogenic activity of isoflavonoids from *Onobrychis ebenoides*. *Planta Med.* **2006**, *72*, 488–493. [CrossRef]
7. Paterni, I.; Granchi, C.; Katzenellenbogen, J.A.; Minutolo, F. Estrogen receptors alpha (ERα) and beta (ERβ): Subtype-selective ligands and clinical potential. *Steroids* **2014**, *90*, 13–29. [CrossRef]
8. Sathyapalan, T.; Aye, M.; Rigby, A.S.; Thatcher, N.J.; Dargham, S.R.; Kilpatrick, E.S.; Atkin, S.L. Soy isoflavones improve cardiovascular disease risk markers in women during the early menopause. *Nutr. Metab. Cardiovasc. Dis.* **2018**, *28*, 691–697. [CrossRef]
9. Hillman, G.G.; Singh-Gupta, V.; Al-Bashir, A.K.; Yunker, C.K.; Joiner, M.C.; Sarkar, F.H.; Abrams, J.; Mark Haacke, E. Monitoring sunitinib-induced vascular effects to optimize radiotherapy combined with soy isoflavones in murine xenograft tumor. *Transl. Oncol.* **2011**, *4*, 110–121. [CrossRef]
10. Ullah, A.; Munir, S.; Badshah, S.L.; Khan, N.; Ghani, L.; Poulson, B.G.; Emwas, A.H.; Jaremko, M. Important flavonoids and their role as a therapeutic agent. *Molecules* **2020**, *25*, 5243. [CrossRef]

11. Chen, P.X.; Tang, Y.; Zhang, B.; Liu, R.; Marcone, M.F.; Li, X.; Tsao, R. 5-Hydroxymethyl-2-furfural and derivatives formed during acid hydrolysis of conjugated and bound phenolics in plant foods and the effects on phenolic content and antioxidant capacity. *J. Agric. Food Chem.* **2014**, *62*, 4754–4761. [CrossRef]
12. Nemitz, M.C.; Moraes, R.C.; Koester, L.S.; Bassani, V.L.; von Poser, G.L.; Teixeira, H.F. Bioactive soy isoflavones: Extraction and purification procedures, potential dermal use and nanotechnology-based delivery systems. *Phytochem. Rev.* **2015**, *14*, 849–869. [CrossRef]
13. Chang, K.H.; Jo, M.N.; Kim, K.T.; Paik, H.D. Evaluation of glucosidases of *Aspergillus niger* strain comparing with other glucosidases in transformation of ginsenoside Rb1 to ginsenosides Rg3. *J. Ginseng Res.* **2014**, *38*, 47–51. [CrossRef]
14. Feng, C.; Jin, S.; Xia, X.X.; Guan, Y.; Luo, M.; Zu, Y.G.; Fu, Y.J. Effective bioconversion of sophoricoside to genistein from *Fructus sophorae* using immobilized *Aspergillus niger* and yeast. *World J. Microbiol. Biotechnol.* **2015**, *31*, 187–197. [CrossRef]
15. Gaya, P.; Peirotén, Á.; Medina, M.; Landete, J.M. Isoflavone metabolism by a collection of lactic acid bacteria and bifidobacteria with biotechnological interest. *Int. J. Food Sci. Nutr.* **2016**, *67*, 117–124. [CrossRef]
16. Singhvi, M.S.; Zinjarde, S.S. Production of pharmaceutically important genistein and daidzein from soybean flour extract by using β-glucosidase derived from *Penicillium janthinellum* NCIM 1171. *Process Biochem.* **2020**, *97*, 183–190. [CrossRef]
17. Liu, F.; Huang, K.; Zheng, A.; Xiao, F.S.; Dai, S. Hydrophobic solid acids and their catalytic applications in green and sustainable chemistry. *ACS Catal.* **2018**, *8*, 372–391. [CrossRef]
18. Zhou, Y.; Noshadi, I.; Ding, H.; Liu, J.; Parnas, R.S.; Clearfield, A.; Xiao, M.; Meng, Y.; Sun, L. Solid acid catalyst based on single-layer α-zirconium phosphate nanosheets for biodiesel production via esterification. *Catalysts* **2018**, *8*, 17. [CrossRef]
19. Peña, L.; Xu, F.; Hohn, K.L.; Li, J.; Wang, D. Propyl-sulfonic acid functionalized nanoparticles as catalyst for pretreatment of corn stover. *J. Biomater. Nanobiotechnol.* **2014**, *5*, 8–16. [CrossRef]
20. Qi, W.; He, C.; Wang, Q.; Liu, S.; Yu, Q.; Wang, W.; Leksawasdi, N.; Wang, C.; Yuan, Z. Carbon-based solid acid pretreatment in corncob saccharification: Specific xylose production and efficient enzymatic hydrolysis. *ACS Sustain. Chem. Eng.* **2018**, *6*, 3640–3648. [CrossRef]
21. Peña, L.; Ikenberry, M.; Ware, B.; Hohn, K.L.; Boyle, D.; Sun, X.S.; Wang, D. Cellobiose hydrolysis using acid-functionalized nanoparticles. *Biotechnol. Bioprocess Eng.* **2011**, *16*, 1214–1222. [CrossRef]
22. Carlier, S.; Hermans, S. Highly efficient and recyclable catalysts for cellobiose hydrolysis: Systematic comparison of carbon nanomaterials functionalized with benzyl sulfonic acids. *Front. Chem.* **2020**, *8*, 347. [CrossRef]
23. Singhvi, M.S.; Deshmukh, A.R.; Kim, B.S. Cellulase mimicking nanomaterial-assisted cellulose hydrolysis for enhanced bioethanol fermentation: An emerging sustainable approach. *Green Chem.* **2021**, *23*, 5064–5081. [CrossRef]
24. Wang, D.; Ikenberry, M.; Pe, L.; Hohn, K.L. Acid-functionalized nanoparticles for pretreatment of wheat straw. *J. Biomater. Nanobiotechnol.* **2012**, *3*, 342–352. [CrossRef]
25. Ingle, A.P.; Philippini, R.R.; de Souza Melo, Y.C.; da Silva, S.S. Acid-functionalized magnetic nanocatalysts mediated pretreatment of sugarcane straw: An eco-friendly and cost-effective approach. *Cellulose* **2020**, *27*, 7067–7078. [CrossRef]
26. Rajkumari, K.; Kalita, J.; Das, D.; Rokhum, L. Magnetic Fe_3O_4@ silica sulfuric acid nanoparticles promoted regioselective protection/deprotection of alcohols with dihydropyran under solvent-free conditions. *RSC Adv.* **2017**, *7*, 56559–56565. [CrossRef]
27. Melero, J.A.; Stucky, G.D.; van Grieken, R.; Morales, G. Direct syntheses of ordered SBA-15 mesoporous materials containing arenesulfonic acid groups. *J. Mater. Chem.* **2002**, *12*, 1664–1670. [CrossRef]
28. Walker, J.M. The bicinchoninic acid (BCA) assay for protein quantitation. *Methods Mol. Biol.* **1994**, *32*, 5–8. [CrossRef]
29. Maharjan, A.; Singhvi, M.; Kim, B.S. Biosynthesis of a therapeutically important nicotinamide mononucleotide through a phosphoribosyl pyrophosphate synthetase 1 and 2 engineered strain of *Escherichia Coli*. *ACS Synth. Biol.* **2021**, *10*, 3055–3065. [CrossRef]
30. Tanuraghaj, H.M.; Farahi, M. Preparation, characterization and catalytic application of nano-Fe_3O_4@ SiO_2@$(CH_2)_3OCO_2Na$ as a novel basic magnetic nanocatalyst for the synthesis of new pyranocoumarin derivatives. *RSC Adv.* **2018**, *8*, 27818–27824. [CrossRef]
31. Safari, J.; Zarnegar, Z. A magnetic nanoparticle-supported sulfuric acid as a highly efficient and reusable catalyst for rapid synthesis of amidoalkyl naphthols. *J. Mol. Catal. A Chem.* **2013**, *379*, 269–276. [CrossRef]
32. Colilla, M.; Izquierdo-Barba, I.; Sánchez-Salcedo, S.; Fierro, J.L.; Hueso, J.L.; Vallet-Regí, M. Synthesis and characterization of zwitterionic SBA-15 nanostructured materials. *Chem. Mater.* **2010**, *22*, 6459–6466. [CrossRef]
33. Alvaro, M.; Corma, A.; Das, D.; Fornés, V.; García, H. "Nafion"-functionalized mesoporous MCM-41 silica shows high activity and selectivity for carboxylic acid esterification and Friedel–Crafts acylation reactions. *J. Catal.* **2005**, *231*, 48–55. [CrossRef]
34. Ramanathan, G.; Banupriya, S.; Abirami, D. Production and optimization of cellulase from *Fusarium oxysporum* by submerged fermentation. *J. Sci. Ind. Res.* **2010**, *69*, 454–459.
35. De Almeida, M.N.; Falkoski, D.L.; Guimarães, V.M.; de Rezende, S.T. Study of gamba grass as carbon source for cellulase production by *Fusarium verticillioides* and its application on sugarcane bagasse saccharification. *Ind. Crops Prod.* **2019**, *133*, 33–43. [CrossRef]
36. Izumi, T.; Piskula, M.K.; Osawa, S.; Obata, A.; Tobe, K.; Saito, M.; Kataoka, S.; Kubota, Y.; Kikuchi, M. Soy isoflavone aglycones are absorbed faster and in higher amounts than their glucosides in humans. *J. Nutr.* **2000**, *130*, 1695–1699. [CrossRef]
37. Hu, S.; Wang, D.; Hong, J. A simple method for beta-glucosidase immobilization and its application in soybean isoflavone glycosides hydrolysis. *Biotechnol. Bioprocess Eng.* **2018**, *23*, 39–48. [CrossRef]

38. Mei, J.; Chen, X.; Liu, J.; Yi, Y.; Zhang, Y.; Ying, G. A biotransformation process for production of genistein from sophoricoside by a strain of *Rhizopus oryza*. *Sci. Rep.* **2019**, *9*, 6564. [CrossRef]
39. Doan, D.T.; Luu, D.P.; Nguyen, T.D.; Hoang Thi, B.; Pham Thi, H.M.; Do, H.N.; Luu, V.H.; Pham, T.D.; Than, V.T.; Thi, H.H.P.; et al. Isolation of *Penicillium citrinum* from roots of *Clerodendron cyrtophyllum* and application in biosynthesis of aglycone isoflavones from soybean waste fermentation. *Foods* **2019**, *8*, 554. [CrossRef]
40. Wang, H.; Covarrubias, J.; Prock, H.; Wu, X.; Wang, D.; Bossmann, S.H. Acid-functionalized magnetic nanoparticle as heterogeneous catalysts for biodiesel synthesis. *J. Phys. Chem. C* **2015**, *119*, 26020–26028. [CrossRef]

catalysts

Review

Nano-Biochar as a Sustainable Catalyst for Anaerobic Digestion: A Synergetic Closed-Loop Approach

Lalit Goswami [1,†], Anamika Kushwaha [1,†], Anju Singh [2], Pathikrit Saha [1], Yoseok Choi [1], Mrutyunjay Maharana [3], Satish V. Patil [4] and Beom Soo Kim [1,*]

[1] Department of Chemical Engineering, Chungbuk National University, Cheongju 28644, Korea; lalitgoswami660323@gmail.com (L.G.); kushwaha.anamika@gmail.com (A.K.); pathikritsaha89@gmail.com (P.S.); y.choi@chungbuk.ac.kr (Y.C.)
[2] Department of Chemical Engineering, Babu Banarsi Das National Institute of Technology and Management, Lucknow 227105, India; anjusinghch1711@gmail.com
[3] School of Electrical Engineering, Xi'an Jiaotong University, Xi'an 710049, China; maharana@xjtu.edu.cn
[4] School of Life Sciences, Kavayitri Bahinabai Chaudhari North Maharashtra University, Jalgaon 425001, India; satish.patil7@gmail.com
* Correspondence: bskim@chungbuk.ac.kr; Tel.: +82-43-261-2372
† These authors contributed equally to this work.

Abstract: Nowadays, the valorization of organic wastes using various carbon-capturing technologies is a prime research area. The anaerobic digestion (AD) technology is gaining much consideration in this regard that simultaneously deals with waste valorization and bioenergy production sustainably. Biochar, a well-recognized carbonaceous pyrogenic material and possessing a broad range of inherent physical and chemical properties, has diverse applications in the fields of agriculture, health-care, sensing, catalysis, carbon capture, the environment and energy. The nano-biochar-amended anaerobic digestion approach has intensively been explored for the past few years. However, an inclusive study of multi-functional roles of biochar and the mechanism involved for enhancing the biogas production via the AD process still need to be evaluated. The present review inspects the significant role of biochar addition and the kinetics involved, further focusing on the limitations, perspectives, and challenges of the technology. Additionally, the techno-economic analysis and life-cycle assessment of biochar-aided AD process for the closed-loop integration of biochar and AD and possible improvement practices are discussed.

Keywords: biochar-amended process; mechanism involved; kinetics; techno-economic analysis; zero-waste approach

Citation: Goswami, L.; Kushwaha, A.; Singh, A.; Saha, P.; Choi, Y.; Maharana, M.; Patil, S.V.; Kim, B.S. Nano-Biochar as a Sustainable Catalyst for Anaerobic Digestion: A Synergetic Closed-Loop Approach. *Catalysts* **2022**, *12*, 186. https://doi.org/10.3390/catal12020186

Academic Editor: Keith Hohn

Received: 30 December 2021
Accepted: 30 January 2022
Published: 1 February 2022

1. Introduction

Presently, the human population has crossed 7.2 billion and is expected to reach between 9.6–12.3 billion by 2100 [1]. This tremendous population growth is further accompanied by enormous industrial development and unprecedented consumption of energy, enhancing the stress on natural resources at a startling level [2,3]. To meet this rising demand, over-exploitation of fossil-based energy is occurring, with deleterious environmental and societal impacts [4–8]. Renewable energy is recognized as a sustainable option to overcome all these challenging issues. Sources such as wind, hydro, geothermal, solar energy, biogas, microbial fuel cells, bioethanol, biodiesel, biohydrogen, etc. have been explored to find a viable solution [9]. Amidst these available options, some utilize silver, titanium, platinum, ruthenium, nickel, and other metal oxides as catalysts on a huge scale [10,11]. Though these metals are very efficient, there is a hunt for sustainable catalyst materials that are cost-effective, efficient, eco-friendly, and widely available.

Biomass is currently the most sustainable option available, and has been widely explored for synthesizing various sustainable materials such as carbon fibers, biochar, activated carbon, graphene, etc. that show tremendous applicability in the energy sector [12,13].

Biochar is a carbonaceous material that is produced via the thermochemical decomposition of organic materials [14]. It is a highly porous, amorphous material with a good surface area containing various functional groups, while displaying stable physicochemical properties and biocompatibility, and easy to further modify in accordance with the particular need [15,16]. The properties of biochar for various application such as bioremediation, energy storage, catalysis, agriculture, carbon capture, wastewater treatment, pharmaceutical, electrodes, cosmetics, etc. rely on the production process, kind of feedstock and operating parameters used [17–21].

During the past decade, biochar has also been utilized in the anaerobic digestion (AD) process. The AD process has ability to use organic biomass and wastes for the production of biogas (containing ~60% methane) and high quality of bio-fertilizers [22,23]. This conversion is purely dependent on to the synergistic metabolic activities of the prevailing microbial consortia within the digester and has to be further maintained under steady state conditions for the best performance. Various electron transfers amongst the similar partners are required to avoid the longer acclimatization period along with the high substrate consumption rate [24]. The direct interspecies electron transfer (DIET) is recognized as more rapid and stable pathway where the transfer of electrons takes place between the syntrophic bacteria to the methanogenic archaea [25].

Recently, researchers have started focusing on non-biological conductive materials such as magnetite, biochar, granular activated carbon, etc. to enhance the DIET performance. Biochar has the ability to enhance the DIET via a conduction-based mechanism that channels the electron flow between the electron-donor and electron-acceptor ends [22]. In addition, biochar supplementation leads to a simple and efficient microbial community possessing the enriched and equilibrated DIET. Biochar-aided anaerobic digestion mediates the formation and degradation of intermittent acids and leads to the enrichment of methanogenic archaea, shortening the lag phase, and enhancing the methane yield [26].

Henceforth, the present review aims to summarize the recent advancements regarding the utilization of nano-biochar in anaerobic digestion processes. This review's focus is on the application of nano-biochar as a sustainable nano-catalyst for the production of renewable energy, particularly by anaerobic digestion. Here, we have considered the biochar according to its particle size, i.e., macro-biochar (>1 μm), colloidal biochar (1 μm–100 nm), and nano-biochar (<100 nm). The role, kinetics, mechanism(s) involved, and possible improvements along with the closed-loop integration of nano-biochar and AD have also been discussed. The review also covers the involvement of techno-economic and environmental life-cycle assessments for moving forward with the least limitations. A network visualization of terms associated with anaerobic digestion and biochar with at least 10 occurrences of the associated keywords is represented in Figure 1. It depicts the present trends in research and development regarding the application of biochar in association with the anaerobic digestion in the Web of Science. Here, the various colors of the nodes represent the different clusters whereas the size of each bubble depicts its frequency of occurrence.

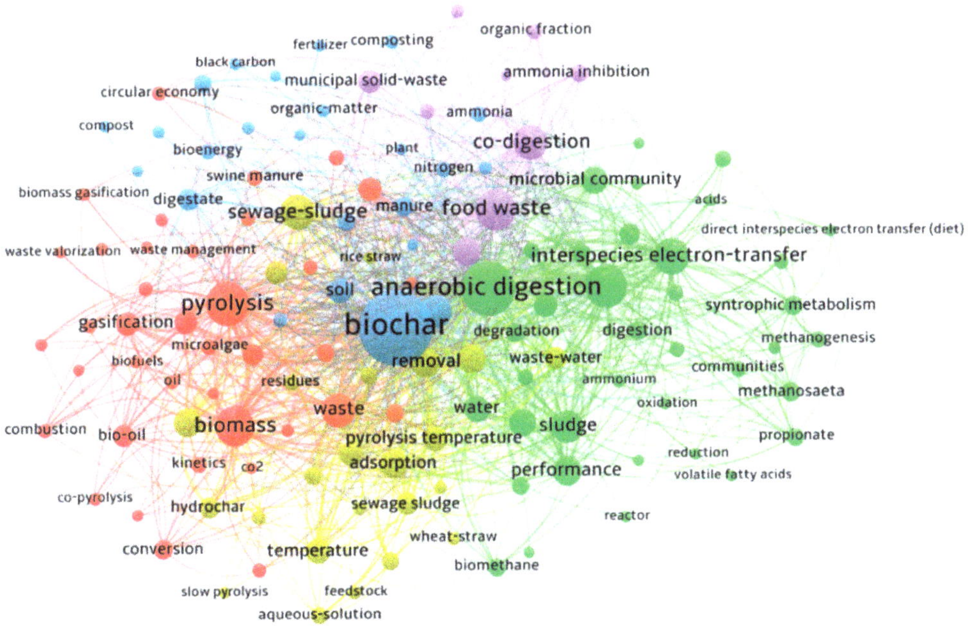

Figure 1. Network visualization of terms associated with anaerobic digestion and biochar.

2. Application of Nano-Biochar in Renewable Energy

Conventionally, biochar has been utilized for soil amendment and bioremediation applications. In this section, we discuss the recent advances in various techniques for biochar application as a catalyst in enhancing renewable energy production. Figure 2 illustrates the emerging applications of biochar for renewable energy. The biomass-derived nano-biochar can be utilized as an electrode in microbial fuel cells (MFCs) and as a catalyst for improving biodiesel and hydrogen generation. The applications of nano-biochar are very dependent on its physicochemical properties such as the biomass composition, biomass-conversion technologies and conditions, pH buffering capability, the presence of various trace elements, etc. [27]. Nano-biochar used in MFC creates a favorable environment for microbial growth and biofilm formation. Further, nano-biochar possessing higher surface area, porosity, and functional groups is more appropriate for microbial film formation, leading to electricity generation in MFC. Owing to the heterogeneous nature of nano-biochar, several techniques are utilized nowadays for its activation for nano-biochar to act as catalyst in a more effective, economical, and reutilizable mode.

2.1. Nano-Biochar for Microbial Fuel Cell

MFCs might be a viable solution for the global energy concerns, containing anodic and cathodic chambers that are further separated via a proton exchange membrane and utilizing microbes as a catalyst for converting chemical energy into electrical energy. Microbes utilize organic matter for their metabolic activities, while simultaneously releasing numerous intermediate metabolites that undergo redox reactions to generate electrons and protons [28,29]. The electrons generated in the anodic chamber under anaerobic conditions move towards the cathode while the proton moves towards the cathode via a proton exchange membrane. Several microorganisms, such as *Shewanella*, *Clostridium*, *Rhodospirillum*, etc., have already been reported in relation to the MFC applications [28]. Here, the bio-electricity generated depends on numerous factors, viz., substrate, rate of electron transfer, electrode performance, rate of oxygen reduction, and external operating conditions. In addition, the performance of the electrode material depends on its nature of physical and chemical properties. During the upscaling of MFCs, low current output and high cost are the major limitations [30]. The electrodes represent 20–50% of the overall cost of a MFC as they are usually made of non-renewable materias, viz., stainless steel, Ni, Cu, etc. These materials further require surface modifications for biofilm formation and electron transfer [31]. Table 1(a) summarizes the utilization of nano-biochar for MFCs. Further, researchers are utilizing sediment MFCs, also known as benthic MFCs (in some cases), applied in the natural systems such as constructed wetland [32]. The energy output from such MFCs is generally very low (e.g., 10–50 mW cm^{-2}). Coconut shell-derived biochar-amended sediment MFC improved the power generation 2–10 times along with the increased total organic carbon (TOC) removal [33]. Also, soil-based MFCs have good performance in low-power continuous energy sources along with the soil remediation application [34]. Li et al. [35] utilized a chicken manure, wheat straw and wood sawdust-derived biochar-amended soil MFC for the biodegradation of petroleum hydrocarbons.

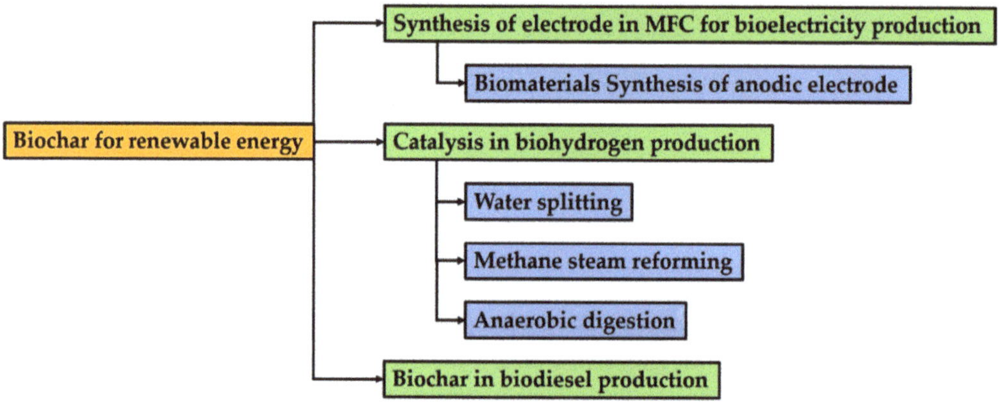

Figure 2. Utilization of biochar for renewable energy production.

Table 1. Application of nano-biochar as catalysts for the production of different renewable energy. (**a**) Nano-biochar for MFC for bioelectricity production; (**b**) Nano-biochar for anaerobic digestion for hydrogen production; (**c**) Nano-biochar for biodiesel production.

(**a**)				
As Anode				
Biomass	**Preparation**	**Comments**	**Power Density**	**References**
Chestnut shell	900 °C, 2 h	Activation with KOH modified microporous structure with reduced O and N content that is beneficial for charge transfer and microbial adhesion	23.6 Wm^{-3}	[36]
Microalgal sludge (MSB)	800 °C, 2 h	Cobalt and chitosan were used as a mediator for electron transfer by immobilization on MSB (MSB/Co/chitosan).	3.1 $mWcm^{-2}$	[37]
Microalgal	900 °C, 1 h	Contains intrinsic N and P	12.8 Wm^{-3}	[38]
Deoiled *Azolla* biomass	600 °C, 3 h	Nano-biochar was activated with KOH at 1:4 ratio at 600 °C for 2 h. Bio-electrode was prepared by using 5% polyvinylidene fluoride (PVDF).	-	[39]

Table 1. *Cont.*

		(a)		
		As Anode		
Biomass	**Preparation**	**Comments**	**Power Density**	**References**
		As Cathode		
Bamboo charcoal	Carbonization at 900 °C under N_2 atmosphere followed by heating at 350 °C under air atmosphere for 2 h	Porous structure of the bamboo derived cathode provides possible channels for oxygen supply and proton transport.	40.4 Wm^{-3}	[40]
Corn cob	650 °C for 2 h	Higher contents of graphitic and pyridinic nitrogen accelerate electron transfer.	458.8 mWm^{-3}	[41]
Balsa wood biochar	800 °C for 1 h	Biochar can be used directly without using the binders and catalysts.	72 mWm^{-2}	[42]
Water hyacinth	900 °C for 2 h	Capable of transferring electrons	24.7 mWm^{-2}	[43]
Eggplant	Pre-treated with $K_3[Fe(C_2O_4)_3]$ and pyrolyzed at 800 °C for 1 h	Possesses hierarchical porous structure with a large specific surface area and high graphitization degree	667 mWm^{-2}	[44]

		(b)		
Biomass	**Synthesis**	**Comment**	**Productivity**	**References**
Corncob	Corncob particles mixed with melamine heated at 121 °C for 2 h. Material soaked with $ZnCl_2$ and pyrolyzed at 700 °C for 3 h	Fabricated biochar promoted the growth of dominant bacteria and the electron transfer rate.	230 mL g^{-1}	[45]
Cornstalk	Pyrolyzed at 600 °C for 2 h	Biochar promotes cellulolytic enzymes activity and leads to increased substrate conversion into hydrogen.	286.1 mL g^{-1}	[46]
Sewage sludge	Pyrolyzed at 600 °C for 3 h	Phosphate-laden biochar used with Ca- and Mg-saturated resin. Both facilitated substrate degradation and reduces the lag phase.	197 mL g^{-1}	[47]
Timber sawdust	Calcinised at 500 °C for 2 h	Biochar along with Fe acts synergistically and results in enhancing the growth and activity of microbes and the utilization of substrate.	50.6 mL g^{-1}	[48]
Woody biomass	Pyrolysis at 400–500 °C	Co-culture of *Enterobacter aerogenes* and *E. coli* was used. Biochar mitigates ammonia inhibitory effect and facilitates biofilm formation for efficient colonization and reduces the lag phase.	96.6 mL g^{-1}	[49]

Table 1. *Cont.*

		(a)		
		As Anode		
Biomass	**Preparation**	**Comments**	**Power Density**	**References**
		(c)		
Biomass	**Oil**	**Catalyst Preparation**	**Yield (%)**	**References**
Peanut shell	Algal oil	400 °C for 1 h followed by H_2SO_4 treatment	94.9	[50]
Brown algae	Waste cooking oil	900 °C for 4 h, calcified with CaO and K_2CO_3 at 500 °C for 3 h	98.8	[51]
Sludge biochar	Palm oil	800 °C for 30 min calcined with CaO at 700 °C	93.7	[52]
Sugarcane bagasse	Palm oil	400 °C for 2 h, sulfonated with $ClSO_3H$ at 300 °C for 5 h	98.6	[53]
Cork biochar	Waste cooking oil	600 °C for 2 h and sulphonated with H_2SO_4 at 100 °C for 10 h	98	[54]

In recent times, bioinspired carbonaceous materials, viz. graphene, biochar, carbon fibers, carbon nanotubes, etc., are gaining much consideration owing to their biocompatible nature for biofilm formation and microbial growth along with high surface area and conductivity. Here, nano-biochar is the renewable carbon-based material produced even by waste-organic materials for electrode production. Various biomass sources have been examined for anode/cathode preparation for efficient biofilm formation, and the properties are further dependent on several factors, i.e., pore size, available surface area, and surface properties. The pyrolyzed carbonaceous heteroatoms function as a natural dopant, delivering admirable electrical conductivity. In addition to the generation of bio-electricity, MFC is also utilized for wastewater treatment and produces low amounts of sludge compared to the traditional anaerobic digestion techniques [31]. For the cathodic performance of the carbonaceous materials, the surface area alone is a weak indicator as the existence of increased acidic functional groups regulates the oxygen reduction capability of the cathode [1].

2.2. Nano-Biochar for Hydrogen Production

Nano-biochar has also been utilized for biohydrogen production via three different processes, namely water splitting, methane steam reforming and anaerobic digestion. A brief discussion regarding all the mentioned technologies is presented in the following section. Table 1(b) summarizes the utilization of nano-biochar for hydrogen generation.

2.2.1. Nano-Biochar for Water Splitting

Water splitting using numerous electro-catalysts of the noble metals and their oxides, e.g., RuO_2 and IrO_2, is the cleanest way of hydrogen production, however, the process is still inefficient owing to the high catalyst cost, instability in the alkaline environment, and overpotential of the H_2 evolution reaction (HER) and O_2 evolution reaction (OER) at the cathode and anode, respectively [55,56]. Therefore, bio-inspired carbonaceous materials are nowadays being explored as electrocatalysts due to their cost-efficient electrical conductivity [1]. Further, to reduce the overpotential and enhance the hydrogen generation ability, transition metals can be doped on carbonaceous materials to create more active sites for rapid electron transference [57]. In addition, the presence of alkali and alkaline earth metals assists in the carbon and porosity activation through the ionic migration effect at the high temperature. Recently, biochar-derived molybdenum carbide (Mo_2C) has gained consideration owing to its Pt-similar stability and structure. Still, the HER efficacy of Mo_2C

electro-catalysts is less because it lacks exposed catalytic sites along with strong Mo-H bonding [1,56].

2.2.2. Nano-Biochar for Methane Steam Reforming

Methane can be generated via the anaerobic digestion of the organic wastes that present the maximum hydrogen to carbon ratio (4:1). Further, the thermo-catalytic decomposition of methane leads to the generation of pure hydrogen at a very high temperature (1200 °C) for the conversion reaction initialization. Thus, metallic catalysts (Ni, Fe, Cu, etc.) are involved in enhancing the methane conversion at a lower temperature. However, these metallic catalysts lose their activity very fast. In addition, the presence of sulphur in the natural atmosphere is further harmful to the catalysts [1]. Carbon materials have greater stability and are resilient to sulphur content. Carbonaceous materials doped with metals in trace amount present enhanced activity owing to the generation of high energy active sites that attract the methane, further helping in the enhanced methane conversion to hydrogen. Nowadays, the nano-biochar prepared from bio-solids recovered from wastewater treatment plants is utilized for the methane steam reforming process resulting in 65% conversion [58].

2.2.3. Nano-Biochar for Biogas Production

Organic bio-waste management using an AD process leads to bioenergy generation. It was stated that the nano-biochar addition during the AD enhances the hydrogen yield with a minimal lag phase. It is predicted that the nano-biochar enables biofilm formation and pH stabilization, and enhances the generation of volatile fatty acids (VFAs) [59]. Further, the minerals existing in the nano-biochar are responsible for providing supplements for microbial metabolism and enzyme synthesis and activity. Biochar-derived from pine dust during the two-phase AD of aqueous carbohydrates resulted in enhanced methane and hydrogen yield by 10% and 31%, respectively [60]. Furthermore, microbes may acclimatize to numerous inhibitors, but that might be time-consuming and affect cellular productivity. The nano-biochar aided AD process tends to remove the inhibitors and reduce the toxicity, thus improving in hydrogen production [1].

2.3. Nano-Biochar for Biodiesel Production

The occurrence of long carbon chain (C_{14}–C_{20}) fatty acids has gained the attention for biodiesel production for the present engines owing to their high energy densities. The transesterified oils from various renewable resources produce biodiesel [61]. During the transesterification process, catalysts playing a major role are further categorized into two types, i.e., homogenous and heterogeneous catalysts. Heterogeneous catalysts can simultaneously perform transesterification and esterification reactions. Table 1(c) lists examples of the utilization of nano-biochar as catalyst for biodiesel production.

Biochar, defined as a heterogeneous catalyst, has clearly shown its potential for biodiesel production. Transesterification reactions were carried out in the presence of nano-biochar that was altered via either acid/alkali as the catalyst. The porosity of the nano-biochar permits the reactants easy access to active sites to enable the transesterification reaction [1]. In addition, the hydrophobic surface of nano-biochar assists in the water elimination during the conversion reaction. The biochar acid-modified via the sulfonation process contains -SO_3H groups on the biochar surface and further acts as the catalyst, while the alkali-modified biochar has basic sites primarily consisting of calcium, potassium, or sodium oxides that are produced during the calcination of the minerals present in the organic substrate [51–53]. During the transesterification reaction, the alcohols and lipids utilize the porous structure to accelerate the reactants' collision frequency at ambient pressure.

3. Nano-Biochar for Anaerobic Digestion

Depending on the conditions, the nano-biochar characteristics can be enhanced for application. Nano-biochar characteristics such as porosity, specific surface area (SSA), cation exchange capacity, electrical conductivity, redox potential, pH, and functional groups play an essential role during the AD process [62].

3.1. Porosity

Nano-biochar porosity is a crucial factor in recognizing the probable association with microbes during AD. The biochar pore size acts as microhabitats for microbes to flourish [63]. The typical microbial size in AD is 0.3–13 μm for bacteria/archaea, 2–80 μm for fungus, and 7–30 μm for protozoa [56]. In addition, the porosity of biochar enables the formation of biofilm, acting as a protection for the amelioration of selective and effective microbes participating in the AD system in an acidic environment [64]. The addition of nano-biochar selectively enhances the numerous bacterial species in the AD system. Most research has documented *Methanolinea*, *Methanobacterium*, and *Methanosarcina* sp. in anaerobic digester supplemented with biochar. Numerous studies reported the spatial dispersion of archaea and bacteria by separating the sludge into several portions [65]. The spatial dispersal of methanogens in the biochar pores is due to their diverse size and morphology [66].

3.2. Specific Surface Area

The SSA of nano-biochar is one of the crucial parameters for environmental pollutants adsorption [63]. The nano-biochar capacity to remove CO_2, H_2S, etc. in a biogas fermenter was studied by Sethupathi et al. [67]. They observed values of 0.208 mmol g^{-1} and 0.126 mmol g^{-1} for CO_2 and H_2S, respectively. The adsorption of CO_2 by biochar (hickory wood and bagasse), the involvement of high SSA and N_2 group of biochar for removal was reported by Creamer et al. [68]. The pore size of biochar (0.5–0.8 nm) and enhanced SSA were reported for sequestration of CO_2 [69], whereas the group I-II A metals and primary functional groups present on the surface play an essential role [70].

3.3. Cation Exchange Capacity

The concentration of ammonium (NH_4^+) and NH_3 are sustained under an optimized threshold for an AD process; it helps in buffer capacity for improved growth of bacteria, whereas the excess amount of total NH_3/free NH_3-N (TAN/FAN) can cause AD catastrophe [71]. It was observed that TAN content (1.7 and 14 g L^{-1}) decreased the generation of CH_4 by 50% and free ammonium nitrogen (FAN) (150 and 1200 mg L^{-1}) content significantly influences the growth of methanogens [71,72]. Biochar could significantly improve the inhibition of NH_3 and increase CH_4 generation by decreasing the microbial lag phase owing to its strong cation exchange capacity (CEC). Shen et al. [73] reported the positive role of biochar in AD of sewage sludge. Su et al. [74] stated that biochar alleviated the NH_3–N (~1500 mg L^{-1}) in food waste AD. Likewise, Lü et al. [66] stated that enhanced NH_4+ content (up to 7 g-N L^{-1}) could be repressed by biochar supplementation during AD. The mechanisms include microbial immobilization, physicochemical adsorption capacity, CEC, surface functional groups (SFG), and DIET advancement [73–75].

3.4. Electrical Conductivity

The microbial syntrophic interaction depends on the electrical conductivity (EC) [63]. However, the EC of nano-biochar is insignificant in contrast to the EC of digestate, which is based on the microbes' metabolism and composition [76]. Nano-biochars' ability to enhance DIET is equal to that of granular activated carbon, despite the lower biochar EC [77] although some nano-biochar has high EC (e.g., basswood nano-biochar) [78]. Barua and Dhar [79] reported increased EC (0.2–36.7 μS cm^{-1}) in numerous microbes due to DIET. Martins et al. [76] proposed that humic substances could act as electron transport media to accelerate DIET.

3.5. *pH*

The pH influences the conductivity of nano-biochar and associated microbes in AD [80]. Due to the ash concentration and acidic functional groups volatilization, the pH values of nano-biochar are alkaline. The rise in biochar's pyrolysis temperature and pH value frequently surges [81]. In addition, the nano-biochar has redox ability, i.e., it can accept or donate the electrons. Further, the presence of aromatics and phenolic groups also aids in the electron transfer [3,82]. Nano-biochar increases the AD alkalinity (minimum pH $\geq$ 6), enhancing the microbial activity for fast CH_4 generation and adaptability to initial the loading shock [83]. Nano-biochar significantly facilitates the methanogenesis stage under acidic conditions (pH 5.3), improving the operating conditions with increased organic loading and total solids [81]. Hence, with the addition of nano-biochar, a continuous AD process can function with a shorter hydraulic retention time and operate under an extra organic loading rate.

3.6. *Surface Functional Groups*

The nano-biochar surface consists of various functional groups such as –OH, C–O, –COOH, CO, $-NH_x$, etc. which assist in nutrient retaining and pollutant removal [84]. Nano-biochar displayed significant outcomes in the adsorption of NH_3 from wastewater and digestate. The nano-biochar's porosity and high SSA aid in the physisorption [85]. However, in some studies, it is not a predominant parameter in adsorption of NH_4+ [86,87]. For instance, amid NH_4+ and nano-biochar acidic functional groups, ion exchange occurs [85], and CEC plays a significant role in enhancing the adsorption capacity of biochar's towards NH_4+ [88]. Sahota et al. [89] employed biochar to remove H_2S from biogas and attained 84.2% elimination efficacy. Likewise, Kanjanarong et al. [90] accomplished 98% H_2S elimination efficacy via biochar and concluded that COOH and OH groups are responsible for the observed H_2S adsorption.

3.7. *Redox Potential*

The nano-biochar redox properties are critical during AD. The biochar redox properties are due to its SFG, free radicals, and metals (M) and metal oxides (MO) [91]. For example, the electron-donating capacity is due to the presence of phenolic C–OH fractions, while electron-accepting ability is due to quinoid C=O fractions [92]. An oxidation process can enhance nano-biochar SFG [93]. Hitherto, the oxidation process should be appropriate to familiarize with novel functioning, but not too robust to trigger the alteration to the COOH group (redox-inactive) or even to eliminate CO_2 [91]. Free radicals influence the nano-biochar redox propensity, such as aryl radicals or as semi-quinoid radicals [94]. Concerning the nano-biochar's inorganic constituent, Fe and Mn oxides (redox-active metals) usually exist in the biomass and diverse oxidation states act as electron acceptors and donors [95].

4. Roles and Mechanisms of Nano-Biochar in Anaerobic Digestion

The intrinsic characteristic of nano-biochar boosts biofilm microbial development (methanogens colonization) and adsorption of NH_3 and acetate (inhibitors) [96]. AD process aids in the development of a defensive layer for microbes that enhances CH_4 generation. Nano-biochar also stabilizes the microbe's nutrient access and eliminates volatile fatty acids and NH_3 [97]. Figure 3 presents a brief schematic of the anaerobic digestion system aided with the nano-biochar for enhanced biogas production.

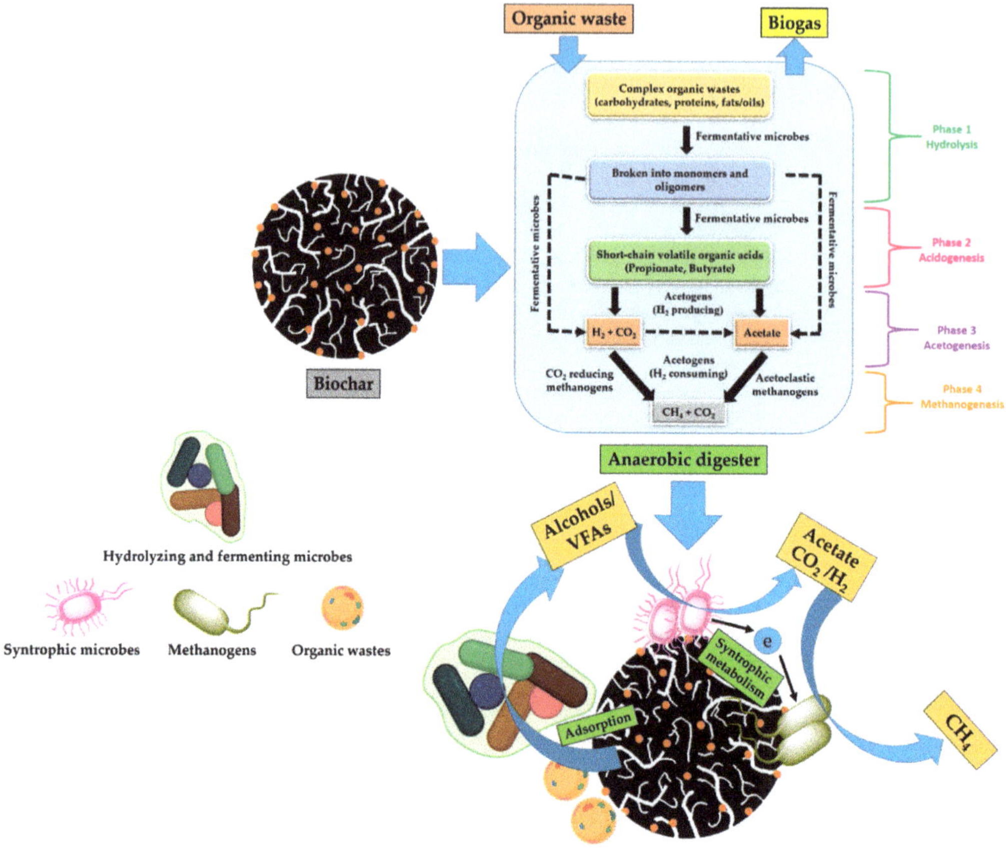

Figure 3. Anaerobic digestion system aided with the nano-biochar for enhanced biogas production.

4.1. Improving the Process Stability

AD stability is crucial for a continuous conversion of biowaste, which can be enhanced by nano-biochar usage via NH_3–N improvement [75]. Nano-biochar improves the methanogenic microbes in the presence of acid and NH_3 impediment, which ultimately helps in nitrogen-rich substrate degradation and reduces NH_3 inhibition, thereby improving AD performance [97]. The nano-biochar addition decreases the AD major inhibitor (free NH_3) by 10.5% and encourages the methanogenesis process under an acidic environment [98]. Nano-biochar significantly increases the AD alkalinity (pH $\geq$ 6), therefore supporting the acclimatization of microbes and enriching their activity in the presence of organic loading for more substantial production of CH_4 [99,100]. However, studies reported the toxic effects of the high biochar concentration (~4.5 g biochar/g of dry sludge), such as decreased microbial activity [98]. Therefore, optimizing the nano-biochar amount and incessant monitoring are crucial to curtailing the adverse effects on the metabolism of microbes and the intermediate metabolite generation.

4.2. Accelerating the Process Rate

AD system process rates can be efficiently enhanced with the addition of nano-biochar. Shanmugam et al. [101] observed a decrease (24 h) in the methanogenic microbe lag phase when algal-derived biochar was used for wastewater AD. Earlier results supported this outcome that the microbial lag phase duration was inversely proportional to the biochar to substrate ratio. The 7.5 days lag phase was the optimized ratio [102]. Likewise, Sunyoto et al. [60] reported 21.4–35.7% and 41–45% decreases in the lag phase of H_2 and CH_4 reactors on biochar, respectively. The magnetic biochar addition during methanogenesis entirely avoided the lag phase or decreased it by 0.9–1.83 days [103,104]. Furthermore, Wang et al. [105] reported a reduction of lag phase from 4.7 to 1.8–3.9 days during co-digestion of food waste and sewage sludge on biochar application. The daily production of CH_4 was improved by 136% after the amendment of biochar (wood-pellet-derived) during the AD process [97].

4.3. Buffering Potency and Alkalinity

The efficiency of AD systems mainly relies on the pH value, where a slight decline in the pH of the solution remarkably hinders microbial growth and functioning [106]. Food wastes with low C:N ratio and enhanced biodegradability result in a rapid acidification rate during AD. During the acidogenesis phase, the fast growth of acidogenic bacteria will affect the activity of methanogen's and result in the VFAs accumulation [107]. Fotidis et al. [108] established that the aforementioned state might arise during high organic load with easily biodegradable biomass. As the AD system has an extended recovery period from gathered VFAs, amendment of nano-biochar can aid in a fast and simple process to hasten the recovery of acidified anaerobic reactors [81]. The buffering capacity and availability of nutrients are significantly influenced by nano-biochar during the co-digestion system [105,109]. However, there are a few opposing results regarding the biochar application in the AD process. Luo et al. [64] proposed that before CH_4 generation, acid inhibition might happen during synthetic wastewater AD, and nano-biochar cannot significantly upsurge the pH buffering capacity. A decrease in pH value (5.0 to 3.0) owing to VFAs accretion even after nano-biochar addition during AD was observed by Sunyoto et al. [60]. Thus, there is no inference about the nano-biochar pH buffering capacity during the AD, which is based on the nano-biochar physicochemical properties and flexible operational parameters.

4.4. Inhibitors Adsorption

One of the key advantages that nano-biochar offers to raise AD effectiveness is the adsorption of inhibitors. During the adsorption process, the biochar aromatic structures enable π-π interaction due to OH and COOH groups [90]. The adsorption of VFAs by nano-biochar and alleviation of acid surges the CH_4 yield [101]. The study showed direct proportionality amid NH_4^+-N adsorption and hydrochar SSA [105]. Linville et al. [110] assessed biochar derived from the walnut shell and found significant removal of CO_2 by coarser (51%) and smaller biochar (61%) in the food waste AD. In the two-stage waste-activated sludge AD by pine wood and corn stover biochar, CH_4 contents of 81–88.6% in corn stover and 72.1–76.6% in pinewood biochar were observed, respectively [73]. The CO_2 was sequestered into carbonate/bicarbonate by base cations released by biochar [96,111].

4.5. Enriched Microbial Functionality

Nano-biochar encourages the extracellular polymeric substance (EPS) secretion from microbes during the formation of biofilm, thereby increasing the adhesion of microbes on the surface of nano-biochar [112]. It is a cost-effective and easy method to evade fast sludge granulation and decrease the methanogens loss during AD. Sun et al. [113] observed the microbial richness in the incidence of biochar carriers. Dang et al. [114] found that *Enterococcus* and *Sporanaerobacter* capacity was improved during biochar addition, which aids in fermentable substrates breakdown for electron transfer to *Methanosarcina*. In addition, the biological interaction between *Methanosaetaceae* and *Methanosarcinales* and biochar efficiently decreases the lag time [66]. Wang et al. [115] found that the hydrochar content is directly proportional to the attachment of methanogenic bacteria. Similarly, in the presence of hydrochar, *Methanosaeta* (acetoclastic methanogen) enrichment was observed [81]. Henceforth, CH_4 generation from VFAs is enhanced by immobilization of methanogens by hydrochar amendment [116]. Furthermore, biochar increases biofilm formation and functional microbial enrichment, thus, improving the AD process [80].

4.6. Electron Transfer Mechanism

The earlier investigations reported the importance of nano-biochar and its electron transfer amid archaea and anaerobic bacteria. AD system efficacy mainly depends upon the syntrophic interactions amid bacteria and methanogens, which help electrons fulfill their energy requirements [76]. Figure 4 gives an overview of proposed electron transfer mechanisms between oxidizing bacteria and methanogens. It is carried out in various ways: DIET via a conductive medium (e.g., granular activated carbon, magnetite, carbon cloth, nano-biochar) [117], membrane-bound transporter proteins [76], electric conductive pili [79], and indirect interspecies electron transfer (IIET) through insoluble (humic compounds) [118] and soluble (acetate, formate) substances [119,120]. In IIET, formate and hydrogen act as electron transport between methanogens and syntrophic-generating bacteria [76]. The microbial metabolite exchange is governed by Fick's Law, occurring via diffusion. Thus, after the cell accumulation is attained, the interspecies hydrogen transfer rate is raised by anaerobic bacteria and methanogenic archaea form the compressed structures [121].

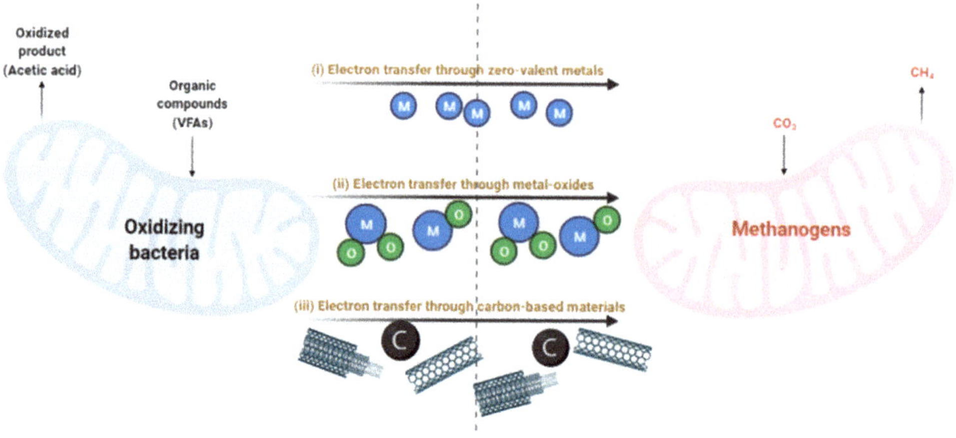

Figure 4. Application of conductive nanoparticles for electron transfer between oxidizing bacteria and methanogens via (i) zero-valent metals; (ii) metal oxides, and (iii) carbon-based materials.

During CH_4 production, IIET and diffusion of soluble metabolites are bottlenecked by decelerating the electron transference and energy transfer rate [77,122]. DIET does not entail the electron transport intercession for generating an electric current amid electron-accepting and donating microbes [76]. DIET was more precise and rapid than IIET [122]. Martínez et al. [123] reported that co-culture enhances CH_4 generation, for instance, enrichment of homoacetogenic bacteria such as *Eubacterium*, *Clostridium*, and *Syntrophomonas*. In synthetic wastewater AD, *Methanosaeta* and *Geobacter* enrichment on biochar in the presence of propionate and butyrate were reported [65]. During conductive biochar amendment, DIET was able to eliminate butyrate and propionate and establish interspecific electron shuttle by *Smithella* and *Syntrophomonas* richness.

5. Kinetics Involved during the AD Process

The kinetic study of AD system is the most important way to evaluate the performance of the reactor for biogas production, mechanism of metabolic pathways, biomass degradation, and the monitoring of the growth rate of microorganisms [124,125]. Based on increasing CH_4 production after the biochemical CH_4 production (BMP assay), the bio-kinetics parameters such as CH_4 generation potential, maximum rate, and its period of lag phase are evaluated to utilize the three diverse kinetics models. In general, the first-order kinetic model such as Chen and Hashimoto model [126], Cone model, and modified Gompertz model [124,127] were applied to mimic the kinetic patterns of biogas production in different condition. The first-order kinetic model and modified Gompertz model are used for biomass degradation rate and biomethane production rate in batch and continuous reactors [128,129]. The obtained cumulative biomethane productivity results through the modified Gompertz kinetic model (fitting error 0.7–13.7%) were more authentic than first-order kinetic (fitting error 9.2–37.1%) for organic waste material [129]. In addition, the quality of products depends on the type of substrates responsible for acidogenesis, acetoclastic, and methanogenesis by microorganisms in the digestion system. Here, the modified Gompertz and logistic function obey the sigmoidal process that correlates the methanogenic archaea growth with the CH_4 generation in an anaerobic reactor. In contrast, the transference function follows the first-order curve to correlate the CH_4 generation with the microbial activity [130].

$$\text{Modified Gompertz Model;} \quad y = M.\exp\left\{-\exp\left[\frac{R_m e}{M}(\lambda - t) + 1\right]\right\} \tag{1}$$

$$\text{Logistic Function Model;} \quad y = \frac{M}{1 + \exp\left(\frac{4R_m(\lambda - t)}{M} + 2\right)} \tag{2}$$

$$\text{Transference Function Model;} \quad y = M\left\{1 - \exp\left(-\frac{R_m(t - \lambda)}{M}\right)\right. \tag{3}$$

where y is the accumulated CH_4 (mL) at time t (h), M, R_m, λ, and e are the potential CH_4 generation (mL CH_4 g^{-1}), maximum rate of CH_4 production (mL CH_4 g^{-1} h^{-1}), lag phase time (h), and base and constant of the natural logarithms, respectively.

Kinetic models can be divided into structured (complex degradation and fermentation mechanism analysis) and unstructured (substrate consumption, growth rate along with production evaluation) models. The structured models are used for unsteady-state balance, while the unstructured models are used to assess steady-state balance conditions during the anaerobic process [127,129]. Some researchers have reported that the maximum growth rate value in the exponential phase is minimum at low substrate concentration. In the case of acetate inhibition at a higher substrate concentration, Andrew's kinetic models for AD (Equations (2) and (3)) that express the acetolactic methanogenesis stage can be used [124,131].

To examine the theoretical CH_4 yield (TMY) from the organic wastes on the basis of the elemental contents of the substrates, the following equations are given (Buswell formula [132]):

$$C_nH_aO_bN_c + \left(n - \frac{a}{4} - \frac{b}{2} + \frac{3c}{4}\right)H_2O = \left(\frac{n}{2} + \frac{a}{8} - \frac{b}{4} - \frac{3c}{8}\right)CH_4 + \left(\frac{n}{2} - \frac{a}{8} + \frac{b}{4} + \frac{3c}{8}\right)CO_2 + cNH_3 \tag{4}$$

$$TMY\left(\frac{\text{mLCH}_4}{gVS}\right) = \frac{(22.4)(1000)\left(\frac{n}{2} + \frac{a}{8} - \frac{b}{4} - \frac{3c}{8}\right)}{12n + a + 16b + 14c} \tag{5}$$

In addition, the anaerobic biodegradability of the organic feedstocks can be evaluated by dividing the experimental CH_4 yield by theoretical CH_4 yield. Furthermore, the surface-related models (Contois kinetics) for anaerobic digestion have not been much developed. Several researchers reported hydrolysis as a first-order kinetic model reaction that depends on various substrates and particle size. These limitations of pertinent simulation in the biogas generation mainly rely on the data availability and accuracy of the process. The methanogenesis effect with the kinetic limitation of the substrate is used in diverse kinds of inhibition models in the AD system. AD process is considered as H_2 and H_2S inhibition with the ionic balance and pH level in the biogas process system. The standard CH_4 formation rate is based on the un-dissociated hydrogen and sulphur concentration. Furthermore, the occurrence of oxygen is very sensitive; it breaks the degradation process in the AD system [133].

6. Closed-Loop Integration of Biochar and Anaerobic Digestion

The inherent complication in AD process is related to the different groups of microbes. The conventional ways such as pre-treatments (reducing the retardation rate) and co-substrates (the balance of nutrients, very low accumulation of toxicity), do not perform well to deal with such challenges [62,134]. In this respect, several new approaches and productive concepts are required, some of which are enumerated below:

(i) a synergetic integration of different technologies and proceed in methodical ways
(ii) development of the zero liquid discharge (ZLD) through cascade system
(iii) processing biomasses in concurrence with the closed-loop integrated system

These methodical routes of advanced blueprinted bio-energetic system offer several advantages, viz. supporting maximum recycling of biomass through managing material and energy which needed from the same feedstock (low energy consumption) [135], promoting the new income of source. Therefore, it also favors the ecological environment [136,137]. This integrated system uses the solid biomass residue for pyrolysis, which produces nano-biochar and is used as an additive in the AD reactor to enhance bio-methane production. Further, nano-biochar can be used as an adsorbent to enrich the nutrients in the slurry, or can also be used for agriculture as a fertilizer and soil conditioner. In this process, methane gas is produced, and it can further be used to generate power and heat. Hence, this integrated closed-loop system promotes crucial techniques for producing biomass during AD [138].

Deng et al. [139] achieved 17% yield increase of CH_4 and 10% bio-oil yield increase through the integrated system of AD and pyrolysis, and obtained 26% reduction in digestate biomass from seaweed. Similarly, Sen et al. [140] reported that in a study of a closed-loop system using the additive nano-biochar, the biomethane yield could be enhanced by 7%, the bio-methanation rate constant by 8.1%, and the maximum methane production rate by 27.6%, as well as increased alkalinity and mitigated NH_3 inhibition for AD. AD with pyrolysis favored both pyrolysis of liquor and pyrolysis of solid digestate, where nano-biochar acts as a catalyst for AD by supporting the mitigated NH_3, buffering, and alkalinity. Therefore, an integrated system of AD with pyrolysis is a promising way for the amalgamation of biological (AD) and thermochemical processes (pyrolysis) [141].

To improve circular economics and the sustainability effect of the AD effluent, alternative methods (converting digestate to pyrogenic carbon) are required, which may be considered as a feasible pathway because nowadays nano-biochar is considered as a carbon reservoir/carbon-negative technology, that is, contributing to greenhouse gases (GHG) emission [142], due to its properties such as having a high amount of highly stable carbon, and long-term carbon segregation capacity especially in an integrated system with AD [135,137]. Monlau et al. [135] revealed that solid-digestate (containing about 30–50% of organic carbon, on dry basis) was treated close to the pyrolysis in terms of excess amount of energy, similar to the pyrolysis nano-biochar. It was evident that the integrated system (AD & pyrolysis) anticipated a net 42% increase in electricity production. The AD system has various limitations such as inhibition of NH_3, quality of digestate, and low quality of biogas. However, by using this vigorous closed-loop integration (AD-pyrogenic carbon), more green and clean energy can be produced.

Further, the accomplishment of nano-biochar on anaerobic reactors such as acidogenesis, acetogenesis, hydrolysis, and methanogenesis is to be carried out for improving the buffering capacity and increasing acid pressure [62]. Nano-biochar is a vital way to aid in bio-shielding of archaea under acidic pressure and its reactive mechanism in AD reactors. Nano-biochar has proven its role as a stabilizer to enhance the syntrophic metabolism of VFAs and alcohol, enabling one to shorten retention times in AD [143].

The productivity for CH_4 during bio-methanation can be improved by improving anaerobic methanogens, which are promoted through a consortium of methanosarcina and syntrophic bacteria. Selective colonization is observed and had an ammonium adsorption capacity up to 17.6 mg g^{-1} by biochar [144]. DIET has been recognized as an alternative pathway for hydrogen interspecies transfer of syntrophic electrons between *Geobacter* and *Methanosarcina* to accelerate the syntrophic metabolism of ethanol by incorporating nano viruses. After which, methanogenic electron movement can be increased due to acetic acid formation [143], amplifying methanogenesis, directing to acidogenesis-acetogenesis, supporting the minimal corrosion and eccentric colonization of methanogenesis for improvement in selective productivity.

The most interesting factor for nano-biochar use as an AD additive is its minimal complexity, low cost, and very low risk of formation of second-hand pollutants. Meanwhile, porous material biochar contains a specific molecular structure [66], which stimulates AD at advanced levels. Many researchers have reported the performance of biochar such as hydrothermal carbonization and pyrolytic biochar for their low effects on ammonia inhibition and microbial growth [63]. Some issues related to using biochar in AD are still unsolved. Thus, there is a need to model the operational procedure, evaluation via advanced techniques with references at an earlier stage. Hence, nano-biochar may be a suitable replacement for conventional approaches.

7. Techno-Economic and Environmental Life Cycle Assessment

The techno-economic and environmental assessment of nano-biochar application in AD is imperative according to the production cost, vending cost, operating cost, profits, low carbon emissions, reduced secondary pollution, and less global warming potential at the commercial level [62]. These evaluations were also performed to raise the economic feasibility of nano-biochar application in full-scale along with the long-term operation of the AD process. In addition, the characteristic properties of the nano-biochar, feed type, production conditions, etc. all play a vital role in the AD performance that is mentioned in the current section.

7.1. Techno-Economic Analysis

Various researchers have performed the economic assessment of nano-biochar production to determine the investment that might be further balanced via the biochar trade price (470 € t^{-1}) [145]. Nano-biochar yield is mostly affected via oxygen to carbon (O/C), molecular ratio during pyrolysis and the ash content. A high O/C ratio in the feed substrate reduces the nano-biochar selectivity and enhances the formation of bio-oil, while a low ash content leads to an increase in the nano-biochar yield and decreases the bio-oil yield [62]. Adding biological and inorganic elements within the nano-biochar as additives in AD is usually carried out to increase methane production. These additions further lead to an increase in the overall production of the biochar; for instance, nutrients and enzymes add in 13–16 and 3.6–4.1 € L^{-1}, respectively [146].

The trending circular economy model was also applied for managing the digestate produced via the AD process of the organics [147]. The researchers have followed the "back to Earth" concept to deal with the digestate produced via AD of food and municipal solid wastes. The simultaneous integrated utilization of the biochar-aided AD and further recycling the digestate for the nano-biochar synthesis is consistent with the "zero-waste" concept approach. The digestate can also be used for fuel, energy, fertilizer, and chemical production for the industries [148]. In addition, the installation cost, pre-treatment cost (covering 43%), and hydrogen purification cost (22%) is chiefly involved in the overall cost. The average operating cost for nano-biochar derived from woody and straw biomasses are 0.68 and 0.86 € L^{-1}, respectively, in comparison to the additives, viz., nutrients (13–16 € L^{-1}) and enzymes (4.10 € L^{-1}) [121]. Traditional biochar (5–25 g L^{-1}) can be utilized as an additive in the AD process and is reused many times, making the process more economically feasible and the additional gain of by-products [149]. González et al. [150]

reported a case study of the economic feasibility of an integrated system and *NPV* could be calculated using the following equation:

$$Net\,present\,value\,(NPV) = -TCC + \sum_{t=1}^{n} \frac{CF_t}{(1+r)^{t'}} \quad (6)$$

where *TCC* is the total capital cost of the investment for the digestion and pyrolysis plant, CF_t is the cash flow expected at time *t*, and *r* is the discount rate.

Nevertheless, this analysis for the cost evaluation of nano-biochar aided AD process on a large scale is still at its embryonic stage. Henceforth, many researchers have further approached the life-cycle assessment (LCA) to analyze the total input energy required for CH_4 production [151].

7.2. Life-Cycle Assessment

An optimistic energy balance should be attained for the addition of nano-biochar for the anaerobic digestion process in the LCA. This "cradle to grave approach" must also be beneficial from the environmental and economic perspective [152]. LCA of the biochar produced via the lignocellulosic wastes showed the GHG emission in between 20–50 g CO_2–eq MJ^{-1}; whereas for pit/shell/husk, it ranges amid 120–250 g CO_2–eq MJ^{-1} [62]. The feedstocks for biochar possessing an ash content in the range of 0–2% and having a high O/C ratio are mainly related to the enhanced GHG emissions [62,84]. In addition, compared to the petro-derived fuels, the utilization of biofuels leads to >85% reduction in GHG release, equivalent to 93 g CO_2–eq MJ^{-1}. The biochar-aided AD process offers encouraging effects in contrast with AD alone. However, further LCA studies are needed to associate and integrate the waste conversion and resource recovery processes [153]. The methodical outcomes primarily depend on the type of biomass, compositions, reaction condition, and reactor, but lack the LCA for biofuels production [154–158].

Moreover, landfills, carbon sources, and other ambiguous factors (infrastructure, transportation, and waste management) should also be considered in the LCA [5]. ReCiPe and Tool for Reduction and Assessment of Chemicals and other environmental Impacts (TRACI) is a method for the life-cycle impact analysis (LCIA) and environmental impact assessment, respectively [5]. Hence, more studies of LCA are needed to assist the optimization of methodical, financial, and environmental performances of the additive nano-biochar and its integration in AD processes.

8. Conclusions and Prospects for Future Research

The properties of nano-biochar are primarily dependent on the organic feedstock used and its processing. Further, it is extensively being used as a catalyst for enhancing renewable energy production. Our comprehensive evaluation of the recent literature available on biochar-amended anaerobic processes concluded the credible importance of nano-biochar from the economical, simple processing, and enhanced biogas yield. Further, the metal-doped/impregnated nano-biochar composites show more magnetization properties and recycling ability, which can reduce the costs of biochar addition. The recycling of nano-biochar composites during the AD process further leads to the loss of the methanogen population due to the digestate disposal. In addition, Fe_2O_3 and Fe_3O_4 impregnated biochar perform as electron channels for promoting the interspecies electron transfer. The enhancement in the stability and reliability of anaerobic digestion via nano-biochar addition is much significant and depicts a novel paradigm for the generation of renewable energy, resource recovery, and waste management. The biochar-aided AD process mitigates the acidification impact caused via VFAs accumulation, encouraging the electron species and microbial growth. Further, in comparison to the traditional AD process, the nano-biochar addition helps reduce the environmental impacts and cost-involved. In addition, the following are some proposals that can further improve this feature:

(i) Numerous references report on the fed-batch operation of nano-biochar-amended AD processes. Further efforts should be made to develop continuous or semi-continuous operational modes, nano-biochar recycling, and reusability.
(ii) Prudent procedures should be developed and followed to determine the quantitative inhibition exhibited via the nano-biochar for the AD process.
(iii) Effective microbial metabolic pathways should be tracked along with the prime attention to the nano-biochar-microbe interactions, and mechanistic insights.
(iv) Techno-economic and the socio-economic analyses of the pilot- and industrial-scale plants, including the mass and energy balance assessments, are essential for the nano-biochar-amended AD process. The life-cycle and supply chain management further needs to be monitored for the overall impact of the integrated process.
(v) Exploration of DIET/IIET is highly encouraged for electron-based elucidation to enhance biogas production and further establish a pioneer avenue of research in renewable energy research.
(vi) Nano-biochar-amended co-digestion approaches need to be explored for the reduction in the reactor volume, zero-waste approach, carbon capturing, and encouraging the circular bioeconomy concept.

Author Contributions: Conceptualization and supervision: L.G., A.K. and B.S.K.; writing—original draft preparation: L.G., A.K., A.S., P.S. and Y.C.; review and editing, artwork and schemes: L.G., A.K., M.M., S.V.P. and B.S.K. All authors have read and agreed to the published version of the manuscript.

Funding: This research was supported by Chungbuk National University BK (Brain Korea) 21 FOUR (2021).

Data Availability Statement: This study did not report any original data.

Conflicts of Interest: All authors declare no competing interest with the work presented in the manuscript.

References

1. Bhatia, S.K.; Palai, A.K.; Kumar, A.; Bhatia, R.K.; Patel, A.K.; Thakur, V.K.; Yang, Y.-H. Trends in renewable energy production employing biomass-based biochar. *Bioresour. Technol.* **2021**, *340*, 125644. [CrossRef]
2. Hussain, C.M.; Singh, S.; Goswam, L. (Eds.) Chapter 13 - Biohythane production from organic waste: Challenges and techno-economic perspective. In *Waste-to-Energy Approaches Towards Zero Waste: Interdisciplinary Methods of Controlling Waste*; Elsevier: Amsterdam, The Netherlands, 2022.
3. Ramanayaka, S.; Vithanage, M.; Alessi, D.S.; Liu, W.-J.; Jayasundera, A.C.A.; Ok, Y.S. Nanobiochar: Production, properties, and multifunctional applications. *Environ. Sci. Nano* **2020**, *7*, 3279–3302. [CrossRef]
4. Kushwaha, A.; Goswami, S.; Sultana, A.; Katiyar, N.K.; Athar, M.; Dubey, L.; Goswami, L.; Hussain, C.M.; Kareem, M.A. Waste biomass to biobutanol: Recent trends and advancements. In *Waste-to-Energy Approaches Towards Zero Waste*; Elsevier: Amsterdam, The Netherlands, 2022; pp. 393–423.
5. Kushwaha, A.; Mishra, V.; Gupta, V.; Goswami, S.; Gupta, P.K.; Singh, L.K.; Gupt, C.B.; Rakshit, K.; Goswami, L. Anaerobic digestion as a sustainable biorefinery concept for waste to energy conversion. In *Waste-to-Energy Approaches Towards Zero Waste*; Elsevier: Amsterdam, The Netherlands, 2022; pp. 129–163.
6. Kushwaha, A.; Yadav, A.N.; Singh, B.; Dwivedi, V.; Kumar, S.; Goswami, L.; Hussain, C.M. Life cycle assessment and techno-economic analysis of algae-derived biodiesel: Current challenges and future prospects. In *Waste-to-Energy Approaches Towards Zero Waste*; Elsevier: Amsterdam, The Netherlands, 2022; pp. 343–372.
7. Kumar, M.; Kushwaha, A.; Goswami, L.; Singh, A.K.; Sikandar, M. A review on advances and mechanism for the phycoremediation of cadmium contaminated wastewater. *Clean. Eng. Technol.* **2021**, *5*, 100288. [CrossRef]
8. Gupt, C.B.; Kushwaha, A.; Prakash, A.; Chandra, A.; Goswami, L.; Sekharan, S. Mitigation of groundwater pollution: Heavy metal retention characteristics of fly ash based liner materials. In *Fate and Transport of Subsurface Pollutants*; Springer: Singapore, 2021; pp. 79–104.
9. Gautam, R.; Nayak, J.K.; Daverey, A.; Ghosh, U.K. Emerging sustainable opportunities for waste to bioenergy: An overview. In *Waste-to-Energy Approaches Towards Zero Waste*; Elsevier: Amsterdam, The Netherlands, 2022; pp. 1–55.
10. Borah, S.N.; Goswami, L.; Sen, S.; Sachan, D.; Sarma, H.; Montes, M.; Narayan, M. Selenite bioreduction and biosynthesis of selenium nanoparticles by *Bacillus paramycoides* SP3 isolated from coal mine overburden leachate. *Environ. Pollut.* **2021**, *285*, 117519. [CrossRef]
11. Sachan, D.; Ramesh, A.; Das, G. Green synthesis of silica nanoparticles from leaf biomass and its application to remove heavy metals from synthetic wastewater: A comparative analysis. *Environ. Nanotechnol. Monit. Manag.* **2021**, *16*, 100467. [CrossRef]

12. Gautam, R.; Nayak, J.K.; Talapatra, K.N.; Ghosh, U.K. Assessment of different organic substrates for Bio-Electricity and Bio-Hydrogen generation in an Integrated Bio-Electrochemical System. *Mater. Today Proc.* **2021**. [CrossRef]
13. Lata, K.; Kushwaha, A.; Ramanathan, G. Chapter 23 - Bacterial enzymatic degradation and remediation of 2,4,6-trinitrotoluene. In *Microbial and Natural Macromolecules*; Academic Press: Cambridge, MA, USA, 2021; pp. 623–659. [CrossRef]
14. Sachan, D.; Das, G. Fabrication of Biochar-Impregnated MnO_2 Nanocomposite: Characterization and Potential Application in Copper (II) and Zinc (II) Adsorption. *J. Hazard. Toxic Radioact. Waste* **2022**, *26*, 04021049. [CrossRef]
15. Goswami, L.; Kushwaha, A.; Goswami, S.; Sharma, Y.C.; Kim, T.; Tripathi, K.M. Nanocarbon-based-ZnO nanocomposites for supercapacitor application. In *Nanostructured Zinc Oxide*; Elsevier: Amsterdam, The Netherlands, 2021; pp. 553–573.
16. Goswami, S.; Kushwaha, A.; Goswami, L.; Singh, N.; Bhan, U.; Daverey, A.; Hussain, C.M. Biological treatment, recovery, and recycling of metals from waste printed circuit boards. In *Environmental Management of Waste Electrical and Electronic Equipment*; Elsevier: Amsterdam, The Netherlands, 2021; pp. 163–184.
17. Kushwaha, A.; Goswami, L.; Lee, J.; Sonne, C.; Brown, R.J.C.; Kim, K.-H. Selenium in soil-microbe-plant systems: Sources, distribution, toxicity, tolerance, and detoxification. *Crit. Rev. Environ. Sci. Technol.* **2021**, 1–38. [CrossRef]
18. Kushwaha, A.; Goswami, S.; Hans, N.; Goswami, L.; Devi, G.; Deshavath, N.N.; Yadav, M.K.; Lall, A.M. An Insight into Biological and Chemical Technologies for Micropollutant Removal from Wastewater. In *Fate and Transport of Subsurface Pollutants*; Springer: Singapore, 2020; pp. 199–226.
19. Devi, G.; Goswami, L.; Kushwaha, A.; Sathe, S.S.; Sen, B.; Sarma, H.P. Fluoride distribution and groundwater hydrogeochemistry for drinking, domestic and irrigation in an area interfaced near Brahmaputra floodplain of North-Eastern India. *Environ. Nanotechnol. Monit. Manag.* **2021**, *16*, 100500. [CrossRef]
20. Goswami, L.; Namboodiri, M.T.; Kumar, R.V.; Pakshirajan, K.; Pugazhenthi, G. Biodiesel production potential of oleaginous Rhodococcus opacus grown on biomass gasification wastewater. *Renew. Energy* **2017**, *105*, 400–406. [CrossRef]
21. Goswami, L.; Manikandan, N.A.; Pakshirajan, K.; Pugazhenthi, G. Simultaneous heavy metal removal and anthracene biodegradation by the oleaginous bacteria *Rhodococcus opacus*. *3 Biotech* **2017**, *7*, 37. [CrossRef]
22. Ma, J.; Chen, F.; Xue, S.; Pan, J.; Khoshnevisan, B.; Yang, Y.; Liu, H.; Qiu, L. Improving anaerobic digestion of chicken manure under optimized biochar supplementation strategies. *Bioresour. Technol.* **2021**, *325*, 124697. [CrossRef]
23. Zhang, T.; Yang, Y.; Liu, L.; Han, Y.; Ren, G.; Yang, G. Improved Biogas Production from Chicken Manure Anaerobic Digestion Using Cereal Residues as Co-substrates. *Energy Fuels* **2014**, *28*, 2490–2495. [CrossRef]
24. Gautam, A.; Kushwaha, A.; Rani, R. Reduction of Hexavalent Chromium [Cr (VI)] by Heavy Metal Tolerant Bacterium Alkalihalobacillus clausii CRA1 and Its Toxicity Assessment Through Flow Cytometry. *Curr. Microbiol.* **2022**, *79*, 33. [CrossRef]
25. Bi, S.; Qiao, W.; Xiong, L.; Mahdy, A.; Wandera, S.M.; Yin, D.; Dong, R. Improved high solid anaerobic digestion of chicken manure by moderate in situ ammonia stripping and its relation to metabolic pathway. *Renew. Energy* **2020**, *146*, 2380–2389. [CrossRef]
26. Pan, J.; Ma, J.; Zhai, L.; Luo, T.; Mei, Z.; Liu, H. Achievements of biochar application for enhanced anaerobic digestion: A review. *Bioresour. Technol.* **2019**, *292*, 122058. [CrossRef]
27. Deng, C.; Lin, R.; Kang, X.; Wu, B.; Wall, D.M.; Murphy, J.D. What physicochemical properties of biochar facilitate interspecies electron transfer in anaerobic digestion: A case study of digestion of whiskey by-products. *Fuel* **2021**, *306*, 121736. [CrossRef]
28. Do, M.H.; Ngo, H.H.; Guo, W.; Chang, S.W.; Nguyen, D.D.; Sharma, P.; Pandey, A.; Bui, X.T.; Zhang, X. Performance of a dual-chamber microbial fuel cell as biosensor for on-line measuring ammonium nitrogen in synthetic municipal wastewater. *Sci. Total Environ.* **2021**, *795*, 148755. [CrossRef]
29. Yadav, A.P.S.; Dwivedi, V.; Kumar, S.; Kushwaha, A.; Goswami, L.; Reddy, B.S. Cyanobacterial Extracellular Polymeric Substances for Heavy Metal Removal: A Mini Review. *J. Compos. Sci.* **2020**, *5*, 1. [CrossRef]
30. Trapero, J.; Horcajada, L.; Linares, J.J.; Lobato, J. Is microbial fuel cell technology ready? An economic answer towards industrial commercialization. *Appl. Energy* **2017**, *185*, 698–707. [CrossRef]
31. Slate, A.J.; Whitehead, K.A.; Brownson, D.A.; Banks, C.E. Microbial fuel cells: An overview of current technology. *Renew. Sustain. Energy Rev.* **2019**, *101*, 60–81. [CrossRef]
32. Xu, B.; Ge, Z.; He, Z. Sediment microbial fuel cells for wastewater treatment: Challenges and opportunities. *Environ. Sci. Water Res. Technol.* **2015**, *1*, 279–284. [CrossRef]
33. Chen, S.; Tang, J.; Fu, L.; Yuan, Y.; Zhou, S. Biochar improves sediment microbial fuel cell performance in low conductivity freshwater sediment. *J. Soils Sediments* **2016**, *16*, 2326–2334. [CrossRef]
34. Gong, L.; Amirdehi, M.A.; Miled, A.; Greener, J. Practical increases in power output from soil-based microbial fuel cells under dynamic temperature variations. *Sustain. Energy Fuels* **2020**, *5*, 671–677. [CrossRef]
35. Li, X.; Li, Y.; Zhang, X.; Zhao, X.; Sun, Y.; Weng, L.; Li, Y. Long-term effect of biochar amendment on the biodegradation of petroleum hydrocarbons in soil microbial fuel cells. *Sci. Total Environ.* **2019**, *651*, 796–806. [CrossRef]
36. Chen, Q.; Pu, W.; Hou, H.; Hu, J.; Liu, B.; Li, J.; Cheng, K.; Huang, L.; Yuan, X.; Yang, C.; et al. Activated microporous-mesoporous carbon derived from chestnut shell as a sustainable anode material for high performance microbial fuel cells. *Bioresour. Technol.* **2018**, *249*, 567–573. [CrossRef]
37. Lee, J.H.; Kim, D.S.; Yang, J.H.; Chun, Y.; Yoo, H.Y.; Han, S.O.; Lee, J.; Park, C.; Kim, S.W. Enhanced electron transfer mediator based on biochar from microalgal sludge for application to bioelectrochemical systems. *Bioresour. Technol.* **2018**, *264*, 387–390. [CrossRef]

38. Chakraborty, I.; Bhowmick, G.D.; Ghosh, D.; Dubey, B.; Pradhan, D.; Ghangrekar, M. Novel low-cost activated algal biochar as a cathode catalyst for improving performance of microbial fuel cell. *Sustain. Energy Technol. Assess.* **2020**, *42*, 100808. [CrossRef]
39. Hemalatha, M.; Sravan, J.S.; Min, B.; Mohan, S.V. Concomitant use of Azolla derived bioelectrode as anode and hydrolysate as substrate for microbial fuel cell and electro-fermentation applications. *Sci. Total Environ.* **2020**, *707*, 135851. [CrossRef]
40. Yang, W.; Li, J.; Zhang, L.; Zhu, X.; Liao, Q. A monolithic air cathode derived from bamboo for microbial fuel cells. *RSC Adv.* **2017**, *7*, 28469–28475. [CrossRef]
41. Li, M.; Zhang, H.; Xiao, T.; Wang, S.; Zhang, B.; Chen, D.; Su, M.; Tang, J. Low-cost biochar derived from corncob as oxygen reduction catalyst in air cathode microbial fuel cells. *Electrochim. Acta* **2018**, *283*, 780–788. [CrossRef]
42. Chang, H.-C.; Gustave, W.; Yuan, Z.-F.; Xiao, Y.; Chen, Z. One-step fabrication of binder-free air cathode for microbial fuel cells by using balsa wood biochar. *Environ. Technol. Innov.* **2020**, *18*, 100615. [CrossRef]
43. Allam, F.; Elnouby, M.; El-Khatib, K.; El-Badan, D.E.; Sabry, S.A. *Water hyacinth* (*Eichhornia crassipes*) biochar as an alternative cathode electrocatalyst in an air-cathode single chamber microbial fuel cell. *Int. J. Hydrogen Energy* **2020**, *45*, 5911–5927. [CrossRef]
44. Zha, Z.; Zhang, Z.; Xiang, P.; Zhu, H.; Zhou, B.; Sun, Z.; Zhou, S. One-step preparation of eggplant-derived hierarchical porous graphitic biochar as efficient oxygen reduction catalyst in microbial fuel cells. *RSC Adv.* **2021**, *11*, 1077–1085. [CrossRef]
45. Zhang, J.; Yang, M.; Zhao, W.; Zhang, J.; Zang, L. Biohydrogen Production Amended with Nitrogen-Doped Biochar. *Energy Fuels* **2021**, *35*, 1476–1487. [CrossRef]
46. Zhao, L.; Wang, Z.; Ren, H.-Y.; Chen, C.; Nan, J.; Cao, G.-L.; Yang, S.-S.; Ren, N.-Q. Residue cornstalk derived biochar promotes direct bio-hydrogen production from anaerobic fermentation of cornstalk. *Bioresour. Technol.* **2021**, *320*, 124338. [CrossRef]
47. Rezaeitavabe, F.; Saadat, S.; Talebbeydokhti, N.; Sartaj, M.; Tabatabaei, M. Enhancing bio-hydrogen production from food waste in single-stage hybrid dark-photo fermentation by addition of two waste materials (exhausted resin and biochar). *Biomass Bioenergy* **2020**, *143*, 105846. [CrossRef]
48. Yang, G.; Wang, J. Synergistic enhancement of biohydrogen production from grass fermentation using biochar combined with zero-valent iron nanoparticles. *Fuel* **2019**, *251*, 420–427. [CrossRef]
49. Sharma, P.; Melkania, U. Biochar-enhanced hydrogen production from organic fraction of municipal solid waste using co-culture of Enterobacter aerogenes and *E. coli*. *Int. J. Hydrogen Energy* **2017**, *42*, 18865–18874. [CrossRef]
50. Behera, B.; Dey, B.; Balasubramanian, P. Algal biodiesel production with engineered biochar as a heterogeneous solid acid catalyst. *Bioresour. Technol.* **2020**, *310*, 123392. [CrossRef]
51. Foroutan, R.; Mohammadi, R.; Razeghi, J.; Ramavandi, B. Biodiesel production from edible oils using algal biochar/CaO/K_2CO_3 as a heterogeneous and recyclable catalyst. *Renew. Energy* **2021**, *168*, 1207–1216. [CrossRef]
52. Wang, Y.; Li, D.; Zhao, D.; Fan, Y.; Bi, J.; Shan, R.; Yang, J.; Luo, B.; Yuan, H.; Ling, X.; et al. Calcium-Loaded Municipal Sludge-Biochar as an Efficient and Stable Catalyst for Biodiesel Production from Vegetable Oil. *ACS Omega* **2020**, *5*, 17471–17478. [CrossRef] [PubMed]
53. Akinfalabi, S.-I.; Rashid, U.; Ngamcharussrivichai, C.; Nehdi, I.A. Synthesis of reusable biobased nano-catalyst from waste sugarcane bagasse for biodiesel production. *Environ. Technol. Innov.* **2020**, *18*, 100788. [CrossRef]
54. Bhatia, S.K.; Gurav, R.; Choi, T.-R.; Kim, H.J.; Yang, S.-Y.; Song, H.-S.; Park, J.Y.; Park, Y.-L.; Han, Y.-H.; Choi, Y.-K.; et al. Conversion of waste cooking oil into biodiesel using heterogenous catalyst derived from cork biochar. *Bioresour. Technol.* **2020**, *302*, 122872. [CrossRef] [PubMed]
55. Yang, Z.; Yang, R.; Dong, G.; Xiang, M.; Hui, J.; Ou, J.; Qin, H. Biochar Nanocomposite Derived from Watermelon Peels for Electrocatalytic Hydrogen Production. *ACS Omega* **2021**, *6*, 2066–2073. [CrossRef]
56. Jiang, C.; Yao, M.; Wang, Z.; Li, J.; Sun, Z.; Li, L.; Moon, K.-S.; Wong, C.-P. A novel flower-like architecture comprised of 3D interconnected Co–Al-Ox/Sy decorated lignosulfonate-derived carbon nanosheets for flexible supercapacitors and electrocatalytic water splitting. *Carbon* **2021**, *184*, 386–399. [CrossRef]
57. Raut, S.D.; Shinde, N.M.; Nakate, Y.T.; Ghule, B.G.; Gore, S.K.; Shaikh, S.F.; Pak, J.J.; Al-Enizi, A.M.; Mane, R.S. Coconut-Water-Mediated Carbonaceous Electrode: A Promising Eco-Friendly Material for Bifunctional Water Splitting Application. *ACS Omega* **2021**, *6*, 12623–12630. [CrossRef]
58. Patel, S.; Kundu, S.; Halder, P.; Marzbali, M.H.; Chiang, K.; Surapaneni, A.; Shah, K. Production of hydrogen by catalytic methane decomposition using biochar and activated char produced from biosolids pyrolysis. *Int. J. Hydrogen Energy* **2020**, *45*, 29978–29992. [CrossRef]
59. Sugiarto, Y.; Sunyoto, N.M.; Zhu, M.; Jones, I.; Zhang, D. Effect of biochar in enhancing hydrogen production by mesophilic anaerobic digestion of food wastes: The role of minerals. *Int. J. Hydrogen Energy* **2021**, *46*, 3695–3703. [CrossRef]
60. Sunyoto, N.M.; Zhu, M.; Zhang, Z.; Zhang, D. Effect of biochar addition on hydrogen and methane production in two-phase anaerobic digestion of aqueous carbohydrates food waste. *Bioresour. Technol.* **2016**, *219*, 29–36. [CrossRef]
61. Goswami, L.; Kumar, R.V.; Manikandan, N.A.; Raja, V.K.; Pugazhenthi, G. Simultaneous polycyclic aromatic hydrocarbon degradation and lipid accumulation by *Rhodococcus opacus* for potential biodiesel production. *J. Water Process Eng.* **2017**, *17*, 1–10. [CrossRef]
62. Kumar, M.; Dutta, S.; You, S.; Luo, G.; Zhang, S.; Show, P.L.; Sawarkar, A.D.; Singh, L.; Tsang, D.C. A critical review on biochar for enhancing biogas production from anaerobic digestion of food waste and sludge. *J. Clean. Prod.* **2021**, *305*, 127143. [CrossRef]
63. Luz, F.C.; Cordiner, S.; Manni, A.; Mulone, V.; Rocco, V. Biochar characteristics and early applications in anaerobic digestion-a review. *J. Environ. Chem. Eng.* **2018**, *6*, 2892–2909.

64. Luo, C.; Lü, F.; Shao, L.; He, P. Application of eco-compatible biochar in naerobic digestion to relieve acid stress and promote the selective colonization of functional microbes. *Water Res.* **2015**, *68*, 710–718. [CrossRef]
65. Zhao, Z.; Zhang, Y.; Holmes, D.E.; Dang, Y.; Woodard, T.L.; Nevin, K.P.; Lovley, D.R. Potential enhancement of direct interspecies electron transfer for syntrophic metabolism of propionate and butyrate with biochar in up-flow anaerobic sludge blanket reactors. *Bioresour. Technol.* **2016**, *209*, 148–156. [CrossRef]
66. Lü, F.; Luo, C.; Shao, L.; He, P. Biochar alleviates combined stress of ammonium and acids by firstly enriching Methanosaeta and then Methanosarcina. *Water Res.* **2016**, *90*, 34–43. [CrossRef]
67. Sethupathi, S.; Ming, Z.; Rajapaksha, A.U.; Sang, R.L.; Nor, N.M.; Mohamed, A.R.; Al-Wabel, M.; Lee, S.S.; Ok, Y.S. Biochars as potential adsorbers of CH_4, CO_2 and H_2S. *Sustainability* **2017**, *9*, 121. [CrossRef]
68. Creamer, A.E.; Gao, B.; Zhang, M. Carbon dioxide capture using biochar produced from sugarcane bagasse and hickory wood. *Chem. Eng. J.* **2014**, *249*, 174–179. [CrossRef]
69. Fiore, S.; Berruti, F.; Briens, C. Investigation of innovative and conventional pyrolysis of ligneous and herbaceous biomasses for biochar production. *Biomass Bioenergy* **2018**, *119*, 381–391. [CrossRef]
70. Chen, T.; Zhang, Y.; Wang, H.; Lu, W.; Zhou, Z.; Zhang, Y.; Ren, L. Influence of pyrolysis temperature on characteristics and heavy metal adsorptive performance of biochar derived from municipal sewage sludge. *Bioresour. Technol.* **2014**, *164*, 47–54. [CrossRef]
71. Rajagopal, R.; Massé, D.I.; Singh, G. A critical review on inhibition of anaerobic digestion process by excess ammonia. *Bioresour. Technol.* **2013**, *143*, 632–641. [CrossRef] [PubMed]
72. Poirier, S.; Madigou, C.; Bouchez, T.; Chapleur, O. Improving anaerobic digestion with support media: Mitigation of ammonia inhibition and effect on microbial communities. *Bioresour. Technol.* **2017**, *235*, 229–239. [CrossRef] [PubMed]
73. Shen, Y.; Forrester, S.; Koval, J.; Urgun-Demirtas, M. Yearlong semi-continuous operation of thermophilic two-stage anaerobic digesters amended with biochar for enhanced biomethane production. *J. Clean. Prod.* **2017**, *167*, 863–874. [CrossRef]
74. Su, C.; Zhao, L.; Liao, L.; Qin, J.; Lu, Y.; Deng, Q.; Chen, M.; Huang, Z. Application of biochar in a CIC reactor to relieve ammonia nitrogen stress and promote microbial community during food waste treatment. *J. Clean. Prod.* **2019**, *209*, 353–362. [CrossRef]
75. Lü, F.; Hua, Z.; Shao, L.; He, P. Loop bioenergy production and carbon sequestration of polymeric waste by integrating biochemical and thermochemical conversion processes: A conceptual framework and recent advances. *Renew. Energy* **2018**, *124*, 202–211. [CrossRef]
76. Martins, G.; Salvador, A.F.; Pereira, L.; Alves, M.M. Methane Production and Conductive Materials: A Critical Review. *Environ. Sci. Technol.* **2018**, *52*, 10241–10253. [CrossRef]
77. Park, J.-H.; Kang, H.-J.; Park, K.-H.; Park, H.-D. Direct interspecies electron transfer via conductive materials: A perspective for anaerobic digestion applications. *Bioresour. Technol.* **2018**, *254*, 300–311. [CrossRef]
78. Liu, M.; Xu, M.; Xue, Y.; Ni, W.; Huo, S.; Wu, L.; Yang, Z.; Yan, Y.-M. Efficient Capacitive Deionization Using Natural Basswood-Derived, Freestanding, Hierarchically Porous Carbon Electrodes. *ACS Appl. Mater. Interfaces* **2018**, *10*, 31260–31270. [CrossRef]
79. Barua, S.; Dhar, B.R. Advances towards understanding and engineering direct interspecies electron transfer in anaerobic digestion. *Bioresour. Technol.* **2017**, *244*, 698–707. [CrossRef]
80. Yin, C.; Shen, Y.; Yuan, R.; Zhu, N.; Yuan, H.; Lou, Z. Sludge-based biochar-assisted thermophilic anaerobic digestion of waste-activated sludge in microbial electrolysis cell for methane production. *Bioresour. Technol.* **2019**, *284*, 315–324. [CrossRef]
81. Ren, S.; Usman, M.; Tsang, D.C.W.; O-Thong, S.; Angelidaki, I.; Zhu, X.; Zhang, S.; Luo, G. Hydrochar-Facilitated Anaerobic Digestion: Evidence for Direct Interspecies Electron Transfer Mediated through Surface Oxygen-Containing Functional Groups. *Environ. Sci. Technol.* **2020**, *54*, 5755–5766. [CrossRef] [PubMed]
82. Goswami, L.; Manikandan, N.A.; Taube, J.C.R.; Pakshirajan, K.; Pugazhenthi, G. Novel waste-derived biochar from biomass gasification effluent: Preparation, characterization, cost estimation, and application in polycyclic aromatic hydrocarbon biodegradation and lipid accumulation by *Rhodococcus opacus*. *Environ. Sci. Pollut. Res.* **2019**, *26*, 25154–25166. [CrossRef] [PubMed]
83. Li, W.; Dang, Q.; Brown, R.C.; Laird, D.; Wright, M.M. The impacts of biomass properties on pyrolysis yields, economic and environmental performance of the pyrolysis-bioenergy-biochar platform to carbon negative energy. *Bioresour. Technol.* **2017**, *241*, 959–968. [CrossRef] [PubMed]
84. Kumar, M.; Xiong, X.; Sun, Y.; Yu, I.K.M.; Tsang, D.C.W.; Hou, D.; Gupta, J.; Bhaskar, T.; Pandey, A. Critical Review on Biochar-Supported Catalysts for Pollutant Degradation and Sustainable Biorefinery. *Adv. Sustain. Syst.* **2020**, *4*, 1900149. [CrossRef]
85. Yin, Q.; Zhang, B.; Wang, R.; Zhao, Z. Biochar as an adsorbent for inorganic nitrogen and phosphorus removal from water: A review. *Environ. Sci. Pollut. Res.* **2017**, *24*, 26297–26309. [CrossRef] [PubMed]
86. Kizito, S.; Wu, S.; Kirui, W.K.; Lei, M.; Lu, Q.; Bah, H.; Dong, R. Evaluation of slow pyrolyzed wood and rice husks biochar for adsorption of ammonium nitrogen from piggery manure anaerobic digestate slurry. *Sci. Total Environ.* **2015**, *505*, 102–112. [CrossRef]
87. Takaya, C.A.; Fletcher, L.A.; Singh, S.; Anyikude, K.U.; Ross, A.B. Phosphate and ammonium sorption capacity of biochar and hydrochar from different wastes. *Chemosphere* **2016**, *145*, 518–527. [CrossRef]
88. Zhang, Y.; Li, Z.; Mahmood, I.B. Recovery of NH_4^+ by corn cob produced biochars and its potential application as soil conditioner. *Front. Environ. Sci. Eng.* **2014**, *8*, 825–834. [CrossRef]

89. Sahota, S.; Vijay, V.K.; Subbarao, P.; Chandra, R.; Ghosh, P.; Shah, G.; Kapoor, R.; Vijay, V.; Koutu, V.; Thakur, I.S. Characterization of leaf waste based biochar for cost effective hydrogen sulphide removal from biogas. *Bioresour. Technol.* **2018**, *250*, 635–641. [CrossRef]
90. Kanjanarong, J.; Giri, B.S.; Jaisi, D.P.; de Oliveira, F.R.; Boonsawang, P.; Chaiprapat, S.; Singh, R.; Balakrishna, A.; Khanal, S.K. Removal of hydrogen sulfide generated during anaerobic treatment of sulfate-laden wastewater using biochar: Evaluation of efficiency and mechanisms. *Bioresour. Technol.* **2017**, *234*, 115–121. [CrossRef]
91. Chacón, F.J.; Sanchez-Monedero, M.; Lezama, L.; Cayuela, M.L. Enhancing biochar redox properties through feedstock selection, metal preloading and post-pyrolysis treatments. *Chem. Eng. J.* **2020**, *395*, 125100. [CrossRef]
92. Klüpfel, L.; Keiluweit, M.; Kleber, M.; Sander, M. Redox Properties of Plant Biomass-Derived Black Carbon (Biochar). *Environ. Sci. Technol.* **2014**, *48*, 5601–5611. [CrossRef] [PubMed]
93. Kumar, M.; Xiong, X.; Wan, Z.; Sun, Y.; Tsang, D.C.; Gupta, J.; Gao, B.; Cao, X.; Tang, J.; Ok, Y.S. Ball milling as a mechanochemical technology for fabrication of novel biochar nanomaterials. *Bioresour. Technol.* **2020**, *312*, 123613. [CrossRef] [PubMed]
94. Joseph, S.; Husson, O.; Graber, E.R.; Van Zwieten, L.; Taherymoosavi, S.; Thomas, T.; Nielsen, S.; Ye, J.; Pan, G.; Chia, C.; et al. The Electrochemical Properties of Biochars and How They Affect Soil Redox Properties and Processes. *Agronomy* **2015**, *5*, 322–340. [CrossRef]
95. Dieguez-Alonso, A.; Anca-Couce, A.; Frišták, V.; Moreno-Jiménez, E.; Bacher, M.; Bucheli, T.D.; Cimò, G.; Conte, P.; Hagemann, N.; Haller, A.; et al. Designing biochar properties through the blending of biomass feedstock with metals: Impact on oxyanions adsorption behavior. *Chemosphere* **2019**, *214*, 743–753. [CrossRef] [PubMed]
96. Masebinu, S.; Akinlabi, E.; Muzenda, E.; Aboyade, A. A review of biochar properties and their roles in mitigating challenges with anaerobic digestion. *Renew. Sustain. Energy Rev.* **2019**, *103*, 291–307. [CrossRef]
97. Indren, M.; Birzer, C.H.; Kidd, S.P.; Hall, T.; Medwell, P.R. Effects of biochar parent material and microbial pre-loading in biochar-amended high-solids anaerobic digestion. *Bioresour. Technol.* **2020**, *298*, 122457. [CrossRef]
98. Shen, Y.; Linville, J.L.; Ignacio-de Leon, P.A.A.; Schoene, R.P.; Urgun-Demirtas, M. Towards a sustainable paradigm of waste-to-energy process: Enhanced anaerobic digestion of sludge with woody biochar. *J. Clean. Prod.* **2016**, *135*, 1054–1064. [CrossRef]
99. Li, H.; Feng, K. Life cycle assessment of the environmental impacts and energy efficiency of an integration of sludge anaerobic digestion and pyrolysis. *J. Clean. Prod.* **2018**, *195*, 476–485. [CrossRef]
100. Li, H.; Dong, X.; da Silva, E.B.; de Oliveira, L.M.; Chen, Y.; Ma, L.Q. Echanisms of metal sorption by biochars: Biochar characteristics and modifications. *Chemosphere* **2017**, *178*, 466–478. [CrossRef]
101. Shanmugam, S.R.; Adhikari, S.; Nam, H.; Sajib, S.K. Effect of bio-char on methane generation from glucose and aqueous phase of algae liquefaction using mixed anaerobic cultures. *Biomass Bioenergy* **2018**, *108*, 479–486. [CrossRef]
102. Fagbohungbe, M.; Herbert, B.M.; Hurst, L.; Li, H.; Usmani, S.Q.; Semple, K.T. Impact of biochar on the anaerobic digestion of citrus peel waste. *Bioresour. Technol.* **2016**, *216*, 142–149. [CrossRef]
103. Viggi, C.C.; Simonetti, S.; Palma, E.; Pagliaccia, P.; Braguglia, C.; Fazi, S.; Baronti, S.; Navarra, M.A.; Pettiti, I.; Koch, C.; et al. Enhancing methane production from food waste fermentate using biochar: The added value of electrochemical testing in pre-selecting the most effective type of biochar. *Biotechnol. Biofuels* **2017**, *10*, 303. [CrossRef] [PubMed]
104. Qin, Y.; Wang, H.; Li, X.; Cheng, J.J.; Wu, W. Improving methane yield from organic fraction of municipal solid waste (OFMSW) with magnetic rice-straw biochar. *Bioresour. Technol.* **2017**, *245*, 1058–1066. [CrossRef]
105. Wang, C.; Liu, Y.; Gao, X.; Chen, H.; Xu, X.; Zhu, L. Role of biochar in the granulation of anaerobic sludge and improvement of electron transfer characteristics. *Bioresour. Technol.* **2018**, *268*, 28–35. [CrossRef] [PubMed]
106. Qiu, L.; Deng, Y.; Wang, F.; Davaritouchaee, M.; Yao, Y. A review on biochar-mediated anaerobic digestion with enhanced methane recovery. *Renew. Sustain. Energy Rev.* **2019**, *115*, 109373. [CrossRef]
107. Ren, Y.; Yu, M.; Wu, C.; Wang, Q.; Gao, M.; Huang, Q.; Liu, Y. A comprehensive review on food waste anaerobic digestion: Research updates and tendencies. *Bioresour. Technol.* **2018**, *247*, 1069–1076. [CrossRef]
108. Fotidis, I.; Karakashev, D.B.; Kotsopoulos, T.; Martzopoulos, G.G.; Angelidaki, I. Effect of ammonium and acetate on methanogenic pathway and methanogenic community composition. *FEMS Microbiol. Ecol.* **2013**, *83*, 38–48. [CrossRef]
109. Jang, H.M.; Choi, Y.K.; Kan, E. Effects of dairy manure-derived biochar on psychrophilic, mesophilic and ther-mophilic anaerobic digestions of dairy manure. *Bioresour. Technol.* **2018**, *250*, 927–931. [CrossRef]
110. Linville, J.L.; Shen, Y.; Leon, P.A.I.-D.; Schoene, R.P.; Urgun-Demirtas, M. In-situ biogas upgrading during anaerobic digestion of food waste amended with walnut shell biochar at bench scale. *Waste Manag. Res. J. A Sustain. Circ. Econ.* **2017**, *35*, 669–679. [CrossRef]
111. Pan, J.; Ma, J.; Zhai, L.; Liu, H. Enhanced methane production and syntrophic connection between microorganisms during semi-continuous anaerobic digestion of chicken manure by adding biochar. *J. Clean. Prod.* **2019**, *240*, 118178. [CrossRef]
112. Zhang, D.; Li, W.; Hou, C.; Shen, J.; Jiang, X.; Sun, X.; Li, J.; Han, W.; Wang, L.; Liu, X. Aerobic granulation accelerated by biochar for the treatment of refractory wastewater. *Chem. Eng. J.* **2017**, *314*, 88–97. [CrossRef]
113. Sun, D.; Hale, L.; Crowley, D. Nutrient supplementation of pinewood biochar for use as a bacterial inoculum carrier. *Biol. Fertil. Soils* **2016**, *52*, 515–522. [CrossRef]
114. Dang, Y.; Holmes, D.E.; Zhao, Z.; Woodard, T.L.; Zhang, Y.; Sun, D.; Wang, L.-Y.; Nevin, K.P.; Lovley, D.R. Enhancing anaerobic digestion of complex organic waste with carbon-based conductive materials. *Bioresour. Technol.* **2016**, *220*, 516–522. [CrossRef] [PubMed]

115. Wang, G.; Li, Q.; Gao, X.; Wang, X.C. Sawdust-Derived Biochar Much Mitigates VFAs Accumulation and Improves Microbial Activities To Enhance Methane Production in Thermophilic Anaerobic Digestion. *ACS Sustain. Chem. Eng.* **2018**, *7*, 2141–2150. [CrossRef]
116. Xu, F.; Li, Y.; Ge, X.; Yang, L.; Li, Y. Anaerobic digestion of food waste—Challenges and opportunities. *Bioresour. Technol.* **2018**, *247*, 1047–1058. [CrossRef]
117. Zhang, J.; Zhao, W.; Zhang, H.; Wang, Z.; Fan, C.; Zang, L. Recent achievements in enhancing anaerobic digestion with carbon-based functional materials. *Bioresour. Technol.* **2018**, *266*, 555–567. [CrossRef]
118. Roden, E.E.; Kappler, A.; Bauer, I.; Jiang, J.; Paul, A.; Stoesser, R.; Konishi, H.; Xu, H. Extracellular electron transfer through microbial reduction of solid-phase humic substances. *Nat. Geosci.* **2010**, *3*, 417–421. [CrossRef]
119. McGlynn, S.E.; Chadwick, G.; Kempes, C.P.; Orphan, V. Single cell activity reveals direct electron transfer in methanotrophic consortia. *Nature* **2015**, *526*, 531–535. [CrossRef]
120. Schink, B.; Montag, D.; Keller, A.; Müller, N. Hydrogen or formate: Alternative key players in methanogenic deg-radation. *Environ. Microbiol. Rep.* **2017**, *9*, 189–202. [CrossRef]
121. Chiappero, M.; Norouzi, O.; Hu, M.; Demichelis, F.; Berruti, F.; Di Maria, F.; Fiore, S. Review of biochar role as ad-ditive in anaerobic digestion processes. *Renew. Sustain. Energy Rev.* **2020**, *131*, 110037. [CrossRef]
122. Lovley, D.R. Happy together: Microbial communities that hook up to swap electrons. *ISME J.* **2017**, *11*, 327–336. [CrossRef] [PubMed]
123. Martínez, E.J.; Rosas, J.G.; Sotres, A.; Moran, A.; Cara-Jiménez, J.; Sánchez, M.E.; Gómez, X. Codigestion of sludge and citrus peel wastes: Evaluating the effect of biochar addition on microbial communities. *Biochem. Eng. J.* **2018**, *137*, 314–325. [CrossRef]
124. Jadhav, P.; Muhammad, N.; Bhuyar, P.; Krishnan, S.; Razak, A.S.A.; Zularisam, A.; Nasrullah, M. A review on the impact of conductive nanoparticles (CNPs) in anaerobic digestion: Applications and limitations. *Environ. Technol. Innov.* **2021**, *23*, 101526. [CrossRef]
125. Lin, R.; Deng, C.; Cheng, J.; Xia, A.; Lens, P.N.L.; Jackson, S.A.; Dobson, A.D.; Murphy, J.D. Graphene Facilitates Biomethane Production from Protein-Derived Glycine in Anaerobic Digestion. *Iscience* **2018**, *10*, 158–170. [CrossRef]
126. Li, P.; Li, W.; Sun, M.; Xu, X.; Zhang, B.; Sun, Y. Evaluation of Biochemical Methane Potential and Kinetics on the Anaerobic Digestion of Vegetable Crop Residues. *Energies* **2019**, *12*, 26. [CrossRef]
127. Kumar, S.S.; Ghosh, P.; Kataria, N.; Kumar, D.; Thakur, S.; Pathania, D.; Singh, L. The role of conductive nano-particles in anaerobic digestion: Mechanism, current status and future perspectives. *Chemosphere* **2021**, *280*, 130601. [CrossRef] [PubMed]
128. Pramanik, S.K.; Suja, F.B.; Porhemmat, M.; Pramanik, B.K. Performance and kinetic model of a single-stage an-aerobic digestion system operated at different successive operating stages for the treatment of food waste. *Processes* **2019**, *7*, 600. [CrossRef]
129. Budiyono, I.S.; Sumardiono, S. Kinetic model of biogas yield production from vinasse at various initial pH: Com-parison between modified Gompertz model and first order kinetic model. *Res. J. Appl. Sci. Eng. Technol.* **2014**, *7*, 2798–2805.
130. Huiliñir, C.; Quintriqueo, A.; Antileo, C.; Montalvo, S. Methane production from secondary paper and pulp sludge: Effect of natural zeolite and modeling. *Chem. Eng. J.* **2014**, *257*, 131–137. [CrossRef]
131. Muloiwa, M.; Nyende-Byakika, S.; Dinka, M. Comparison of unstructured kinetic bacterial growth models. *S. Afr. J. Chem. Eng.* **2020**, *33*, 141–150. [CrossRef]
132. Buswell, A.M.; Hatfield, W.D. Anaerobic Fermentations. Bulletin No. 32. 1939. Available online: https://www.ideals.illinois.edu/bitstream/handle/2142/94555/ISWSB-32.pdf?sequence=1 (accessed on 26 January 2021).
133. Abuabdou, S.M.; Ahmad, W.; Aun, N.C.; Bashir, M.J. A review of anaerobic membrane bioreactors (AnMBR) for the treatment of highly contaminated landfill leachate and biogas production: Effectiveness, limitations and future perspectives. *J. Clean. Prod.* **2020**, *255*, 120215. [CrossRef]
134. González, J.; Sánchez, M.E.; Gómez, X. Enhancing Anaerobic Digestion: The Effect of Carbon Conductive Materials. *C* **2018**, *4*, 59. [CrossRef]
135. Monlau, F.; Francavilla, M.; Sambusiti, C.; Antoniou, N.; Solhy, A.; Libutti, A.; Zabaniotou, A.; Barakat, A.; Monteleone, M. Toward a functional integration of anaerobic digestion and pyrolysis for a sustainable resource management. Comparison between solid-digestate and its derived pyrochar as soil amendment. *Appl. Energy* **2016**, *169*, 652–662. [CrossRef]
136. Elsayed, M.; Ran, Y.; Ai, P.; Azab, M.; Mansour, A.; Jin, K.; Zhang, Y.; Abomohra, A.E.-F. Innovative integrated approach of biofuel production from agricultural wastes by anaerobic digestion and black soldier fly larvae. *J. Clean. Prod.* **2020**, *263*, 121495. [CrossRef]
137. Giwa, A.S.; Xu, H.; Chang, F.; Zhang, X.; Ali, N.; Yuan, J.; Wang, K. Pyrolysis coupled anaerobic digestion process for food waste and recalcitrant residues: Fundamentals, challenges, and considerations. *Energy Sci. Eng.* **2019**, *7*, 2250–2264. [CrossRef]
138. Song, J.; Wang, Y.; Zhang, S.; Song, Y.; Xue, S.; Liu, L.; Lvy, X.; Wang, X.; Yang, G. Coupling biochar with anaerobic digestion in a circular economy perspective: A promising way to promote sustainable energy, environment and agriculture development in China. *Renew. Sustain. Energy Rev.* **2021**, *144*, 110973. [CrossRef]
139. Deng, C.; Lin, R.; Kang, X.; Wu, B.; O'Shea, R.; Murphy, J.D. Improving gaseous biofuel yield from seaweed through a cascading circular bioenergy system integrating anaerobic digestion and pyrolysis. *Renew. Sustain. Energy Rev.* **2020**, *128*, 109895. [CrossRef]
140. Shen, Y.; Linville, J.L.; Urgun-Demirtas, M.; Schoene, R.P.; Snyder, S.W. Producing pipeline-quality biomethane via anaerobic digestion of sludge amended with corn stover biochar with in-situ CO_2 removal. *Appl. Energy* **2015**, *158*, 300–309. [CrossRef]

141. Hübner, T.; Mumme, J. Integration of pyrolysis and anaerobic digestion—Use of aqueous liquor from digestate pyrolysis for biogas production. *Bioresour. Technol.* **2015**, *183*, 86–92. [CrossRef] [PubMed]
142. Li, S.; Chan, C.Y.; Sharbatmaleki, M.; Trejo, H.; Delagah, S. Engineered Biochar Production and Its Potential Benefits in a Closed-Loop Water-Reuse Agriculture System. *Water* **2020**, *12*, 2847. [CrossRef]
143. Zhao, Z.; Zhang, Y.; Woodard, T.L.; Nevin, K.P.; Lovley, D.R. Enhancing syntrophic metabolism in up-flow an-aerobic sludge blanket reactors with conductive carbon materials. *Bioresour. Technol.* **2015**, *191*, 140–145. [CrossRef]
144. Khalid, Z.B.; Siddique, M.N.I.; Nayeem, A.; Adyel, T.M.; Ismail, S.B.; Ibrahim, M.Z. Biochar application as sustainable precursors for enhanced anaerobic digestion: A systematic Review. *J. Environ. Chem. Eng.* **2021**, *9*, 105489. [CrossRef]
145. Sahoo, K.; Bilek, E.; Bergman, R.; Mani, S. Techno-economic analysis of producing solid biofuels and biochar from forest residues using portable systems. *Appl. Energy* **2019**, *235*, 578–590. [CrossRef]
146. Kumar, A.N.; Dissanayake, P.D.; Masek, O.; Priya, A.; Lin, C.S.K.; Ok, Y.S.; Kim, S.-H. Recent trends in biochar integration with anaerobic fermentation: Win-win strategies in a closed-loop. *Renew. Sustain. Energy Rev.* **2021**, *149*, 111371. [CrossRef]
147. Peng, W.; Pivato, A. Sustainable Management of Digestate from the Organic Fraction of Municipal Solid Waste and Food Waste Under the Concepts of Back to Earth Alternatives and Circular Economy. *Waste Biomass Valorization* **2019**, *10*, 465–481. [CrossRef]
148. Opatokun, S.A.; Kan, T.; Al Shoaibi, A.; Srinivasakannan, C.; Strezov, V. Characterization of Food Waste and Its Digestate as Feedstock for Thermochemical Processing. *Energy Fuels* **2016**, *30*, 1589–1597. [CrossRef]
149. Zhao, W.; Yang, H.; He, S.; Zhao, Q.; Wei, L. A review of biochar in anaerobic digestion to improve biogas production: Performances, mechanisms and economic assessments. *Bioresour. Technol.* **2021**, *341*, 125797. [CrossRef]
150. González, R.; González, J.; Rosas, J.; Smith, R.; Gómez, X. Biochar and Energy Production: Valorizing Swine Manure through Coupling Co-Digestion and Pyrolysis. *C* **2020**, *6*, 43. [CrossRef]
151. Ambaye, T.G.; Rene, E.R.; Nizami, A.S.; Dupont, C.; Vaccari, M.; van Hullebusch, E.D. Beneficial role of biochar addition on the anaerobic digestion of food waste: A systematic and critical review of the operational parameters and mech-anisms. *J. Environ. Manag.* **2021**, *290*, 112537. [CrossRef] [PubMed]
152. Moreira, M.T.; Noya, I.; Feijoo, G. The prospective use of biochar as adsorption matrix—A review from a lifecycle perspective. *Bioresour. Technol.* **2017**, *246*, 135–141. [CrossRef] [PubMed]
153. Deshavath, N.N.; Mogili, N.V.; Dutta, M.; Goswami, L.; Kushwaha, A.; Veeranki, V.D.; Goud, V.V. Role of lig-nocellulosic bioethanol in the transportation sector: Limitations and advancements in bioethanol production from lignocellulosic biomass. In *Waste-to-Energy Approaches Towards Zero Waste*; Elsevier: Amsterdam, The Netherlands, 2022; pp. 57–85.
154. Bajgai, R.C.; Tamang, D.T.; Kushwaha, A.; Goswami, L. Strategic consideration as feedstock resource for biofuel production as a holistic approach to control invasive plant species. In *Waste-to-Energy Approaches Towards Zero Waste*; Elsevier: Amsterdam, The Netherlands, 2022; pp. 245–268.
155. Gautam, A.; Kushwaha, A.; Rani, R. Microbial remediation of hexavalent chromium: An eco-friendly strategy for the remediation of chromium-contaminated wastewater. In *The Future of Effluent Treatment Plants*; Elsevier: Amsterdam, The Netherlands, 2021; pp. 361–384.
156. Goswami, L.; Pakshirajan, K.; Pugazhenthi, G. Biological treatment of biomass gasification wastewater using hy-drocarbonoclastic bacterium Rhodococcus opacus in an up-flow packed bed bioreactor with a novel waste-derived nano-biochar based bio-support material. *J. Clean. Prod.* **2020**, *256*, 120253. [CrossRef]
157. Singh, A.; Kushwaha, A.; Goswami, S.; Tripathi, A.; Bhasney, S.M.; Goswami, L.; Hussain, C.M. Roadmap from microalgae to biorefinery: A Circular Bioeconomy approach. In *Emerging Trades to Approaching Zero Waste*; Elsevier: Amsterdam, The Netherlands, 2022; pp. 373–392.
158. Anupama; Khare, P. A comprehensive evaluation of inherent properties and applications of nano-biochar prepared from different methods and feedstocks. *J. Clean. Prod.* **2021**, *320*, 128759. [CrossRef]

catalysts

MDPI

Article

Rapid Detection of Mercury Ions Using Sustainable Natural Gum-Based Silver Nanoparticles

Samie Yaseen Sharaf Zeebaree [1,*], Osama Ismail Haji [2,3], Aymn Yaseen Sharaf Zeebaree [1], Dunya Akram Hussein [1] and Emad Hameed Hanna [1]

1 Deptartment of Medical Laboratory Technology, College of Health and Medical Technology, Duhok Polytechnic University, Kurdistan Region, Dahuk 42001, Iraq
2 Medical Biochemical Analysis Department, College of Health Technology, Cihan University-Erbil, Kurdistan Region, Erbil 44008, Iraq
3 General Science Department, Faculty of Education, Soran University, Soran 44008, Iraq
* Correspondence: samie.yasin@dpu.edu.krd

Abstract: Fabrication of metal nanostructures using natural products has attracted scientists and researchers due to its renewable and environmentally benign availability. This work has prepared an eco-friendly, low-cost, and rapid colorimetric sensor of silver nanoparticles using tree gum as a reducing and stabilizing agent. Several characterization techniques have been exploited to describe the synthesized nanosensor morphology and optical properties. Ultraviolet−Visible (UV−Vis) spectroscopy has been used for monitoring the localized plasmon surface area. High-resolution transmission electron microscopy (HR-TEM) illustrated the size and shape of silver nanoparticles. X-ray diffraction spectra showed the crystallography and purity of the product. Silver nanoparticles decorated with almond gum molecules (AgNPs@AG) demonstrated high sensitivity and colorimetric detection of mercury ions in water samples. The method is based on the aggregation of AgNPs and the disappearing yellow color of AgNPs via a spectrophotometer. The detection limit of this method was reported to be 0.5 mg/L. This work aimed to synthesize a rapid, easy-preparation, eco-friendly, and efficient naked-eye colorimetric sensor to detect toxic pollutants in aqueous samples.

Keywords: natural gum; silver nanoparticles; mercury ion detection

Citation: Sharaf Zeebaree, S.Y.; Haji, O.I.; Zeebaree, A.Y.S.; Hussein, D.A.; Hanna, E.H. Rapid Detection of Mercury Ions Using Sustainable Natural Gum-Based Silver Nanoparticles. *Catalysts* **2022**, *12*, 1464. https://doi.org/10.3390/catal12111464

Academic Editors: Beom Soo Kim and Pritam Kumar Dikshit

Received: 13 August 2022
Accepted: 10 November 2022
Published: 18 November 2022

1. Introduction

Population growth worldwide has led to increased demand for drinking water in recent years. However, water resources have suffered from pollution due to industrial sectors' discharging various kinds of pollutants and harmful substances. Many kinds of pollutants, such as herbicides, pesticides, dyes, plastics, oil derivatives, and heavy metals [1–4], have been found in water effluents, which is the main reason behind the development of health issues among people. Mercury and its compounds are among the most harmful species reported and adversely affect human health [5]. Mercury affects the central nervous system, liver, and kidney. Further, it can disturb the immune system due to impaired hearing, vision, paralysis, and emotional instability. Many analytical techniques are employed for detecting mercury ions, such as inductively coupled plasma atomic emission and mass spectroscopy, cold vapor atomic absorption spectroscopy, cold vapor atomic fluorescence spectroscopy, and high-performance liquid chromatography with UV−Vis detection [6]. However, these techniques suffer from several limitations, for instance, time consumption, harmful solvents, equipment costs, and complex procedures [7]. Therefore, developing new approaches to detecting and determining mercury ions becomes crucial and an urgent need.

Metal nanoparticles possess several unique optical, photothermal, and electrical properties. These distinctive characteristics make scientists more curious to investigate and uncover the potential applications of metal nanoparticles [8]. Gold, silver, and copper nanoparticles have been used widely in many fields such as medicine, industry, agriculture, and the environment. The vast spectrum of applications of these nanostructured metals come from the nanoscale size, which has a variant behavior for its bulk state [9]. However, it must undergo chemical, physical, or biological processing to convert a metal from a bulk state to nanoscale particles. Biological methods are considered an emerging technique used to fabricate metal nanoparticles. It is classified into two main subclasses, that is, plant and microorganism. The plant's parts include roots, stems, flowers, leaves, and seeds [10]. Some trees' stems release exudate (gum) as a defense weapon against microorganisms and insects [11]. However, the gum has been proven to contain many active chemical components. Almond gum is a natural product released from the tree stem. It contains fats (0.85%), protein (2.45%), and carbohydrates (92.36%). Carbohydrates include glucose, galactose, arabinose, mannose, rhamnose, and uronic acid. These bioactive compounds possess active functional groups such as hydroxyl groups (OH), carboxylic groups (COOH), and carbonyl groups (CO). It is considered a rich site electron donor and has a crucial role in reducing action [12].

Prior studies have been performed to prepare silver nanoparticles using plant exudate. For instance, *Mimosa pudica* gum was successfully employed in the synthesis of silver nanoparticles as an antimicrobial agent [13]. Gum *Tragacanth* has been used as a reducing and stabilizing agent for fabricating silver nanoparticles [14]. The aqueous solution of gum *Acacia* was utilized efficiently in the preparation of silver nanoparticles [15]. Arabic gum has served a dual function for synthesizing silver nanoparticles and evaluating its activity against pathogen bacteria [16]. *Cashew* gum has been exploited for the production of silver nanoparticles to study its activity as an anti-bacterium [17]. Biogenic silver nanoparticles were prepared with gum *ghatti* and gum *olibanum*. Further, inhibition action against certain bacteria was observed and was biocompatible with other bacteria [18]. Catalytic activity has been evaluated in silver nanoparticles synthesized by gum *Karaya* [19]. Plant gums of *Araucaria heterophylla*, *Azadirachta indica*, and *Prosopis chilensis* mediated have been used for the fabrication of silver nanoparticles and assisted in the removal of heavy metal, anticancer, and antibacterial agents [20]. Most of the mentioned studies fabricated distinctive and efficient silver nanoparticles. However, they have used complex apparatuses, high temperature, high pressure, and long-time reactions. Further, they concentrated on biological applications.

The current study has employed almond gum as a reducing and capping agent to synthesize a eco-friendly, low-cost, easy-preparation and rapid naked eye colorimetric sensor for silver nanoparticles. Further, AgNps@AG showed rapid and high sensitivity to mercury (II) ions in an aqueous medium.

2. Results and Discussion

2.1. Fabrication of AgNPs@AG

Nanotechnology's essence is synthesizing nanoparticles according to simple, easy, inexpensive, and environmentally friendly steps, including the use of green natural resources. However, validating the principle of the synthesis of nanoparticles using bio-waste extracts, including plants, is considered to be of paramount importance in the scientific and research community. Over the past years, many green metallic nanoparticles, including copper, selenium, titanium, and silver, have been synthesized according to the green reduction mechanism through active biological components operating in plants. Most of the plant body comprises multiple vital compounds, such as proteins, carboxylic acids, flavonoids, glycosides, and polysaccharides, in addition to a percentage of metals and hydrating compounds. Due to its high biosignificance, the gum contents have been investigated by Hossein and Ali's group using HPLC and FTIR analysis [21]. The HPLC analysis revealed that the gum contains an excellent ratio of many active chemical compounds, such as proteins, polysaccharides, uronic acid, fats, and ash.

In contrast, the FTIR study for these compounds showed that their structures consist of various active functional groups such as C=O, CO_2H, OH, and NH_2. According to the analysis, the high value of the almond gum return to the polysaccharide, which involve several monosaccharides such as rhamnose, mannose, glucose, xylose, arabinose, and galactose existing in the branched chain of structure. However, due to the foundation of hydroxy−flavonoid structure in these saccharides containing a high OH groups ratio, it is responsible for reducing many metal ions to nanoparticles. Therefore, the formation of AgNPs@AG can be attributed to the poly- and monosaccharides molecules, which are considered active chelating compounds because they possess multiple hydroxyl groups located at carbon atoms of the flavone rings (Scheme 1). This hypothesis has been confirmed by several published reports on reducing various metal ions, such as Cu^{2+}, Se^{2+}, Ti^{4+} Fe^{3+}, and Zn^{2+}, using plant extracts [22].

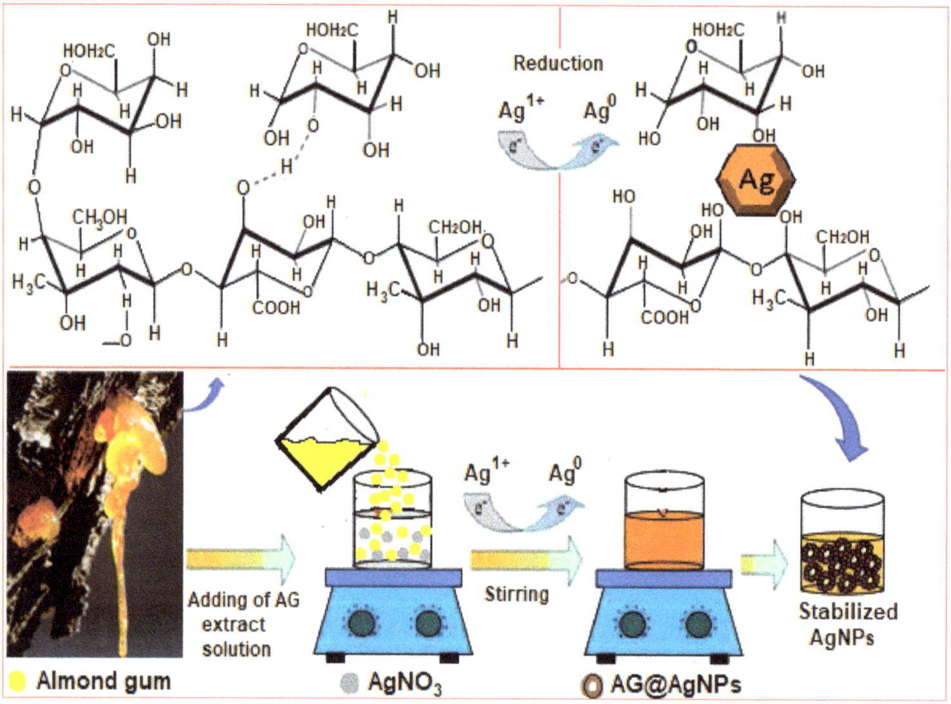

Scheme 1. The supposed mechanism for the fabrication of AgNPs@AG by an almond gum extract.

2.2. UV−Vis Analysis Measurement

Employing ultra violet-visible spectroscopy to identify nanoparticles is a powerful and valuable method for the preliminary diagnosis of bio-fabricated AgNPs@AG. However, after mixing the required colorless components for the preparation of the AgNPs@AG catalyst, dark brown color was formed. The formation of this color was an initial confirmation of the synthesis of AgNPs@AG. Nevertheless, for more support, the observed color value of the catalyst was read on a nanodrop 2000/2000c spectrophotometer at the scale of 250−700 nm, where a robust and broad curve at 410 nm was recorded on the formation of nanoparticles with LSPR properties. The appearance of this result is a confirmation point on the formation of NPs in nanosizes, which agrees with a previous published report' data [23]. The brown−yellow color formation was returned to the reduction process between the Ag^+ ions and active functional groups present in the gum extract components and eventually formed Ag^0. Moreover, it was due to the excitation of the surface plasmon resonance (SPR) vibrations of electrons in AgNPs (Figure 1—red line). The valence band and the conduction band of the AgNPs were very close, making a good reason for electrons moving freely, giving a sharp SPR absorption band at 410 nm [24]. In contrast, this band was monitored after adding a brown−yellow sensor solution to the mercury ions sample (Figure 1—blue line). The analysis revealed excellent suppression for the sensor color to be colorless during 13 seconds at ambient conditions. Furthermore, this color's disappearance and formation of a colorless phase were approved on the well unit of Ag with Hg ions to form a Ag−Hg phase in the aqueous solution. This result has been proven in many literature and research works [25].

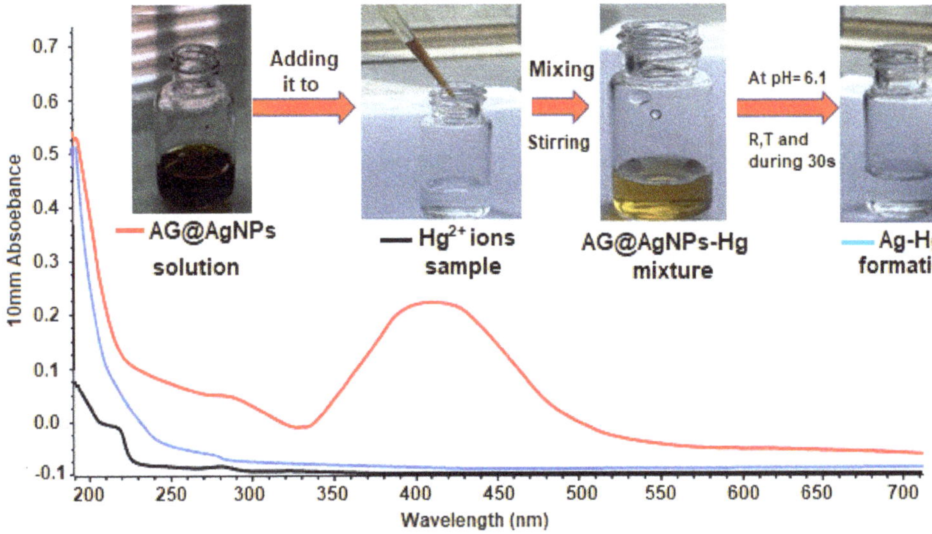

Figure 1. Measured UV−Vis spectra of Hg^{2+} (black line), AgNPs@AG (red line), and Ag−Hg (blue line).

2.3. *Surface Morphology Diagnosis*

Investigation of the oxidation states and nature of the surface of the prepared nanomaterials is crucial and widely needed in multiple applications and fields. For this requirement, XPs analysis was utilized to diagnose the orbitals type formed for AG surfaces and encapsulated AgNps as well as the reaction mixture sample (AgNPS−Hg or Ag−Hg) as shown in Figure 2a,b. For the AG extract (Figure 2a—blue line), the XPS analysis chart showed two firm peaks located at binding energies of 283 and 539 eV, returning to the carbon atom (C 1s) and an oxygen atom (O 1s), respectively. However, the appearance of these atoms (i.e., C and O) are generally back to the presence of multi-organic compounds in gum bio content such as flavones, phenols, and carboxylic acids, which especially possess multi-functional groups in their structures such as C−C, C−O−C, C=O, OH, and COOH. The presence of these active groups came from the existence of hydroxyl and carbonyl groups that built the polymeric structure of polysaccharides. Additionally, these single bands (C 1s and O 1s) were extensive supports and evidence of the stability of cyclic C−C atoms of ketonic polysaccharides rings in the resin structure. The measured results of previous works strongly confirm the recorded data in this regard [26]. In contrast, after encapsulating the AG−Ag NPs by resin, the analysis survey of XPs (Figure 2a—black line) showed excellent three satellite peaks located at 288, 400, and 533 eV, respectively, which are attributed to the core levels of C 1s, Ag 3d, and O 1s, respectively. Noticing and recording these orbitals is a good indication of the coating of synthesized NPs by gum components. This diagnosis corresponds well with the previously published reports for the fabrication of a zerovalent silver state [27]. No presence or appearance of other elements in the spectrum analysis is a well proof of the use of almond gum as a sustainable reducing and stabilizing agent for designing silver nanosensors. However, the sensor activity for the prepared AgNPs@AG was tested, and XPs analysis survey was applied to show the sensor efficiency in detecting Hg ions to prove the reaction process between AgNPs and Hg ions. XPs diagnosis for the mixture (Figure 2b—brown line) revealed significant photoelectron peaks values at 100 and 120 eV, which returned to the orbital states of Ag 3d5/2 and Hg 4f, respectively. These types of orbitals were formed due to the adsorption of mercury ions on the gum surface. The redox process by the transition of electrons between the Ag^0 and Hg^{2+} ions occurred, thus forming the Ag−Hg state. These obtained results agree with the previous literature in forming the Ag−Hg phase [28]. The stability and contacting effects for each AgNPs@AG and formatted Ag−Hg solutions with ambient conditions were evaluated for more support. The process was performed by opening the bottle head at different times (24 to 240 h), as shown in Figure 2b (black line). The fabricated sensor recorded no notable change in color or their stability to produce another state, indicating the perfect role of the gum in protecting the silver nanoparticles and using it for long periods.

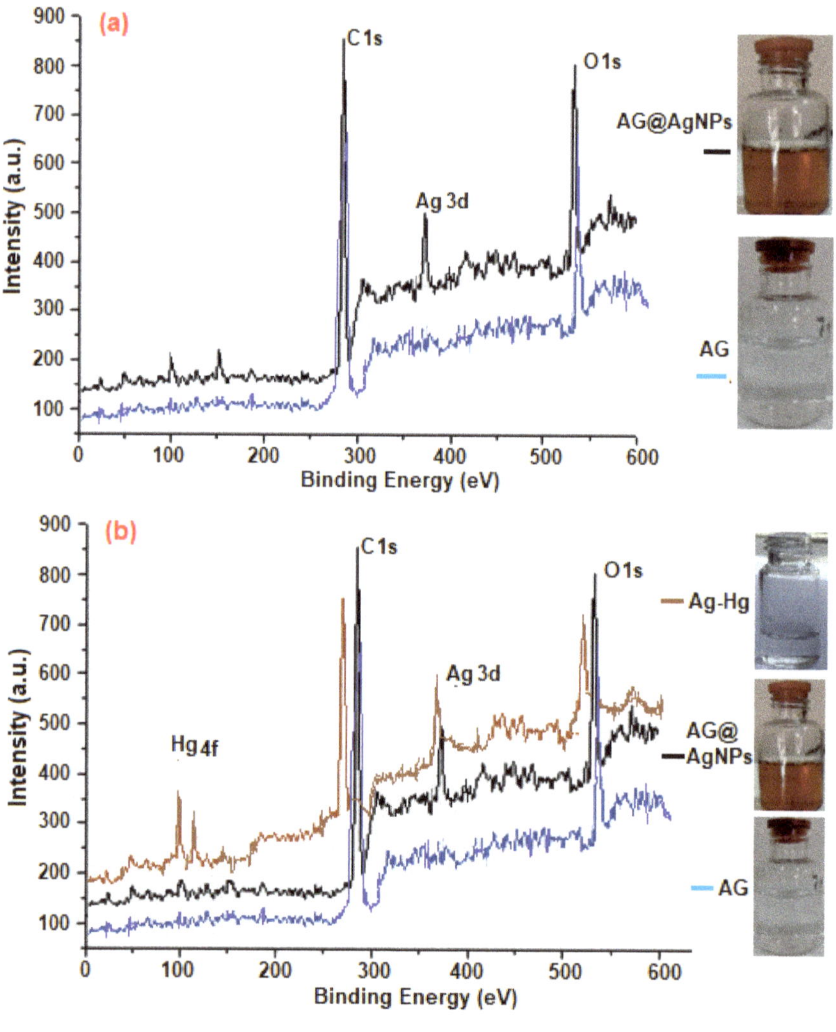

Figure 2. (**a**) XPs spectra of AG and AgNPs@AG; (**b**) XPs analysis of AG, AgNPs@AG, and Ag−Hg.

2.4. XRD Measurement

XRD is a unique tool used to determine the crystalline structure of the fabricated materials at different scales. However, XRD analysis has been applied for each AG, AgNPs@AG, and Ag−Hg. The spectrum line pattern of AG (black line) revealed a significant broad peak located at 2θ, 20°, which assumed back to the polysaccharide present in the gum composition [29]. This value was recorded besides the formation of AgNPs, after they encapsulated the AgNPs, where the analysis spectrum of AgNPs@AG (brown line) revealed excellent four peaks localized at 2θ values of 38.27°, 44.48°, 64.59°, and 77.46°, respectively, which had the corresponding planes (hkl) (111), (200), (220), and (311), respectively, as shown in Figure 3. The recorded results confirmed that the prepared AgNPs@AG possessed face-centered cubic lattices [15]. The appearance of sharp peaks in the X-ray spectrum at the base level strongly indicated that the synthesized AgNPs were prepared in a good form. The average size (*D*) for the as-prepared AgNPs@AG was calculated by employing the Scherrer equation ($D = [0.9\lambda/\beta\cos\theta]$), where λ and β represent the wavelength source and the breadth of the diffraction line at its half intensity maximum recorded in XRD spectrum, respectively. The calculations showed that the average size of the prepared AgNPs@AG crystalline was found to be 19 nm. However, after treating the prepared sensor with the sample of mercury ions, the XRD spectrum pattern (blue line) recorded the absence of AgNPs@AG peaks with the appearance of new interesting sharp peaks at locations 25.79°, 32.51°, 56.26°, and 66.11°, which returned to the formation of the Ag−Hg status. These measured values agree with the previous work's results [30].

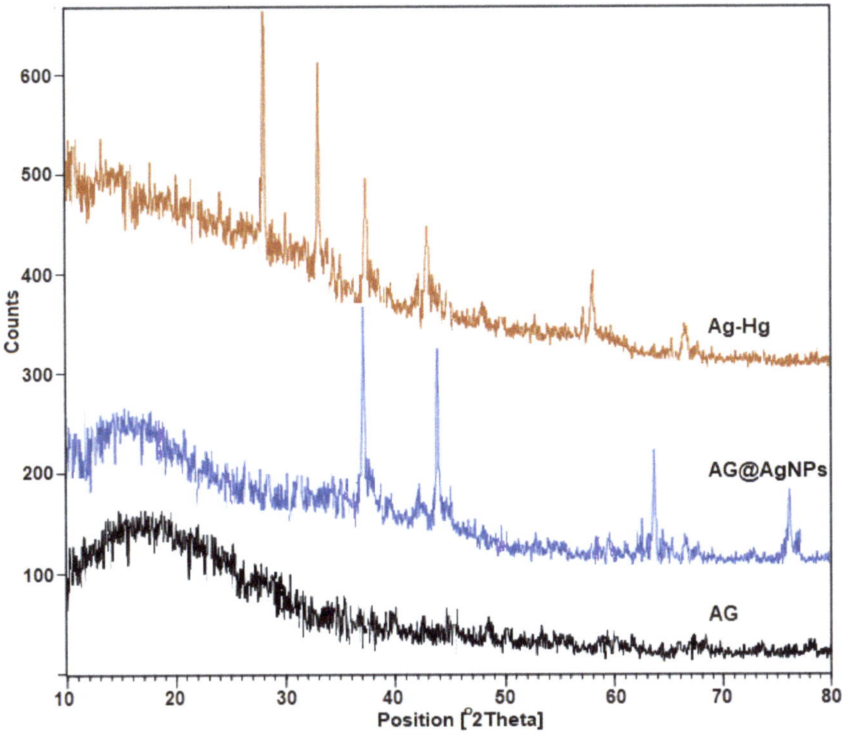

Figure 3. XRD studies of AG (black line), AgNPs@AG (blue line), and Ag−Hg (brown line).

2.5. FESEM and HR-TEM Photograph Analysis

The external and internal structures of the prepared nanoparticles has been explored and determined by FESEM and HR-TEM microanalysis. It is essential for its contribution to characterizing the particle type involved in the colorimetric response process. The analysis of FE-SEM photos (Figure 4a) revealed that AgNPs@AG formed in tiny and nonagglomerated spherical shapes due to the oily biomass nature of the gum extract. In addition, the image analysis (Figure 4b) revealed that the AgNPs@AG particles formed with perfect sizes ranging from 15 to 17 nm, which indicated the efficiency of the bio fabrication method. However, after adding these small nanosize AgNPs to the Hg ions sample, FESEM images showed excellent aggregation of precipitate, attributed to the formation of Ag−Hg (Figure 4c). The formation of the mercury−silver state occurred through the adsorption of mercury ions on the active surface of the AgNPs. Further, it could be due to its adhesion to the resin covering the AgNPs surface, which then interacted with the active AgNPs. There was a consistent parallel agreement between the images acquired from the HR-TEM and the data collected from the FESEM regarding the formation of small nanosphere particles. (Figure 4d). In contrast, Figure 4e proved the good trapping of mercury ions by the sensor with perfect aggregation. XRD and XPs have confirmed this hypothesis analysis results, in addition to the results and explanations of the published research.

2.6. EDS and EDC Identification

Synthesizing and producing good, clean, cheap, and sustainable sensors using green methods for colorimetric response is considered one of the most important challenges for researchers. For this purpose, the purity of fabricated AgNPs@AG was evaluated by determining their components using elements distribution spectroscopy (EDS) supported by elements spread diagram (ESD). The EDS spectrum (Figure 5a) revealed good weights ratios of elements for silver (Ag = 34.2%), oxygen (O = 17.5%), and carbon (C = 47.4%). No emergence or recording of other elements in the EDS chart is strong evidence of the clean methodology for this process. However, this advantage was exploited for the detection of Hg ions by the prepared sensor, where the analysis revealed the excellent formation of the Ag−Hg compound with ratios of 45, 67, 77, and 98, which belonged to the Ag, O, C and Hg, respectively (Figure 5b). EDS analysis clearly supported the XRD, XPs, FESEM, and HR-TEM analysis results, as it proved that the agglomeration of components and the formation of the Ag−Hg structure were due to the role of the bio-active surface of AgNPs and their resin coating that acted as a linking and reducing agent.

2.7. Factors Affecting the Sensing Process

2.7.1. Effect of pH on the Reaction Medium

It was recorded that the pH of AgNPs@AG was slightly acidic (6.1). This could promote that the as-prepared silver nanosensor was coated with proton donor functional groups such as hydroxyl, carboxyl, and carbonyl groups, which remarkably exist in carbohydrate molecules mentioned in almond gum constituents [31]. However, by examining the impact of acidity and basicity on the medium of the reaction, we observed that the absorbance value of the mixture was increased. This could be attributed to the formation of turbidity. Moreover, the assumed reason we expected was the presence of a small percentage of protein participating in the capping process of silver nanoparticles. However, the optimum pH of the mixture was assigned at 6.1, as shown in Figure 6.

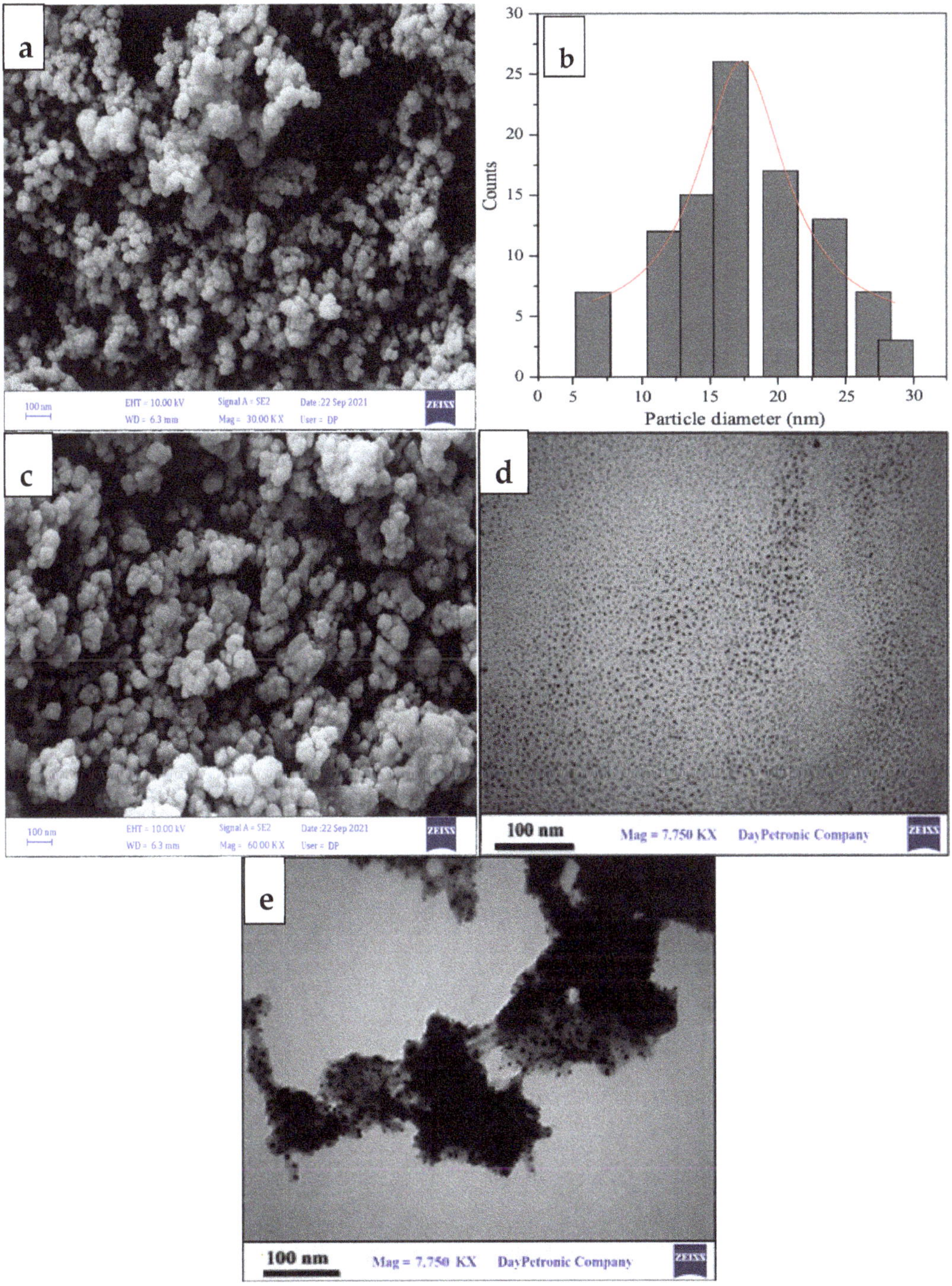

Figure 4. FESEM images of AgNPs@AG and size distribution(**a**,**b**) and Ag−Hg (**c**); HR-TEM images of AgNPs@AG (**d**) and Ag−Hg (**e**).

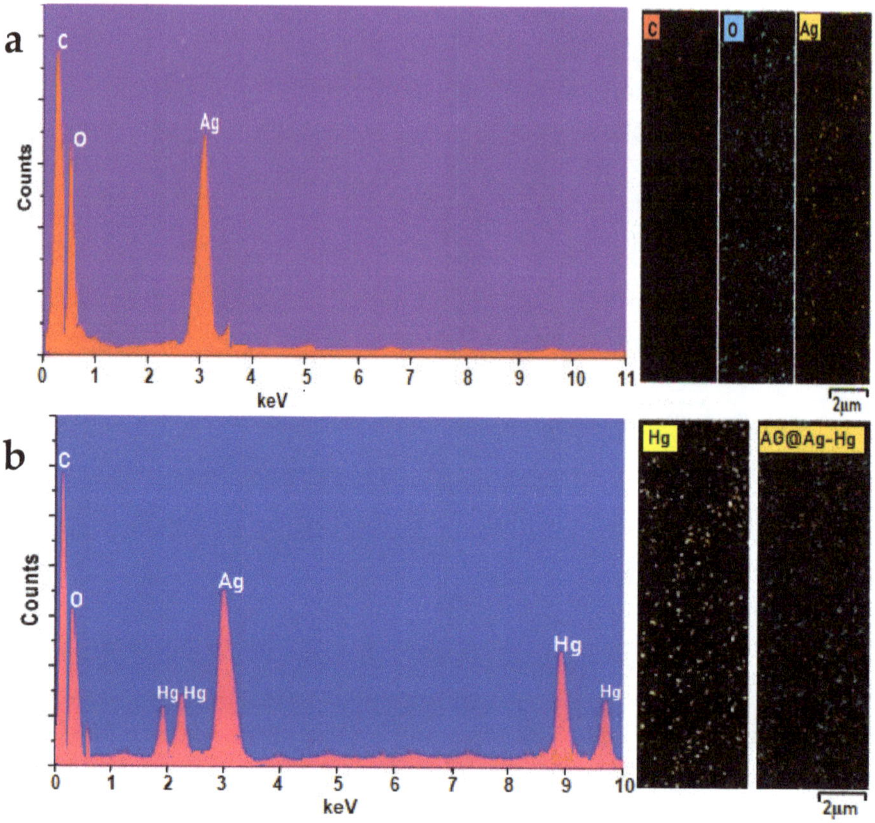

Figure 5. EDS measurements of AgNPs@AG supported with ESD (**a**) and Ag−Hg with supported ESD (**b**).

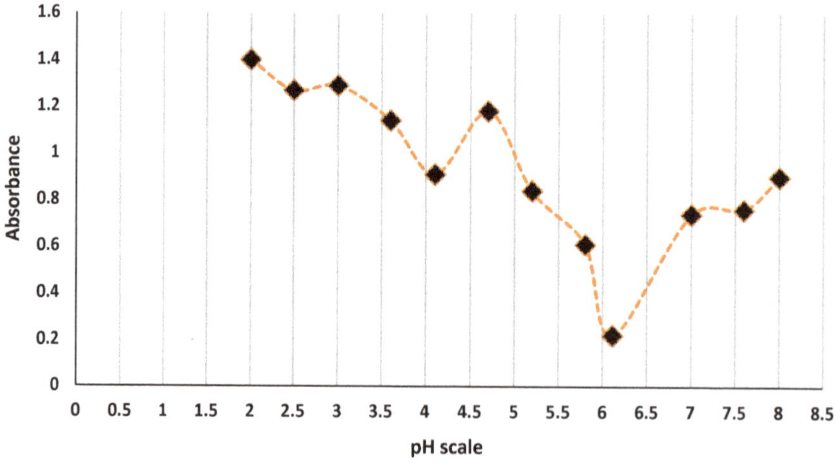

Figure 6. pH scale and its effect on the AgNPs@AG reaction medium.

2.7.2. Calibration Curve of the Method

The validity of the method was examined after standardizing the optimum conditions. The change in ΔA value of AgNPs@AG was monitored at 410 nm (Figure 7a) upon mixing the silver nanosensor with Hg^{2+} ions. The method observed good linearity achieved over a concentration range of 10–180 mg/L (see Figure 7b) by plotting the absorbance values against Hg^{2+} concentrations. Further, the correlation of determination (R^2) was 0.989. Therefore, the limit of detection (LOD) was 0.5 mg/L, and the limit of quantification (LOQ) was calculated to be 1.69 mg/L. The limit of detection (LOD) and the limit of quantification (LOQ) were determined from the equation (3 ∂/slope) and (10 ∂/slope), respectively.

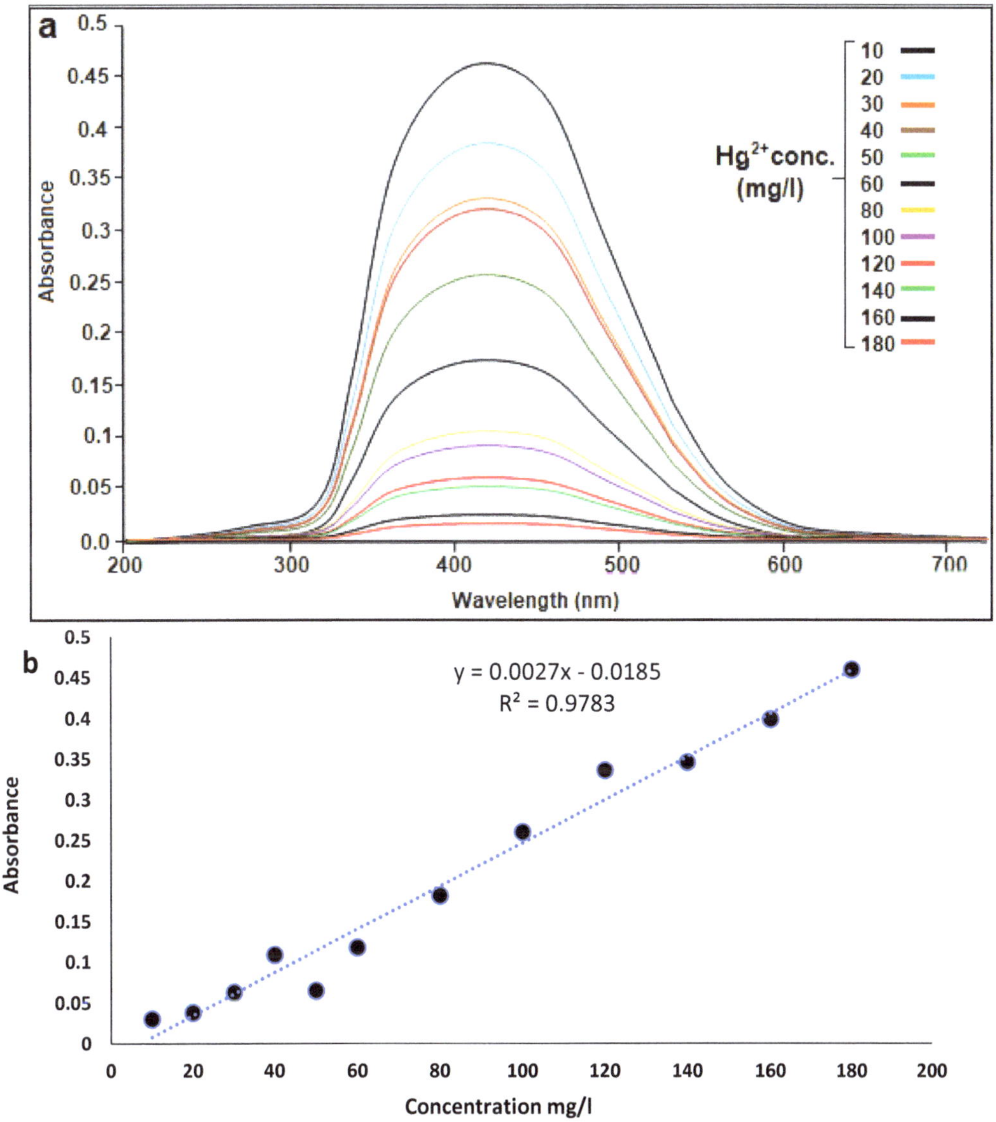

Figure 7. UV–Vis analysis (**a**) and calibration calculation (**b**) of AgNPs@AG performance validity at different concentrations (10–180 mg/L).

2.7.3. Effect of Time on the Sensing Process

The time effect was studied by monitoring the color change of AgNPs@AG from yellow color to colorless. Further, the sensing behavior was tracked using a spectrophotometer following the decreasing absorbance value against blank. It was shown that the sensing of mercury ions in an aqueous medium by silver nanoparticles was very rapid and the yellow color of AgNPs@AG disappeared instantly after the addition of silver nanoparticles with Hg^{2+} ions. The recorded time for this action was 30 s, indicating that the sensing process was swift and efficient.

2.7.4. Comparative Study between AgNPs@AG and Other Catalysts in Hg Detection

Comparing and evaluating the preparation approach, ease of conduct, and activity of any newly prepared catalyst with previously synthesized catalysts is crucial to achieving the same purpose. Here, the activity and efficiency of our fabricated sensor based on the LSPR of AgNPs@AG to detect Hg^{2+} ions were compared with those obtained by previous methods and works (Table 1). Furthermore, the prepared catalyst was compared with many different nano metallic sensors for detecting Hg^{2+} ions. The comparison table shows that our novel fabricated AgNPs@AG revealed good results for Hg^{2+} ions detection compared to the different synthesized green sensors and chemically synthesized ones. The good benefits of AgNPs@AG sensor are represented in the quick sensing of Hg^{2+} ion with the color change and noticing it with the naked eye in a short time. This result confirmed that the surface of our catalyst had excellent activity. In addition, it had high efficiency with low cost and was environment-friendly. Another advantage was that our procedure for estimating Hg^{2+} ions was compared with the different approaches. The recorded data for our work principle proved that it is easy to perform, sensitive, and cheap without needing any further enhancement or using large complex apparatuses or relatively complex detection steps that need a long time.

Table 1. Comparison of AgNPs@AG activity against detection of mercury (Hg^{2+}) with the previous published literature and reports.

No.	Principles and Sensor Used	Target Ion	Limit of Detection (LOD)	Limit Time	Reference
	(A) Greenly synthesized				
1	Spectrophotometry: AgNPs	Mercury	16×10^{-2} mg/L	-	[32]
2	Spectrophotometry: AgNPs	Mercury	44×10^{-2} mg/L	5 min	[33]
3	Spectrophotometry: AgNPs	Mercury	74×10^{-7} mg/L	20 min	[34]
4	Spectrophotometry: AgNPs	Mercury	8×10^{-6} mg/L	5 min	[35]
	(B) Chemically synthesized				
5	Spectrophotometry: Ag NPs	Mercury	16×10^{-4} mg/L	6 min	[36]
6	Spectrophotometry: Ag NPs	Mercury	6×10^{-1} mg/L	8 min	[37]
	(C) Other nano metals				
7	Spectrophotometry: Au NPs	Mercury	2×10^{-4} mg/L	10 min	[38]
8	Spectrophotometry: Pt NPs	Mercury	7×10^{-5} mg/L	30 min	[39]
	(D) Other principles				
9	HPLC	Mercury	8×10^{-4} mg/L	12 min	[40]
10	Electrochemistry	Mercury	12×10^{-3} mg/L	1.5 min	[41]
11	Electrophoresis	Mercury	1.4 mg/L	5 min	[42]
12	Fluorimetry	Mercury	8.4×10^{-5} mg/L	5min	[43]
13	GC-FI	Mercury	1×10^{-5} mg/L	60 min	[44]
14	Amperometry	Mercury	2×10^{-6} mg/L	1 min	[45]
15	AgNPs@AG: Spectrophotometry	Mercury	5×10^{-1} mg/L	30 s	This work

2.8. Evaluation of the Sensor Selectivity

The performance sensitivity, selectivity, and the naked eye noticeable for AgNPs@AG sensor against groups of different anions and cations were evaluated. The test was performed with the same optimum conditions (a mixture of 3 mL Hg^{2+} and 50 μL of AgNPs@AG) against several cations and anions. The recommended method was tracked to evaluate the effect of the interferences on determining the Hg^{2+} ions. However, interferences for I^-, Cl^-, Na^+, and Li^+ were at 1000 ppm and those for Mg^{2+}, Ba^{2+}, and $NO3^-$ were at 250 ppm, whereas interferences for Zn^{2+}, Fe^{2+} and Co^{2+} were at 50 ppm. For adequate consideration, an error of 5% in concentration was recorded. The sensor behavior against mercury ions revealed excellent yellow color with a clear naked eye seen (Figure 8a). In contrast, the foreign ions did not overlap with the required ion (Hg ions) during the sensing process, even with the high concentrations, with no color change. However, based on the monitoring process of the reaction mixture by the naked eye (Figure 8a), no change in the yellow color solution was observed in less than 13 seconds and even 10 minutes for the common ions of Zn^{2+}, Co^{2+}, Fe^{2+}, Mg^{2+}, Ba^{2+}, Na^+, K^+, I^-, Cl^-, and NO_3^- even at concentrations 500 times higher than for mercury. UV−Vis analysis supported this behavior more (Figure 8b,c). The variation detail of UV−Vis measurements revealed that the sensor's selectivity worked only with mercury ions with a significant change in the absorbance curve, ranging from 0 to 0.19 nm. In contrast, the analysis revealed low absorbance changes with the other ions (Zn^{2+}, Co^{2+}, Fe^{2+}, Mg^{2+}, Ba^{2+}, Na^+, Li^+, I^-, Cl^-, and NO_3^- for the ranges of 0.01, 0.03, 0.02, 0.04, 0.04, 0.04, 0.02, 0.04, 0.04, and 0.05 nm, respectively). The results confirmed that this novel sensor had excellent detection performance for mercury ions in the effluent.

2.9. Performance of the AgNPs@AG Sensor in Tap Water Samples

In this part, the performance of the developed AgNPs@AG sensor was evaluated by utilizing it with collected actual water samples for the detection and determination of mercury ions. Water samples (home drinking water) were collected from different places in Shekan area in Duhok city and in Kurdistan region in Iraq during different times. This test was performed to prove the reliability of the efficient colorimetric detection response method of the AgNPs@AG sensor in practical applications. First, the principle of pretreatment for water samples was applied according to the protocol of Section 2.4. According to the obtained data in Table 2, it was reported that the sensor behavior revealed no color transformation with the samples of tap water (i.e., before dealing it with mercury ions) even with high addition concentrations of the sensor. Furthermore, it indicated no mercury ions in tap water samples or overlapping water total contents with the sensor solution. In contrast, after the treatment of the same water samples with mercury ions, the recovery experiments for mercury ions were carried out, and the results were recorded as shown in Table 2. The recovery process results revealed that the prepared silver catalyst had good recovery activity for mercury ions and with ratios of 91.5–101.7% at different concentrations (10–50 ppm), which proved that the presented method is feasible for the mercury ions detection in real environmental water samples.

Table 2. Detection of Hg ions in real water samples.

Sample	Added Analyte (mg/L)	Found (mg/L)	Recovery Ratios * (%, n = 3)
Before treatment	0	0	0
After treatment	10	9.1	91 ± 0.89
	20	19.4	97 ± 1.59
	30	29.7	99 ± 0.67
	40	40.2	100.5 ± 0.93
	50	50.7	101.7 ± 1.88

* Mean ± standard deviation (n = 3).

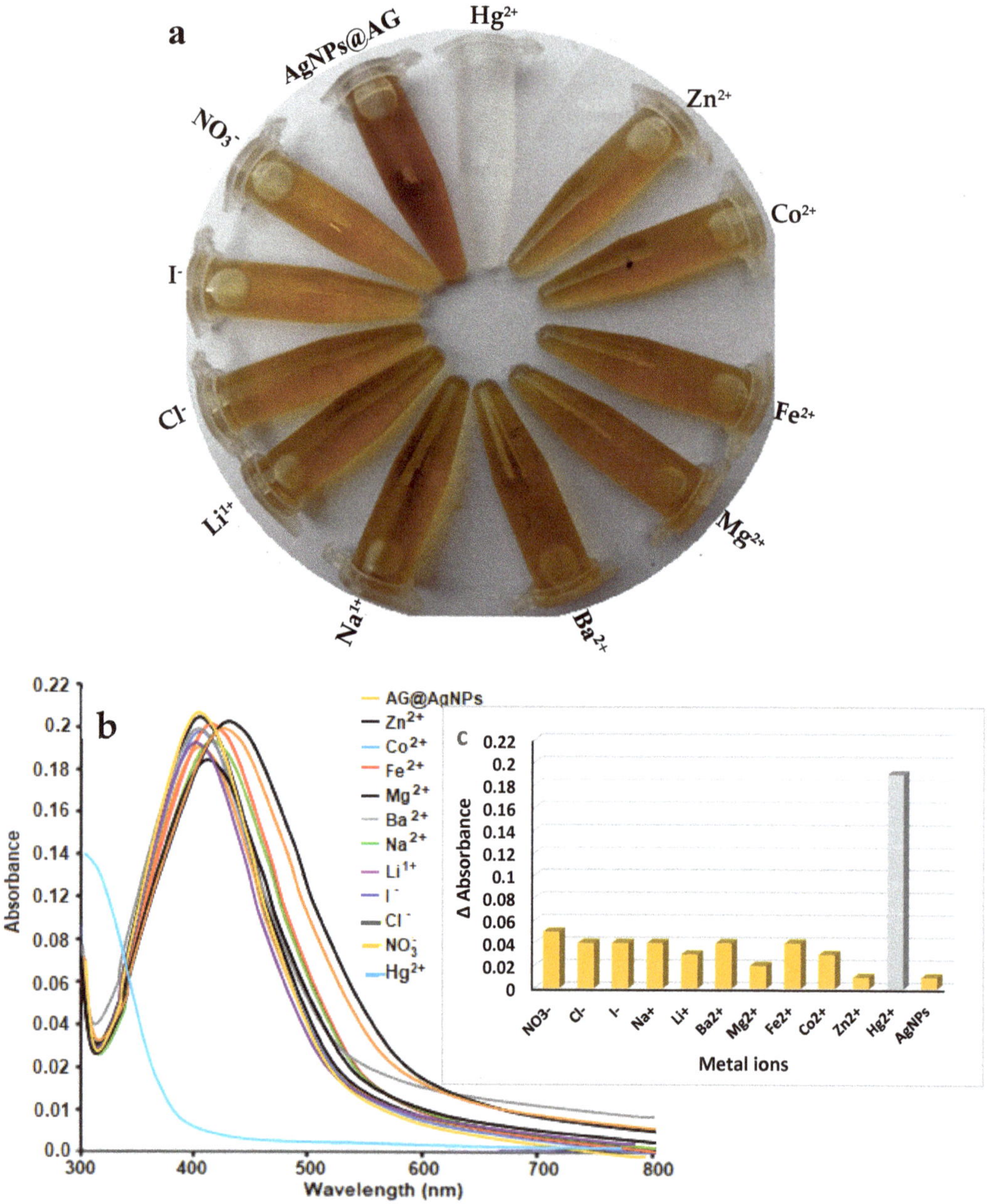

Figure 8. (**a**) Photo of the AgNPs@AG sensor performance against different groups of anions and cations; (**b**) absorbances of anionic and cationic foreign ions at certain wavelengths; (**c**) absorbance of different anions and cations at different wavelengths.

2.10. Assumed Colorimetric Response of AgNPs@AG against Hg^{2+} Ions

The proposed mechanism for the response of the as-prepared nano silver catalyst (AgNPs@AG) against sensing and detection of mercury ions [Hg^{2+}] is predicted, as the illustrated pathway (Scheme 2): firstly, mercury ion [Hg^{2+}] coordinates with almond gum by hydroxyl [−OH] and carboxylic [−CO2H] groups, as confirmed and proved by the XPs study, leading to the formation of Hg-almond gum complex. Furthermore, during an increase in Hg^{2+} ion concentration, Hg^{2+} ions quickly adsorb on the active surface of the AgNPs, after the redox reaction between Ag0 and Hg^{2+} and production of Ag^{+} and Hg^{+}. After that, the state of Ag^{0} is changed to the Ag^{+} state through the electron transfer process from almond gum to Ag^{+}, thus initializing the process between both Ag^{+} and Hg^{+} to form of the Ag−Hg phase. Therefore, in this process, almond gum plays a significant role in reducing, stabilizing, and being a good electron donor in preparation for each AgNP and Ag−Hg state. Furthermore, the electron transmission from almond gum to Ag^{+} gains support from the suggested synthetic procedure without additional reducing agents used during the AgNPs fabrication [30]. The fabrication of AgNPs@AG and the formation of Ag-Hg were confirmed and supported by UV−Vis measurement, XRD study, XPs analysis, and an increase in the nanoparticles' sizes for both silver nanoparticles and the amalgamate that appeared by the FESEM and HR-TEM images analysis. The increase in nanoparticle size was due to the well-active surface of the prepared NPs, causing perfect agglomeration and a "swelling effect" that occurs, which is attributed to the diffusion of Hg atoms, thus increasing the total nanoparticle sizes. Another vital piece of evidence, the high decrease in the wavelength value of AgNPs@AG at a constant position (410 nm) at the addition of mercury ions, could be attributed to the displacement of almond gum on the AgNPs surface by mercury ions due to the higher affinity of Ag toward the Hg ions. Hence, this leads to the fabrication of Ag−Hg with the aqueous phase.

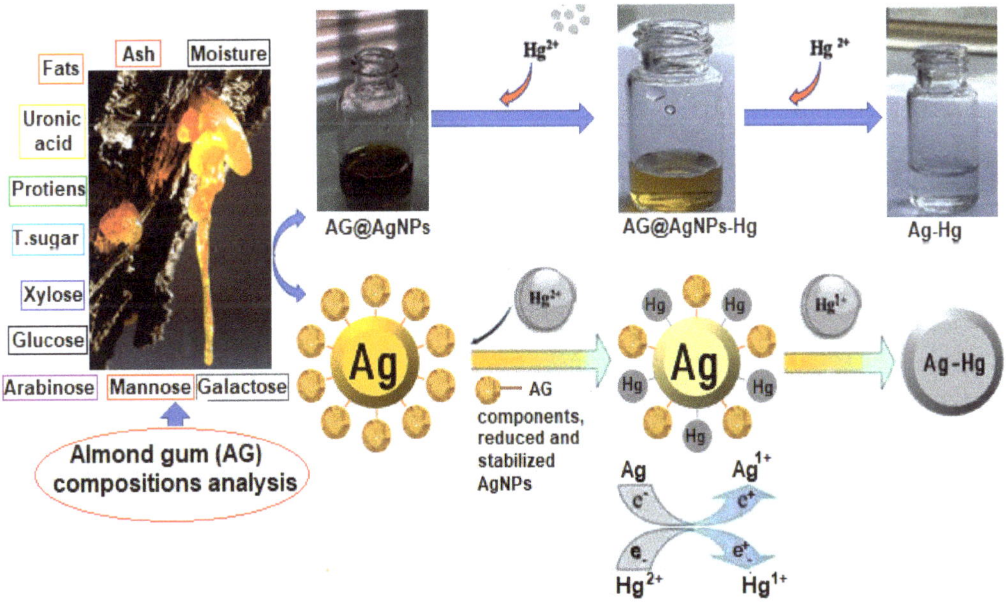

Scheme 2. Assumed colorimetric response of AgNPs@AG against Hg^{2+} ions.

3. Methodology

3.1. Materials and Methods

All chemical materials were pure-grade analytical reagents, including silver nitrate (AgNO3, FW = 169.87 g/mol, CAS No: 7761-88-8MP) and sodium hydroxide (NaOH, FW = 40 g/mol, CAS No:1310-73-2), which were provided from Scharlue comp. in Barcelona Spain.

A healthy almond gum was collected from Charboot village, Zeebar district in Akre and Duhok−Kurdistan region of Iraq in August 2021. A mercury chloride stock solution

(100 mg/L) was prepared daily due to dissolving a certain weight of mercury chloride salt in distilled water. The stock solution of mercury chloride was then stored in a dark container in a cool place for the subsequent experiments. All experiments were performed in the laboratory under ambient conditions.

3.2. Preparation of the Almond Gum Solution

The solution of 1% almond gum was prepared following these steps. First, the solid particles of almond gum were crushed to a fine powder by a mortar. Next, 1 g of the gum powder was transferred to a conical flask, dissolved with 100 mL distilled water and stirred vigorously at 50 °C for 30 min. Finally, the mixture was filtered with a gauze and a filter paper to separate insoluble solid fractions and stored in a refrigerator at 2 °C to be used for the reduction process of silver ions.

3.3. Fabrication of Silver Nanoparticles (AgNPs@AG)

The synthesis process included mixing 3 mL of a 0.1 M AgNO3 solution with 25 μL of 0.3 M NaOH. The mixture was stirred at 70 °C for 20 min. Afterwards, 2 mL of 1% almond gum solution were added to the mixture, followed by stirring for 45 min. A yellow−brown color appeared 15 min after the addition of the almond gum solution. The yellow−brown color continuously developed, until the whole mixture became brown, indicating the reduction process of silver ions and the formation of silver nanoparticles.

3.4. Detection of Hg^{2+} Ions by AgNPs@AG

An easy and short procedure was conducted to investigate the efficiency of AgNPs@AG as a rapid colorimetric sensor to detect mercury ions in an aqueous solution. A group of 5 mL glass vials was arranged and filled with 3 mL of different concentrations of mercury ions (10, 20, 30, 40, 50, 60, 70, 80, 90, and 100 mg/L). It was shown that the yellow color of AgNPs@AG disappeared rapidly after each addition of silver nanoparticles to the mercury solution, indicating that AG@NPs worked as an efficient probe against mercury ions in an aqueous solution. The procedure was monitored by a UV−Vis device to follow the decrease in absorbance at 410 nm to confirm the sensing process.

3.5. Detection of Mercury Ions in Real Water Samples

Various drinking water samples were collected from different locations in Shekhan District and Duhok in Iraq. The sensing of mercury ions in real water samples was conducted according to the optimum conditions of the procedure.

3.6. Diagnosis Techniques Used

The synthesized AgNPs@AG sensor was diagnosed with several techniques, including using an ultraviolet−visible (UV–Vis) spectrophotometer (nanodrop 2000/2000 c) to record the predicted LSPR band for the brown−yellow solution of AgNPs. X-ray diffraction (XRD) measurement was performed for AgNPs@AG by a PAN analytical X'PERT PRO model X-ray diffractometer operated at 30 kV and 45 mA, to reveal the crystalline nature of AgNPs@AG. The chemical surface nature of AgNPs@AG was diagnosed with an X-ray photoelectron spectroscopy (XPS) tool (PHI 5000 Versa Prob II: Supplier Model, japan), supported with an Al anode set at 140 W and operated at a pass energy of 50 eV. Field-emission scanning electron microscopy (FESEM) analysis (SEM-ZEISS-Singapore, Malaysia) with an accelerating voltage of 20 kV, supported with energy-dispersive spectroscopy (EDS) measurement, was used to predict the shape nature of the prepared nanoparticles. High-resolution-transmission electron microscopy (HR-TEM) measurements were carried out for AgNPs@AG by a JOEL apparatus (model 1200 EX; Cu grid), with an accelerating voltage of 40 kV, supported with a elements spread diagram (ESD) using (ZEISS-TEM instrument), operating at a 200 magnification.

4. Conclusions

In this study, silver nanoparticles have been synthesized employing a green strategy at ambient conditions. Natural exudate (almond gum) has been used as a reducing and stabilizing agent in order to produce a stable silver nanostructure. The distinctive optical properties and sensing behaviors of the as-prepared silver nanoparticles were observed due to reaction with mercury ions (Hg^{2+}) in an aqueous solution. Rapid and high sensitivity have been reported when mixing the silver nanosensor with a proper quantity of the target ions. In addition, it has been shown that the synthesized silver nanoparticles were stable and lasted for two months and no decolorization was observed.

Author Contributions: S.Y.S.Z.: conceiving the main idea, writing, and results analysis; A.Y.S.Z.: conceptualization, writing, and results analysis; O.I.H.: writing and editing; D.A.H.: viewing and editing; E.H.H.: viewing and editing. All authors have read and agreed to the published version of the manuscript.

Funding: This research received no external funding.

Data Availability Statement: All data and details regarding the work of this original article have been provided in this paper without further information.

Conflicts of Interest: The authors declare that there is no conflict of interest or any personal circumstances or interest that may be perceived as inappropriately influencing the representation or interpretation of reported research results.

References

1. Alshawi, J.M.S.; Mohammed, M.Q.; Alesary, H.F.; Ismail, H.K.; Barton, S. Voltammetric Determination of Hg^{2+}, Zn^{2+}, and Pb^{2+} Ions Using a PEDOT/NTA-Modified Electrode. *ACS Omega* **2022**, *7*, 20405–20419. [CrossRef] [PubMed]
2. Mohammed, M.Q.; Ismail, H.K.; Alesary, H.F.; Barton, S. Use of a Schiff base-modified conducting polymer electrode for electrochemical assay of Cd(II) and Pb(II) ions by square wave voltammetry. *Chem. Pap.* **2022**, *76*, 715–729. [CrossRef]
3. Ali, L.I.A.; Ismail, H.K.; Alesary, H.F.; Aboul-Enein, H.Y. A nanocomposite based on polyaniline, nickel and manganese oxides for dye removal from aqueous solutions. *Int. J. Environ. Sci. Technol.* **2021**, *18*, 2031–2050. [CrossRef]
4. Ismail, H.K.; Ali, L.I.A.; Alesary, H.F.; Nile, B.K.; Barton, S. Synthesis of a poly(p-aminophenol)/starch/graphene oxide ternary nanocomposite for removal of methylene blue dye from aqueous solution. *J. Polym. Res.* **2022**, *29*, 159. [CrossRef]
5. Pomal, N.C.; Bhatt, K.D.; Modi, K.M.; Desai, A.L.; Patel, N.P.; Kongor, A.; Kolivoška, V. Functionalized Silver Nanoparticles as Colorimetric and Fluorimetric Sensor for Environmentally Toxic Mercury Ions: An Overview. *J. Fluoresc.* **2021**, *31*, 635–649. [CrossRef]
6. Suvarapu, L.N.; Baek, S.O. Recent Studies on the Speciation and Determination of Mercury in Different Environmental Matrices Using Various Analytical Techniques. *Int. J. Anal. Chem.* **2017**, *2017*, 1–27. [CrossRef] [PubMed]
7. Sarkar, P.K.; Polley, N.; Chakrabarti, S.; Lemmens, P.; Pal, S.K. Nanosurface Energy Transfer Based Highly Selective and Ultrasensitive "turn on" Fluorescence Mercury Sensor. *ACS Sens.* **2016**, *1*, 789–797. [CrossRef]
8. Yasin, S.A.; Abbas, J.A.; Ali, M.M.; Saeed, I.A.; Ahmed, I.H. Methylene blue photocatalytic degradation by TiO2 nanoparticles supported on PET nanofibres. *Mater. Today Proc.* **2020**, *20*, 482–487. [CrossRef]
9. Tauran, Y.; Brioude, A.; Coleman, A.W.; Rhimi, M.; Kim, B. Molecular recognition by gold, silver and copper nanoparticles. *World J. Biol. Chem.* **2013**, *4*, 35. [CrossRef]
10. Sabri, M.A.; Umer, A.; Awan, G.H.; Hassan, M.F.; Hasnain, A. Selection of suitable biological method for the synthesis of silver nanoparticles. *Nanomater. Nanotechnol.* **2016**, *6*, 1–20. [CrossRef]
11. Naidoo, S.; Külheim, C.; Zwart, L.; Mangwanda, R.; Oates, C.N.; Visser, E.A.; Wilken, F.E.; Mamni, T.B.; Myburg, A.A. Uncovering the defence responses of eucalyptus to pests and pathogens in the genomics age. *Tree Physiol.* **2014**, *34*, 931–943. [CrossRef]
12. Mahfoudhi, N.; Chouaibi, M.; Donsì, F.; Ferrari, G.; Hamdi, S. Chemical composition and functional properties of gum exudates from the trunk of the almond tree (*Prunus dulcis*). *Food Sci. Technol. Int.* **2012**, *18*, 241–250. [CrossRef] [PubMed]
13. Mumtaz, A.; Munir, H.; Zubair, M.T.; Arif, M.H. Mimosa pudica gum based nanoparticles development, characterization, and evaluation for their mutagenicity, cytotoxicity and antimicrobial activity. *Mater. Res. Express* **2019**, *6*, 105308. [CrossRef]
14. Kora, A.J.; Arunachalam, J. Green fabrication of silver nanoparticles by gum tragacanth (*Astragalus gummifer*): A dual functional reductant and stabilizer. *J. Nanomater.* **2012**, *2012*, 69. [CrossRef]
15. Venkatesham, M.; Ayodhya, D.; Madhusudhan, A.; Veerabhadram, G. Synthesis of stable silver nanoparticles using gum acacia as reducing and stabilizing agent and study of its microbial properties: A novel green approach. *Int. J. Green Nanotechnol. Biomed.* **2012**, *4*, 199–206. [CrossRef]

16. El-Adawy, M.M.; Eissa, A.E.; Shaalan, M.; Ahmed, A.A.; Younis, N.A.; Ismail, M.M.; Abdelsalam, M. Green synthesis and physical properties of Gum Arabic-silver nanoparticles and its antibacterial efficacy against fish bacterial pathogens. *Aquac. Res.* **2021**, *52*, 1247–1254. [CrossRef]
17. Quelemes, P.V.; Araruna, F.B.; de Faria, B.E.F.; Kuckelhaus, S.A.S.; da Silva, D.A.; Mendonça, R.Z.; Eiras, C.; Soares, M.J.d.S.; Leite, J.R.S.A. Development and antibacterial activity of cashew gum-based silver nanoparticles. *Int. J. Mol. Sci.* **2013**, *14*, 4969–4981. [CrossRef]
18. Kora, A.J.; Sashidhar, R.B. Antibacterial activity of biogenic silver nanoparticles synthesized with gum ghatti and gum olibanum: A comparative study. *J. Antibiot.* **2015**, *68*, 88–97. [CrossRef]
19. Venkatesham, M.; Ayodhya, D.; Madhusudhan, A.; Santoshi Kumari, A.; Veerabhadram, G.; Girija Mangatayaru, K. A Novel Green Synthesis of Silver Nanoparticles Using Gum Karaya: Characterization, Antimicrobial and Catalytic Activity Studies. *J. Clust. Sci.* **2014**, *25*, 409–422. [CrossRef]
20. Samrot, A.V.; Angalene, J.L.A.; Roshini, S.M.; Raji, P.; Stefi, S.M.; Preethi, R.; Selvarani, A.J.; Madankumar, A. Bioactivity and Heavy Metal Removal Using Plant Gum Mediated Green Synthesized Silver Nanoparticles. *J. Clust. Sci.* **2019**, *30*, 1599–1610. [CrossRef]
21. Rezaei, A.; Nasirpour, A.; Tavanai, H. Fractionation and some physicochemical properties of almond gum (*Amygdalus communis* L.) exudates. *Food Hydrocoll.* **2016**, *60*, 461–469. [CrossRef]
22. Yaseen Sharaf Zeebaree, A.; Yaseen Sharaf Zeebaree, S.; Rashid, R.F.; Ismail Haji Zebari, O.; Albarwry, A.J.S.; Ali, A.F.; Yaseen Sharaf Zebari, A. Sustainable engineering of plant-synthesized TiO2 nanocatalysts: Diagnosis, properties and their photocatalytic performance in removing of methylene blue dye from effluent. A review. *Curr. Res. Green Sustain. Chem.* **2022**, *5*, 100312. [CrossRef]
23. Roy, K.; Sarkar, C.K.; Ghosh, C.K. Rapid colorimetric detection of Hg2+ ion by green silver nanoparticles synthesized using Dahlia pinnata leaf extract. *Green Process. Synth.* **2015**, *4*, 455–461. [CrossRef]
24. Kumar, V.; Singh, D.K.; Mohan, S.; Bano, D.; Gundampati, R.K.; Hasan, S.H. Green synthesis of silver nanoparticle for the selective and sensitive colorimetric detection of mercury (II) ion. *J. Photochem. Photobiol. B Biol.* **2017**, *168*, 67–77. [CrossRef] [PubMed]
25. FIrdaus, M.L.; Itriani, I.F.; Yantuti, S.W.; Artati, Y.W.H.; Haydarov, R.K.; Calister, J.A.M.; Bata, H.O.; Amo, T.G. Colorimetric Detection of Mercury (II) Ion in Aqueous Solution Using Silver Nanoparticle. *Anal. Sci.* **2017**, *8*, 831–837. [CrossRef]
26. Abu-Dalo, M.A.; Othman, A.A.; Al-Rawashdeh, N.A.F. Exudate gum from acacia trees as green corrosion inhibitor for mild steel in acidic media. *Int. J. Electrochem. Sci.* **2012**, *7*, 9303–9324.
27. Han, J.K.; Madhiisudhan, A.; Bandi, R.; Park, C.W.; Kim, J.C.; Lee, Y.K.; Lee, S.H.; Wona, J.M. Green synthesis of AgNPs using lignocellulose nanofibrils as a reducing and supporting agent. *BioResources* **2020**, *15*, 2119–2132. [CrossRef]
28. Schiesaro, I.; Burratti, L.; Meneghini, C.; Fratoddi, I.; Prosposito, P.; Lim, J.; Scheu, C.; Venditti, I.; Iucci, G.; Battocchio, C. Hydrophilic Silver Nanoparticles for Hg(II) Detection in Water: Direct Evidence for Mercury-Silver Interaction. *J. Phys. Chem. C* **2020**, *124*, 25975–25983. [CrossRef]
29. Zeebaree, S.Y.S.; Ismail Haji, O.; Farooq Rashid, R.; Yasin, S.A.; Zeebaree, A.Y.S.; Albarwary, A.J.S.; Zebari, A.Y.S.; Gerjees, H.A. Novel natural exudate as a stabilizing agent for fabrication of copper nanoparticles as a colourimetric sensor to detect trace pollutant. *Surf. Interfaces* **2022**, *32*, 102131. [CrossRef]
30. Abbasi, A.; Hanif, S.; Shakir, M. Gum acacia-based silver nanoparticles as a highly selective and sensitive dual nanosensor for Hg(ii) and fluorescence turn-off sensor for S2- and malachite green detection. *RSC Adv.* **2020**, *10*, 3137–3144. [CrossRef]
31. Bashir, M.; Haripriya, S. Assessment of physical and structural characteristics of almond gum. *Int. J. Biol. Macromol.* **2016**, *93*, 476–482. [CrossRef] [PubMed]
32. Ahmed, F.; Kabir, H.; Xiong, H. Dual Colorimetric Sensor for Hg^{2+}/Pb^{2+} and an Efficient Catalyst Based on Silver Nanoparticles Mediating by the Root Extract of Bistorta amplexicaulis. *Front. Chem.* **2020**, *8*, 591958. [CrossRef] [PubMed]
33. Demirezen Yılmaz, D.; Aksu Demirezen, D.; Mıhçıokur, H. Colorimetric detection of mercury ion using chlorophyll functionalized green silver nanoparticles in aqueous medium. *Surf. Interfaces* **2021**, *22*, 100840. [CrossRef]
34. Azimpanah, R.; Solati, Z.; Hashemi, M. Green synthesis of silver nanoparticles and their applications as colorimetric probe for determination of Fe and Hg ions. *IET Nanobiotechnology* **2018**, *12*, 673–677. [CrossRef] [PubMed]
35. Memon, R.; Memon, A.A.; Sirajuddin Balouch, A.; Memon, K.; Sherazi, S.T.H.; Chandio, A.A.; Kumar, R. Ultrasensitive colorimetric detection of Hg^{2+} in aqueous media via green synthesis by Ziziphus mauritiana Leaf extract-based silver nanoparticles. *Int. J. Environ. Anal. Chem.* **2020**, 1–16. [CrossRef]
36. Annadhasan, M.; Rajendiran, N. Highly selective and sensitive colorimetric detection of Hg(II) ions using green synthesized silver nanoparticles. *RSC Adv.* **2015**, *5*, 94513–94518. [CrossRef]
37. Prosposito, P.; Burratti, L.; Bellingeri, A.; Protano, G.; Faleri, C.; Corsi, I.; Battocchio, C.; Iucci, G.; Tortora, L.; Secchi, V.; et al. Bifunctionalized silver nanoparticles as Hg^{2+} plasmonic sensor in water: Synthesis, characterizations, and ecosafety. *Nanomaterials* **2019**, *9*, 1353. [CrossRef]
38. Yang, M.; Yao, J.; Liu, Y.; Duan, Y. Sensitive detection of mercury (II) ion using wave length-tunable visible-emitting gold nanoclusters based on protein-templated synthesis. *J. Mater. Res.* **2014**, *29*, 2416–2424. [CrossRef]
39. Li, H.; Liu, H.; Zhang, J.; Cheng, Y.; Zhang, C.; Fei, X.; Xian, Y. Platinum Nanoparticle Encapsulated Metal-Organic Frameworks for Colorimetric Measurement and Facile Removal of Mercury(II). *ACS Appl. Mater. Interfaces* **2017**, *9*, 40716–40725. [CrossRef]

40. Ichinoki, S.; Kitahata, N.; Fujii, Y. Selective determination of mercury(II) ion in water by solvent extraction followed by reversed-phase HPLC. *J. Liq. Chromatogr. Relat. Technol.* **2004**, *27*, 1785–1798. [CrossRef]
41. Gold, G.; Nkosi, D.; Pillay, K.; Arotiba, O. Electrochemical detection of Hg(II) in water using self-assembled single walled carbon nanotube-poly(m-amino benzene sulfonic acid) on gold electrode. Sens. *Bio-Sens. Res.* **2016**, *10*, 27–33. [CrossRef]
42. Sukesan, R.; Chen, Y.; Shahim, S.; Wang, S.; Sarangadharan, I.; Wang, Y. Instant Mercury Ion Detection in Industrial Waste Water with a Microchip Using Extended Gate Field-E ff ect Transistors and a Portable Device. *Sensors* **2019**, *19*, 2209. [CrossRef] [PubMed]
43. Kaewprom, C.; Areerob, Y.; Oh, W. Simultaneous determination of Hg (II) and Cu (II) in water samples using fluorescence quenching sensor of N-doped and N, K co-doped graphene quantum dots. *Arab. J. Chem.* **2020**, *13*, 3714–3723. [CrossRef]
44. Sarafraz-yazdi, A.; Fatehyan, E.; Amiri, A. Determination of Mercury in Real Water Samples Using in situ Derivatization Followed by Sol-Gel—Solid-Phase Microextraction with Gas Chromatography—Flame Ionization Detection. *J. Chromatogr. Sci.* **2014**, *52*, 81–87. [CrossRef] [PubMed]
45. Gumpu, M.B.; Krishnan, U.M. Design and development of amperometric biosensor for the detection of lead and mercury ions in water matrix—A permeability approach. *Anal. Bioanal. Chem.* **2017**, *409*, 4257–4266. [CrossRef]

catalysts

Review

Recent Advances on Metal Oxide Based Nano-Photocatalysts as Potential Antibacterial and Antiviral Agents

Jai Prakash [1,2,*], Suresh Babu Naidu Krishna [3], Promod Kumar [2], Vinod Kumar [4], Kalyan S. Ghosh [1], Hendrik C. Swart [2], Stefano Bellucci [5] and Junghyun Cho [6]

1 Department of Chemistry, National Institute of Technology Hamirpur, Hamirpur 177005, Himachal Pradesh, India
2 Department of Physics, University of the Free State, Bloemfontein 9300, South Africa
3 Department of Biomedical and Clinical Technology, Durban University of Technology, Durban 4000, South Africa
4 Department of Physics, College of Natural and Computational Science, Dambi Dollo University, Dambi Dollo P.O. Box 260, Ethiopia
5 INFN Laboratori Nazionali di Frascati, Via Enrico Fermi 40, 00044 Frascati, Italy
6 Department of Mechanical Engineering & Materials Science and Engineering Program, State University of New York (SUNY), Binghamton, NY 13902-6000, USA
* Correspondence: jaip@nith.ac.in

Abstract: Photocatalysis, a unique process that occurs in the presence of light radiation, can potentially be utilized to control environmental pollution, and improve the health of society. Photocatalytic removal, or disinfection, of chemical and biological species has been known for decades; however, its extension to indoor environments in public places has always been challenging. Many efforts have been made in this direction in the last two–three years since the COVID-19 pandemic started. Furthermore, the development of efficient photocatalytic nanomaterials through modifications to improve their photoactivity under ambient conditions for fighting with such a pandemic situation is a high research priority. In recent years, several metal oxides-based nano-photocatalysts have been designed to work efficiently in outdoor and indoor environments for the photocatalytic disinfection of biological species. The present review briefly discusses the advances made in the last two to three years for photocatalytic viral and bacterial disinfections. Moreover, emphasis has been given to the tailoring of such nano-photocatalysts in disinfecting surfaces, air, and water to stop viral/bacterial infection in the indoor environment. The role of such nano-photocatalysts in the photocatalytic disinfection of COVID-19 has also been highlighted with their future applicability in controlling such pandemics.

Keywords: antibacterial; antiviral; air/water disinfection; surface decontamination; metal oxide semiconductors; nano-photocatalysts

Citation: Prakash, J.; Krishna, S.B.N.; Kumar, P.; Kumar, V.; Ghosh, K.S.; Swart, H.C.; Bellucci, S.; Cho, J. Recent Advances on Metal Oxide Based Nano-Photocatalysts as Potential Antibacterial and Antiviral Agents. *Catalysts* **2022**, *12*, 1047. https://doi.org/10.3390/catal12091047

Academic Editors: Beom Soo Kim and Pritam Kumar Dikshit

Received: 15 August 2022
Accepted: 8 September 2022
Published: 14 September 2022

1. Introduction

Even with the fast growing technology and industrial developments, the modern world is still lacking in the control of environmental and health issues. The best example is the current COVID-19 pandemic, which has made us realize that the modern world should also take care of the development of novel technologies, materials and medical innovations to control such health- and environment-related issues [1]. Various unwanted components present in the environment affect human health directly or indirectly. In this context in particular, different microbial pathogens such as viruses, bacteria, protozoa, etc. present in the environment may sometimes threaten human health and cause dangerous infectious illnesses [1,2]. Recent developments suggest that nanotechnology-based methods and materials could be alternate options with the huge potential for controlling such bacterial/viral outbreaks [3–6] which have been a serious issue and increased at a disquieting rate over the past decades [2].

Photocatalysis, which uses nano-photocatalysts, is one of the unique processes occurs in the presence of solar radiation [7,8]. This process is promising for the control of environmental issues and for improving the health of the society due to the presence of unspent solar energy on the Earth [9,10]. Photocatalysis has multifunctional applications in the field of environmental studies, including the photocatalytic degradation of toxic/harmful organic compounds and gases [11–13], and the photocatalytic viral and bacterial disinfection of water, air, or on surfaces, which ultimately protects the environment and improves human health [14–17]. Photocatalytic removal or disinfection of such species is a promising and environmentally friendly process using suitable photocatalysts under the influence of solar radiation. Furthermore, it is also very cost-effective and promising in the open environment [1,9]. In recent years, several metal oxide semiconductor photocatalysts such as TiO_2, ZnO, CuO, WO_3, etc. have been designed as visible active photocatalysts. Their properties have been improved through some modifications which enable them to work efficiently in solar light towards photocatalytic degradation and disinfection of chemical and biological species [18,19], respectively. These are found to be very useful for disinfecting surfaces, air, and water by killing several microorganisms i.e., bacteria and fungi, and inactivating several viruses including influenza virus, hepatitis C virus, coronavirus, etc., [20]. These photocatalysts exhibit oxidative capabilities via the photocatalytic production of cytotoxic reactive oxygen species (ROS) as shown in Figure 1 for photo-degradation/inactivation of such species in outdoor as well as indoor environment. It has been found to be very beneficial for the treatment of various bacterial/viral diseases such as measles, influenza, herpes, Ebola, current COVID-19, etc., [1,2] as is shown schematically in Figure 2.

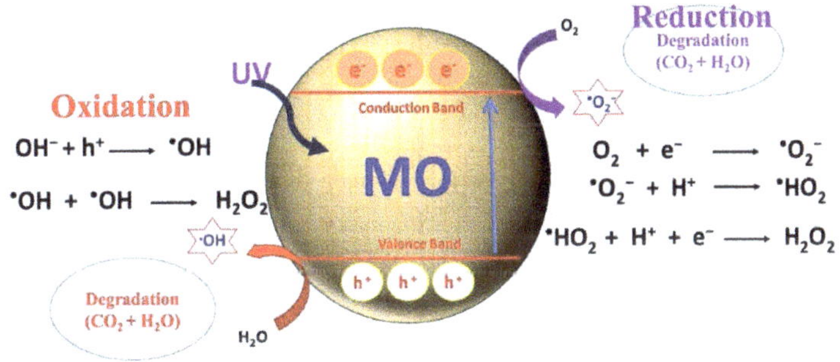

Figure 1. Photocatalytic mechanism of metal oxide nano-photocatalysts towards photocatalytic degradation of chemical species [12].

These semiconductor nano-photocatalysts are potential candidates as next-generation antibiotics and antiviral agents to deal with multi-drug-resistant pathogens and viruses, respectively, owing to their outstanding antibacterial/viral performance. The action of photocatalytic inactivation/degradation of these nano-photocatalysts on various bacterial and viruses has been successfully explained by several authors; however, the proposed mechanisms are still under debate and continuous investigations are going on by the scientific community [11,21–23]. This review covers briefly recent advances carried out in this field using metal oxide nano-photocatalysts with an emphasis on the understanding of photomechanism processes and potential applications in the environment. The role of such nano-photocatalysts in photocatalytic disinfection of COVID-19 has also been highlighted, along with their future applicability in controlling such pandemic situations.

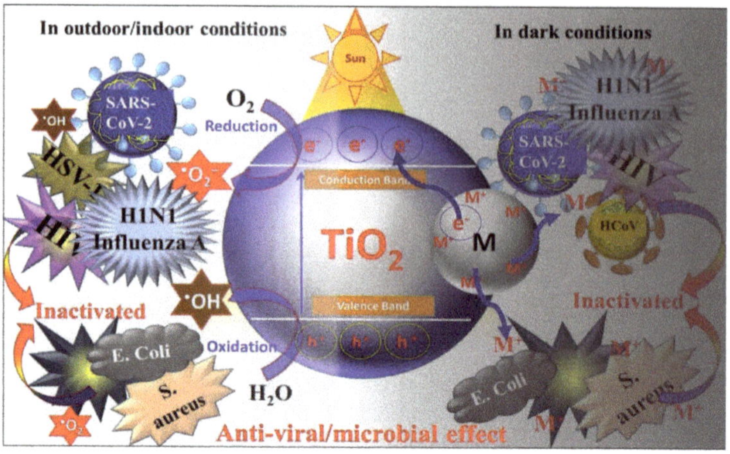

Figure 2. Schematic representation of photodegradation of viruses/microbes in outdoor as well as indoor environment using metal oxide nano-photocatalysts [1].

2. Metal Oxide Based Nano-Photocatalysts: Antibacterial and Antiviral Mechanisms

The photocatalysis method of disinfection using metal oxide semiconductors shows great potential in outdoor and indoor environments as compared to the conventional methods for the removal of bacteria or viruses. These nano-photocatalysts can effectively inactivate the bacteria and viruses in the presence of light radiation under ambient conditions without producing any other by-products as compared to that of using chemicals [1,24].

Metal oxide semiconductor-based nano-photocatalysts such as TiO_2, and ZnO have been extensively investigated for inactivation of several bacteria and viruses. The basic mechanism behind their photoinduced inactivation involves the photocatalytic production of short-lived but effective biocidal ROS, i.e., hydroxyl radicals (•OH), superoxide (•O_2^-), and hydrogen peroxide (H_2O_2), through photochemical redox reactions under light irradiation. [25,26]. The formation mechanism of various ROS in various cases is shown in Figure 3. Such an effectively biocidal ROS further inactivates the bacteria and viruses by damaging deoxyribonucleic acid (DNA), Ribonucleic acid (RNA), proteins, and lipids [2,17,26,27]. The generation of ROS and the disinfection of bacteria and virus, including severe acute respiratory syndrome coronavirus 2 (SARS-CoV-2) virus inactivation, are shown schematically in Figure 3a–c.

Under the influence of ultra-violet (UV) light radiation, these nano-photocatalysts absorb the radiation resulting in excitation and promotion of valance band (VB) electrons into the conduction band (CB). The holes in VB interact with the adsorbed water molecules and produce active OH and H_2O_2 free radicals. These free radicals are powerful oxidants which generally oxidize the components/chemical in the shell and capsid of the bacteria and viruses [27]. Subsequently, whereas electrons in CB generally reduce the atmospheric O_2 (or available from the medium) and produce •O_2^- radicals [27,28]. Similarly, •O_2^- radicals produced in photocatalysis are effective in rupturing the capsid shell that results in the leakage and rapid destruction of capsid proteins and RNA [29] (Figure 3b,c).

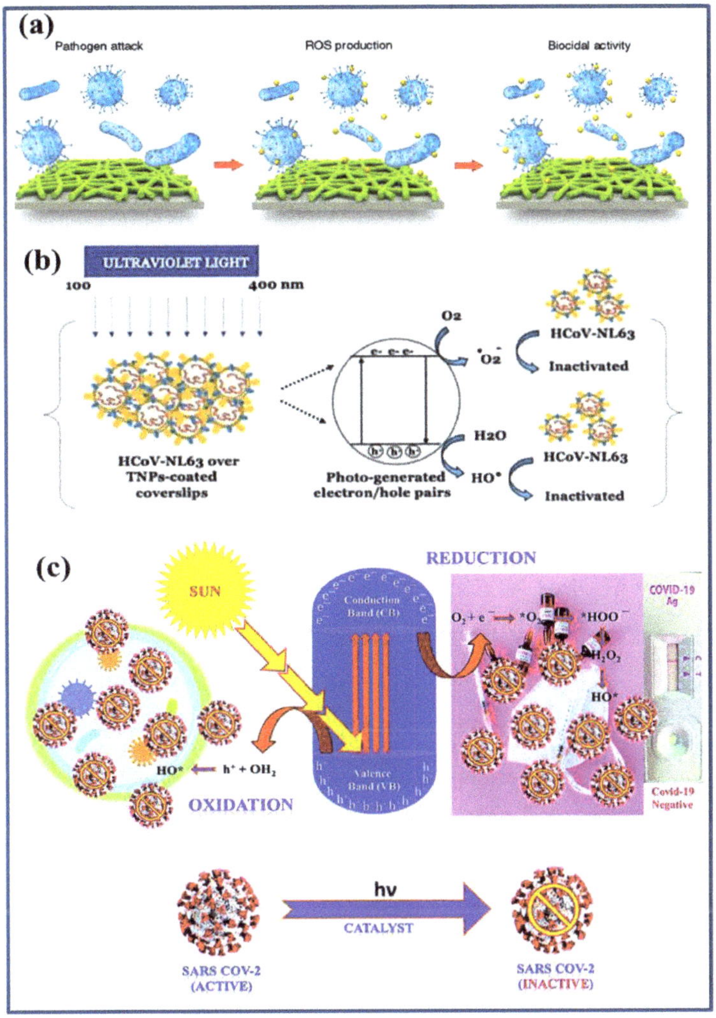

Figure 3. Schematic representation of photocatalytic disinfection of: (**a**) bacteria [26]; (**b**) HCoV-NL63 virus [28]; and (**c**) SARS-CoV-2 virus [27]. Under the influence of light irradiation, the photocatalysts produce electrons and holes that undergo oxidation and reduction processes with O_2 and H_2O generating strong free radical on their surfaces. These radicals interact with the adsorbed bacteria or viruses and inactivate them.

The ROS as produced generally attack or interact with the cytoplasmic membrane and cell wall of the bacteria or viruses during the inactivation mechanism [30]. However, the rate of photo inactivation/disinfection depends on the photocatalyst used and the amount of ROS produced under the influence of the available wavelength of light irradiation, and also depends on the internal as well as external cell structures of the type of pathogens, because all the bacteria or viruses do not have similar cell wall and membrane structures. These components may have complicated layered structures and contain various types of RNA/DNA, proteins, or enzymes [1,27,30]. For instance, cyclobutene pyrimidine dimers (CPDs) and pyrimidine-6,4-pyrimidone (6.4 PP) photoproducts, together with the Dewar-valence isomers, are the most studied and best described UV-induced photoreactions between and within nucleic acids [30,31]. Following UV light exposure, pyrimidine dimers (see Scheme 1) are formed. CPD and 6,4PP dimers are mainly responsible for bending the double helix 7–9 and 44 degrees, respectively, once they are formed. DNA replication is stopped because of these alterations.

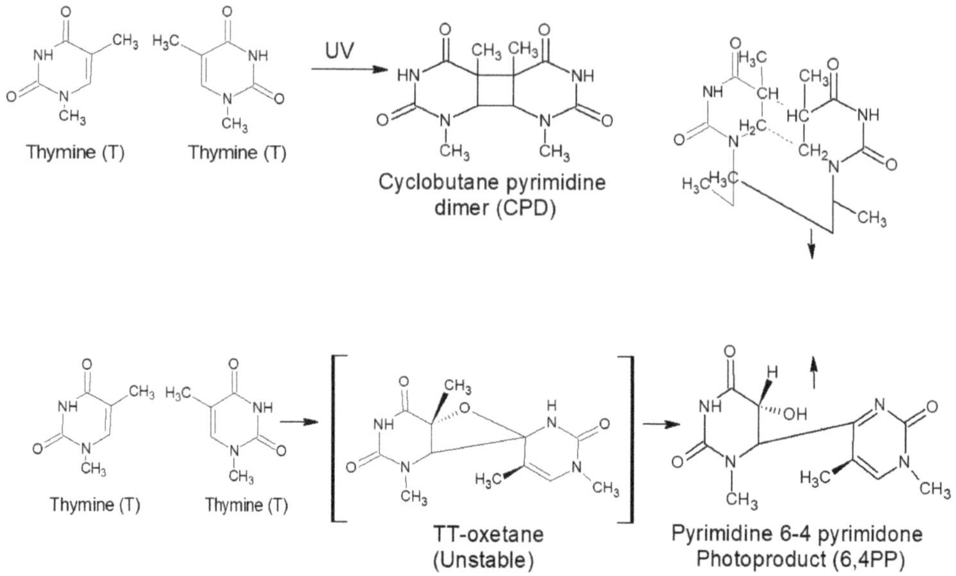

Scheme 1. Schematic diagram of how pyrimidine dimers form after DNA is exposed to UV light. between two adjacent thymine (T) nitrogenous bases, the production of cyclobutane pyrimidine dimers (CPDs); and pyrimidine-6.4-pyrimidone photoproducts (6.4 PP). Similar reactions for uracil in the case of RNA could take place (U) [30].

Because of their wide band gap, the TiO_2 (3.2 eV) and ZnO (3.37 eV) nano-photocatalysts absorb the high energy UV radiation. This limits their potential photocatalytic applicability more efficiently in outdoor environments under the sunlight because it has only 3–5% UV radiation. Furthermore, photocatalytic disinfection processes in indoor environments are challenging, and modifications of these metal oxide nano-photocatalysts to make them visible light active photocatalysts need to be explored [9,25]. There are several ways to modify these nano-photocatalysts, such as doping with metals/non-metals [32], surface modification via sensitizing or heterojunction formation [10,33] with other functional nanomaterials such as noble metals, carbon based nanomaterials (i.e., graphene, carbon nanotubes, graphene oxide, etc.) [8,34] other metal oxides, etc. [35–37] Emphasis has been given to enhance the surface area, prevent the recombination of photogenerated charge carriers, and bandgap modification to extend into visible light absorption for effective use in ambient conditions [11,38]. For example, Yu et al. [39,40] demonstrated the enhanced photocatalytic activity of mesoporous TiO_2 via F doping attributed to the stronger absorption in UV-visible region with a red shift in the band gap transition. Fe doped TiO_2 were found to be very effective in visible region with excellent antifungal activities under natural environment [41]. Similarly, Ag doped ZnO [23] and TiO_2 NPs [11] showed better antibacterial activities in normal room conditions due to Ag ion-induced visible light activity in these nano-photocatalysts. These nano-photocatalysts, modified with plasmonic noble metals [42–44], are effective antibacterial and antiviral agents in dark conditions [1,45]. Interestingly, such nano-photocatalysts have also been used as memory catalysis because of their unique talent to retain the catalytic performance in dark conditions [33,45,46]. For example, Tatsuma et al. [47] demonstrated that TiO_2–WO_3 heterojuction nanocomposite photocatalyst films could be charged by UV light irradiation which showed good antibacterial effect on Escherichia coli in dark environment. Similarly, Ag-modified TiO_2 films were also shown to exhibit disinfection memory activity [45].

As discussed above, a great deal of research has been performed in real practical applications of such nano-photocatalysts. Recent developments show that modified metal oxide nano-photocatalysts are promising disinfection agents in indoor environments in ambient room conditions when applied in the form of surface coating/thin films on commonly used surfaces in hospitals, offices, home, etc. Additionally, potential practical applications have been carried out which show excellent results while using these nanomaterials for the disinfection of polluted water and air (in indoor as well as outdoor environments) which show their potential to combat pandemics such as COVID-19. The disinfection applications of such nano-photocatalysts in air, water and on surfaces have been discussed in the next sections, with an emphasis on their mechanism of actions.

3. Recent Advances on Metal Oxide Based Nano-Photocatalysts as Potential Antibacterial and Antiviral Agents

3.1. TiO_2 Based Nano-Photocatalysts as Potential Antibacterial and Antiviral Agents

The sudden pandemic spread due to the novel SARS-CoV-2 virus has made the world realize that the development of simple and cost-effective nanomaterials-based technology is needed. Meanwhile, a great deal of research has been focused on developing various types of nanomaterials to fight against disease-causing bacteria and viruses which could be effective in water, air and on surfaces [30]. More emphasis has been given to developing promising materials to be used in indoor spaces such as hospitals, homes, and other public places to minimize the health- and environment-related risks [1]. As discussed in the previous section, metal oxide-based nano-photocatalysts are promising mainly in outdoor circumstances due to their light absorption activity cum disinfection ability. However, these nanomaterials have been modified to function in many ways [48–53], and have been explored in the case of indoor disinfection activities, including surface decontamination. TiO_2 is one of the most used and studied metal oxide semiconductor nano-photocatalysts for disinfection caused by bacteria and viruses in the last decades. UV and visible light activation techniques have been used as promising approaches for the outdoor and indoor disinfection applications of TiO_2. TiO_2 has been researched for last many decades owing to its promising photocatalytic activity. The sole TiO_2 is known to be an active photocatalyst under UV light radiation and transparent to visible light due to its bandgap of 3.2 eV [32,54,55]. It is a fascinating material because of its tunable characteristics and the existence of its different forms. Several efforts have been made to enhance its photocatalytic activity towards removal or disinfection under visible light, which is still challenging under ambient conditions [56–61].

Krumdieck et al. [56] developed a TiO_2 based nanocomposite coating for stainless steel-made surfaces such as door handles, bed rails and other high touch surfaces which could be the main source of spreading germs/infections in hospitals. The coating was made of nanostructured anatase, rutile dendrites and carbon (NsARC), with promising structural and adhesive properties (Figure 4a). The NsARC coating showed excellent antibacterial activity under sunlight along with a significant disinfection effect in dark conditions (Figure 4b). The enhanced photocatalytic activity (greater than 3-log reduction in viable *E coli*) was attributed to the rutile-anatase heterojunctions, nanostructured single crystals with high surface area and the low migration path length of the charge carriers (Figure 4c).

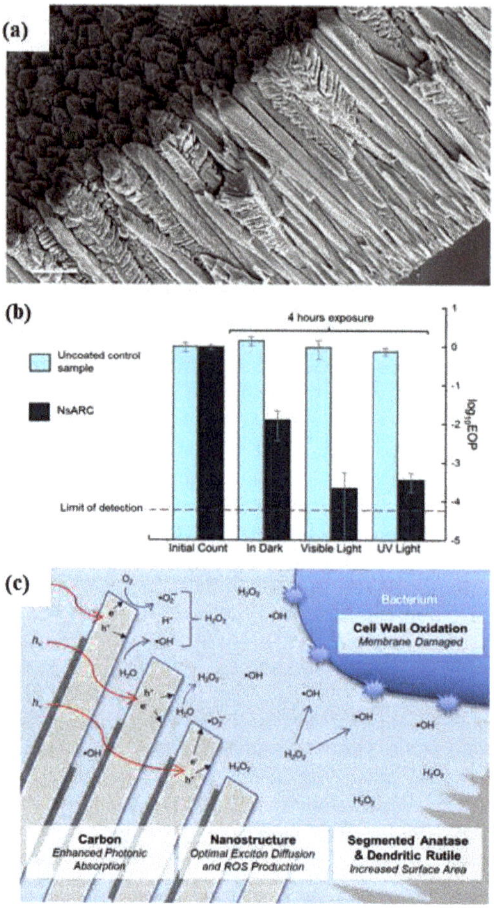

Figure 4. (**a**) Secondary electron microscopy (SEM) image of NsARC nanocomposite; (**b**) antibacterial performance of NsARC coated stainless steel on E. coli in different conditions including in dark conditions also; and (**c**) mechanisms of disinfection of antibacterial activity of NsARC on *E.coli* [56].

Reduced graphene oxide (rGO) modified TiO_2 composite photocatalysts under artificial solar light showed enhanced antibacterial performance as compared to only TiO_2 against *E. coli* that was attributed to a change on the surface properties due to rGO. Figure 5a shows the SEM micrograph of *E. coli* bacterial cell and Figure 5b shows the *E. coli* covered with rGO-TiO_2 nanocomposites. After sunlight exposure, bacterial cells were damaged as shown in Figure 5c,d [57]. Similarly, Zhou et al. [58] reported the excellent antibacterial behavior of rGO-TiO_2 nanocomposites under UV and visible light environment against *E. hormaechei*. Dhanasekar et al. [60] reported on rGO-Cu doped TiO_2 nanocomposites that showed excellent ambient light antibacterial performance against several bacteria (Figure 5e). The material modification i.e., doping [62,63] and forming the composite with rGO, enhanced the visible light activity in ambient conditions by reducing the recombination of photogenerated charge carriers and their transport, respectively, for powerful ROS production for activation. From a practical application point of view, the nanocomposite was explored as a coating embedded in a polymer film/coating [37,64] which exhibited equal performance, showing the potential for its use as an antibacterial coating for different applications in surface protection. Similarly, complete disinfection under low intensity-simulated solar light irradiation was shown by Bonnefond et al. [61] using hybrid acrylic/TiO_2 films. This kind of photocatalytic film with a self-cleaning ability [54] and high disinfection performance could be applied for surfaces in ambient conditions in the indoor environment to stop the spreading of bacterial germs/viruses, etc. (Figure 5f).

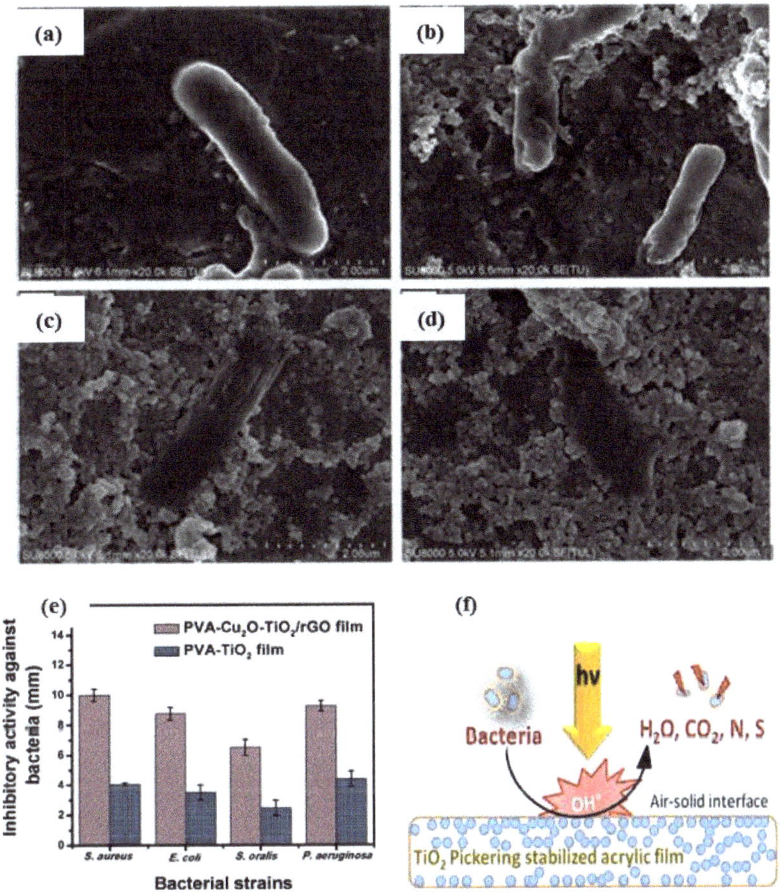

Figure 5. SEM images of (**a**) *E. coli* bacterial cell; (**b**) *E.coli* covered with TiO_2-rGO before sunlight irradiation; (**c**,**d**) damaged bacterial cells after sunlight irradiation (reproduced with permission from Ref. [57] Copyright 2018-Elsevier); (**e**) antibacterial effect of rGO-Cu doped TiO_2 nanocomposites in ambient light against several microbes (reproduced with permission from Ref. [60] Copyright 2018-Elsevier) and (**f**) self-cleaning ability hybrid acrylic/TiO_2 films (reproduced with permission from Ref. [61]. Copyright 2015-Elsevier).

The usse of photocatalysts in airborne bacterial disinfection is very important in indoor environment but very challenging [65–67]. Hernández-Gordillo et al. [65] demonstrated that CuO- and CdS-modified TiO_2 monoliths surfaces could be efficient for the removal and inactivation of airborne bacteria under visible light irradiation in a continuous flow photoreactor. The flow cytometry and SEM micrograph results were used to study the damaging of the bacteria. The SEM micrographs exhibited the collapsing of the cell membrane of bacteria due to the lipid peroxidation because of interaction with ROS generated on the monolith surface as shown in Figure 6a–g. The photocatalytic inactivation of TiO_2/CdS monoliths, was found to be maximum (99.9%). Similarly, Valdez-Castillo et al. [68] demonstrated bioaerosols disinfection using perlite-supported ZnO and TiO_2. It was found that these photocatalysts showed better disinfection performance leading to cell death. It was concluded that these photocatalytic systems could be potentially used for practical applications in removing viable airborne microorganisms from air.

Various TiO_2 based floating photocatalysts were proposed to efficiently disinfect the polluted water [69,70]. Varnagiris et al. [69] reported on the disinfection application of C-doped TiO_2 films supported on high-density polyethylene (HDPE) beads in polluted water as floating photocatalysts. Interesting results of microbial disinfection i.e., 95% of *Salmonella typhimurium* bacteria in water environment along with recyclability of the substrates were observed. Similarly, they used polystyrene (PS) beads to coat with photocatalytic TiO_2 and used for the photocatalytic disinfection of *E. coli* in antibacterial contaminated water as shown in Figure 6h [70]. More than 90% of inactivation was observed. Recently, Sboui et al. [71] proposed TiO_2 and Ag_2O nanocomposites immobilized on cellulose paper as an excellent floating photocatalytic antibacterial system against *E. coli*.

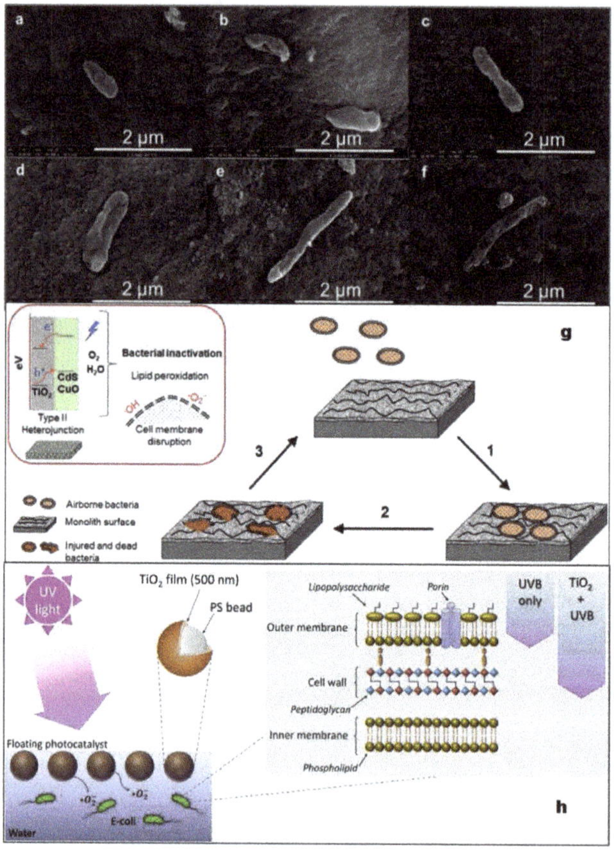

Figure 6. SEM images of bioaerosols adsorbed on TiO_2 based monoliths: (**a**,**b**) m-TiO_2/CdS; (**c**,**d**) m-TiO_2/CuO and (**e**,**f**) m-TiO_2, before and after photocatalysis, respectively; (**g**) Bioaerosol inactivation mechanism in the continuous flow photoreactor. (1) airborne bacteria are adsorbed on the monoliths. (2) photocatalysis produces ROS affecting the adsorbed bacteria. (3) inactive bioaerosols are released, the monolith surface becomes available to adsorb new airborne bacteria. Inset shows schematically the generation of reactive oxygen species from the monolith surface under blue light, causing lipid peroxidation (reproduced with permission from Ref. [65] Copyright 2021-Springer Nature); (**h**) schematic of *E. coli* bacteria hoto-decomposition using floating photocatalyst (reproduced with permission from Ref. [70]. Copyright 2020-Elsevier).

Tong et al. [29] demonstrated that TiO_2 coating could be efficiently used to inhibit *hepatitis C virus* (HCV) infection under the effect of weak indoor light. It was explained that OH radicals (as compared to $\bullet O_2^-$ radicals) produced by photo-activated TiO_2 were mainly responsible for damaging RNA genome by inactivating the virus without destroying the virion global structure and contents. The overall experiment has been described in Figure 7a. More details can be found elsewhere [29]. Figure 7b shows the effect of photoactivated TiO_2 on the inhibition of HCV infection in different conditions (light/off). Figure 7c,d show that there is no change in the quantity of HCV protein and RNA in the virions which proves that there is no destruction by photoactivated TiO_2-produced ROS on the virion structures. Figure 7e,f show the effect of photocatalytic treatment on the cells post-transfection. It signified that there were no viral amplifications after the photocatalytic TiO_2 inactivation suggesting that although there was no change in the amount of viral RNA after interaction with photocatalytic TiO_2 but their biological activity was impaired. This resulted in the realization of the damaging effect of RNA genome for the inactivation without destroying the virion structure. The formation of ROS was studied by electron spin-resonance (ESR) spectroscopy measurements as shown in Figure 7g,h. The ESR results exhibited presence of both $\bullet OH$ and $\bullet O_2^-$ radicals under the effect of light in indoor environments, whereas no peaks corresponding to these radicals were observed in dark conditions [29]. These results show that ROS were produced and played a major role in the inactivation of the HCV viruses.

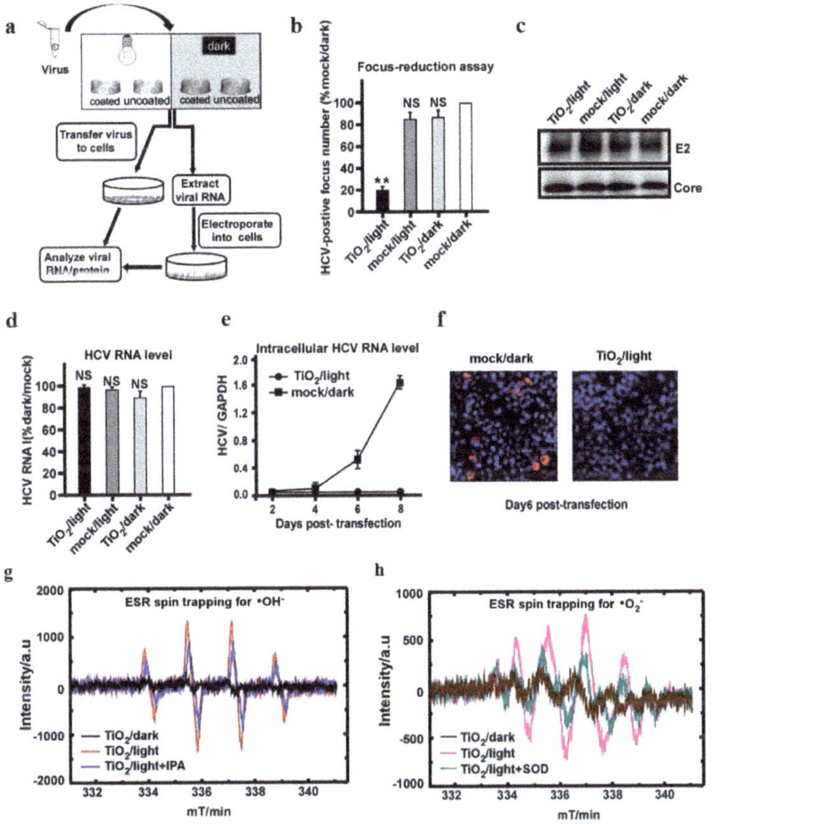

Figure 7. (**a**) Schematic representation of TiO_2 photocatalytic inhibition of HCV infection in dark and light; (**b**) the HCV-positive foci as determined by immunofluorescence of HCV NS3 proteins and

expressed as the percentage of the mock/dark group; (**c**,**d**) HCV E2 and Core proteins (**c**) and genome RNA (**d**) in the input virions after the TiO2/light treatment (as quantified by Western blotting and RT-qPCR, respectively); (**e**) the intracellular HCV RNA levels; (**f**) immunofluorescence of NS3 proteins (red) transfected with the extracted RNA from TiO_2/light treated virions on day six post-transfection. Nuclei (blue) were stained with Hoechst dye; (**g**,**h**) ESR results show the presence of both •OH and $•O_2-$ radicals under the effect of light in indoor environment whereas, no peaks corresponding to these radicals were observed in the dark conditions (reproduced with permission from Ref. [29] Copyright 2021-Elsevier).

Similarly, Khaiboullina et al. [28] studied the photocatalytic TiO_2 coating effect on HCoV-NL63 virus, which is known to be an alpha coronavirus causing acute respiratory distress symptoms. It was found that TiO_2 coating effectively inactivated HCoV-NL63 viruses by reducing the viral genomic RNA stability and virus infectivity. The HCoV-NL63 viruses, placed on TiO_2 nanoparticles (NPs) coated surfaces, were treated with UV, and complete disinfection was observed even after one minute of UV light exposure. In order to make sure of the disinfection activity, a virus infectivity assay was carried out by adding the recovered viruses (from TiO_2 coated and uncoated surfaces after UV light exposure) onto the monolayer of HEK293L cells. As observed in immunofluorescent signals from HEK293L cells as shown in Figure 8 (when compared A–D to E–H), it was found that those HCoV-NL63 viruses taken from TiO_2 coated surfaces after UV exposure were completely inactivated, while those on uncoated TiO_2 surface were not fully inactivated and live infectious viruses were detected. The overall experiment concluded that the viruses on the TiO_2 coated surface could be disinfected very quickly as compared to that of the uncoated surface under the UV light exposure [28].

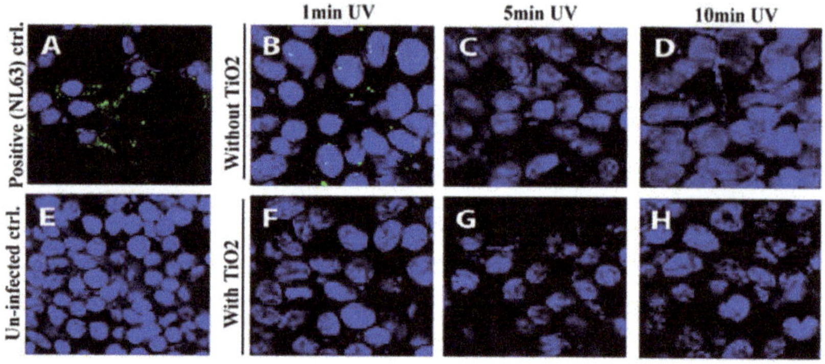

Figure 8. Post-UV treatment, HCoV-NL63 virus was collected and subjected to the infection of HEK293L monolayer. Viral protein, indicator of virus infectivity and replication, was detected by immune localization through IFA (green fluorescent dots): (**A**) NL63 infection (positive control); (**B**) 1 min UV light exposure without TiO_2; (**C**) 5 min UV light exposure without TiO_2; (**D**) 10 min UV light exposure without TiO_2; (**E**) uninfected control (negative control); (**F**) 1 min UV light exposure with TiO_2; (**G**) 5 min UV light exposure with TiO_2; and (**H**) 10 min UV light exposure with TiO_2 [28].

Mathur et al. [72] demonstrated that the UV activation killing mechanism of TiO_2 could be effectively used for purifying the air in a closed cabin by killing living organic germs, bacteria, pathogen viruses, etc., from the cabin air in recirculation mode. It was suggested that by using this model, various air condition systems could be designed to make the cabin free of bacterial and viral infections. Similarly, Matsuura et al. [20] studied the disinfection of SARS-CoV-2 in liquid/aerosols using coated TiO_2 nano-photocatalyst activated by a light-emitting diode in an air cleaner. The photocatalytic disinfection was explained by observing the virion morphology using transmission electron microscopy (TEM) by detecting damaged viral RNA and proteins as shown in Figure 9a. These nano-photocatalysts are shown to be very promising antiviral coatings for protecting surfaces against the spread of viruses such as COVID-19 [73–75]. Recently Uppal et al. [74] performed the study of a TiO_2 photocatalytic coating for virucidal activity against the HCoV-OC43 virus which is a member of the beta coronaviruses family just like SARS-CoV-2 under the influence of UV irradiation using T-qPCR and virus infectivity assays. It was found that the glass surface coated with TiO_2 exhibited better antiviral response as compared to uncoated glass against the virus. UV exposure reduced the viral RNA copies and the infectious virus with increasing exposure time and a 60 min exposure completely disinfected the viruses, as shown in Figure 9b,c.

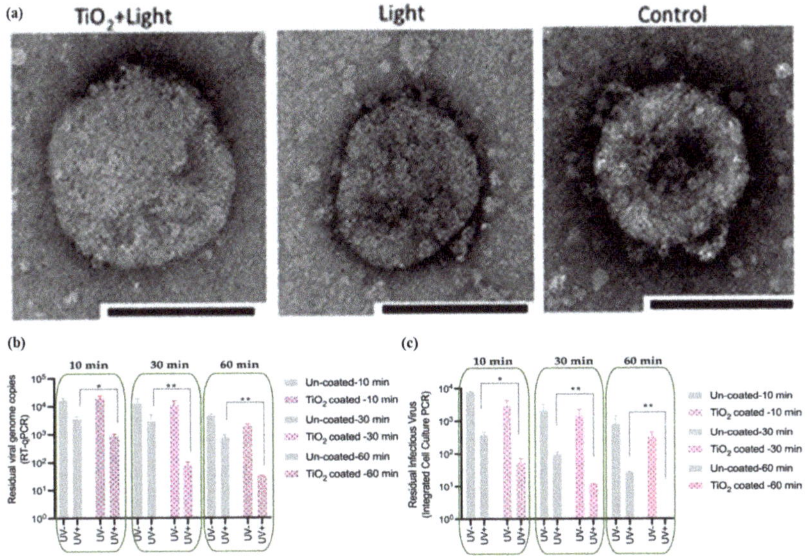

Figure 9. (**a**) TEM images SARS-CoV-2 virus showing the effect of TiO_2 and UV irradiation on the stability of human coronavirus [20]; TiO_2 photocatalytic coating for virucidal activity against the HCoV-OC43 virus (**b**) followed by exposure to UVA light (Post UV-irradiation, total viral RNA was extracted for the detection of intact viral genome copies via RT-qPCR.) (**c**) followed by infection of A549-hACE2 cells (total viral RNA was extracted to detect intracellular viral genome copies following infection and replication) [74].

Only UV exposure has also been reported to be effective against corona and other viruses spreading but has limited applications in surgical instruments, respiratory masks, and indoor environments that are also very challenging [76]. TiO_2 coating on the tile surface was also studied to investigate the effect of the coating against virus contamination keeping in mind public places such as hospitals. It was found that ambient light could not have any effect on the viral viability but when the tiles were coated with TiO_2/Ag–TiO_2, complete inactivation of the viruses under the same ambient light conditions in 5 h was exhibited. The coating was also found very stable after 4 months and showed similar disinfection response in indoor environment [77]. Nakano et al. [73] reported on antiviral effect of Cu_xO/TiO_2 induced inactivation of SARS-CoV-2 virus, including its Delta variant [78] under dark conditions as well as light irradiation using a normal white fluorescent bulb. Figure 10a,b show the morphology of the nanocomposites of Cu_xO/TiO_2 photocatalyst and the enhanced visible light absorption spectra respectively. Role of CuxO/TiO_2 photocatalyst was examined against SARS-CoV-2 virus which revealed the inactivation through damaging proteins and RNAs in the virus even under the dark conditioned. The damaging action was enhanced further under the white bulb illumination in presence of visible light active

Cu_xO/TiO_2 photocatalyst as shown in the graph of Figure 10c. These results provide an important and promising implication of such photocatalysts for virus disinfection in dark and indoor as well as outdoor environments. This is evidenced by the presence of virus (white spots in Figure 10d) in the digital image of active SARS-CoV-2 virus without Cu_xO/TiO_2 photocatalysts while when the photocatalysis was conducted under visible light irradiation using a white fluorescence bulb, no virus was seen (Figure 10e [73]). It is evident that the TiO_2 coating could be a potential nanomaterial for limiting the virus spread in poorly ventilated as well as in high-traffic public places [28,74,77].

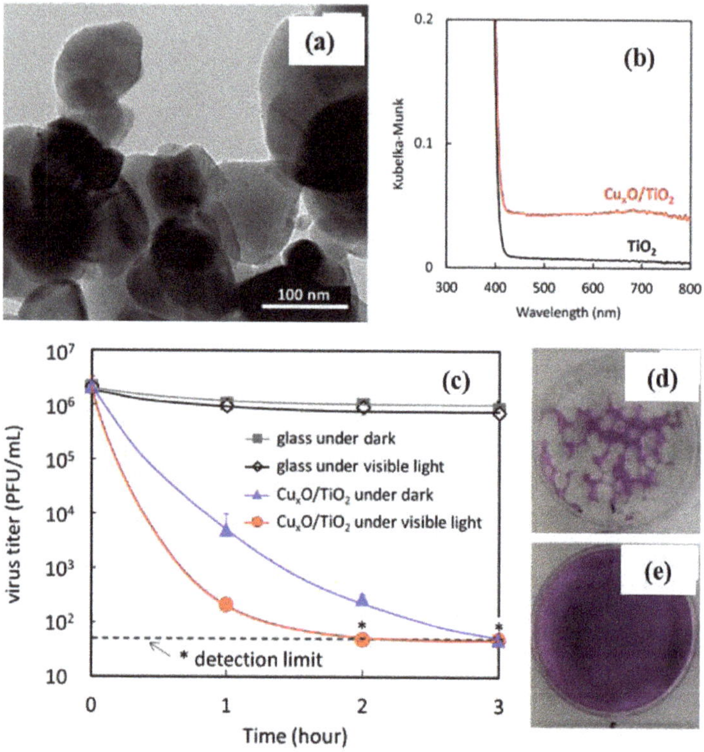

Figure 10. (**a**) Morphology of Cu_xO/TiO_2 and (**b**) UV–Vis spectra of $CuxO/TiO_2$ showing the visible light active photocatalyst. Inactivation of wild-type strain of SARS-CoV-2 by photocatalyst; (**c**) experimental results of inactivation of virus titer of SARS-CoV-2; photograph of (**d**) viral plaques infected by SARS-CoV-2 without photocatalyst; and (**e**) viral plaques infected by SARS-CoV-2 for $CuxO/TiO_2$ photocatalyst under visible light irradiation [73].

3.2. *ZnO Based Nano-Photocatalysts as Potential Antibacterial and Antiviral Agents*

Zinc oxide (ZnO) is the most capable wide band gap (3.37 eV) semiconductor inorganic material with a broad range of applications in the field of sensors, UV laser, photocatalysis and photovoltaics [79]. ZnO have also been used in different biological and environmental applications. The food and drug administration of United States has already provided the safety confirmation about ZnO [80]. Many biological applications of ZnO have been reported by the research in the field of biosensors, glucose, biomedical imaging, estimations of enzyme, etc. [81]. Particularly, as a potential antibacterial and antiviral agent, ZnO shows promising disinfection activities due to the creation of ROS and zinc ions which leads to the cell membrane disintegration, cell lysis, membrane protein damage, resulting in cell death [82]. The largely valuable antibacterial mechanism is the production of ROS, which is directly related to cell death by damaging functional cellular components, such as proteins and DNA.

As discussed in the last section, use of nano-photocatalysts in environmental disinfection is promising for the health of society and can be achieved by producing efficient nanomaterials with excellent antimicrobial and antiviral agents that can work in disinfecting air, water and common surfaces [83]. ZnO is another highly researched nano-photocatalysts which has been investigated in previous years during the pandemic period for such purposes in the environmental and healthcare applications. The ZnO based nanostructures have been modified by different approaches such as tailoring of size, surface modifications by doping, annealing, or forming heterojunctions [12,19,23,84–86] for making an efficient antimicrobial, and antiviral nanomaterials. For example, Silva et al. [19] reported that the surface modification of the ZnO NPs using (3-glycidyloxypropyl) trimethoxy silane which allowed the dispersion of ZnO NPs in water providing an important pathway for uniform water disinfection without further contamination. Similarly, use of ZnO NPs was demonstrated for the bacterial decontamination on the surfaces as well as in drinking water, remotely, and most importantly, in the absence of sunlight by Milionis et al. [85]. ZnO based superhydrophobic, and self-cleaning surfaces were prepared by depositing highly conformal, biodegradable, and water-soluble fluoroalkylsilane (FAS) or ethanol-soluble stearic acid (SA) as shown in Figure 11a–c. The low bacterial adhesion was recorded on superhydrophobic ZnO substrates due to the self-cleaning properties. Hence, inhibition of surface bacterial contamination was observed. The wetting properties were studied using contact angle measurements as shown in Figure 11a. It was found that optimal concentrations for SA and FAS resulted in advancing contact angle (ACA) of 160° and CAH of 5° for the case of SA and ACA of 158° and contact angle hysteresis (CAH) of 3° for the FAS and corresponding morphologies were studied with SEM micrographs shown in Figure 11b (SA) and Figure 11c (FAS). Mechanism of interaction with ZnO nanostructures with microbes was proposed as shown in Figure 11d–e. When microbes were in contact with the ZnO surface, antibacterial action took place through the released zinc ions (Zn^{2+}) and ROS as shown schematically in Figure 11d(i,ii). When interaction took place with superhydrophilic ZnO nanostructures, due to the sharp edges of the ZnO nanostructures via killing mechanism puncturing of the cell walls occurred as shown in Figure 11d(iii). These mechanisms of antibacterial actions of ZnO nanosctures on *E. coli* are shown in SEM micrographs on different surfaces as shown in Figure 11e(i–iii). They applied these ZnO superhydrophilic surfaces in inactivation of *E.coli* in water disinfection under static and shaking conditions as shown in Figure 11f–g. It was found that these ZnO nanostructures showed the highest efficacy.

The size and concentration change of the ZnO NPs are another important factors which influence the antimicrobial/viral activities. The smaller sizes are more effective and promising antimicrobial agents, however, the structural differences in various pathogens are also important to consider the effectiveness of the NPs. Raj et al. [84] studied the disinfection activities with emphasis on effect of size and concentration of ZnO NPs on microbes with different structures. It was found that the bacterial strains are important factor and due to the structural differences between *E. coli* and *P. aeruginosa*, the antibacterial efficiency of ZnO NPs was also not same. It was also found that the zone of inhibition was increased with increasing the concentration and decreasing the size of ZnO NPs. The smaller ZnO NPs showed better bactericidal activity whereas the larger NPs presented the bacteriostatic activity. The doped ZnO nanostructures show enhanced properties such as surface area and charge transfer which are beneficial for the disinfection activities. A number of investigations have been carried out on doped ZnO with an emphasis on improving its surface, optical and charge transport properties for particularly photocatalysis and antibacterial activities [87–89]. These all contribute to the production of greater ROS which enhances disinfection activities. Naskar et al. [90] reported that the antibacterial activity of ZnO NPs was improved with Ni doping because of the increased surface area and decreased crystallite size. The morphological changes occurred in multidrug-resistant strains of *E. coli* and *A. baumannii* were investigated by SEM technique before and after exposure to 5% Ni doped ZnO (NZO) as shown in Figure 12a–d. It can be seen in the SEM micrographs that untreated *E. coli* and *A. baumannii* cells are with smooth and intact surfaces (Figure 12a,c respectively). However, variations in morphologies of the cells were observed as a result of membrane corrugations after treatment with 5% doped NZO due to wrinkling and damaging of the cell membranes as shown by red circles indicating the areas of cell membrane disruption (Figure 12b,d corresponding to the Figure 12a,c respectively). The 5% doped NZO NPs demonstrated better antibacterial activity with respect to other concentration which was attributed to the high production of ROS which was studied and quantified using fluorescence intensity at 520 nm of *E. coli* ATCC 25922 cells as shown Figure 12e.

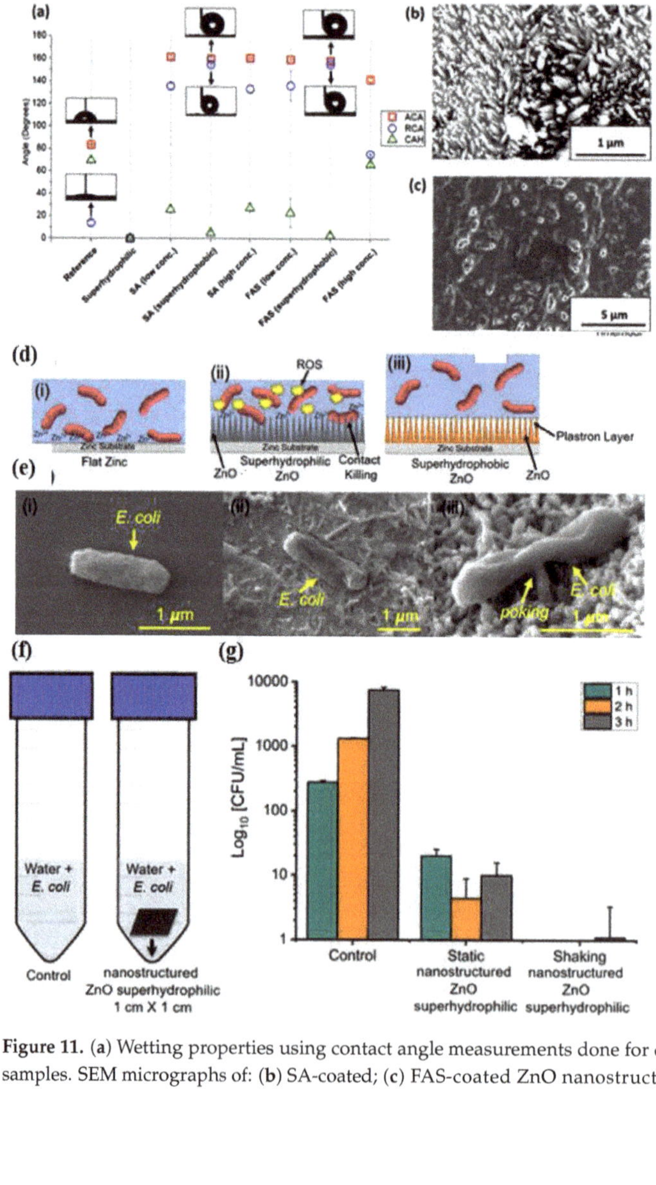

Figure 11. (**a**) Wetting properties using contact angle measurements done for different ZnO based samples. SEM micrographs of: (**b**) SA-coated; (**c**) FAS-coated ZnO nanostructures; (**d**) schematic

representation mode of interactions of surface–bacteria on flat Zn substrates, nanostructured superhydrophilic ZnO surfaces, and nanostructured hydrophobized ZnO surfaces; (**e**) SEM micrographs of E. coli on (i) glass coverslip, (ii) flat zinc, and (iii) nanostructured ZnO superhydrophilic surface; and (**f**,**g**) ZnO superhydrophilic surfaces in inactivation of E.coli in water disinfection under static and shaking conditions. (reproduced with permission from Ref. [85] copyright-2020 American Chemical Society).

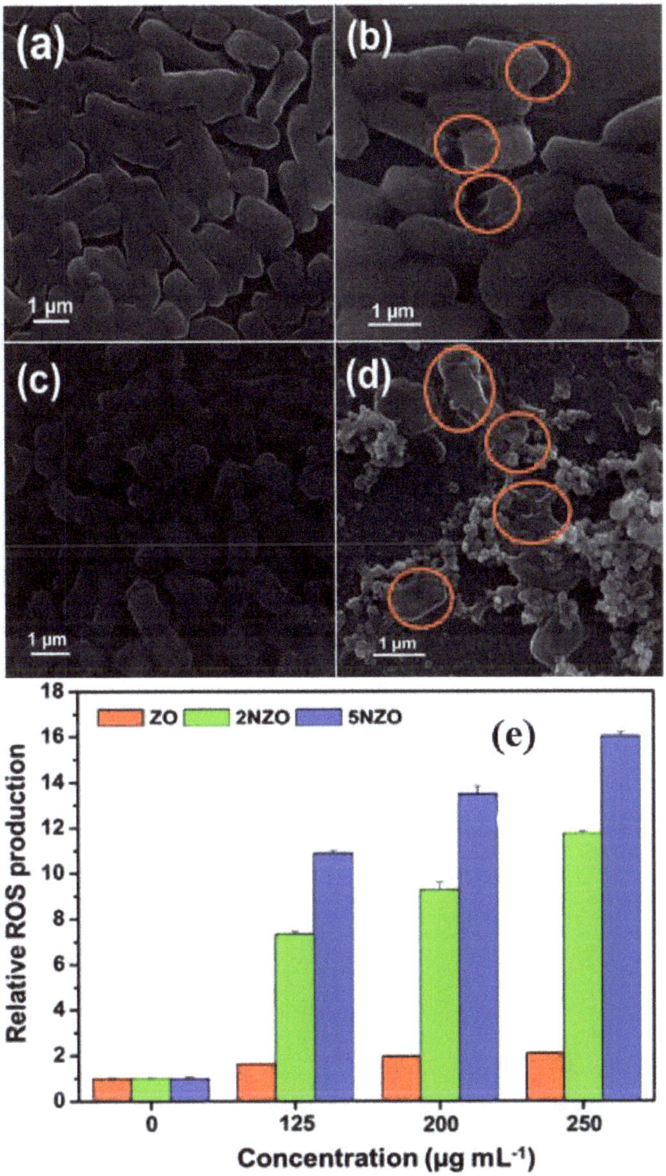

Figure 12. The surface morphology of bacterial cells: (**a**) untreated samples of *E. coli;* (**b**) treated sample of *E coli* by 5% doped NZO; (**c**) untreated samples of *A. baumannii;* (**d**) treated samples of *A. baumannii* with 5% doped NZO; and (**e**) quantification of ROS production using fluorescence dye [90].

Stability of such functionalized ZnO nanostructures is one of the important concerns for real practical applications. In this context, different approaches have been applied. For example, concerning air pollution with bacterial contaminants which really causes several type of health problems, Geetha et al. [91] produced stable and low cost ZnO nanofibers based polymer nanocomposites as a promising coating for face masks, with an aim to filter both particulate and bacterial contaminants. ZnO NPs were dispersed homogeneously in the mixed PVA/PVP polymer blend solution and using electrospinning system, ZnO nanofilbers-polymer nanocomposites were produced with different ZnO NPs concentrations as shown in Figure 13a–d. Figure 13a Shows the SEM micrographs of the ZnO NPs which are agglomerated having NPs of different sizes. The morphology of the only polymer nanofibers shows uniform and smooth morphology as shown in Figure 13b. Whereas, ZnO based polymer nanocomposites are shown in Figure 13c,d which exhibit rough surface morphology attributed to the embedded ZnO NPs. ZnO based polymer nanofibers are also shown to be of increased size attributed to the increased concentration of ZnO NPs. Several bacteria were treated with these ZnO based polymer nanocomposites and their antibacterial effects were studied. The microbial growth inhibition efficiency of ZnO NPs, only polymer, and ZnO-polymer nanocomposite fibers for the *S. aureus*, *E. coli*, *K. pneumonia* and *S. aeruginosa* bacteria are shown in Figure 13e. It was found that around the electrospun circular mask, there was no bacterial growth observed for a particular distance when nanocomposite was applied with different concentration of ZnO NPs in nanocomposites to the mask, whereas there was no any disinfection/inhibition zone effect observed for only polymer nanofibers. These results show good results, indicating the usefulness of the prepared nanofibrous material for antimicrobial face masks as shown in Figure 13f.

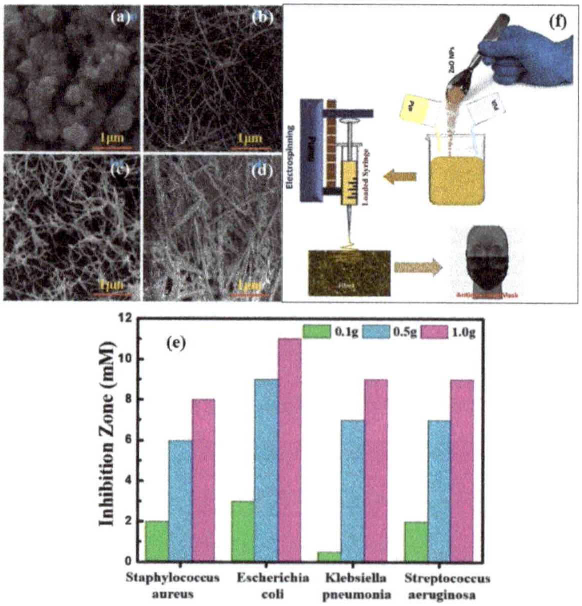

Figure 13. SEM micrographs of (**a**) ZnO NPs; (**b**) PVA/PVA; (**c**) ZnO (0.1 g)/PVA/PVP; (**d**) ZnO (1 g)/PVA/PVP composites; (**e**) the microbial growth inhibition efficiency of ZnO/PVA/PVP nanocomposite fiber against *S.aureus*, *E. coli*, *K. pneumonia* and *S. aeruginosa;* and (**f**) schematic of synthesis of ZnO-polymer based composite nanofibers for the application of face mask (eproduced with permission from Ref. [91]. Copyright 2022-Elsevier).

ZnO NPs are exploited as promising antiviral nanomaterials against different kind of viruses including SARS-CoV-2 [83,92]. ZnO NPs have been used against herpes simplex virus type 111 and H1N1 influenza virus [93–97]. Usually, zinc is an important element that observed in bone, muscle, brain, and skin of human. This vital element is also concerned in different enzyme processes like protein and metabolism [98]. Zn is observed to impede both SARS-CoV and retrovirus in vitro RNA polymerase activity, as well as zinc ionophores which hinder the viruses replication in cell culture [99]. Zn was also established to reduce the viral replication of different RNA viruses like respiratory syncytial virus influenza virus, and numerous picornaviruses [97,100]. ZnO NPs show better performance in antiviral activity as compared to the other antiviral materials because they have also emanated from their good compatibility to biological systems, high safety, low price, and good stability [101]. Additionally, visible fluorescence of ZnO is used in bioimaging and control monitoring of the drug. All the premises of ZnO NPs presented show potential in the design of nanomedical viral-targeting therapies.

Hamdi et al. [93] reported that ZnO could be used for vitro characterizations and cellular uptakes in human lung fibroblast cells and then proposed a mechanism of ZnO against COVID-19. They synthesized ZnO NPs and studied its surface, structural and morphological properties in view of its interaction with virus. The estimation of surface charge of ZnO NPs was found to be important to understand the interaction with the biological membranes. The value of zeta potential for optimized ZnO NPs was recorded from +25.32 to −18.78 and found to be highly dependent on the dispersant medium pH. The release profiles of Zn^{2+} at different pH was optimized and it was found that higher Zn^{2+} solubility in the acidic medium was the reason for the higher Zn release at pH 5.5. In silico molecular docking was investigated to speculate the possible interaction between ZnO NPs and COVID-19 targets including the ACE2 receptor, COVID-19 RNA-dependent RNA polymerase, and main protease. Interestingly, felicitous binding of these ZnO NPs with the three tested COVID-19 targets, via hydrogen bond formation, was observed and an enhanced dose-dependent cellular uptake was proposed. The proposed mechanism of interaction of ZnO NPs with SARS-CoV-2 has been shown schematically in Figure 14.

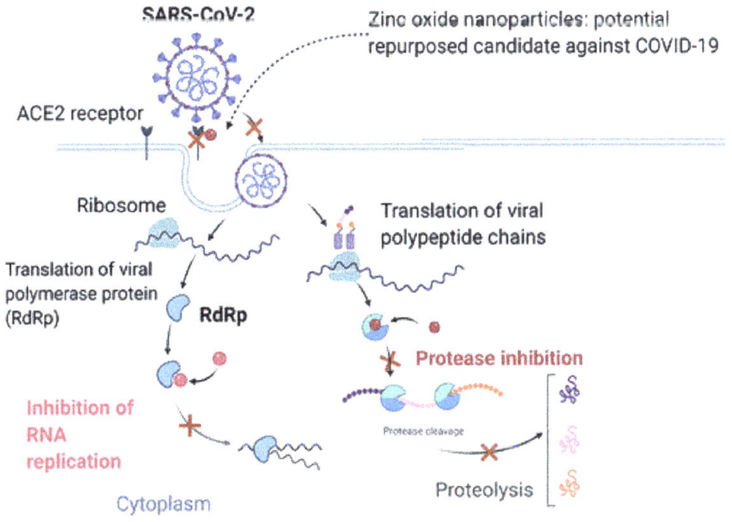

Figure 14. Schematic of interaction of ZnO NPs with SARS-CoV-2 [93].

3.3. CuO and Other Metal Oxides Based Nano-Photocatalysts as Potential Antimicrobial and Antiviral Agents

Due to their narrow band structure and open surface sites, structurally modified inorganic metals NPs and their oxide forms possess special physical and chemical characteristics that make them more suitable to interact with pathogens and microorganisms. In this regard, a number of antiviral medicines made of transitional metal-oxide such as ZnO, CuO, and TiO_2 were found. It has been discovered that after entering the host cells, viruses attach, invade, duplicate, and branch out. While the antiviral properties of the metal-based nanomaterials may interact with the virus to restrict the insertion of the virus into the cell body. Additionally, the metal-oxides generate extremely potent free radical species that damage the nucleus of the virus and prevent it from killing the organelles of living cells.

Due to its functionalized surfaces and electronic band structures, CuO has been investigated as an antibacterial and antiviral agent among the many metal oxides [102–104]. As one of the most effective nanocatalysts for fighting bacteria, fungus, and viruses at the right concentration, it is also one of the least poisonous and expensive nanomaterials [103]. The antimicrobial activity of CuO NPs is due to their close interaction with bacterial membranes and the release of metal ions. When the NPs are near the lipid membrane, they slowly oxidize, release copper ions, and generate harmful hydroxyl free radicals. These free radicals separate the lipids from the cell membrane through oxidation and destroy the membrane [105]. The entire genome order of the virus is frequently destroyed in the nuclei by these reactive ions or radicals [106]. Mechanistically, the free radical ions damage the host's proteins, lipids, and viral glycoproteins in the outermost receptor surface. The damaged cell membranes (coenzyme A) therefore, disrupted metabolic processes (respiration), leading to cell lysis or virus destruction [107,108]. Consequently, in the case of CuO, the free Cu ions produced from Cu_2O/CuO inspire the formation of free radicals that prompt the destruction of the outer capsid level of the virus. It eventually destroys their genomic sequences and the replication process is stopped at a non-cytotoxic amount [109]. In order to remove organic pollutants and different pathogens from waste water, it has also been employed as a better photocatalyst [110,111]. The band gap of CuO is ~1.7 eV [110]. It has demonstrated excellent promise for use in the creation of antibacterial compounds as well as the photocatalytic destruction of organic contaminants. For instance, Akhavan et al. [112] showed that the Cu and CuO NPs covering had antibacterial action both in the dark and when exposed to light.

The information above demonstrates the potential of CuO NPs to function as effective antibacterial or antiviral agents in indoor or outdoor settings, i.e., in the absence of sunshine or underwater. A composite coating made of Cu_2O NPs and linear low-density polyethylene, for instance, showed antibacterial characteristics for the disinfection of water as studied by Gurianov et al. [113]. The composite exhibited no or very low leaching of copper ions into the aqueous phase, showing good antibacterial activity against *S. aureus* and *E. coli*. Similarly, Domagala et al. [114] investigated the antiviral effectiveness of Cu_2O nanostructures encapsulated on multi-walled carbon nanotubes (MWCNTs) based filters and their stability for virus removal from water. Three different procedures were followed to produce different nanocomposites with MWCNTs such as (1) direct Cu ion attachment, (2) $Cu(OH)_2$ extraction, and (3) $[Cu(NH_3)_4]^{2+}$ complex bonding with MWCNTs. The formation of nanocomposites and distribution of Cu_2O on MWCNTs were confirmed by TEM images as shown in Figure 15a–c. The MS2 bacteriophages elimination assays were carried out twice, once before (day 1) and again (day 2) post treating filters using water at pH of 5 and 7. (24 h) as shown in Figure 15d,e. The effective removal outcomes showed how important the methodology is for the removal of viruses, the efficiency of disinfection in Cu_2O-coated MWCNTs, and the potential for virus elimination due to copper's antibacterial capabilities as well as retaining viral electrostatic adsorption in MWCNTs. Similarly, Mazukow et al. [115] developed a spray based on alumina granules deposited with CuO NPs filters for virus removal from water as schematically depicted in Figure 15f and also investigated the effect of copper oxidation state on virus removal capacity. It was found that an alumina support of this kind offered a porous structure to ensure extended interaction with water-harbouring viruses. It was discovered that the principal virus-removing phases were copper (I) oxide and metallic copper, and 99.9% of MS2 bacteriophages could be eliminated [115].

Antimicrobial property of thermocycled polymethyl methacrylate denture base resin reinforced with CuO NPs was studied by Giti et al. [104]. The CuO NPs may also be effective as a potential control agent or candidate for avoiding dental infections or caries [116]. In a study, it was found that CuO NPs doped with Fe were effective against biofilm forming bacteria and fungi [103]. These Fe doped CuO NPs showed better photocatalytic and antimicrobial efficacy because of reduced band gap of 1 eV. By co-culturing CuO NPs with HSV-1-infected cells at a certain concentration of CuO NPs (100 g/mL), Ahmad Tavakoli and colleagues [109] showed that the CuO NPs had excellent anti-HSV-1 viral efficacy. This disinfection level of the cell was subsequently increased to 83.3%. ROS were discovered to be mostly produced by semiconducting CuO NPs, which were then bound to the HSV-1 and destructed the virus's DNA. The HCV was also discovered to be resistant to Cu_2O NPs [117] (Figure 15g–k). According to this study, Cu_2O NPs exhibited inhibitory effect on virus infection on the target cells by blocking the virus infection both at the attachment and entry stages.

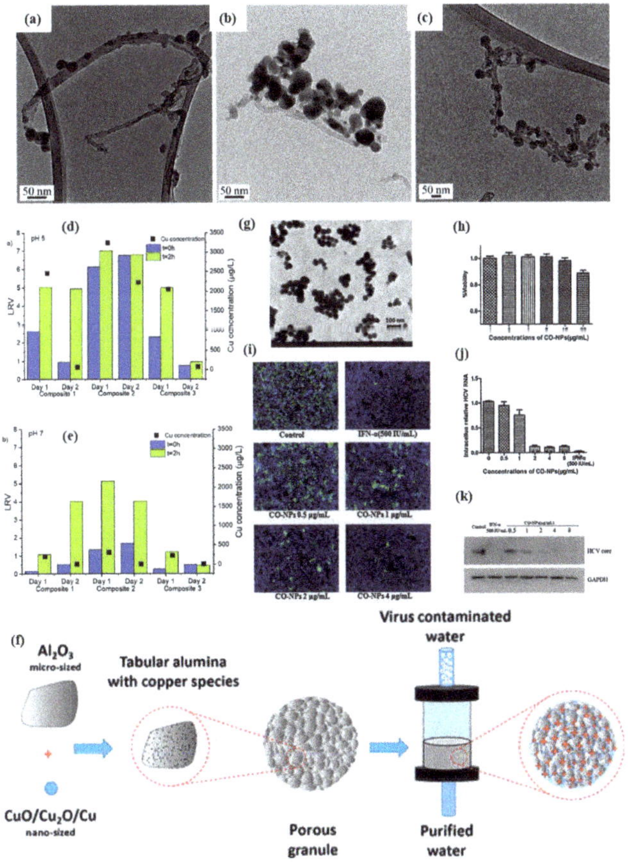

Figure 15. TEM images of Cu_2O-MWCNTs nanocomposites: (**a**) composite 1; (**b**) composite 2; (**c**) composite 3. MS2 removal test results for composite 1, composite 2 and composite 3 at (**d**) pH 5

and (**e**) pH 7, and the associated copper concentrations detected in the permeate (reproduced with permission from Ref. [114] Copyright 2020-Elsevier); (**f**) Schematic representation of nano-sized copper (Oxide) on alumina granules filters for water filtration. Reproduced with permission from Ref. [115] Copyright 2020, American Chemical Society. Cu_2O (CO) NPs inhibited HCV infection. HCVcc (MOI = 0.1) virions were added with the indicated concentrations of Cu_2O NPs or interferon-(IFN-, 500 IU/mL) to Huh7.5.1 cells. After incubation at 37 °C for 2 h, the cells were washed three times to remove the virus and treated with the indicated concentrations of Cu_2O NPs for additional 72 h; (**g**) the TEM image of Cu_2O NPs; (**h**) Huh7.5.1 cells were treated with Cu_2O NPs at the indicated doses for 72 h. The effect of Cu_2O NPs on cell viability was measured by a CCK8 assay kit; (**i**) Huh7.5.1 cells at 72 h post-infection were stained with HCV-positive serum from patients and with DAPI. Representative fluorescence images are shown (80× magnification); (**j**) The intracellular HCV RNA content was determined by qRT-PCR; and (**k**) Western blot analysis of the harvested cell lysates using the anti-core or anti-GAPDH antibodies. Results are represented as the mean ± SD from 3 independent experiments ($p < 0.001$) (reproduced with permission from Ref. [117]. Copyright 2015-Elsevier).

The aforementioned examples demonstrate the potent antiviral and antibacterial characteristics of CuO and Cu_2O NPs, as well as how these properties can be maintained even when embedded in layers of polymeric covering. This may hold promise for effective real-world applications against viral and microbial diseases on surfaces used for public usage. Without altering the mask's standard filtration procedure, Borkow et al. [118] demonstrated that CuO impregnated masks could be helpful in lowering the danger of pathogen transmission/contamination in the air. It was proposed that such masks containing CuO NPs also provide protection from any kind of pathogens and are very important for combatting against the spread of and infection by dangerous pathogens attributed to the potent antiviral and antibacterial properties of CuO. As shown in Figure 16a–c [118], a respiration face mask was prepared containing CuO as an antiflu treatment. It was found that these CuO containing masks effectively filtered more than 99% of inhalational influenza viruses such as H1N1 and the H9N2. Very recently, Leung et al. also proved a high efficacy of CuO NPs decorated respiratory masks that efficiently decrease viral infections [119].

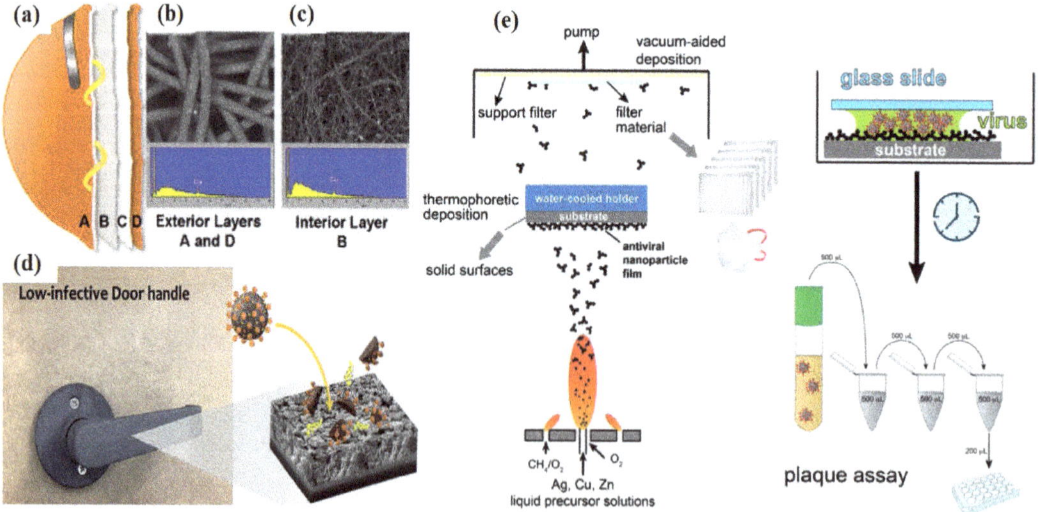

Figure 16. (**a**) CuO impregnated masks; (**b**) outer coating of mask covering CuO NPs; (**c**) inner coating of mask covering CuO NPs. [118]; (**d**) schematic of cupric Oxide coating on the common

surfaces such as door locks that rapidly reduces infection by SARS-CoV-2 (reproduced with permission from Ref. [120] Copyright 2021-American Chemical Society); and (**e**) schematic of the flame aerosol deposition process of antiviral nanoparticle coatings on solid flat substrates as well as on porous filter materials. The as-prepared nanoparticle coatings are then incubated with SARS-CoV-2, and their antiviral activity is examined by the plaque assay [121].

The primary cause of the coronavirus illness, COVID-19, is the SARS-CoV-2 virus, which spreads through repeated contact or aspiration of respiratory secretions. CuO and Cu_2O are ionic forms of the Cu ions, and their antiviral, antifungal, and antibacterial properties imply that they may also be useful against the SARS-CoV-2 virus. The effects of a Cu_2O coating on a SARS-CoV-2 virus disseminated solution in an aqueous droplet were investigated by Hosseini et al. [120]. In the experiment, they used a live SARS-CoV-2 strain rather than a proxy virus on BSL-3 environments. It was discovered that the developed coating was quite helpful in reducing COVID-19 transmission and contacting concerns. To reduce disease transmissions, the cost-effective and stable coating was created and applied to places where the public can touch, like door handles (Figure 16d). The Cu_2O NPs were compacted into a strong coating followed by thermal annealing which allows forming the crystalline solid phase of CuO crystals. The CuO coated on SARS-CoV-2 infectivity was lowered by 99.8% in 30 min and then 99.9% in an hour, while the coating stayed hydrophilic for around 5 months. Merkl et al. [121] investigated the effect of antiviral coating of various nanomaterials such as Ag, CuO and ZnO by depositing on both solid flat surfaces as well as porous filter media. The antiviral studies were carried out against SARS-CoV-2 viability and were compared with a viral plaque assay. The coatings were prepared by aerosol nanoparticle self-assembly during their flame synthesis as shown in Figure 16e. It was found that as compared to ZnO, other coatings were more effective including CuO showing their potential as antiviral coatings on variety of surfaces to reduce the transmission of viruses. Recently, Delumeau et al. [122] demonstrated that Cu and Cu_2O thin-film coatings deposited on glass had very strong antiviral effect on human coronavirus HCoV-229E. Additionally, the coating was placed to a N95 mask, and it was discovered through droplet studies that the coating decreased viral infectivity by 1–2 orders of magnitude in just over an hour.

In addition to everything mentioned above, CuO has demonstrated superior antiviral and antimicrobial photocatalyst properties, as well as being a better supporting material to increase the photocatalytic disinfection efficiency of other wide band gap photocatalysts like TiO_2 NPs in both indoor and outdoor environments [123]. Inactivation of various variant types of SARS-CoV-2 by indoor-light-sensitive TiO_2-based photocatalyst [110]. For example, Farah et al. [123] studied the *E. coli* disinfection in indoor air under the photoreactor and found that CuO-TiO_2 nanocomposite acted as excellent photocatalyst. Nakano et al. [73] studied antiviral behaviour of Cu_xO/TiO_2 photocatalyst under the indoor light environment and found the excellent antiviral effect against COVID-19. Kanako et al. [124] demonstrated the photocatalytic antibacterial activity of TiO_2, TiO_2 + CuO, and WO_3 + CuO and found that composite with CuO enhanced the photocatalytic antibacterial activity of TiO_2 and WO_3 photocatalysts. In addition, WO_3 has been investigated for its potential photocatalytic properties for disinfection activity [125,126]. It is an n-type semiconductor capable of absorbing the visible light up to 480 nm (bandgap energy ranges between 2.5–3.0 eV) with an excellent photostability and surface transport properties [127]. The potential application of WO_3 based nano-photocatalyst coating in inactivation of SARS-CoV-2 under the visible light was demonstrated by Uema et al. [125] in different environmental conditions. A low-cost multiphase photocatalyst from industrial waste and WO_3 was proposed by Hojamberdiev et al. [127] for photocatalytic removal of SARS-CoV-2 antiviral drugs (lopinavir and ritonavir) in real wastewater for practical applications.

4. Challenges, Future Prospects and Summary

In conclusion, this paper provides a concise overview of the research background and significance of recent advances in metal oxide-based nano-photocatalysts as potential antimicrobial and antiviral agents. Emphasis is placed on understanding photomechanism processes and potential applications in the outdoor and indoor environment including water and air contamination [128]. To advance current technologies, many research groups have focused on the development of new photoreactors based on metal oxides. Using nanomaterial components not only enables more efficient and speedier regulation of the spread of dangerous viruses such as H1N1, SARS-CoV, and SARS-CoV-2, but also tends to make protective masks, fabrics, and screens reproducible in community settings. More research into the antiviral properties of transition metal oxides is required to reduce the severity of viral infections and prevent potential pandemics. As a result, it is reasonable to assume that in the coming years, such novel transitional metal oxide photocatalysts will undoubtedly pave the way toward an effective way to completely overcome the dangerous SARS-CoV-2. Metal oxide semiconductor photocatalysts have been extensively investigated for their ability to inactivate a variety of viruses and microorganisms. Examples of these photocatalysts include TiO_2 and ZnO. The photocatalytic inactivation of microorganisms is a synergistic bactericidal and virucidal effect of electromagnetic radiation at a given wavelength and the oxidative radicals produced by the photocatalyst when exposed to UV light. In this context, TiO_2 has been the subject of extensive research and has been successfully implemented in various well-known disinfection technologies. The crystallinity and concentration of the photocatalyst, as well as the appropriate combination of the intensity of the light applied, and the irradiation time, play a major role in determining the efficiency of UV-induced TiO_2 photocatalysis. The efficiency is a function of several parameters.

At present copper oxide impregnated masks safely reduce the risk of influenza virus environmental contamination without altering the filtration capacities of the masks. Due to the potent antiviral and antibacterial properties of copper oxide, we believe that these masks also confer protection from additional pathogens, and, as such, are an important additional armament in the combat against the spread of and infection by dangerous pathogens. The production of the mask layers with copper oxide and the manufacture of the mask using these materials do not add any significant costs to the price of the masks. It is suggested that copper oxide should be also included in other personal protective equipment to further confer protection to the wearer and to the environment [118]. Pathogens and certain other infectious germs cannot enter the wearer's nose or mouth using defensive breathing face masks. Nevertheless, improper handling and maintenance of masks, particularly when they are used by non-professionals like the common person, could probably induce pathogen transmission. The ROS-based concepts have already been reported by several different types of studies for the improved antibacterial and antifungal properties using semiconducting metal oxides nanomaterials. This improvement is attributed to the fast productions of ROS and the slow recombination rate of the electron-hole charges in the nanomaterials. Therefore, to reduce the spread and infections of viruses like SARS-CoV-2, numerous reasonable precautions including antiviral medications, passive vaccination, many antiseptic solutions, UV irradiation, antibodies are in practice [129–131].

As a result, CuO might find use in medical research and in the design of healthy environments in the form of NPSs, thin films, or functionalized nanostructures. Overall, the semiconducting metal-oxides (photocatalyst) like CuO, TiO_2, and ZnO have been appreciated for enhanced antiviral activities due to their structural, optical, and surface engineering at the nanoscale. The best method for inactivating encapsulated viruses is thought to be photocatalytic nanomaterials, which need illumination as their power source. Therefore, photocatalytic reactions take place because of a cumulative impact of solar energy at a fixed frequency and the photoactive substance, which captures a diverse range of sunlight wavelengths. These could be more practically applicable for the betterment of society and the environment.

5. Challenges

1. Due to the wide bandgap that it possesses, natural TiO_2 can only be excited by the near-UV photons that are present in the solar spectrum (390 nm). However, visible light accounts for approximately 42% of solar radiation [132], despite the fact that UV makes up only 4% of solar light. Because of its wide bandgap, it is not useful for applications involving the environment. As a result, boosting the photocatalytic activity of TiO_2 is a difficult problem;
2. It is common knowledge that photocatalytic nanomaterials could generate ROS, which can then destroy the structural components of viruses. However, the light source has a significant impact on their performance, which may result in an increase in the cost of their application;
3. It is important to tailor the development and choice of antiviral nanomaterials to their intended uses. The high flow rate of air purifiers makes it easier for viruses to gain momentum when antiviral materials are used. Because of this, nanomaterials' electrostatic effects on their surfaces should be amplified to improve their adsorption capacities toward viruses, and their physical structures should be strengthened to break viruses;
4. The biodistribution of metal oxide NPs is influenced by their interactions with proteins, which take place through a process known as opsonization. As a result, the NP's properties are altered [133];
5. When metal NPs are utilized for in vivo applications, it is imperative that the potentially harmful effects of these particles be taken into consideration. Nanotoxicity can be explained by two factors: (i) the potential release of toxic ions from metallic nanoparticles, and (ii) the oxidative stress caused by the inherent properties of the nanoparticles themselves (morphology, surface charge, size, and chemical surface composition) [134].

To sum up, it is crucial to integrate the safety assessment of the metal oxide-based nano-photocatalysts from the earliest stages of material design, synthesis, and development in order to implement the full potential of antimicrobial or antiviral nanomaterials, to consider both environmental and human health risks at each stage of the product life cycle, and, as is especially important, in the case of antibacterial nanomaterials, to consider the potential exposure effects on human commensal microorganisms. Many nanomaterials used to treat bacterial infections also have antiviral properties, so studying them could lead to novel approaches in treating and preventing the spread of viruses.

Funding: This research received no external funding.

Acknowledgments: Author (J.P.) acknowledges Department of Science and Technology (DST), India for providing INSPIRE Faculty award. Authors (P.K. & H.C.S.) acknowledge support provided by the South African Research Chairs Initiative of the DST and University of the Free State, Bloemfontein, SA.

Conflicts of Interest: The authors declare no conflict of interest.

References

1. Prakash, J.; Cho, J.; Mishra, Y.K. Photocatalytic TiO2 nanomaterials as potential antimicrobial and antiviral agents: Scope against blocking the, SARS-CoV-2 spread. *Micro Nano Eng.* **2022**, *14*, 100100. [CrossRef]
2. Soni, V.; Khosla, A.; Singh, P.; Nguyen, V.-H.; Van Le, Q.; Selvasembian, R.; Hussain, C.M.; Thakur, S.; Raizada, P. Current perspective in metal oxide based photocatalysts for virus disinfection: A review. *J. Environ. Manag.* **2022**, *308*, 114617. [CrossRef] [PubMed]
3. Guerra, F.D.; Attia, M.F.; Whitehead, D.C.; Alexis, F. Nanotechnology for Environmental Remediation: Materials and Applications. *Molecules* **2018**, *23*, 1760. [CrossRef] [PubMed]
4. Talebian, S.; Wallace, G.G.; Schroeder, A.; Stellacci, F.; Conde, J. Nanotechnology-based disinfectants and sensors for, SARS-CoV-2. *Nat. Nanotechnol.* **2020**, *15*, 618–621. [CrossRef]
5. Aghalari, Z.; Dahms, H.-U.; Sillanpää, M. Investigating the effectiveness of nanotechnologies in environmental health with an emphasis on environmental health journals. *Life Sci. Soc. Policy* **2021**, *17*, 8. [CrossRef]

6. Soliman, A.M.; Khalil, M.; Ali, I.M. Novel and Facile Method for Photocatalytic Disinfection and Removal of Organic Material from Water Using Immobilized Copper Oxide Nano Rods. *J. Water Process Eng.* **2021**, *41*, 102086. [CrossRef]
7. Chakraborty, A.; Samriti Ruzimuradov, O.; Gupta, R.K.; Cho, J.; Prakash, J. TiO_2 nanoflower photocatalysts: Synthesis, modifications and applications in wastewater treatment for removal of emerging organic pollutants. *Environ. Res.* **2022**, *212*, 113550. [CrossRef]
8. Samriti, M.; Chen, Z.; Sun, S.; Prakash, J. Design and engineering of graphene nanostructures as independent solar-driven photocatalysts for emerging applications in the field of energy and environment. *Mol. Syst. Des. Eng.* **2022**, *7*, 213–238. [CrossRef]
9. Prakash, J.; Sun, S.; Swart, H.C.; Gupta, R.K. Noble metals-TiO_2 nanocomposites: From fundamental mechanisms to photocatalysis, surface enhanced, Raman scattering and antibacterial applications. *Appl. Mater. Today* **2018**, *11*, 82–135. [CrossRef]
10. Gupta, T.; Cho, J.; Prakash, J. Hydrothermal synthesis of TiO_2 nanorods: Formation chemistry, growth mechanism, and tailoring of surface properties for photocatalytic activities. *Mater. Today Chem.* **2021**, *20*, 100428. [CrossRef]
11. Prakash, J.; Kumar, P.; Harris, R.A.; Swart, C.; Neethling, J.H.; van Vuuren, A.J.; Swart, H.C. Synthesis, characterization and multifunctional properties of plasmonic Ag–TiO_2 nanocomposites. *Nanotechnology* **2016**, *27*, 355707. [CrossRef]
12. Singh, N.; Prakash, J.; Gupta, R.K. Design and engineering of high-performance photocatalytic systems based on metal oxide–Graphene—Noble metal nanocomposites. *Mol. Syst. Des. Eng.* **2017**, *2*, 422–439. [CrossRef]
13. Rajput, V.; Gupta, R.K.; Prakash, J. Engineering metal oxide semiconductor nanostructures for enhanced charge transfer: Fundamentals and emerging SERS applications. *J. Mater. Chem. C* **2022**, *10*, 73–95.
14. Ahmadi, Y.; Bhardwaj, N.; Kim, K.-H.; Kumar, S. Recent advances in photocatalytic removal of airborne pathogens in air. *Sci. Total Environ.* **2021**, *794*, 148477. [CrossRef]
15. Channegowda, M. Functionalized Photocatalytic Nanocoatings for Inactivating COVID-19 Virus Residing on Surfaces of Public and Healthcare Facilities. *Coronaviruses* **2021**, *2*, 3–11. [CrossRef]
16. Saravanan, A.; Kumar, P.S.; Jeevanantham, S.; Karishma, S.; Kiruthika, A.R. Photocatalytic disinfection of micro-organisms: Mechanisms and applications. *Environ. Technol. Innov.* **2021**, *24*, 101909. [CrossRef]
17. Liu, Y.; Huang, J.; Feng, X.; Li, H. Thermal-Sprayed Photocatalytic Coatings for Biocidal Applications: A Review. *J. Therm. Spray Technol.* **2021**, *30*, 1–24. [CrossRef]
18. Kumar, P.; Mathpal, M.C.; Prakash, J.; Viljoen, B.C.; Roos, W.; Swart, H. Band gap tailoring of cauliflower-shaped CuO nanostructures by Zn doping for antibacterial applications. *J. Alloys Compd.* **2020**, *832*, 154968. [CrossRef]
19. da Silva, B.L.; Caetano, B.L.; Chiari-Andréo, B.G.; Pietro, R.C.L.R.; Chiavacci, L.A. Increased antibacterial activity of ZnO nanoparticles: Influence of size and surface modification. *Colloids Surf. B. Biointerfaces* **2019**, *177*, 440–447. [CrossRef]
20. Matsuura, R.; Lo, C.-W.; Wada, S.; Somei, J.; Ochiai, H.; Murakami, T.; Saito, N.; Ogawa, T.; Shinjo, A.; Benno, Y.; et al. SARS-CoV-2 Disinfection of Air and Surface Contamination by TiO_2 Photocatalyst-Mediated Damage to Viral Morphology, RNA, and Protein. *Viruses* **2021**, *13*, 942. [CrossRef]
21. Gold, K.; Slay, B.; Knackstedt, M.; Gaharwar, A.K. Antimicrobial Activity of Metal and Metal-Oxide Based Nanoparticles. *Adv. Ther.* **2018**, *1*, 1700033. [CrossRef]
22. Liaqat, F.; Khazi, M.I.; Awan, A.S.; Eltem, R.; Li, J. 15-Antimicrobial studies of metal oxide nanomaterials. In *Metal Oxide-Carbon Hybrid Materials*; Chaudhry, M.A., Hussain, R., Butt, F.K., Eds.; Elsevier: Amsterdam, The Netherlands, 2022; pp. 407–435.
23. Kumar, V.; Prakash, J.; Singh, J.P.; Chae, K.H.; Swart, C.; Ntwaeaborwa, O.M.; Swart, H.C.; Dutta, V. Role of silver doping on the defects related photoluminescence and antibacterial behaviour of zinc oxide nanoparticles. *Colloids Surf. B Biointerfaces* **2017**, *159*, 191–199. [CrossRef]
24. Zacarías, S.M.; Satuf, M.L.; Vaccari, M.C.; Alfano, O.M. Photocatalytic inactivation of bacterial spores using TiO_2 films with silver deposits. *Chem. Eng. J.* **2015**, *266*, 133–140. [CrossRef]
25. Park, G.W.; Cho, M.; Cates, E.L.; Lee, D.; Oh, B.-T.; Vinjé, J.; Kim, J.-H. Fluorinated TiO_2 as an ambient light-activated virucidal surface coating material for the control of human norovirus. *J. Photochem. Photobiol. B Biol.* **2014**, *140*, 315–320. [CrossRef]
26. Si, Y.; Zhang, Z.; Wu, W.; Fu, Q.; Huang, K.; Nitin, N.; Ding, B.; Sun, G. Daylight-driven rechargeable antibacterial and antiviral nanofibrous membranes for bioprotective applications. *Sci. Adv.* **2018**, *4*, eaar5931. [CrossRef]
27. Kumar, A.; Soni, V.; Singh, P.; Khan, A.A.P.; Nazim, M.; Mohapatra, S.; Saini, V.; Raizada, P.; Hussain, C.M.; Shaban, M.; et al. Green aspects of photocatalysts during corona pandemic: A promising role for the deactivation of COVID-19 virus. *RSC Adv.* **2022**, *12*, 13609–13627. [CrossRef]
28. Khaiboullina, S.; Uppal, T.; Dhabarde, N.; Subramanian, V.R.; Verma, S.C. Inactivation of Human Coronavirus by Titania Nanoparticle Coatings and UVC Radiation: Throwing Light on SARS-CoV-2. *Viruses* **2021**, *13*, 19. [CrossRef]
29. Tong, Y.; Shi, G.; Hu, G.; Hu, X.; Han, L.; Xie, X.; Xu, Y.; Zhang, R.; Sun, J.; Zhong, J. Photo-catalyzed TiO_2 inactivates pathogenic viruses by attacking viral genome. *Chem. Eng. J.* **2021**, *414*, 128788. [CrossRef]
30. Bono, N.; Ponti, F.; Punta, C.; Candiani, G. Effect of UV Irradiation and TiO_2-Photocatalysis on Airborne Bacteria and Viruses: An Overview. *Materials* **2021**, *14*, 1075. [CrossRef]
31. Mullenders, L.H.F. Solar UV damage to cellular DNA: From mechanisms to biological effects. *Photochem. Photobiol. Sci.* **2018**, *17*, 1842–1852. [CrossRef]
32. Prakash, J.; Samriti, K.A.; Dai, H.; Janegitz, B.C.; Krishnan, V.; Swart, H.C.; Sun, S. Novel rare earth metal–doped one-dimensional TiO_2 nanostructures: Fundamentals and multifunctional applications. *Mater. Today Sustain.* **2021**, *13*, 100066. [CrossRef]

33. Cai, T.; Liu, Y.; Wang, L.; Zhang, S.; Ma, J.; Dong, W.; Zeng, Y.; Yuan, J.; Liu, C.; Luo, S. "Dark Deposition" of Ag Nanoparticles on TiO_2: Improvement of Electron Storage Capacity To Boost "Memory Catalysis" Activity. *ACS Appl. Mater. Interfaces* **2018**, *10*, 25350–25359. [CrossRef] [PubMed]
34. Prakash, J. Mechanistic Insights into Graphene Oxide Driven Photocatalysis as Co-Catalyst and Sole Catalyst in Degradation of Organic Dye Pollutants. *Photochem* **2022**, *2*, 651–671. [CrossRef]
35. Verma, S.; Mal, D.S.; de Oliveira, P.R.; Janegitz, B.C.; Prakash, J.; Gupta, R.K. A facile synthesis of novel polyaniline/graphene nanocomposite thin films for enzyme-free electrochemical sensing of hydrogen peroxide. *Mol. Syst. Des. Eng.* **2022**, *7*, 158–170. [CrossRef]
36. Sharma, P.; Kherb, J.; Prakash, J.; Kaushal, R. A novel and facile green synthesis of SiO_2 nanoparticles for removal of toxic water pollutants. *Appl. Nanosci.* **2021**. [CrossRef]
37. Prakash, J.; Harris, R.A.; Swart, H.C. Embedded plasmonic nanostructures: Synthesis, fundamental aspects and their surface enhanced Raman scattering applications. *Int. Rev. Phys. Chem.* **2016**, *35*, 353–398. [CrossRef]
38. Qi, K.; Cheng, B.; Yu, J.; Ho, W. A review on TiO_2-based Z-scheme photocatalysts. *Chin. J. Catal.* **2017**, *38*, 1936–1955. [CrossRef]
39. Yu, J.; Wang, W.; Cheng, B.; Su, B.-L. Enhancement of Photocatalytic Activity of Mesporous TiO_2 Powders by Hydrothermal Surface Fluorination Treatment. *J. Phys. Chem. C* **2009**, *113*, 6743–6750. [CrossRef]
40. Yu, J.C.; Yu, J.; Ho, W.; Jiang, Z.; Zhang, L. Effects of F- Doping on the Photocatalytic Activity and Microstructures of Nanocrystalline TiO_2 Powders. *Chem. Mater.* **2002**, *14*, 3808–3816. [CrossRef]
41. Li, J.; Ren, D.; Wu, Z.; Huang, C.; Yang, H.; Chen, Y.; Yu, H. Visible-light-mediated antifungal bamboo based on Fe-doped TiO_2 thin films. *RSC Adv.* **2017**, *7*, 55131–55140. [CrossRef]
42. Kumar, P.; Mathpal, M.C.; Prakash, J.; Jagannath, G.; Roos, W.D.; Swart, H.C. Plasmonic and nonlinear optical behavior of nanostructures in glass matrix for photonics application. *Mater. Res. Bull.* **2020**, *125*, 110799. [CrossRef]
43. Prakash, J.; Tripathi, A.; Gautam, S.; Chae, K.H.; Song, J.; Rigato, V.; Tripathi, J.; Asokan, K. Phenomenological understanding of dewetting and embedding of noble metal nanoparticles in thin films induced by ion irradiation. *Mater. Chem. Phys.* **2014**, *147*, 920–924. [CrossRef]
44. Kumar, P.; Mathpal, M.C.; Jagannath, G.; Prakash, J.; Maze, J.-R.; Roos, W.D.; Swart, H.C. Optical limiting applications of resonating plasmonic Au nanoparticles in a dielectric glass medium. *Nanotechnology* **2021**, *32*, 345709. [CrossRef]
45. Li, J.; Ma, R.; Wu, Z.; He, S.; Chen, Y.; Bai, R.; Wang, J. Visible-Light-Driven Ag-Modified TiO_2 Thin Films Anchored on Bamboo Material with Antifungal Memory Activity against *Aspergillus niger*. *J. Fungi* **2021**, *7*, 592. [CrossRef]
46. Ma, R.; Li, J.; Han, S.; Wu, Z.; Bao, Y.; He, S.; Chen, Y. Solar-driven $WO_3 \cdot H_2O/TiO_2$ heterojunction films immobilized onto bamboo biotemplate: Relationship between physical color, crystal structure, crystal morphology, and energy storage ability. *Surf. Interfaces* **2022**, *31*, 102028. [CrossRef]
47. Tatsuma, T.; Takeda, S.; Saitoh, S.; Ohko, Y.; Fujishima, A. Bactericidal effect of an energy storage TiO_2–WO_3 photocatalyst in dark. *Electrochem. Commun.* **2003**, *5*, 793–796. [CrossRef]
48. Singh, J.P.; Chen, C.L.; Dong, C.L.; Prakash, J.; Kabiraj, D.; Kanjilal, D.; Pong, W.F.; Asokan, K. Role of surface and subsurface defects in MgO thin film: XANES and magnetic investigations. *Superlattices Microstruct.* **2015**, *77*, 313–324. [CrossRef]
49. Pathak, T.K.; Kumar, V.; Prakash, J.; Purohit, L.P.; Swart, H.C.; Kroon, R.E. Fabrication and characterization of nitrogen doped p-ZnO on n-Si heterojunctions. *Sens. Actuators A. Phys.* **2016**, *247*, 475–481. [CrossRef]
50. Patel, S.P.; Chawla, A.K.; Chandra, R.; Prakash, J.; Kulriya, P.K.; Pivin, J.C.; Kanjilal, D.; Kumar, L. Structural phase transformation in ZnS nanocrystalline thin films by swift heavy ion irradiation. *Solid State Commun.* **2010**, *150*, 1158–1161. [CrossRef]
51. Chen, Z.; Zhang, G.; Prakash, J.; Zheng, Y.; Sun, S. Rational Design of Novel Catalysts with Atomic Layer Deposition for the Reduction of Carbon Dioxide. *Adv. Energy Mater.* **2019**, *9*, 1900889. [CrossRef]
52. Prakash, J.; Tripathi, A.; Khan, S.A.; Pivin, J.C.; Singh, F.; Tripathi, J.; Kumar, S.; Avasthi, D.K. Ion beam induced interface mixing of Ni on PTFE bilayer system studied by quadrupole mass analysis and electron spectroscopy for chemical analysis. *Vacuum* **2010**, *84*, 1275–1279. [CrossRef]
53. Prakash, J.; Swart, H.C.; Zhang, G.; Sun, S. Emerging applications of atomic layer deposition for the rational design of novel nanostructures for surface-enhanced Raman scattering. *J. Mater. Chem. C* **2019**, *7*, 1447–1471. [CrossRef]
54. Prakash, J. Fundamentals and applications of recyclable SERS substrates. *Int. Rev. Phys. Chem.* **2019**, *38*, 201–242. [CrossRef]
55. Singh, N.; Prakash, J.; Misra, M.; Sharma, A.; Gupta, R.K. Dual Functional Ta-Doped Electrospun TiO_2 Nanofibers with Enhanced Photocatalysis and SERS Detection for Organic Compounds. *ACS Appl. Mater. Interfaces* **2017**, *9*, 28495–28507. [CrossRef]
56. Krumdieck, S.P.; Boichot, R.; Gorthy, R.; Land, J.G.; Lay, S.; Gardecka, A.J.; Polson, M.I.J.; Wasa, A.; Aitken, J.E.; Heinemann, J.A.; et al. Nanostructured TiO_2 anatase-rutile-carbon solid coating with visible light antimicrobial activity. *Sci. Rep.* **2019**, *9*, 1883. [CrossRef]
57. Wanag, A.; Rokicka, P.; Kusiak-Nejman, E.; Kapica-Kozar, J.; Wrobel, R.J.; Markowska-Szczupak, A.; Morawski, A.W. Antibacterial properties of TiO_2 modified with reduced graphene oxide. *Ecotoxicol. Environ. Saf.* **2018**, *147*, 788–793. [CrossRef]
58. Zhou, X.; Zhou, M.; Ye, S.; Xu, Y.; Zhou, S.; Cai, Q.; Xie, G.; Huang, L.; Zheng, L.; Li, Y. Antibacterial activity and mechanism of the graphene oxide (rGO)- modified TiO_2 catalyst against Enterobacter hormaechei. *Int. Biodeterior. Biodegrad.* **2021**, *162*, 105260. [CrossRef]
59. Rokicka-Konieczna, P.; Wanag, A.; Sienkiewicz, A.; Kusiak-Nejman, E.; Morawski, A.W. Antibacterial effect of TiO_2 nanoparticles modified with APTES. *Catal. Commun.* **2020**, *134*, 105862. [CrossRef]

60. Dhanasekar, M.; Jenefer, V.; Nambiar, R.B.; Babu, S.G.; Selvam, S.P.; Neppolian, B.; Bhat, S.V. Ambient light antimicrobial activity of reduced graphene oxide supported metal doped TiO_2 nanoparticles and their PVA based polymer nanocomposite films. *Mater. Res. Bull.* **2018**, *97*, 238–243. [CrossRef]
61. Bonnefond, A.; González, E.; Asua, J.M.; Leiza, J.R.; Kiwi, J.; Pulgarin, C.; Rtimi, S. New evidence for hybrid acrylic/TiO_2 films inducing bacterial inactivation under low intensity simulated sunlight. *Colloids Surf. B. Biointerfaces* **2015**, *135*, 1–7. [CrossRef]
62. Lin, Y.-P.; Ksari, Y.; Prakash, J.; Giovanelli, L.; Valmalette, J.-C.; Themlin, J.-M. Nitrogen-doping processes of graphene by a versatile plasma-based method. *Carbon* **2014**, *73*, 216–224. [CrossRef]
63. Komba, N.; Zhang, G.; Wei, Q.; Yang, X.; Prakash, J.; Chenitz, R.; Rosei, F.; Sun, S. Iron (II) phthalocyanine/N-doped graphene: A highly efficient non-precious metal catalyst for oxygen reduction. *Int. J. Hydrog. Energy* **2019**, *44*, 18103–18114. [CrossRef]
64. Prakash, J.; Pivin, J.C.; Swart, H.C. Noble metal nanoparticles embedding into polymeric materials: From fundamentals to applications. *Adv. Colloid Interface Sci.* **2015**, *226*, 187–202. [CrossRef] [PubMed]
65. Hernández-Gordillo, A.; Arriaga, S. Mesoporous TiO_2 Monoliths Impregnated with CdS and CuO Nanoparticles for Airborne Bacteria Inactivation Under Visible Light. *Catal. Lett.* **2022**, *152*, 629–640. [CrossRef]
66. Islam, M.A.; Ikeguchi, A.; Naide, T. Effectiveness of an air cleaner device in reducing aerosol numbers and airborne bacteria from an enclosed type dairy barn. *Environ. Sci. Pollut. Res.* **2022**, *29*, 53022–53035. [CrossRef]
67. Lee, M.; Koziel, J.A.; Macedo, N.R.; Li, P.; Chen, B.; Jenks, W.S.; Zimmerman, J.; Paris, R.V. Mitigation of Particulate Matter and Airborne Pathogens in Swine Barn Emissions with Filtration and UV-A Photocatalysis. *Catalysts* **2021**, *11*, 1302. [CrossRef]
68. Valdez-Castillo, M.; Saucedo-Lucero, J.O.; Arriaga, S. Photocatalytic inactivation of airborne microorganisms in continuous flow using perlite-supported ZnO and TiO_2. *Chem. Eng. J.* **2019**, *374*, 914–923. [CrossRef]
69. Urbonavicius, M.; Varnagiris, S.; Sakalauskaite, S.; Demikyte, E.; Tuckute, S.; Lelis, M. Application of Floating TiO_2 Photocatalyst for Methylene Blue Decomposition and *Salmonella typhimurium* Inactivation. *Catalysts* **2021**, *11*, 794. [CrossRef]
70. Varnagiris, S.; Urbonavicius, M.; Sakalauskaite, S.; Daugelavicius, R.; Pranevicius, L.; Lelis, M.; Milcius, D. Floating TiO_2 photocatalyst for efficient inactivation of *E. coli* and decomposition of methylene blue solution. *Sci. Total Environ.* **2020**, *720*, 137600. [CrossRef]
71. Sboui, M.; Lachheb, H.; Bouattour, S.; Gruttadauria, M.; La Parola, V.; Liotta, L.F.; Boufi, S. TiO_2/Ag_2O immobilized on cellulose paper: A new floating system for enhanced photocatalytic and antibacterial activities. *Environ. Res.* **2021**, *198*, 111257. [CrossRef]
72. Mathur, G. COVID Killing Air Purifier Based on UV & Titanium Dioxide Based Photocatalysis System. *SAE Int. J. Adv. Curr. Pract. Mobil.* **2021**, *4*, 143–150.
73. Nakano, R.; Yamaguchi, A.; Sunada, K.; Nagai, T.; Nakano, A.; Suzuki, Y.; Yano, H.; Ishiguro, H.; Miyauchi, M. Inactivation of various variant types of SARS-CoV-2 by indoor-light-sensitive TiO2-based photocatalyst. *Sci. Rep.* **2022**, *12*, 5804. [CrossRef]
74. Uppal, T.; Reganti, S.; Martin, E.; Verma, S.C. Surface Inactivation of Human Coronavirus by MACOMA^{TM} UVA-TiO_2 Coupled Photocatalytic Disinfection System. *Catalysts* **2022**, *12*, 690. [CrossRef]
75. Zan, L.; Fa, W.; Peng, T.; Gong, Z.-K. Photocatalysis effect of nanometer TiO_2 and TiO_2-coated ceramic plate on Hepatitis B virus. *J. Photochem. Photobiol. B Biol.* **2007**, *86*, 165–169. [CrossRef]
76. Bhardwaj, S.K.; Singh, H.; Deep, A.; Khatri, M.; Bhaumik, J.; Kim, K.-H.; Bhardwaj, N. UVC-based photoinactivation as an efficient tool to control the transmission of coronaviruses. *Sci. Total Environ.* **2021**, *792*, 148548. [CrossRef]
77. Micochova, P.; Chadha, A.; Hesseloj, T.; Fraternali, F.; Ramsden, J.; Gupta, R. Rapid inactivation of SARS-CoV-2 by titanium dioxide surface coating. *Wellcome Open Res.* **2021**, *6*, 56. [CrossRef]
78. Kupferschmidt, K.; Wadman, M. Delta variant triggers new phase in the pandemic. *Science* **2021**, *372*, 1375–1376. [CrossRef]
79. Sharma, D.K.; Shukla, S.; Sharma, K.K.; Kumar, V. A review on ZnO: Fundamental properties and applications. *Mater. Today Proc.* **2020**, *49*, 3028–3035. [CrossRef]
80. Yamada, H.; Suzuki, K.; Koizumi, S. Gene expression profile in human cells exposed to zinc. *J. Toxicol. Sci.* **2007**, *32*, 193–196. [CrossRef]
81. Umar, A.; Rahman, M.; Vaseem, M.; Hahn, Y.-B. Ultra-sensitive cholesterol biosensor based on low-temperature grown ZnO nanoparticles. *Electrochem. Commun.* **2009**, *11*, 118–121. [CrossRef]
82. Sharmila, G.; Muthukumaran, C.; Sandiya, K.; Santhiya, S.; Pradeep, R.S.; Kumar, N.M.; Suriyanarayanan, N.; Thirumarimurugan, M. Biosynthesis, characterization, and antibacterial activity of zinc oxide nanoparticles derived from Bauhinia tomentosa leaf extract. *J. Nanostruct. Chem.* **2018**, *8*, 293–299. [CrossRef]
83. Sportelli, M.C.; Izzi, M.; Loconsole, D.; Sallustio, A.; Picca, R.A.; Felici, R.; Chironna, M.; Cioffi, N. On the Efficacy of ZnO Nanostructures against SARS-CoV-2. *Int. J. Mol. Sci.* **2022**, *23*, 3040. [CrossRef]
84. Raj, N.B.; PavithraGowda, N.; Pooja, O.; Purushotham, B.; Kumar, M.A.; Sukrutha, S.; Ravikumar, C.; Nagaswarupa, H.; Murthy, H.A.; Boppana, S.B. Harnessing ZnO nanoparticles for antimicrobial and photocatalytic activities. *J. Photochem. Photobiol.* **2021**, *6*, 100021. [CrossRef]
85. Milionis, A.; Tripathy, A.; Donati, M.; Sharma, C.S.; Pan, F.; Maniura-Weber, K.; Ren, Q.; Poulikakos, D. Water-based scalable methods for self-cleaning antibacterial ZnO-nanostructured surfaces. *Ind. Eng. Chem. Res.* **2020**, *59*, 14323–14333. [CrossRef]
86. Prakash, J.; Kumar, V.; Erasmus, L.J.B.; Duvenhage, M.M.; Sathiyan, G.; Bellucci, S.; Sun, S.; Swart, H.C. Phosphor Polymer Nanocomposite: ZnO:Tb3+ Embedded Polystyrene Nanocomposite Thin Films for Solid-State Lighting Applications. *ACS Appl. Nano Mater.* **2018**, *1*, 977–988. [CrossRef]

87. Kumar, R.S.; Dananjaya, S.H.S.; De Zoysa, M.; Yang, M. Enhanced antifungal activity of Ni-doped ZnO nanostructures under dark conditions. *RSC Adv.* **2016**, *6*, 108468–108476. [CrossRef]
88. Iqbal, G.; Faisal, S.; Khan, S.; Shams, D.F.; Nadhman, A. Photo-inactivation and efflux pump inhibition of methicillin resistant Staphylococcus aureus using thiolated cobalt doped ZnO nanoparticles. *J. Photochem. Photobiol. B Biol.* **2019**, *192*, 141–146. [CrossRef]
89. Vijayalakshmi, K.; Sivaraj, D. Enhanced antibacterial activity of Cr doped ZnO nanorods synthesized using microwave processing. *RSC Adv.* **2015**, *5*, 68461–68469. [CrossRef]
90. Naskar, A.; Lee, S.; Kim, K.-S. Antibacterial potential of Ni-doped zinc oxide nanostructure: Comparatively more effective against Gram-negative bacteria including multi-drug resistant strains. *RSC Adv.* **2020**, *10*, 1232–1242. [CrossRef]
91. Geetha, K.; Sivasangari, D.; Kim, H.-S.; Murugadoss, G.; Kathalingam, A. Electrospun nanofibrous ZnO/PVA/PVP composite films for efficient antimicrobial face masks. *Ceram. Int.* **2022**, *48*, 29197–29204. [CrossRef]
92. Attia, G.H.; Moemen, Y.S.; Youns, M.; Ibrahim, A.M.; Abdou, R.; El Raey, M.A. Antiviral zinc oxide nanoparticles mediated by hesperidin and in silico comparison study between antiviral phenolics as anti-SARS-CoV-2. *Colloids Surf. B Biointerfaces* **2021**, *203*, 111724. [CrossRef] [PubMed]
93. Hamdi, M.; Abdel-Bar, H.M.; Elmowafy, E.; El-Khouly, A.; Mansour, M.; Awad, G.A. Investigating the internalization and COVID-19 antiviral computational analysis of optimized nanoscale zinc oxide. *ACS Omega* **2021**, *6*, 6848–6860. [CrossRef] [PubMed]
94. El-Megharbel, S.M.; Alsawat, M.; Al-Salmi, F.A.; Hamza, R.Z. Utilizing of (zinc oxide nano-spray) for disinfection against "SARS-CoV-2" and testing its biological effectiveness on some biochemical parameters during (COVID-19 pandemic)—"ZnO nanoparticles have antiviral activity against (SARS-CoV-2)". *Coatings* **2021**, *11*, 388. [CrossRef]
95. Ishida, T. Anti-viral vaccine activity of Zinc (II) for viral prevention, entry, replication, and spreading during pathogenesis process. *Curr. Trends. Biomed. Eng. Biosci.* **2019**, *19*, MS.ID.556012. [CrossRef]
96. Antoine, T.E.; Mishra, Y.K.; Trigilio, J.; Tiwari, V.; Adelung, R.; Shukla, D. Prophylactic, therapeutic and neutralizing effects of zinc oxide tetrapod structures against herpes simplex virus type-2 infection. *Antivir. Res.* **2012**, *96*, 363–375. [CrossRef]
97. Siddiqi, K.S.; Rahman, A.U.; Tajuddin, H.A. Properties of Zinc Oxide Nanoparticles and Their Activity Against Microbes. *Nanoscale Res. Lett.* **2018**, *13*, 141. [CrossRef]
98. Prasad, A.S. Zinc in human health: Effect of zinc on immune cells. *Mol. Med.* **2008**, *14*, 353–357. [CrossRef]
99. Te Velthuis, A.J.; van den Worm, S.H.; Sims, A.C.; Baric, R.S.; Snijder, E.J.; van Hemert, M.J. Zn^{2+} inhibits coronavirus and arterivirus RNA polymerase activity in vitro and zinc ionophores block the replication of these viruses in cell culture. *PLoS Pathog.* **2010**, *6*, e1001176. [CrossRef]
100. Kalsi, T.; Mitra, H.; Roy, T.K.; Godara, S.K.; Kumar, P. Comprehensive Analysis of Band Gap and Nanotwinning in Cd1–xMgxS QDs. *Cryst. Growth Des.* **2020**, *20*, 6699–6706. [CrossRef]
101. Jiang, J.; Pi, J.; Cai, J. The advancing of zinc oxide nanoparticles for biomedical applications. *Bioinorg. Chem. Appl.* **2018**, *2018*, 1062562. [CrossRef]
102. Momeni, M.; Mirhosseini, M.; Nazari, Z.; Kazempour, A.; Hakimiyan, M. Antibacterial and photocatalytic activity of CuO nanostructure films with different morphology. *J. Mater. Sci. Mater. Electron.* **2016**, *27*, 8131–8137. [CrossRef]
103. Pugazhendhi, A.; Kumar, S.S.; Manikandan, M.; Saravanan, M. Photocatalytic properties and antimicrobial efficacy of Fe doped, CuO nanoparticles against the pathogenic bacteria and fungi. *Microb. Pathog.* **2018**, *122*, 84–89. [CrossRef]
104. Giti, R.; Zomorodian, K.; Firouzmandi, M.; Zareshahrabadi, Z.; Rahmannasab, S. Antimicrobial activity of thermocycled polymethyl methacrylate resin reinforced with titanium dioxide and copper oxide nanoparticles. *Int. J. Dent.* **2021**, *2021*, 6690806. [CrossRef]
105. Wang, L.; Hu, C.; Shao, L. The antimicrobial activity of nanoparticles: Present situation and prospects for the future. *Int. J. Nanomed.* **2017**, *12*, 1227. [CrossRef]
106. Soni, V.; Xia, C.; Cheng, C.K.; Nguyen, V.-H.; Nguyen, D.L.T.; Bajpai, A.; Kim, S.Y.; Van Le, Q.; Khan, A.A.P.; Singh, P. Advances and recent trends in cobalt-based cocatalysts for solar-to-fuel conversion. *Appl. Mater. Today* **2021**, *24*, 101074. [CrossRef]
107. Bandala, E.R.; Kruger, B.R.; Cesarino, I.; Leao, A.L.; Wijesiri, B.; Goonetilleke, A. Impacts of COVID-19 pandemic on the wastewater pathway into surface water: A review. *Sci. Total Environ.* **2021**, *774*, 145586. [CrossRef]
108. Ghotekar, S.; Pansambal, S.; Bilal, M.; Pingale, S.S.; Oza, R. Environmentally friendly synthesis of Cr2O3 nanoparticles: Characterization, applications and future perspective—A review. *Case Stud. Chem. Environ. Eng.* **2021**, *3*, 100089. [CrossRef]
109. Tavakoli, A.; Hashemzadeh, M.S. Inhibition of herpes simplex virus type 1 by copper oxide nanoparticles. *J. Virol. Methods* **2020**, *275*, 113688. [CrossRef]
110. Dulta, K.; Ağçeli, G.K.; Chauhan, P.; Jasrotia, R.; Ighalo, J.O. Multifunctional CuO nanoparticles with enhanced photocatalytic dye degradation and antibacterial activity. *Sustain. Environ. Res.* **2022**, *32*, 1–15. [CrossRef]
111. Akhavan, O.; Azimirad, R.; Safa, S.; Hasani, E. CuO/$Cu(OH)_2$ hierarchical nanostructures as bactericidal photocatalysts. *J. Mater. Chem.* **2011**, *21*, 9634–9640. [CrossRef]
112. Akhavan, O.; Ghaderi, E. Cu and CuO nanoparticles immobilized by silica thin films as antibacterial materials and photocatalysts. *Surf. Coat. Technol.* **2010**, *205*, 219–223. [CrossRef]
113. Gurianov, Y.; Nakonechny, F.; Albo, Y.; Nisnevitch, M. Antibacterial composites of cuprous oxide nanoparticles and polyethylene. *Int. J. Mol. Sci.* **2019**, *20*, 439. [CrossRef]

114. Domagała, E. Stability evaluation of Cu_2O/MWCNTs filters for virus removal from water. *Water Res.* **2020**, *179*, 115879. [CrossRef]
115. Mazurkow, J.M.; Yüzbasi, N.S.; Domagala, K.W.; Pfeiffer, S.; Kata, D.; Graule, T. Nano-sized copper (oxide) on alumina granules for water filtration: Effect of copper oxidation state on virus removal performance. *Environ. Sci. Technol.* **2019**, *54*, 1214–1222. [CrossRef] [PubMed]
116. Amiri, M.; Etemadifar, Z.; Daneshkazemi, A.; Nateghi, M. Antimicrobial effect of copper oxide nanoparticles on some oral bacteria and candida species. *J. Dent. Biomater.* **2017**, *4*, 347. [PubMed]
117. Hang, X.; Peng, H.; Song, H.; Qi, Z.; Miao, X.; Xu, W. Antiviral activity of cuprous oxide nanoparticles against Hepatitis C Virus in vitro. *J. Virol. Methods* **2015**, *222*, 150–157. [CrossRef] [PubMed]
118. Borkow, G.; Zhou, S.S.; Page, T.; Gabbay, J. A novel anti-influenza copper oxide containing respiratory face mask. *PLoS ONE* **2010**, *5*, e11295. [CrossRef] [PubMed]
119. Leung, N.H.; Chu, D.K.; Shiu, E.Y.; Chan, K.H.; McDevitt, J.J.; Hau, B.J.; Cowling, B.J. Respiratory virus shedding in exhaled breath and efficacy of face masks. *Nat. Med.* **2020**, *26*, 676–680. [CrossRef]
120. Hosseini, M.; Chin, A.W.; Behzadinasab, S.; Poon, L.L.; Ducker, W.A. Cupric oxide coating that rapidly reduces infection by SARS-CoV-2 via solids. *ACS Appl. Mater. Interfaces* **2021**, *13*, 5919–5928. [CrossRef]
121. Merkl, P.; Long, S.; McInerney, G.M.; Sotiriou, G.A. Antiviral activity of silver, copper oxide and zinc oxide nanoparticle coatings against SARS-CoV-2. *Nanomaterials* **2021**, *11*, 1312. [CrossRef]
122. Delumeau, L.V.; Asgarimoghaddam, H.; Alkie, T.; Jones, A.J.B.; Lum, S.; Mistry, K.; Musselman, K.P. Effectiveness of antiviral metal and metal oxide thin-film coatings against human coronavirus 229E. *APL Mater.* **2021**, *9*, 111114. [CrossRef]
123. Farah, J.; Ibadurrohman, M.; Slamet. Synthesis of CuO-TiO_2 nano-composite for Escherichia coli disinfection and toluene degradation. *AIP Conf. Proc.* **2020**, *2237*, 020050.
124. Kanako, Y.; Yui, S.; Momo, I.; Takakiyo, T.; Toshikazu, S.; Jin-ichi, S. Photocatalytic Antibacterial Activity of TiO_2, TiO_2+CuO, and WO_3 +CuO -Evaluation of Codoping Effect. *Technol. Innov. Pharm. Res.* **2021**, *1*, 130–139.
125. Uema, M.; Yonemitsu, K.; Momose, Y.; Ishii, Y.; Tateda, K.; Inoue, T.; Asakura, H. Effect of the Photocatalyst under Visible Light Irradiation in SARS-CoV-2 Stability on an Abiotic Surface. *Biocontrol Sci.* **2021**, *26*, 119–125. [CrossRef]
126. Lin, H.; Li, T.; Janani, B.J.; Fakhri, A. Fabrication of Cu_2MoS_4 decorated WO_3 nano heterojunction embedded on chitosan: Robust photocatalytic efficiency, antibacterial performance, and bacteria detection by peroxidase activity. *J. Photochem. Photobiol. B Biol.* **2022**, *226*, 112354. [CrossRef]
127. Hojamberdiev, M.; Czech, B.; Wasilewska, A.; Boguszewska-Czubara, A.; Yubuta, K.; Wagata, H.; Daminova, S.S.; Kadirova, Z.C.; Vargas, R. Detoxifying SARS-CoV-2 antiviral drugs from model and real wastewaters by industrial waste-derived multiphase photocatalysts. *J. Hazard. Mater.* **2022**, *429*, 128300. [CrossRef]
128. Czech, B.; Krzyszczak, A.; Boguszewska-Czubara, A.; Opielak, G.; Jośko, I.; Hojamberdiev, M. Revealing the toxicity of lopinavir- and ritonavir-containing water and wastewater treated by photo-induced processes to Danio rerio and Allivibrio fischeri. *Sci. Total Environ.* **2022**, *824*, 153967. [CrossRef]
129. Teymoorian, T.; Teymourian, T.; Kowsari, E.; Ramakrishna, S. Direct and indirect effects of SARS-CoV-2 on wastewater treatment. *J. Water Process Eng.* **2021**, *42*, 102193. [CrossRef]
130. Yang, J.Z.Y.; Gou, X.; Pu, K.; Chen, Z.; Guo, Q.; Ji, R.; Wang, H.; Wang, Y.; Zhou, Y. Prevalence of comorbidities and its effects in patients infected with SARS-CoV-2: A systematic review and meta-analysis. *Int. J. Infect. Dis.* **2020**, *94*, 91–95. [CrossRef]
131. Huang, F.L.Y.; Leung, E.L.; Liu, X.; Liu, K.; Wang, Q.; Lan, Y.; Li, X.; Yu, H.; Cui, L.; Luo, H.; et al. A review of therapeutic agents and Chinese herbal medicines against SARS-COV-2 (COVID-19). *Pharm. Res.* **2020**, *158*, 104929. [CrossRef]
132. Elgohary, E.A.; Mohamed, Y.M.A.; El Nazer, H.A.; Baaloudj, O.; Alyami, M.S.S.; El Jery, A.; Assadi, A.A.; Amrane, A. A Review of the Use of Semiconductors as Catalysts in the Photocatalytic Inactivation of Microorganisms. *Catalysts* **2021**, *11*, 1498. [CrossRef]
133. Sanvicens, N.; Marco, M.P. Multifunctional nanoparticles–properties and prospects for their use in human medicine. *Trends Biotechnol.* **2008**, *26*, 425–433. [CrossRef]
134. Seabra, A.B.; Durán, N. Nanotoxicology of metal oxide nanoparticles. *Metals* **2015**, *5*, 934–975. [CrossRef]

catalysts

MDPI

Review

A Review on Green Synthesis of Nanoparticles and Their Diverse Biomedical and Environmental Applications

Melvin S. Samuel [1], Madhumita Ravikumar [1], Ashwini John J. [2], Ethiraj Selvarajan [2,*], Himanshu Patel [3], P. Sharath Chander [4], J. Soundarya [4], Srikanth Vuppala [5,6,*], Ramachandran Balaji [7] and Narendhar Chandrasekar [8]

1 Department of Chemistry, Indian Institute of Technology, Kharagpur 721302, West Bengal, India; melvinsamuel08@gmail.com (M.S.S.); madhumita.lifi@gmail.com (M.R.)
2 Department of Genetic Engineering, School of Bioengineering, SRM Institute of Science and Technology, Kattankulathur 603203, Tamil Nadu, India; ashwinijohn97@gmail.com
3 Applied Science and Humanities Department, Pacific School of Engineering, Kadodara, Palasana Road, Surat 394305, Gujarat, India; hjpatel123@yahoo.in
4 Department of Computer Science and Engineering, SSN College of Engineering, Kalavakkam 603110, Tamil Nadu, India; sharathchander.p@gmail.com (P.S.C.); jayakum3@uwm.edu (J.S.)
5 Department of Civil and Environmental Engineering, Politecnico di Milano, Piazza Leonardo da Vinci, 32, 20133 Milan, Italy
6 Seamthesis Srl, Via IV Novembre, 156, 9122 Piacenza, Italy
7 Department of Chemical Engineering and Biotechnology, National Taipei University of Technology, Taipei 10608, Taiwan; balajiyashik@gmail.com
8 Department of Nanoscience and Technology, Sri Ramakrishna Engineering College, Coimbatore 641022, Tamil Nadu, India; narendhar.nano@gmail.com
* Correspondence: selrajan@gmail.com (E.S.); srikanth.vuppala22@gmail.com (S.V.)

check for updates

Citation: Samuel, M.S.; Ravikumar, M.; John J., A.; Selvarajan, E.; Patel, H.; Chander, P.S.; Soundarya, J.; Vuppala, S.; Balaji, R.; Chandrasekar, N. A Review on Green Synthesis of Nanoparticles and Their Diverse Biomedical and Environmental Applications. *Catalysts* **2022**, *12*, 459. https://doi.org/10.3390/catal12050459

Academic Editors: Pritam Kumar Dikshit and Beom Soo Kim

Received: 18 March 2022
Accepted: 13 April 2022
Published: 20 April 2022

Abstract: In recent times, metal oxide nanoparticles (NPs) have been regarded as having important commercial utility. However, the potential toxicity of these nanomaterials has also been a crucial research concern. In this regard, an important solution for ensuring lower toxicity levels and thereby facilitating an unhindered application in human consumer products is the green synthesis of these particles. Although a naïve approach, the biological synthesis of metal oxide NPs using microorganisms and plant extracts opens up immense prospects for the production of biocompatible and cost-effective particles with potential applications in the healthcare sector. An important area that calls for attention is cancer therapy and the intervention of nanotechnology to improve existing therapeutic practices. Metal oxide NPs have been identified as therapeutic agents with an extended half-life and therapeutic index and have also been reported to have lesser immunogenic properties. Currently, biosynthesized metal oxide NPs are the subject of considerable research and analysis for the early detection and treatment of tumors, but their performance in clinical experiments is yet to be determined. The present review provides a comprehensive account of recent research on the biosynthesis of metal oxide NPs, including mechanistic insights into biological production machinery, the latest reports on biogenesis, the properties of biosynthesized NPs, and directions for further improvement. In particular, scientific reports on the properties and applications of nanoparticles of the oxides of titanium, cerium, selenium, zinc, iron, and copper have been highlighted. This review discusses the significance of the green synthesis of metal oxide nanoparticles, with respect to therapeutically based pharmaceutical applications as well as energy and environmental applications, using various novel approaches including one-minute sonochemical synthesis that are capable of responding to various stimuli such as radiation, heat, and pH. This study will provide new insight into novel methods that are cost-effective and pollution free, assisted by the biodegradation of biomass.

Keywords: biosynthesis; biocompatible; remediation; degradation; metal oxide; nanoparticles

1. Introduction

The exquisiteness of nanomaterials was reflected upon by Feynman (1960) as "there is plenty of room at the bottom" [1]. True to his speculation, the technology and science behind miniaturization has opened up innovative avenues for dealing with the synthesis and characterization of nanomaterials and their employment in society. The resulting scientific interest in NPs can be attributed to the fact that these entities serve as bridges to manage the gap between bulk constituents and atomic or molecular assemblies. Several well-characterized bulk materials possess interesting properties at the nanoscale. NPs have a high aspect ratio, facilitating improved reactivity as well as effectiveness compared to the majority of materials. Over time, researchers have demonstrated their competency and developed nano-sized complements for composites, along with exclusive nano-based materials [2–4]. Significant and important applications of nanotechnology include capturing higher resolution images, many nano-sized sensors for ecological contamination, a high quantity of optoelectronics strategies, and nano-engineered solar applications. Nanotechnology deals with the nanoscale range. There is evidence of the existence of nanostructures dating from the beginning of life [5–7]. A significant need for nanotechnology has arisen due to the cumulative claims for nanostructured materials in several fields such as catalysis. In the past few centuries, materials experts have discovered carbon-based materials and mineral elemental blends exhibiting potential optoelectronic and dimensional qualities that are greater than the majority of their complements [8–11]. Organic NPs include carbon in the arrangement of liposomes, fullerenes, dendrimers, and polymeric micelles, and inorganic NPs consisting of magnetic, noble metal, and semiconductor NPs [12–15]. Metallic NPs are important in research, due to the fact that their precise properties are not easily accessible in isolated molecules [16]. The development of metallic NPs serves as an active area in theoretical and, more importantly, "applied research" in nanotechnology [17]. This review focuses on contemporary research activities that deal with the green synthesis of inorganic NPs, which has advantages over traditional approaches that use chemical agents that are detrimental to the environment. The current article looks at traditional synthetic procedures, with a focus on recent developments of greener routes to manufacturing metal, metal oxide, and other important NPs. It then goes on to discuss formation mechanisms and the conditions that control the surface morphology, dispersity, and other properties of these biosynthesized NPs. The report finishes with a discussion of the current situation and future forecasts for nanoparticle production via various green techniques. Briefly, nanomaterials used for various applications ranging from biomedical to bioenergy are in very high demand, due to the fact that the nano size is accompanied by a high surface area that can facilitate loading of the molecule of interest for various scientific applications including drug delivery systems for various disease conditions, especially cancer. When using nanomaterials as a drug carrier, it is very important to analyze the toxicity of the carrier; this concept gave rise to the introduction of green synthesis, which can replace the chemical methods that produce toxic nanocarriers. A synthesis of a metal oxide that responds to multiple stimuli can be an effective way to target drug delivery to the required site. Other than drug delivery applications, these nanomaterials can also be efficiently used in bioremediation as they can degrade the pollutant without affecting the ecosystem, since the nanocarriers are synthesized from natural products. This study mainly focuses on the unique and advanced green synthesis methods for metal oxide nanoparticles that are sensitive to many stimuli, resulting in cost-effective and prominent nanomaterials that can be used for a wide variety of applications, along with their biodegrading capacity, which serves as the novelty of this work. This green synthesis not only produces highly efficient nanocarriers but also performs its specified work without disturbing living organisms or the environment.

2. Synthesis of NPs via Bio/Green Synthesis

Earlier investigations provided two methods for the development of metallic NPs: the top-down method and the bottom-up method. In top-down methods, the nanoscopic features are etched onto a substrate using electron rays, and subsequently by using appropriate engraving and deposition processes. The commonly adopted top-down approaches are physical methods such as evaporation–condensation and the technique of laser ablation.

In this technique, the major resources, i.e., most of the initial metal materials are evaporated using a radiator, and the evaporated vapor subsequently cools at a suitably high rate with the assistance of the steep temperature gradient in the vicinity of the heater surface. The rapid heating and cooling result in unstable NPs at high concentrations. While evaporation–condensation methods are carried out employing an inert gas, the laser ablation technique uses a laser to target a metallic material in solution. For example, silver nano-spheroids (20–50 nm) can be produced by laser ablation in water with femtosecond laser pulses at 800 nm. A major drawback is the inadequacy of the surface construction. Such flaws can have a substantial influence on physical properties and the exterior interactions of the metallic NPs, owing to the high feature relation [2–4]. The most popular approach is chemical reduction utilizing a variety of carbon-based and mineral-reducing mediators. In general, various reducing mediators such as sodium citrate, ascorbate, elemental hydrogen, sodium borohydride ($NaBH_4$), polyols, Tollen's reagent, N, N-dimethylformamide (DMF), and poly (ethylene glycol)-block copolymers are employed for the reduction of metal ions in aqueous as well as non-aqueous solutions, leading to the formation of zerovalent metal, followed by agglomeration into oligomeric clusters. These clusters eventually form metallic colloidal particles. It is also notable that most of these approaches employ protecting mediators (polymers) as stabilizers, to avoid the accumulation of NPs. The presence of surfactants and polymers (e.g., thiols, amines, acids, and alcohols) affects the functionalities for interactions within the particle surfaces, stabilizing particle growth and protecting particles from agglomeration, sedimentation, or loss of their surface properties. Most of these methods persist in the development stages, as the extraction and purification of the produced NPs for further applications still represent important hurdles [5–7]. Several mechanical and irradiation-assisted techniques have been employed for the synthesis of metallic NPs. Recently, green synthesis of metal oxides by the sonochemical method has gained popularity, as this is the only method that facilitates the mixing of the chemical constituents at the atomic level, as a result of an unusual chemical reaction caused by cavitation in aqueous media at a temperature of 5000 °C and a pressure of 1800 kpa. In 2021, Pérez-Beltrán synthesized a magnetic iron oxide nanoparticle using a high-energy sonochemical approach, considering an amplitude of 2826 J and time of 1 min as major factors. This novel one-minute green synthesis by sonochemistry produced nanoparticles of 11 ± 2 nm particle size and was used for the biosensing of mercury in water [8]. In another study, conducted by Goudarzi, it is stated that copper oxide nanoparticles can be ultrasonically synthesized using Dactylopius coccus and can be further thermally decomposed at 60 °C for drug release in breast cancer applications [9].

The sono-electrochemistry technique employs alternating sonic and electric pulses, ultrasonic power, and electrolyte configuration for the mechanical manipulation of the material. Recent advances in the synthesis of metallic NPs include photoinduced or photocatalytic reduction methods [10,11]. Table 1 reviews some of the common traditional approaches reported for the synthesis of metallic NPs.

Table 1. Traditional approaches reported for the synthesis of metallic nanoparticles.

Method	Characteristics	Nanoparticle	Size	Morphology	Advantages	Disadvantages	Reference
Physical Methods							
Plasma Synthesis	Gas–liquid interfacial plasma is produced in ionic liquid. Plasmas provide a rich source of chemically active species that react with a surface or react with each other to produce secondary, short-lived chemical precursors needed for thin film deposition.	Pd	20 nm	Nanorods	Low-temperature operation, non-destructive materials treatment capability.	High-pressure limit. Economic constraint.	[12]
Ball Milling	Arc melting followed by grinding. The milling process and handling of the starting powders and the milled particles are carried out in an oxygen-free inert environment.	FeCo	30 nm	Nanorods	Adaptable for toxic and abrasive materials.	Contamination of product.	[13]
Pulsed Laser Desorption	The precursor (liquid or gas) is ionized, dissociated, sublimated, or evaporated using a laser and then condensed.	Au	5.5 nm	Nanorods	Fewer defects, Homogenous chemical composition, narrow size distribution.	Scale-up is difficult, economic concerns.	[14]
Lithographic Techniques	Uses light or electron beam to selectively remove micron-scale structures from a precursor material called a resist.	Au	50 nm	Nanorods	Simple to implement, low cost. Large surface patterning.	Patterning accuracy and nanoparticle size variation due to diffraction effects.	[15]
Molecular Beam Epitaxy	Ultra-pure elements are heated in separate quasi-Knudsen effusion cells until they begin to slowly evaporate. The evaporated elements then condense on the wafer	Pa	250 nm	Nanowires	Precisely controllable operating conditions.	Expensive, complicated system.	[16]
Chemical Methods							
Electrodeposition	Deposition of metal nanoparticles on supported material performed in acidic or basic baths containing metal salts. Nanoparticle synthesis is accomplished by scanning between a few voltage ranges.	Pt	NA	Nanotubes	Porosity-free finished products. Low initial capital investment. High production rates with few shape and size limitations.	Complex operational conditions.	[17]
Chemical Vapor Deposition	Solid is deposited on a heated surface via a chemical reaction from the vapor or gas phase.	Ru	3.1 nm	NA	Is self-cleaning—extremely high purity deposits (>99.995% purity). Conforms homogeneously to contours of the substrate surface.	Chemical and particle contamination.	[18]

NA—Not assessed.

It can be seen that physical and chemical schemes for metallic NPs synthesis are exceedingly diverse, and the findings show that process parameters such as temperature and concentration, etc., greatly affect the morphology, stability, and physicochemical properties of the NPs. Moreover, the synthesis of NPs employing conventional methods involves expensive chemical and physical processes with the potential hazards of ecological damage, cellular toxicity, and carcinogenicity [19,20]. These arise due to the use of harmful materials such as organic solvents, reducing agents, and stabilizers for the prevention of unwanted agglomeration of the colloids. Certain NPs are lethal, owing to features such as their magnitude, chemical composition, form, and external interactions, resulting in the incidence of lethal agents in the manufactured NPs possibly preventing their use in clinical and biomedical applications. As a result, there is a requirement for evolving new, biologically compatible, and eco-friendly green processes for manufacturing NPs [21–23].

Biological agents that have extensively been used for metallic NPs synthesis include unicellular and multicellular organisms. A few notable examples are bacteria, fungi, plant extracts, algae, diatoms, viruses, yeast, and a few higher organisms such as earthworms. Numerous sources in the literature have elaborated on the various attempts to synthesize metallic NPs in biofactories. The biological factories act as clean, non-toxic, and environmentally friendly systems for synthesizing biocompatible NPs over a wide range of sizes, shapes, compositions, and physicochemical natures. Most biological entities act as templates that assist in the stabilization of the nanostructures with the aid of biological polymers. The biopolymers enhance the biocompatibility of these NPs and prevent their agglomeration into clusters. However, plant extracts provide a plethora of enzymes and reducing agents that assist in the straightforward synthesis of NPs. Figure 1 shows a schematic representation of the synthesis of NPs using plant extracts. Compared with microorganisms, the plant method is highly beneficial as it does not require separate, complex, or numerous procedures such as isolation, culture development, and culture preservation. In addition, synthesis using plants is quicker, more cost-effective, and easy to scale up for the manufacture of bulk quantities of NPs.

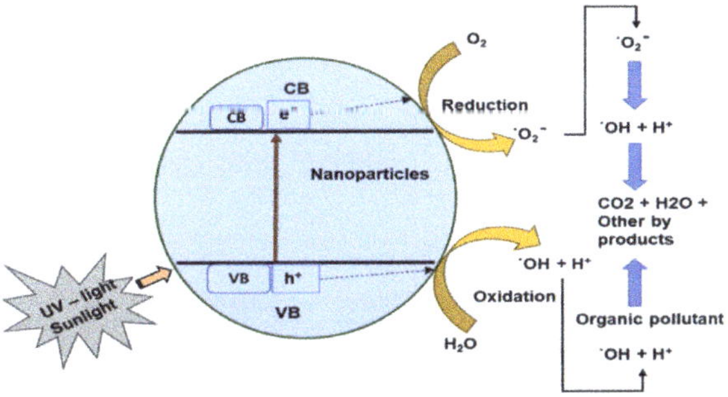

Figure 1. A schematic diagram for the production of reactive oxygen and hydroxyl radicals.

Table 2 shows a list of a few metal oxide nanoparticles synthesized from various plants and having various applications. In a relatively new report, quantum dots have been synthesized using the enzyme milieu in the midgut of earthworms. In summary, the utilization of biological resources for metallic NPs synthesis has increased exponentially over the past few years. The following sections elaborate on the interplay of operational conditions in bio-systems for the synthesis of NPs [24–26].

Table 2. Green synthesis of metal oxide nanoparticles using various plants, with applications.

Plant	Source of Nanoparticles	Metal Oxide	Size	Application	Reference
Ficus carica	Leaf	Fe_3O_4	43–57 nm	Antioxidant activity	[27]
Azadirachta indica	Leaf	CuO	NA	Anticancer property	[28]
Peltophorum pterocarpum	Leaf	Fe_3O_4	85 nm	Rhodomine degradation	[29]
Terminalia chebula	Seed	Fe_3O_4	NA	Methylene blue degradation	[30]
Punica granatum	Peel	ZnO	118.6 nm	Antibacterial property	[31]
Lactuca serriols	Seed	NiO	NA	Degradation of dye	[32]
Vitis rotundifolia	Fruit	CoO	NA	Degradation of acid blue dye	[33]

2.1. *Influence of Various Parameters on the Synthesis of Nanoparticles*

Several features control the nucleation and construction of stabilized NPs. A variety of claims for properties such as antioxidant, antimicrobial, anticancer, larvicidal, and antibiofilm properties have been made for crystalline NPs with different shapes and controlled sizes. These features (form and magnitude) are mostly reliant on the process limitations of the extract, along with the metal salt's response, pH, time of reaction, temperature, and ratio of plant extract to metal salts [34]. The following sections briefly discuss each of these factors in detail for the growth phase of the organism. Experimental efforts to optimize and enhance the synthesis of NPs have been reported by several authors. In 2011, Kalimuthu studied the effect of the growth phase of biomass on the synthesis of Ag NPs [21]. It was observed that during the stationary phase, the organism (*Bacillus* sp.) produced a relatively high number of NPs compared with the biomass obtained from other phases. Sweeney et al. demonstrated intracellular dense packing of NPs in E.coli in the stationary phase of bacterial growth [35]. According to the literature, the metal tolerance of fungus is enhanced during the stationary phase due to the release of enzymes and other chemical metabolites that reduce the metal stress. Furthermore, the metal tolerance capacity is reported to vary with the type of microbe and the metal under consideration. For instance, the presence of nickel in the growth medium has been shown to result in an extended mid-log phase in *Aspergillus* sp. However, the presence of chromium in the medium was reported to extend the stationary phase for the same organism [36]. Nevertheless, most of the studies in the literature suggest the preferential use of microbes in their stationary phase for NPs synthesis.

2.2. *pH and Precursor Concentration*

The molar ratios of reactants have also been reported to be important parameters that influence the NP size in chemical synthesis protocols. It is known that the concentration of reactants can directly influence the products in chemical synthesis. In this regard, Perumal Karthiga demonstrated that the shape of silver nanocrystals biosynthesized using silver nitrate and citrus leaf extract can be controlled systematically by varying the reactant concentration [37]. According to the authors, a $AgNO_3$: citric acid ratio of 1:4 (vol:vol) yielded spherical NPs. However, it was also reported that the production of bio-organics from plant extract increased the particle size of Ag NPs. Although a definite relationship between the precursor concentrations and the shape of the nanocrystal was not found, it could be noted that precursors at a higher molar ratio had a significant effect on the shape of the NPs. The pH was also stated to have a profound effect on the reduction reaction of the metallic ions. Pandian analyzed the effect of varied pH conditions on the synthesis of CdS nanocrystallites by *Brevibacterium species* [38]. The pH of the incubation mixtures was subjected to adjustments using 1 M HCl and 1 M NaOH solutions. It was observed that the size of the NPs varied greatly with pH. In general, an alkaline pH assisted the possibility of accessible functional groups in the reaction mixture, which in turn aided nucleation and NPs formation. The alkaline environment was previously found to aid the synthesis of various NPs in association with protein molecules [25]. Kowshik checked the pH stability of the biosynthesized NPs. It was observed that acidification of the nanocrystallites from pH 7 to 6 led to particle agglomeration [39]. In another study, Ag NPs synthesized from extract of *Cinnamon zeylanicum* bark increased in number with cumulative concentrations of bark extract and at greater pH values (pH 5 and above). Furthermore, pH values below 6 resulted in the precipitation of nanocrystals out of the solution. Due to their low toxicity, lower production of pollutants, and energy conservation, biomanufacturing methods for metal/metallic nanomaterials with ordered micro/nanostructures and programmable functionalities is critical in both fundamental investigations and practical applications. Microorganisms, as effective biofactories, have a significant ability to biomineralize and bioreduce metal ions, which can be obtained as nanocrystals of varied morphologies and sizes. The advancement of nanoparticle biosynthesis improves the safety and sustainability of nanoparticle production [31].

2.3. *Temperature*

Numerous studies indicate the predominant influence of temperature in the morphology and distribution of nanocrystals. Most of the studies in the literature report that elevated temperature conditions result in a size reduction in NPs. For instance, researchers reported a size reduction in biosynthesized Ag NPs from 35 nm to 10 nm when the reaction temperature was increased from 25 °C to 60 °C [40]. The biosynthesis was initiated using sweet orange peel extract. The reaction rate and particle formation rate increased with an increase in reaction temperature, although the average particle size decreased and the particle change rate progressively increased on increasing the temperature. In this context, it is also important to consider the temperature tolerance profile of the biological entity being considered for the synthesis of the NPs. Many researchers have reported the production of heat shock proteins by microorganisms at the elevated temperature conditions that aid NPs synthesis [41].

3. Applications of Nanoparticles

NPs possess tremendous advantages for use in many areas of day-to-day activities. Therefore, it is important to explore NPs in depth. Figure 2 shows a schematic representation of nanoparticle synthesis methods and the applications of NPs discussed in this review. NPs for use in the human body include biosynthesized noble metal NPs, which have many important applications. They make use of the molecular engine to address medicinal difficulties, and molecular information is used to support and advance human fitness at the molecular scale. This leads to the protection and development of human health. Fernández-Llamosas biosynthesized selenium NPs, which have many benefits for human health, using *Azoarcus* sp. *CIB*, [42]. The classification of different nanoparticle synthesis methods and their applications is depicted in Figure 2.

Regarding uses in biomedical research, the medicinal field still has unsolved issues, and NPs are the key to certain issues. The synthesis of NPs using extracts of leaves (plant) and/or bark provides more extensive applications in biotechnology [43], sensors [44], medicine [45], catalysis [46], optical devices [47], coatings [48], drug delivery [49], water remediation [50], and agriculture [51]. The NPs have micro and/or nanomolar sensitivity and can be detected via imaging instruments, which makes them suitable for imaging, therapy, and the delivery of drugs [52]. NPs of different dimensions have different biomedical uses. NPs have even been loaded onto TiO_2 nanotube implants for use as orthopedic implant materials. The NPs increase the biocompatibility of the implants, ultimately leading to a longer life span and greater effectiveness of the implant.

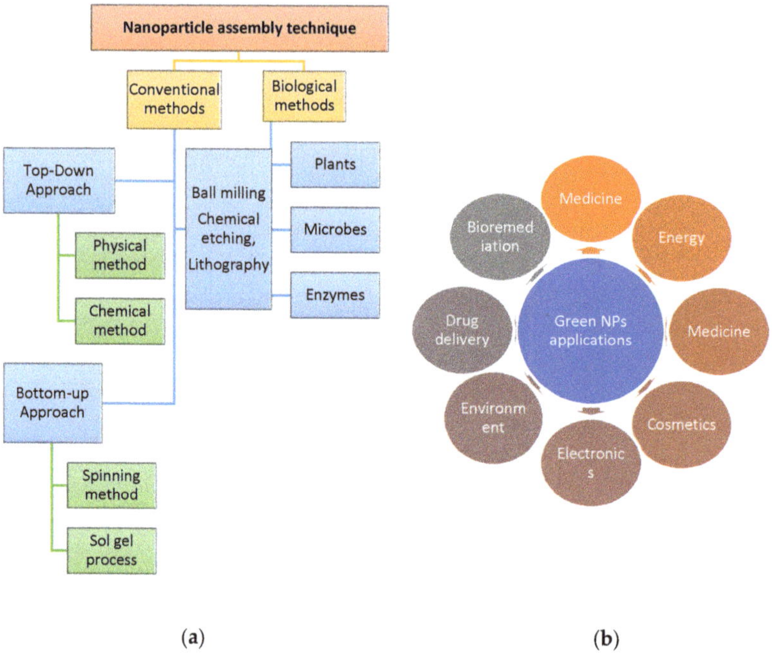

Figure 2. Classification of different nanoparticle synthesis methods (**a**) and their applications (**b**).

3.1. Anti-Inflammatory Properties of Nanoparticles

Nanoparticles have been developed as anti-inflammatory mediators in recent years. NPs have a large surface-area-to-volume ratio and are used for obstructing substances accompanying inflammation such as cytokines and inflammation-supporting enzymes, associated with other complements. Numerous metal-based NPs have been reported with excellent anti-inflammatory properties, such as those based on silver, gold, copper, and iron oxide. In this review, we demonstrate the mechanism for constructing anti-inflammatory properties in NPs. Figure 3 depicts the mechanism of nanoparticles in anti-inflammatory systems. Swelling is the body's instant response to interior damage, contagion, hormone inequity, and failure in the interior structures or external features, such as in an attack by pathogenic microorganisms or an external element. This leads to overweight, food allergies, or interactions with ecological contagions. Distinctive resistant cells possess antigen receptors capable of sensing biochemical signs. Swelling is caused by cellular and tissue injury resulting from an imbalance in the signals controlling the inflammation [53]. Upon injury or infection, muscles invoke an inflammatory response that leads to the deployment of macrophages and killer cells [54,55]. Macrophages have the main role in auto-modifiable inflammatory processes. Macrophages are large, mononucleated phagocytes produced in the bone marrow and originate as moveable white blood cells (WBCs) called monocytes in the bloodstream [56]. These monocytes then drift to various locations in numerous tissues and form macrophages. Macrophages are of two kinds: pro-inflammatory M1 macrophages whose manufacture encourages inflammation and M2 macrophages that are alternatively activated as an anti-inflammatory response and stimulate the remodeling of the swollen tissues and organs. Macrophages are able to sustain the inflammatory process by inducing changes among the two phenotypes contingent on the retarder's disorder [57,58]. Through swelling, the macrophages overwhelm cellular and tissue damage by phagocytosis and lead to inflammation via activation signals stimulating the macrophages.

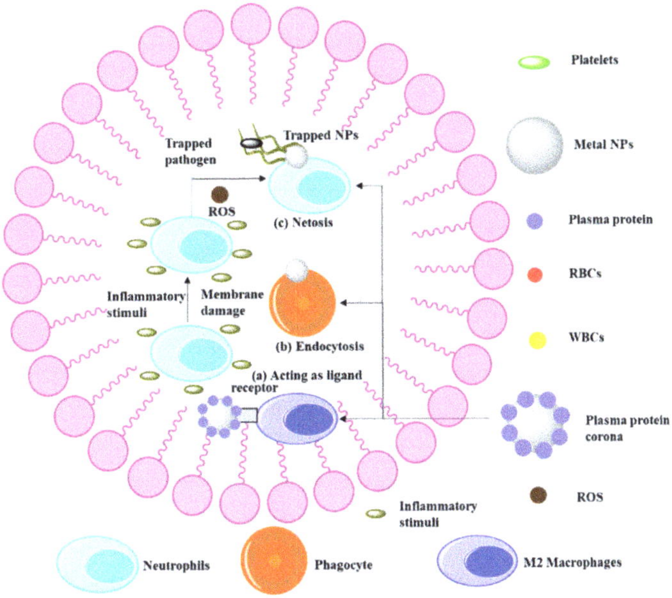

Figure 3. Anti-inflammatory mechanism adopted by various nanoparticles.

3.2. In Therapeutics

NPs are the ultimate platform for biomedical uses and therapeutic interventions. Cancer is a notorious and deadly disease and still stands as one of the principal health issues of the 21st century. Hence, there is an urgent need for anti-cancer medicine. The current advances in therapeutic options for cancer have lagged in differentiating between cancerous and normal cells, failing to produce a complete anti-cancer response [59]. In recent times, researchers have found that metal oxide NPs such as Zn and Ce oxide NPs hold considerable promise as anti-cancer medicines [60,61]. Cerium nanoparticles (CeO NPs) consisting of a cerium core enclosed by an oxygen lattice have shown extensive potential as a therapeutic agent [62]. Silver (Ag) NPs synthesized using *Abelmoschus esculentus* (L.) pulp extract have shown potential therapeutic uses and efficacy in killing Jurkat cells in vitro. The anticancer activity of Ag NPs was found to be strongly associated with higher levels of reactive oxygen species (ROS) and reactive nitrogen species, with a loss of integrity in the mitochondrial membrane [63]. More recently, the anti-cancer activity of Ag NPs synthesized from Punica granatum leaf extract (PGE) was investigated against a liver cancer cell line (HepG2). The results showed that the PGE-AgNPs showed greater efficacy in killing cancer cells. Figure 4 shows a schematic representation of the killing of cancer cells using AgNPs. Yet another report by Saratale showed that AgNPs synthesized from the common medicinal plant dandelion (Taraxacum officinale) had a high cytotoxic effect against HepG2 [64]. It is clear that in the future, NPs could be personalized for patient care. Furthermore, AgNPs developed using Olax scandens leaf extract showed anti-cancer activities with respect to different cancer cells (B16: mouse melanoma cell line, A549: human lung cancer cell lines, and MCF7: human breast cancer cells) [65].

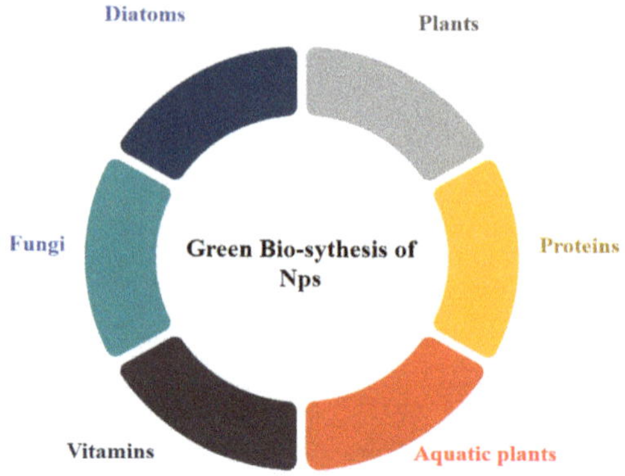

(a)

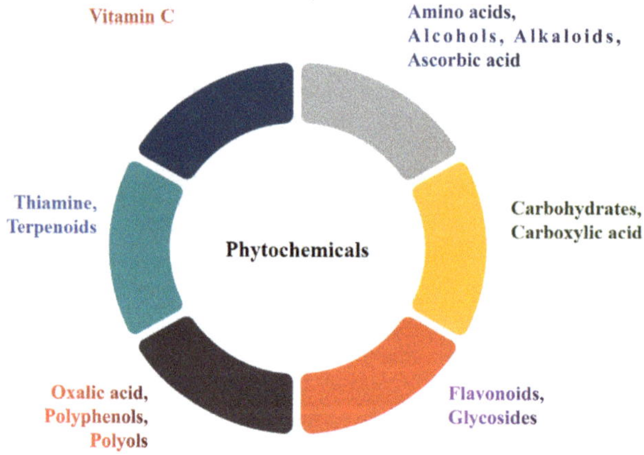

(b)

Figure 4. Several natural resources for synthesizing green NPs (**a**) and a few important bio-reductants found in plant extracts (**b**).

It was reported recently that iron oxide NPs have the dual capacity to act as both magnetic and photothermal agents in cancer therapy. This dual action was found to yield complete apoptosis-mediated cell death. Furthermore, these iron oxide NPs can be combined with laser therapy, showing complete regression of tumor cells in vivo [66]. Studies showed that photothermal therapy using green synthesized iron oxide nanoparticles loaded with the drug temozolomide with near-infrared light irradiation resulted in the death of glioblastoma cancer cells [67]. Bilici stated that superparamagnetic iron oxide nanoparticles act as a highly efficient photothermal therapy agent. Indocyanine-green-coated iron oxide nanoparticles were irradiated using laser treatment at 795 nm. This photothermal effect efficiently reduced the breast cancer cell line MCF7 when the ICG was free [68]. Green synthesized metal oxides play an important role in photothermal therapy as the metal oxide nanoparticles are irradiated by the light and can help in targeted drug release with a controllable dose. In 2013, Geetha synthesized gold NPs from the flower of the tree *C. guianensis* and explored their antileukemic cancer activity [69]. Fazal reported green synthesized anisotropic and cytocompatible gold NPs without any capping agent and studied its effectiveness along with photothermal therapy [70]. Their report found that the anisotropic particles exhibited a good photothermal effect for femtosecond laser exposure at 800 nm on A431 cancer cells, with low influence. Parida prepared metallic gold NPs, stabilizing them with ethanolic extract of clove (*Syzygium aromaticum*) and studying their anti-cancer potential and biomechanism using the human SUDHL-4 cell line. They found that the gold NPs could decrease the growth and viability of the SU-DHL-4 cell line and increase apoptosis [71]. The synthesis protocol and the important bio-reductants found in plant extracts are shown in Figure 4b.

3.3. In Drug Delivery Systems

Management of infections of the frontal section of the eye using commercially obtainable ocular drug delivery schemes has low efficiency. NPs have been designed for employment in preparations for eye drops or injectable solutions. Drugs loaded with NPs possess good drug pharmacokinetics, non-specific toxicity, pharmacodynamics, immunogenicity, and biorecognition, thereby improving the efficacy of the drugs [72]. Chitosan based polymeric NPs can act as drug carriers, paving the way for the growth of numerous dissimilar colloidal delivery vehicles. These NPs can cross biological barriers and protect macromolecules such as peptides, oligonucleotides, proteins, and genes from the degradation of biological media, allowing the delivery of drugs or macromolecules to the target site followed by precise release [73]. NPs are a promising strategy for the controlled delivery of a drug against human immunodeficiency virus (HIV) named lamivudine, which acts as a potent and selective inhibitor of type 1 and type 2 HIV [74]. Superparamagnetic iron oxide NPs (SPIONs), together with the drug, have been used for site-specific delivery of drugs. The drug can readily bind to the SPION surface and can be guided with an external magnetic field to the desired site, where the NPs can enter the target cell and deliver the drug [75]. The SPION is exposed to another cell once the drug is dissolved inside the target cell, which can reduce the absorption time, the quantity of the drug, and the interaction of the drug with non-target cells. Sripriya reported a sophisticated technique for the fabrication of multifunctional polyelectrolyte thin films in the loading and delivery of therapeutic drugs. The Ag NPs biosynthesized from the leaf extract of *Hybanthus enneaspermus* were found to be effective reducing agents with significant potential for remotely activated drug delivery, antibacterial coatings and wound dressings [48].

3.4. Medical Diagnosis, Imaging, and Sensors

In recent times, NPs have played a vital role in multimodal and multifunctional molecular imaging. Owing to the nanoscale sizes, high agent loadings, tailored surface properties, and controlled release patterns, as well as the enhanced permeability and retention effect, nanotechnology has emerged as a promising strategy for cancer diagnosis. Magnetic NPs such as iron oxide have gained tremendous attention in drug delivery systems and magnetic resonance imaging, as well as in magnetic fluid hyperthermia for diagnosis and cancer therapy [76]. Critical information regarding the progress of a deadly cancerous disease can be obtained readily via imaging of the sentinel lymph nodes (SLNs). The naked carbon NPs obtained from food-grade honey can be effectively employed in SLN imaging, which is attributed to their strong optical absorption in the near-infrared region, smaller size, and rapid lymphatic transport. This has great potential for faster resection of the SLN and also decreases complications in axillary investigations using low-resolution imaging techniques [77]. Fluorescent carbon NPs were synthesized using grape juice via chemical-free simple hydrothermal treatment with high water stability, lower toxicity, and excellent stability. These NPs can be employed as excellent fluorescent probes for the cellular imaging process and could be a promising alternative to traditional quantum dots [78]. The medicinal applications of green synthesized metal oxide nanoparticles are shown in Figure 5.

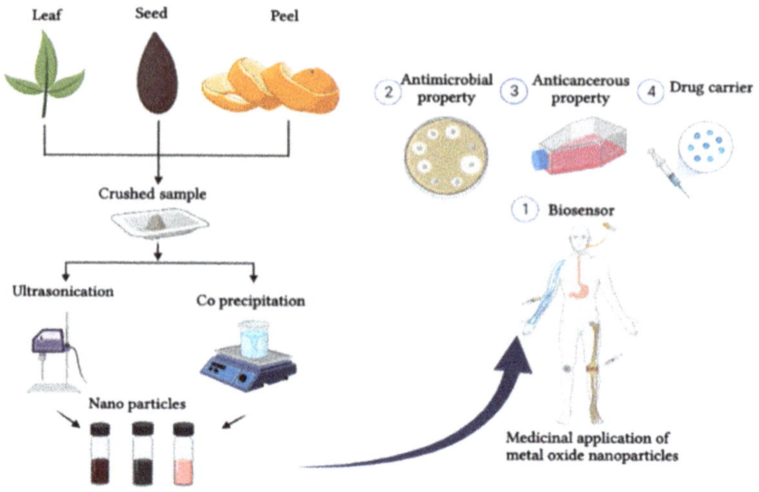

Figure 5. Medicinal applications of green synthesized metal oxides (created on 14 April 2022 by BioRender.com).

Fluorescent-nanoparticle-based imaging probes are equipped with current labeling technology and are also expected to be used in new medical diagnostic tools, due to their superior brightness and photostable properties compared to conventional molecular probes [79]. Raja prepared Ag NPs using *Calliandra haematocephala* leaf extract for the detection of H_2O_2. The results showed that the Ag NPs could be successfully used to detect the concentrations of H_2O_2 present in various samples [80]. Zheng prepared Ag NPs via a green biochemical method employing *Corymbia citriodora* leaf extract as an effective reducing and stabilizing agent and also explored the application of biosynthesized zinc oxide NPs in constructing an H_2O_2 biosensor [81]. The results showed that the fabricated electrochemical H_2O_2 sensor could potentially be employed in the pharmaceutical field and in clinical trials. In recent years, food adulteration has become a serious issue; for instance, adulteration of milk makes it hazardous to drink. Varun synthesized Ag NPs for sensing melamine in milk [82]. Monitoring aquatic ecosystems is important because potentially toxic metal ions such as Cu^{2+} and Hg^{2+} can have severe effects on human health as well as on the environment. Ag NPs synthesized using the juice extract of *Citrullus lanatus* (watermelon) exhibited good ability to detect these ions in aqueous solutions [83]. Moreover, biosynthesized Ag-NPs using *Camellia sinensis* (green tea) aqueous extract possessed good properties for sensing Cu^{2+} and Pb^{2+} ions in aqueous solutions [84]. Gold NPs synthesized from *Osmundaria obtusiloba* extract proved to be an excellent agent with good optical properties that could be employed as a suitable candidate for sensor applications [85]. Polluted water is a major threat to both quality of life and public health. Ag NPs prepared by green synthesis using *Achyranthes aspera* L. extract and protected by chitosan could be employed as a sensor for removing thiocyanate ions present in contaminated water [86].

4. Environment and Energy

Nanomaterials are of prime importance in environmental remediation and green processes. They have potential in cleaning hazardous waste sites as well as in the treatment of pollutants. Figure 6. shows the photocatalytic degradation of Acid Blue-74 by the nanoparticles. The self-cleaning nanoscale surface coatings can eliminate several chemicals for cleaning purposes employed in maintenance routines. Fe NPs have gained interest owing to rapidly developing applications for the disinfection of water, as well as remediation of heavy metals in the soil. NPs serve as alternatives to pesticides in the control and management of plant disease and act as effective fertilizers that are eco-friendly and capable of increasing crop production. Magnetite (Fe_3O_4) and the siliceous material produced by employing bacterial cells and diatoms have been successfully employed in coating optical instruments for solar energy applications [87].

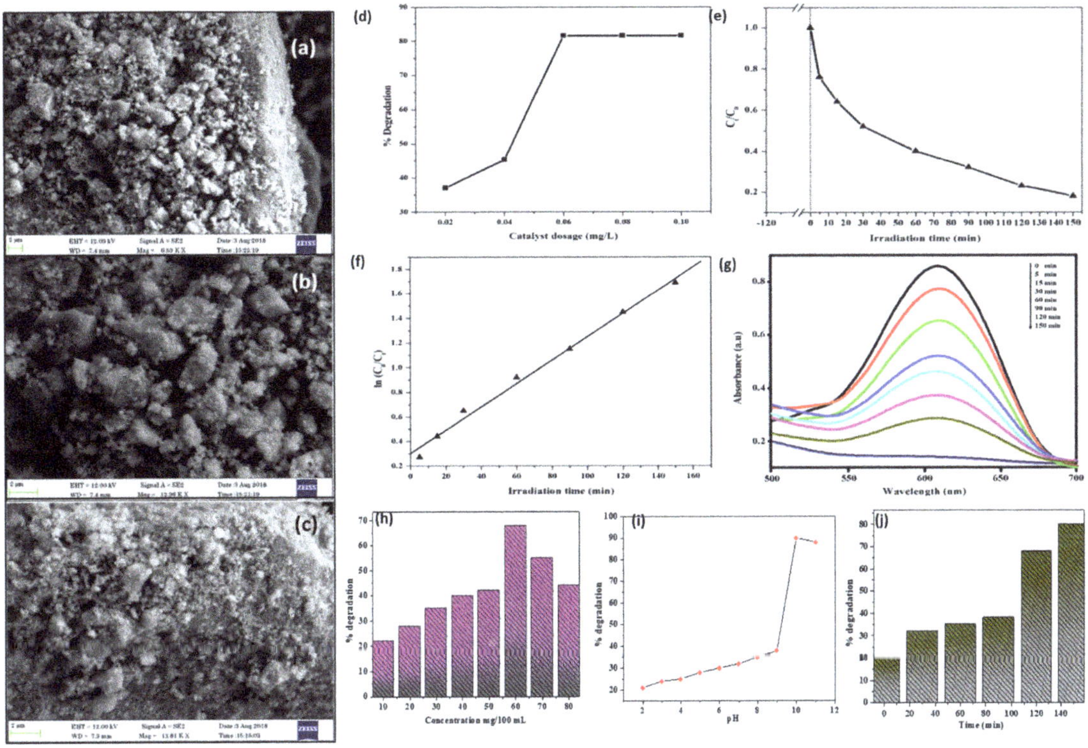

Figure 6. (**a**–**c**) FESEM micrographs of green synthesized GGCo-NPs nanoparticles. (**d**) Effect of initial catalytic dose. (**e**) Photocatalytic degradation of AB-74 by GCo-NPs under irradiation by sunlight. (**f**) Pseudo-first-order reaction kinetic model for GCo-NPs as an NP photocatalyst. (**g**) Absorption spectrum of photocatalytically degraded AB-74 at different time intervals. Photocatalytic degradation of AB-74, with varying initial dye concentrations (10 mg/100 mL–80 mg/100mL) (**h**), pH (2–12) (**i**), contact time (0–150 min) (**j**) [33].

4.1. Remediation and Degradation

Remediation solves the problem. "Bioremediation" refers to a process involving the use of biological agents such as bacteria, fungi, protists, or their enzymes for the degradation of environmental contaminants into less-toxic versions [88,89]. Bioremediation provides many advantages over conventional treatments as it is more economically feasible, has a high competence level, minimizes chemical and biological sludge, is selective to specific metals, has supplementary nutrient requirements, and has the possibility of regeneration of the biosorbent and metal recovery [90,91]. There are multiple reasons for the employment of different NPs in bioremediation; for example, when materials are at the nanoscale, the surface area per unit mass of the material increases, and as a result, a larger amount of the material comes into contact with the surrounding materials, thereby affecting the reactivity. NMs have the potential to exhibit a quantum effect with less activation energy for accessing the chemical feasibility of the reactions. Surface plasmon resonance is another phenomenon that is exhibited by NPs, and this can be employed for the detection of toxic materials [92]. Regarding shape and size, various metallic and nonmetallic NMs with different morphologies have been employed in environmental clean-up processes. For example, various single-metal NPs, carbon-based NMs, and bimetallic NPs can be used. Bioremediation processes use agents in solid waste, groundwater and wastewater management, petroleum and petroleum goods management, uranium remediation, soil remediation, and remediation of heavy metal pollution. The study of the ability of NMs to combat contamination is advancing and could potentially result in revolutionary changes in the ecological field. The various uses of NPs include the following:

- Nanoscale zero-valent iron (nZVI) has been produced and verified for its ability to efficiently remove As (III), which is an exceedingly lethal, mobile, and major arsenic species in anoxic groundwater.
- Engineered polymeric NPs have been employed for the bioremediation of hydrophobic contaminants.
- PAMAM dendrimers with special structures and properties have been employed in water treatment, as they are efficient as well as innoxious as a water treatment agent.
- Engineered polymeric NPs have been employed for soil remediation.

NPs could have a deeper impact on biodegradation. With the increased development of the textile trade, major anxieties have arisen regarding the pollution of the environment with dye contaminants, leading to serious conservational contamination as well as detrimental consequences to health, given their variety, toxic nature, and ability to persist. Most of these dyes have complex compositions and high chemical stability, facilitating their persistence for extended distances in flowing water and thereby retarding photosynthetic action, inhibiting the development of aquatic biota by the blockage of sunlight, and inhibiting the utilization of dissolved oxygen, leading to a decreased recovery rate of the watercourse. Degradation of the dyes in the manufacturing wastewaters has gained considerable attention due to the bulk production, less decoloration, slower biodegradation and high toxicity. Metal oxides can adopt a huge number of physical geometries and have an electronic assembly that can exhibit metallic, semiconducting, or insulating features and can therefore perform efficient roles in many areas of science. In the past few decades, enormous interest has been shown in heterogeneous photocatalysis technology with the incorporation of metal oxides, owing to their possible applications in both ecology and organic synthesis. Several attempts to study the photocatalytic activity of different metal oxides such as ZrO_2, SnO_2, and CdS have been made. Titanium dioxide (TiO_2) and zinc oxide (ZnO) have been characterized as having chemical stability, eco-friendly properties, and a lack of toxicity, and they can be produced relatively cheaply. They have been employed in diverse areas of photochemistry ranging from large-scale products to more advanced applications. For instance, in the case of environmental remediation, they have been used in the photoelectrolysis of water and in dye-sensitized solar cells. Sunlight is an abundantly accessible resource that can be used to irradiate semiconductors in photon-based degradation of polluting agents, and these techniques are economically relatively feasible [93]. Worldwide soil contamination is severe, damaging normal ecological services and preventing human activities Traditional approaches for dealing with dirty soils include excavation followed by discarding or ex-situ action such as soil washing or thermal desorption. However, these approaches can be expensive and time-consuming, and they lead to large quantities of secondary contamination. Therefore, low-influence in situ actions such as inoculation with NPs are increasingly preferred. Numerous effective NPs are stated to have better activity over a wide range of contaminants including heavy metals and organic contaminants.

4.2. Wastewater Treatment

Water contamination is one of the key problematic issues faced by the world today, as the survival of the species relies on the presence of water fit for consumption. Contamination of water has highly detrimental consequences affecting the environment as well as human health, along with multiple impacts on socioeconomic progress. Many commercial and non-commercial techniques are available for combatting this problem, which is increasing due to technological progress [94]. Nanotechnology has also proved to be one of the best and most advanced strategies for the treatment of wastewater. NPs have high adsorption, interaction, and reaction capacities owing to their small size and high proportion of atoms at the surface [95–98]. They can also be suspended in aqueous solutions, to behave as colloids. These particles achieve energy conservation owing to their small size, and this can ultimately lead to cost-effectiveness. NPs possess great advantages for treating water at large depths and in any location that has not been cleared by the available conventional technologies [10,96,99,100]. Green nanomaterials possess a wide range of abilities for the treatment of water that is contaminated by toxic metal ions, organic and inorganic solutes, and pathogenic microorganisms. Advanced research and commercialization of various nanomaterials (nanostructured catalytic membranes, nanosorbents, bioactive NPs, nanocatalysts, biomimetic membranes, and molecularly imprinted polymers (MIPs)) has been undertaken, in order to eradicate toxic metal ions, pathogenic microbes, and organic and inorganic solutes from water [101,102].

5. Nanosorbents

Nanosorbents possessing high and specific sorption potential are widely exploited for the purification of water and for remediation purposes, as well as in treatment processes, e.g., carbon-based nanosorbents such as Captymer™. Nanocatalysts, e.g., silver (Ag) nanocatalyst, AgCCA catalyst, etc., have been widely employed as they can increase the catalytic activity at the surface owing to their special characteristics of possessing a high surface area with a shape-dependent property for the enhancement of the reactivity, as well as the degradation of contaminants. Figure 7 shows the degradation of 4-nitrophenol by the AgNPs. Nanostructured catalytic membranes are employed widely for the purpose of treating contaminated water, and this is facilitated due to several advantages such as high uniformity of the catalytic sites, the potential for optimization, the limited contact time of the catalyst facilitating sequential reactions, and the ease of industrial scale-ups. Examples include immobilization of the metallic NPs onto membranes such as cellulose acetate, chitosan, polyvinylidene fluoride (PVDF), polysulfone, etc.

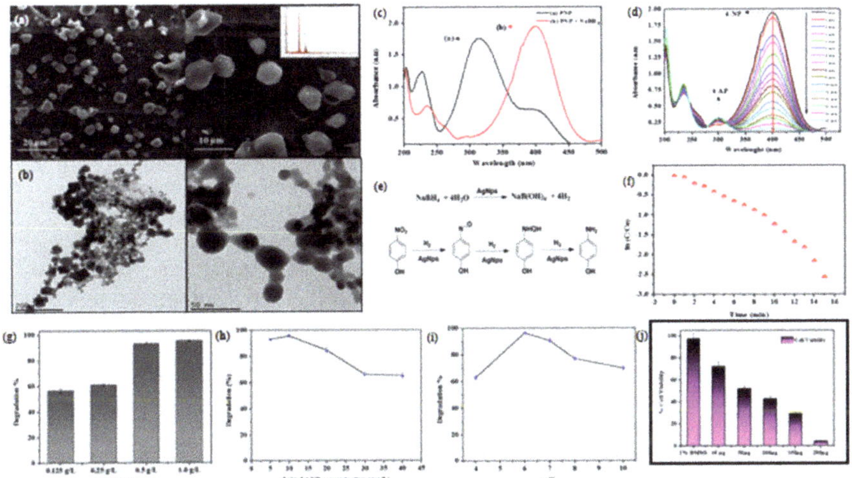

Figure 7. (**a**) SEM analysis. The EDX insert shows that the formed nanoparticles are silver. (**b**) TEM image showing the synthesized AgNPs using B. amyloliquefaciens. (**c**) Absorption spectra of 4-nitrophenol in aqueous medium and 4-nitrophenol+$NaBH_4$ in aqueous solution. (**d**) Absorption

spectra for reduction of 4-nitrophenol by $NaBH_4$ in aqueous medium in presence of biosynthesized AgNPs as catalyst. (**e**) Kinetic modeling of the 4-nitrophenol reduction in the presence of biosynthesized AgNPs. (**f**) Mechanism of reduction of 4-nitrophenol (4-NP) to 4-aminophenol (4-AP) using biosynthesized AgNPs catalyst. (**g**) Effect of adsorbent dosage on the degradation of 4- NP. (**h**) Effect of initial dye concentration on the degradation of 4-NP (experimental conditions: H_2O_2: 40 mM; AgNPs: 0.5 g/L; initial pH: 5.5). (**i**) Effect of initial pH on the degradation of 4-NP (experimental conditions: 4-NP: 10 mg/L; AgNPs: 0.5 g/L; H_2O_2: 40 mM). (**j**) Effect of hydrogen peroxide (H_2O_2) on the degradation of 4-NP (experimental conditions: 4-NP: 10 mg/L; AgNPs: 0.5 g/L; initial pH: 6. Effect of biosynthesized AgNPs on A549 lung carcinoma epithelial cells at 50–200 μg for 24 h [11].

Silver nanoparticles (AgNPs) with a very high antibacterial potential can be synthesized extracellularly by employing the bacterium *Bacillus cereus*. Nanotechnologies have facilitated several sophisticated solutions for counteracting issues of water contamination and are likely to produce many strategies composed of enhancements in the future. Treatments based on nanotechnology offer highly effective, durable, efficient and eco-friendly approaches. These strategies are cost-effective, less time-consuming, and energy-efficient, with much lower waste generation than conventional bulk-materials-based technologies. However certain precautions are necessary for avoiding threats to human health or the environment [103].

6. Cosmetics and Food Industry

Applications of nanotechnology and nanomaterials are widely present in several cosmetic products such as moisturizers, hair care products, makeup accessories, and sunscreen. The main uses for nanotechnology in cosmetics are as follows. NPs are employed in cosmetics as UV filters. TiO_2 and ZnO are the key compounds used in these applications. Organic substitutes for these have also been established. Nanotechnology is also used for delivery. The cosmetic industry takes advantage of liposomes as vehicles for delivery. Novel structures comprising solid lipid NPs and nanostructured carriers composed of lipids have been reported to perform better than liposomes. NPs also enhance and facilitate penetration. Encapsulation or suspension of the key ingredients in nanospheres or nanoemulsions facilitates skin penetration. Regarding NPs in hair-related products, the employment of nanoemulsions for the encapsulation of desired substances facilitates their delivery into the deeper hair shafts. In sunscreen lotions, the employment of zinc and titanium micronized NPs results in transparency, a less greasy texture, and less odor, and makes the lotions highly absorbable into the layers of the skin [103].

Nanotechnology has emerged as an important strategy for several food-related applications. In these types of applications, NPs of a core type are introduced into a specified food-related product for the development of certain desirable properties in the food. Nanotechnology has become an integral part of research and development for the large-scale manufacturing of agricultural products and processed foods, as well as in food packaging sectors across the world. In recent decades, the use of nanotechnology has increased tremendously, revolutionizing technology in the food sector. The emergence of demands from consumers concerning the quality of food and hygienic aspects of health have shifted the focus of researchers to developing strategies for the enhancement of food quality without any implications for the nutritional value. The demand for NP-based materials has increased in the food industry, as most of them contain essential elements and are also non-toxic and stable at high temperatures and pressures [104]. Nanotechnology can offer a wide array of solutions at various stages ranging from the manufacturing of food to processing and packaging. They have the potential to make a great difference not only in terms of food quality and safety but also in the terms of the health benefits that the food provides. Several research and industrial organizations are investigating novel techniques, methodologies, and products involving a direct application of nanotechnology in the food science sector. The applications of nanotechnology can be fitted into two main groups: nanostructured food ingredients and the nanosensing of food. Nanostructured food ingredients are widespread, in areas from the processing of food to the packaging of food. In the processing of food, these nanostructures are employed as additives for the food, antimicrobial agents, carriers for the smart delivery of nutrients, anti-caking agents, fillers that improve the mechanical strength and durability of the packaged material, etc. In the case of food nanosensing, they are employed to achieve a better quality of food and for safety evaluation purposes [104]. Several reports have stated that nanomaterials are possible candidates for improving food safety by enhancing the efficacy of packaging and the shelf life, with no alterations to the nutritional value, along with additives that do not alter the taste and physical features of the food products. Although they have the potential to help create innovative products along with the production processes prevailing in the food sector, nano techniques face a major challenge regarding the employment of cost-effective processing operations for the synthesis of edible and non-toxic nanoscale delivery systems and the efficient development of effective formulations that are considered safe for human consumption. Thus, owing to the increased employment of NSMs, mounting apprehensions in terms of developing biocompatible, safe, and non-toxic nanostructures from food-grade ingredients have emerged with respect to the use of the modest, greener processes as well as the cost-effective processes utilizing layer-by-layer technology [37]. Despite the application of nanotechnology in terms of green synthesized NSMs for numerous technologies in the food sector, the use of NSMs has led to controversies in a few instances, as they are scientifically uncertain and could have a long-term detrimental effect on human health, as well as on the environment. In this context, the complexity and the limitations of nanotechnology in terms of toxicity and accumulation could be overruled by the elucidation of the physiochemical and biological properties of the NSMs through extensive large-scale research.

Unique $CoNi_2S_4$ nanoparticles have been synthesized using a one-step solvothermal technique. When used as supercapacitor electrode materials, $CoNi_2S_4$ nanoparticles, with their lower manufacturing costs, exhibit better electrochemical characteristics such as higher specific capacitance, higher rate capability, and higher energy density, making this a promising candidate electrode material for next-generation supercapacitors.

Porous carbon electrodes are ideal for energy storage systems. A simple in situ reduction approach used gold nanoparticles to improve the electrochemical performance of carbon materials. Scanning electron microscopy, transmission electron microscopy, and the Brunauer–Emmett–Teller method all confirmed that the porous carbon microspheres coated with gold nanoparticles had a 3D honeycomb-like structure with a high specific surface area of roughly 1635 m^2g^{-1}. The electrochemical performance of the as-synthesized porous carbon microspheres as electrode materials for supercapacitors was demonstrated; they were shown to have a high specific capacitance of 440 F g^{-1} at a current density of 0.5 A g^{-1} and excellent cycling stability, with a capacitance retention of 100 percent after 2000 cycles at 10 Ag^{-1} in 6 M KOH electrolyte. The method paved the way for the gold-nanoparticle-decorated synthesis of porous carbon microspheres and could be used to create porous carbon microspheres with a variety of nanoparticle decorations for a variety of applications such as energy storage devices, enhanced absorption materials, and catalytic sites [105,106].

7. Summary and Conclusions

Apprehension over the secondary effects related to the development of NPs and an increasing desire for greener technologies have arisen in the field of green and maintainable remediation. The acceptance of green synthesis promises not only to avoid secondary conservational contamination but also to reduce manufacturing costs. However, there are still gaps in the research that should be addressed to assist the development of the field. For example, an explanation of the precise mechanisms involved in green synthesis remains essential to advance additional expected outcomes. Most studies rely on sensible norms, but detailed assessments of the precise mechanisms remain subtle. The current green synthesis research has produced NPs with several geometric structures, but methods that can produce more composite forms with more detailed surface areas are still required.

Moreover, varying the crystal-like construction of greener NPs, to discover innovative properties that vary from their majority material is an additional future avenue for green synthesis investigators. Nanotechnology has emerged as an attractive tool capable of revolutionizing several fields. It is a technology that functions on the nanometer scale and deals with atoms, molecules, and macromolecules approximately in the 1–100 nm range, to synthesize and employ materials that possess novel properties. Nature is the best coach for teaching us mechanisms for the synthesis of miniaturized functional materials. Despite being a methodological approach, the synthesis of metal oxide NPs with microorganisms or plant extracts, using biological mechanisms, opens up tremendous opportunities to produce biocompatible and cost-effective particles with potential applications in the healthcare sector.

Synthesizing of NPs via the bio-green route has attracted a great deal of attention as it involves no harmful chemicals in its synthesis method. Hence, bio-green synthesized NPs could be promising materials, opening up new prospects in clinical, energy, and environmental research. One of the most important areas calling for attention is cancer therapy and the use of nanotechnology to improve existing therapeutic practices. Cancer is a leading cause of mortality and morbidity worldwide, and the use of traditional chemotherapeutics is often limited by the adverse side effects they cause. The need for a novel strategy to combat this is important for effective cancer therapy. Recent progression in the nanotechnology sector offers many strategies for combating cancer with innovative and personalized treatments that are capable of overcoming the barriers encountered with traditional drugs. Nanomaterials have been known to enhance the efficacy of food processing and its nutritional value as additives, without changing the characteristics of food products. They are also effective agents of bioremediation and have been used in wastewater treatments. Nanotechnology has found applications in a variety of areas and will form an important strategy for solving several problems.

8. Future Perspectives

- Nanotechnology is highly recommended for future perspectives since it replaces dangerous solvents in green synthesis and process approaches, improves catalytic efficiency and selectivity, is cost-effective, and involves less toxic waste disposal.
- The primary advantages of the greener techniques are low cost and the use of antimicrobial nanoparticle combinations, which allows for the use of local plant extracts without harmful chemical reducing agents, as well as additional applications such as antibacterial bandages.
- It is necessary to have a thorough understanding of a variety of microbial/biochemical constituents as well as the various pathways involved in laboratory synthesis, including the isolation and tracing of components that are precisely used in the reduction of several metallic salts to the required materials.
- Future difficulties and current accomplishments linked with green perspectives for nanomaterial production must be addressed, extending laboratory-based compliance to an achievable industrial standard by considering current/past issues in terms of health and environmental repercussions.
- However, a greener approach technique based on bio-derived materials or nanomaterials is required and will be widely used in the field of environmental remediation as well as other broad fields such as the food, cosmetics, and pharmaceutical industries.
- Furthermore, biomaterials made from marine plants and algae found in specific locations remain undiscovered. As a result, there are still many opportunities for the development of novel green pathway strategies based on biogenic synthesis.
- To enable the industrial production of such green nanomaterials, a great deal of scientific research is required. The eventual release of such nanomaterials into the environment might cause odd behavior, and this is a concern that must be investigated further.

- Toxicity evaluation must be undertaken for nanoparticles and effective risk management processes provided for their synthesis, materials handling, storage, and disposal.

Author Contributions: Conceptualization, M.S.S., E.S. and S.V.; software, M.S.S., M.R., A.J.J. and H.P.; resources, M.S.S., E.S., S.V. and N.C.; writing—original draft preparation, M.S.S., A.J.J. and R.B.; writing—review and editing, M.S.S., E.S., S.V., N.C., P.S.C., J.S. and M.R.; supervision, E.S. All authors have read and agreed to the published version of the manuscript.

Funding: This research received no external funding.

Institutional Review Board Statement: Not applicable.

Informed Consent Statement: Not applicable.

Data Availability Statement: Not applicable.

Acknowledgments: The authors are grateful to the SRM Institute of Science and Technology for providing excellent facilities and support to carry out the work.

Conflicts of Interest: The authors declare no conflict of interest.

Abbreviations

J	Joule
M	Molar
Zn	Zinc
Pb	Lead
As	Arsenic
Se	Selenium
Ag	Silver
Cu	Copper
Hg	Mercury
Pd	Palladium
Pt	Platinum
Ru	Ruthenium
Kpa	Kilopascal
Nm	Nanometer
Fe_3O_4	Iron oxide
NPs	Nanoparticles
ZnO	Zinc oxide
SnO_2	Tin oxide
NiO	Nickel oxide
CuO	Copper oxide
CeO	Cerium oxide
CoO	Cobalt oxide
$AgNO_3$	Silver nitrate
NMs	Nanomaterials
TiO_2	Titanium dioxide
ZrO	Zirconium oxide
HCL	Hydrochloric acid
WBCs	White blood cells
NaOH	Sodium hydroxide
H_2O_2	Hydrogen peroxide
SLN	Sentinel lymph nodes
$NaBH_4$	Sodium borohydrate
$CoNi_2S_4$	Cobalt nickel sulfite
KOH	Potassium hydroxide
DMF	N-Dimethylformamide
nZVI	Nanoscale zero-valent iron

PGE	Punica granatum leaf extract
HIV	Human immunodeficiency virus
SPIONS	Superparamagnetic iron oxide
MIPs	Molecularly imprinted polymers
NCMs	Nanostructuured catalytic membranes

References

1. Feynman, R.P. There Is Plenty of Room at the Bottom. *Eng. Sci. Calif. Inst. Technol.* **1960**, *254.5036*, 1300–1301.
2. Liu, J.; Qiao, S.Z.; Hu, Q.H.; Lu, G.Q. Magnetic Nanocomposites with Mesoporous Structures: Synthesis and Applications. *Small* **2011**, *7*, 425–443. [CrossRef] [PubMed]
3. Sinha, A.; Cha, B.G.; Kim, J. Three-Dimensional Macroporous Alginate Scaffolds Embedded with Akaganeite Nanorods for the Filter-Based High-Speed Preparation of Arsenic-Free Drinking Water. *ACS Appl. Nano Mater.* **2018**, *1*, 1940–1948. [CrossRef]
4. Nassar, N.N. Rapid removal and recovery of Pb(II) from wastewater by magnetic nanoadsorbents. *J. Hazard. Mater.* **2010**, *184*, 538–546. [CrossRef]
5. Gupta, V.; Nayak, A. Cadmium removal and recovery from aqueous solutions by novel adsorbents prepared from orange peel and Fe_2O_3 nanoparticles. *Chem. Eng. J.* **2012**, *180*, 81–90. [CrossRef]
6. Mohammadkhani, F.; Montazer, M.; Latifi, M. Microwave absorption characterization and wettability of magnetic nano iron oxide/recycled PET nanofibers web. *J. Text. Inst.* **2019**, *110*, 989–999. [CrossRef]
7. Kumar, S.; Nair, R.R.; Pillai, P.B.; Gupta, S.N.; Iyengar, M.A.R.; Sood, A.K. Graphene Oxide–$MnFe_2O_4$ Magnetic Nanohybrids for Efficient Removal of Lead and Arsenic from Water. *ACS Appl. Mater. Interfaces* **2014**, *6*, 17426–17436. [CrossRef] [PubMed]
8. Pérez-Beltrán, C.H.; García-Guzmán, J.J.; Ferreira, B.; Estevez-Hernandez, O.; Lopez-Iglesias, D.; Cubillana-Aguilera, L.; Link, W.; Stănică, N.; da Costa, A.M.R.; Palacios-Santander, J.M. One-Minute and Green Synthesis of Magnetic Iron Oxide Nanoparticles Assisted by Design of Experiments and High Energy Ultrasound: Application to Biosensing and Immunoprecipitation. *Mater. Sci. Eng. C* **2021**, *123*, 112023. [CrossRef]
9. Goudarzi, M.; Salavati-Niasari, M.; Yazdian, F.; Amiri, M. Sonochemical Assisted Thermal Decomposition Method for Green Synthesis of $CuCo_2O_4$/CuO Ceramic Nanocomposite Using *Dactylopius Coccus* for Anti-Tumor Investigations. *J. Alloy. Compd.* **2019**, *788*, 944–953. [CrossRef]
10. Samuel, M.S.; Sumanb, S.; Venkateshkannanc; Selvarajand, E.; Mathimanie, T.; Pugazhendhif, A. Immobilization of $Cu_3(btc)_2$ on graphene oxide-chitosan hybrid composite for the adsorption and photocatalytic degradation of methylene blue. *J. Photochem. Photobiol. B Biol.* **2020**, *204*, 111809. [CrossRef]
11. Samuel, M.S.; Jose, S.; Selvarajan, E.; Mathimani, T.; Pugazhendhi, A. Biosynthesized silver nanoparticles using Bacillus amyloliquefaciens; Application for cytotoxicity effect on A549 cell line and photocatalytic degradation of p-nitrophenol. *J. Photochem. Photobiol. B Biol.* **2020**, *202*, 111642. [CrossRef] [PubMed]
12. Rather, G.A.; Nanda, A.; Chakravorty, A.; Hamid, S.; Khan, J.; Rather, M.A.; Bhattacharya, T.; Rahman, M.H. Biogenic Green Synthesis of Nanoparticles from Living Sources with Special Emphasis on Their Biomedical Applications. *Res. Sq.* **2021**. [CrossRef]
13. Chakka, V.M.; Altuncevahir, B.; Jin, Z.Q.; Li, Y.; Liua, J.P. Magnetic nanoparticles produced by surfactant-assisted ball milling. *J. Appl. Phys.* **2006**, *99*, 08E912. [CrossRef]
14. Mafuné, F.; Kohno, J.-Y.; Takeda, Y.; Kondow, T. Full Physical Preparation of Size-Selected Gold Nanoparticles in Solution: Laser Ablation and Laser-Induced Size Control. *J. Phys. Chem. B* **2002**, *106*, 7575–7577. [CrossRef]
15. Barbillon, G.; Hamouda, F.; Bartenlian, B. Large Surface Nanostructuring by Lithographic Techniques for Bioplasmonic Applications. In *Manufacturing Nanostructures*; One Central Press: Altrincham, UK, 2014; pp. 244–262.
16. Vossen, J.L.; Kern, W.; Kern, W. *Thin Film Processes II*; Gulf Professional Publishing: Houston, TX, USA, 1991; Volume 2, ISBN 0127282513.
17. Islam, M.; Islam, M.S. Electro-Deposition Method for Platinum Nano-Particles Synthesis. Saidul, Electro-Deposition Method Platin. Nano-Particles Synth. *Eng. Int.* **2013**, *1*, 2. [CrossRef]
18. Ogura, Y.; Sato, K.; Miyahara, S.-I.; Kawano, Y.; Toriyama, T.; Yamamoto, T.; Matsumura, S.; Hosokawa, S.; Nagaoka, K. Efficient ammonia synthesis over a $Ru/La_{0.5}Ce_{0.5}O_{1.75}$ catalyst pre-reduced at high temperature. *Chem. Sci.* **2018**, *9*, 2230–2237. [CrossRef]
19. Yin, S.-J.; Zhang, L.; Zhang, L.; Wan, J.; Song, W.; Jiang, X.; Park, Y.-D.; Si, Y.-X. Metabolic responses and arginine kinase expression of juvenile cuttlefish (*Sepia pharaonis*) under salinity stress. *Int. J. Biol. Macromol.* **2018**, *113*, 881–888. [CrossRef]
20. Samuel, M.S.; Sivaramakrishna, A.; Mehta, A. Degradation and detoxification of aflatoxin B1 by Pseudomonas putida. *Int. Biodeterior. Biodegrad.* **2014**, *86*, 202–209. [CrossRef]
21. Kalishwaralal, K.; Deepak, V.; Pandian, S.R.K.; Nellaiah, H.; Sangiliyandi, G. Extracellular biosynthesis of silver nanoparticles by the culture supernatant of Bacillus licheniformis. *Mater. Lett.* **2008**, *62*, 4411–4413. [CrossRef]
22. Vinoth, V.; Wu, J.J.; Asiri, A.M.; Anandan, S. Sonochemical synthesis of silver nanoparticles anchored reduced graphene oxide nanosheets for selective and sensitive detection of glutathione. *Ultrason. Sonochem.* **2017**, *39*, 363–373. [CrossRef]
23. Luo, L.; Xu, L.; Zhao, H. Biosynthesis of reduced graphene oxide and its in-vitro cytotoxicity against cervical cancer (HeLa) cell lines. *Mater. Sci. Eng. C* **2017**, *78*, 198–202. [CrossRef] [PubMed]
24. Vigneshwaran, N.; Ashtaputre, N.; Varadarajan, P.; Nachane, R.; Paralikar, K.; Balasubramanya, R. Biological synthesis of silver nanoparticles using the fungus Aspergillus flavus. *Mater. Lett.* **2007**, *61*, 1413–1418. [CrossRef]

25. Gurunathan, S.; Kalishwaralal, K.; Vaidyanathan, R.; Venkataraman, D.; Pandian, S.R.K.; Muniyandi, J.; Hariharan, N.; Eom, S.H. Biosynthesis, purification and characterization of silver nanoparticles using Escherichia coli. *Colloids Surf. B Biointerfaces* **2009**, *74*, 328–335. [CrossRef] [PubMed]
26. Ahmed, M.; Abdel-Messih, M.; El-Sherbeny, E.; El-Hafez, S.F.; Khalifa, A.M. Synthesis of metallic silver nanoparticles decorated mesoporous SnO_2 for removal of methylene blue dye by coupling adsorption and photocatalytic processes. *J. Photochem. Photobiol. A Chem.* **2017**, *346*, 77–88. [CrossRef]
27. Üstün, E.; Önbaş, S.C.; Çelik, S.K.; Ayvaz, M.Ç.; Şahin, N. Green Synthesis of Iron Oxide Nanoparticles by Using Ficus Carica Leaf Extract and Its Antioxidant Activity. *Biointerface Res. Appl. Chem.* **2022**, *2021*, 2108–2116.
28. Patil, S.P.; Chaudhari, R.Y.; Nemade, M.S. Azadirachta indica leaves mediated green synthesis of metal oxide nanoparticles: A review. *Talanta Open* **2022**, *5*, 100083. [CrossRef]
29. Shah, Y.; Maharana, M.; Sen, S. *Peltophorum pterocarpum* leaf extract mediated green synthesis of novel iron oxide particles for application in photocatalytic and catalytic removal of organic pollutants. *Biomass-Convers. Biorefin.* **2022**, 1–14. [CrossRef]
30. Singh, P.; Singh, K.R.; Verma, R.; Singh, J.; Singh, R.P. Efficient electro-optical characteristics of bioinspired iron oxide nanoparticles synthesized by Terminalia chebula dried seed extract. *Mater. Lett.* **2021**, *307*, 131053. [CrossRef]
31. Abdelmigid, H.M.; Hussien, N.A.; Alyamani, A.A.; Morsi, M.M.; AlSufyani, N.M.; Kadi, H.A. Green Synthesis of Zinc Oxide Nanoparticles Using Pomegranate Fruit Peel and Solid Coffee Grounds vs. Chemical Method of Synthesis, with Their Biocompatibility and Antibacterial Properties Investigation. *Molecules* **2022**, *27*, 1236. [CrossRef]
32. Ali, T.; Warsi, M.F.; Zulfiqar, S.; Sami, A.; Ullah, S.; Rasheed, A.; Alsafari, I.A.; Agboola, P.O.; Shakir, I.; Baig, M.M. Green nickel/nickel oxide nanoparticles for prospective antibacterial and environmental remediation applications. *Ceram. Int.* **2021**, *48*, 8331–8340. [CrossRef]
33. Samuel, M.S.; Selvarajan, E.; Mathimani, T.; Santhanam, N.; Phuong, T.N.; Brindhadevi, K.; Pugazhendhi, A. Green synthesis of cobalt-oxide nanoparticle using jumbo Muscadine (*Vitis rotundifolia*): Characterization and photo-catalytic activity of acid Blue-74. *J. Photochem. Photobiol. B Biol.* **2020**, *211*, 112011. [CrossRef] [PubMed]
34. Ramkumar, V.S.; Pugazhendhi, A.; Prakash, S.; Ahila, N.; Vinoj, G.; Selvam, S.; Kumar, G.; Kannapiran, E.; Rajendran, R.B. Synthesis of platinum nanoparticles using seaweed Padina gymnospora and their catalytic activity as PVP/PtNPs nanocomposite towards biological applications. *Biomed. Pharmacother.* **2017**, *92*, 479–490. [CrossRef] [PubMed]
35. Sweeney, R.Y.; Mao, C.; Gao, X.; Burt, J.L.; Belcher, A.M.; Georgiou, G.; Iverson, B.L. Bacterial Biosynthesis of Cadmium Sulfide Nanocrystals. *Chem. Biol.* **2004**, *11*, 1553–1559. [CrossRef] [PubMed]
36. Congeevaram, S.; Dhanarani, S.; Park, J.; Dexilin, M.; Thamaraiselvi, K. Biosorption of chromium and nickel by heavy metal resistant fungal and bacterial isolates. *J. Hazard. Mater.* **2007**, *146*, 270–277. [CrossRef]
37. Perumal, A.B.; Nambiar, R.B.; Sellamuthu, P.S.; Sadiku, E.R. Application of Biosynthesized Nanoparticles in Food, Food Packaging and Dairy Industries. In *Biological Synthesis of Nanoparticles and Their Applications*; CRC Press: Boca Raton, FL, USA, 2019; pp. 145–158. ISBN 0429265239.
38. Pandian, S.R.K.; Deepak, V.; Kalishwaralal, K.; Gurunathan, S. Biologically synthesized fluorescent CdS NPs encapsulated by PHB. *Enzym. Microb. Technol.* **2011**, *48*, 319–325. [CrossRef]
39. Kowshik, M.; Vogel, W.; Urban, J.; Kulkarni, S.; Paknikar, K. Microbial Synthesis of Semiconductor PbS Nanocrystallites. *Adv. Mater.* **2002**, *14*, 815–818. [CrossRef]
40. Pordanjani, A.H.; Aghakhani, S.; Afrand, M.; Sharifpur, M.; Meyer, J.P.; Xu, H.; Ali, H.M.; Karimi, N.; Cheraghian, G. Nanofluids: Physical phenomena, applications in thermal systems and the environment effects—A critical review. *J. Clean. Prod.* **2021**, *320*, 128573. [CrossRef]
41. Cuenya, B.R. Synthesis and catalytic properties of metal nanoparticles: Size, shape, support, composition, and oxidation state effects. *Thin Solid Film.* **2010**, *518*, 3127–3150. [CrossRef]
42. Fernández-Llamosas, H.; Castro, L.; Blázquez, M.L.; Díaz, E.; Carmona, M. Biosynthesis of selenium nanoparticles by *Azoarcus* sp. CIB. *Microb. Cell Factories* **2016**, *15*, 109. [CrossRef]
43. Khatoon, N.; Mazumder, J.A.; Sardar, M. Biotechnological Applications of Green Synthesized Silver Nanoparticles. *J. Nanosci. Curr. Res.* **2017**, *2*, 2572-0813. [CrossRef]
44. Zhang, Y.; Chu, W.; Foroushani, A.D.; Wang, H.; Li, D.; Liu, J.; Barrow, C.J.; Wang, X.; Yang, W. New Gold Nanostructures for Sensor Applications: A Review. *Materials* **2014**, *7*, 5169–5201. [CrossRef] [PubMed]
45. Moghaddam, A.B.; Namvar, F.; Moniri, M.; Tahir, P.M.; Azizi, S.; Mohamad, R. Nanoparticles Biosynthesized by Fungi and Yeast: A Review of Their Preparation, Properties, and Medical Applications. *Molecules* **2015**, *20*, 16540–16565. [CrossRef] [PubMed]
46. Xu, L.; Wu, X.-C.; Zhu, J.-J. Green preparation and catalytic application of Pd nanoparticles. *Nanotechnology* **2008**, *19*, 305603. [CrossRef] [PubMed]
47. Sathyavathi, R.; Krishna, M.B.M.; Rao, D.N. Biosynthesis of Silver Nanoparticles Using Moringa oleifera Leaf Extract and Its Application to Optical Limiting. *J. Nanosci. Nanotechnol.* **2011**, *11*, 2031–2035. [CrossRef] [PubMed]
48. Sripriya, J.; Anandhakumar, S.; Achiraman, S.; Antony, J.J.; Siva, D.; Raichur, A.M. Laser receptive polyelectrolyte thin films doped with biosynthesized silver nanoparticles for antibacterial coatings and drug delivery applications. *Int. J. Pharm.* **2013**, *457*, 206–213. [CrossRef] [PubMed]
49. Khan, S.A.; Gambhir, S.; Ahmad, A. Extracellular biosynthesis of gadolinium oxide (Gd_2O_3) nanoparticles, their biodistribution and bioconjugation with the chemically modified anticancer drug taxol. *Beilstein J. Nanotechnol.* **2014**, *5*, 249–257. [CrossRef] [PubMed]

50. Kefeni, K.K.; Mamba, B.; Msagati, T. Application of spinel ferrite nanoparticles in water and wastewater treatment: A review. *Sep. Purif. Technol.* **2017**, *188*, 399–422. [CrossRef]
51. Sabir, S.; Arshad, M.; Chaudhari, S.K. Zinc Oxide Nanoparticles for Revolutionizing Agriculture: Synthesis and Applications. *Sci. World J.* **2014**, *2014*, 925494. [CrossRef]
52. Padmanabhan, P.; Kumar, A.; Kumar, S.; Chaudhary, R.K.; Gulyás, B. Nanoparticles in practice for molecular-imaging applications: An overview. *Acta Biomater.* **2016**, *41*, 1–16. [CrossRef]
53. Brenner, P.S.; Krakauer, T. Regulation of Inflammation: A Review of Recent Advances in Anti-Inflammatory Strategies. *Curr. Med. Chem. Anti-Allergy Agents* **2003**, *2*, 274–283. [CrossRef]
54. Yanai, H.; Matsuda, A.; An, J.; Koshiba, R.; Nishio, J.; Negishi, H.; Ikushima, H.; Onoe, T.; Ohdan, H.; Yoshida, N.; et al. Conditional ablation of HMGB1 in mice reveals its protective function against endotoxemia and bacterial infection. *Proc. Natl. Acad. Sci. USA* **2013**, *110*, 20699–20704. [CrossRef] [PubMed]
55. Yu, D.; Rao, S.; Tsai, L.M.; Lee, S.K.; He, Y.; Sutcliffe, E.L.; Srivastava, M.; Linterman, M.; Zheng, L.; Simpson, N.; et al. The Transcriptional Repressor Bcl-6 Directs T Follicular Helper Cell Lineage Commitment. *Immunity* **2009**, *31*, 457–468. [CrossRef] [PubMed]
56. O'Sullivan, D.; van der Windt, G.J.; Huang, S.C.-C.; Curtis, J.D.; Chang, C.-H.; Buck, M.; Qiu, J.; Smith, A.M.; Lam, W.Y.; DiPlato, L.M.; et al. Memory CD8^{+} T Cells Use Cell-Intrinsic Lipolysis to Support the Metabolic Programming Necessary for Development. *Immunity* **2014**, *41*, 75–88. [CrossRef] [PubMed]
57. Liu, Y.-C.; Zou, X.-B.; Chai, Y.-F.; Yao, Y.-M. Macrophage Polarization in Inflammatory Diseases. *Int. J. Biol. Sci.* **2014**, *10*, 520–529. [CrossRef]
58. Dunster, J.L. The macrophage and its role in inflammation and tissue repair: Mathematical and systems biology approaches. *WIREs Syst. Biol. Med.* **2015**, *8*, 87–99. [CrossRef]
59. Rasmussen, J.W.; Martinez, E.; Louka, P.; Wingett, D.G. Zinc oxide nanoparticles for selective destruction of tumor cells and potential for drug delivery applications. *Expert Opin. Drug Deliv.* **2010**, *7*, 1063–1077. [CrossRef]
60. Bisht, G.; Rayamajhi, S. ZnO Nanoparticles: A Promising Anticancer Agent. *Nanobiomedicine* **2016**, *3*, 9. [CrossRef]
61. Kuddus, S.A. Nanoceria and Its Perspective in Cancer Treatment. *J. Cancer Sci. Ther.* **2017**, *9*, 368–373. [CrossRef]
62. Gao, Y.; Gao, F.; Chen, K.; Ma, J.-L. Cerium oxide nanoparticles in cancer. *OncoTargets Ther.* **2014**, *7*, 835–840. [CrossRef]
63. Mollick, M.M.R.; Rana, D.; Dash, S.K.; Chattopadhyay, S.; Bhowmick, B.; Maity, D.; Mondal, D.; Pattanayak, S.; Roy, S.; Chakraborty, M.; et al. Studies on green synthesized silver nanoparticles using *Abelmoschus esculentus* (L.) pulp extract having anticancer (in vitro) and antimicrobial applications. *Arab. J. Chem.* **2019**, *12*, 2572–2584. [CrossRef]
64. Saratale, R.G.; Shin, H.S.; Kumar, G.; Benelli, G.; Kim, D.-S.; Saratale, G.D. Exploiting antidiabetic activity of silver nanoparticles synthesized using *Punica granatum* leaves and anticancer potential against human liver cancer cells (HepG2). *Artif. Cells Nanomed. Biotechnol.* **2017**, *46*, 211–222. [CrossRef] [PubMed]
65. Mukherjee, S.; Chowdhury, D.; Kotcherlakota, R.; Patra, S.; Vinothkumar, B.; Bhadra, M.P.; Sreedhar, B.; Patra, C.R. Potential Theranostics Application of Bio-Synthesized Silver Nanoparticles (4-in-1 System). *Theranostics* **2014**, *4*, 316–335. [CrossRef] [PubMed]
66. Espinosa, A.; Di Corato, R.; Kolosnjaj-Tabi, J.; Flaud, P.; Pellegrino, T.; Wilhelm, C. Duality of Iron Oxide Nanoparticles in Cancer Therapy: Amplification of Heating Efficiency by Magnetic Hyperthermia and Photothermal Bimodal Treatment. *ACS Nano* **2016**, *10*, 2436–2446. [CrossRef]
67. Kwon, Y.M.; Je, J.; Cha, S.H.; Oh, Y.; Cho, W.H. Synergistic combination of chemo-phototherapy based on temozolomide/ICG-loaded iron oxide nanoparticles for brain cancer treatment. *Oncol. Rep.* **2019**, *42*, 1709–1724. [CrossRef] [PubMed]
68. Bilici, K.; Muti, A.; Sennaroğlu, A.; Acar, H.Y. Indocyanine Green Loaded APTMS Coated SPIONs for Dual Phototherapy of Cancer. *J. Photochem. Photobiol. B Biol.* **2019**, *201*, 111648. [CrossRef] [PubMed]
69. Geetha, R.; Ashokkumar, T.; Tamilselvan, S.; Govindaraju, K.; Sadiq, M.; Singaravelu, G. Green synthesis of gold nanoparticles and their anticancer activity. *Cancer Nanotechnol.* **2013**, *4*, 91–98. [CrossRef]
70. Fazal, S.; Jayasree, A.; Sasidharan, S.; Koyakutty, M.; Nair, S.V.; Menon, D. Green Synthesis of Anisotropic Gold Nanoparticles for Photothermal Therapy of Cancer. *ACS Appl. Mater. Interfaces* **2014**, *6*, 8080–8089. [CrossRef]
71. Parida, U.K.; Biswal, S.K.; Bindhani, B.K. Green Synthesis and Characterization of Gold Nanoparticles: Study of Its Biological Mechanism in Human SUDHL-4 Cell Line. *Adv. Biol. Chem.* **2014**, *04*, 360–375. [CrossRef]
72. Janagam, D.R.; Wu, L.; Lowe, T.L. Nanoparticles for drug delivery to the anterior segment of the eye. *Adv. Drug Deliv. Rev.* **2017**, *122*, 31–64. [CrossRef]
73. Rajan, M.; Raj, V. Potential Drug Delivery Applications of Chitosan Based Nanomaterials. *Int. Rev. Chem. Eng.* **2013**, *5*, 145–155.
74. Dev, A.; Binulal, N.S.; Anitha, A.; Nair, S.V.; Furuike, T.; Tamura, H.; Jayakumar, R. Preparation of Poly (Lactic Acid)/Chitosan Nanoparticles for Anti-HIV Drug Delivery Applications. *Carbohydr. Polym.* **2010**, *80*, 833–838. [CrossRef]
75. Siddiqi, K.S.; Rahman, A.U.; Tajuddin; Husen, A. Biogenic Fabrication of Iron/Iron Oxide Nanoparticles and Their Application. *Nanoscale Res. Lett.* **2016**, *11*, 498. [CrossRef] [PubMed]
76. Malekigorji, M.; Curtis, A.D.M.; Hoskins, C. The Use of Iron Oxide Nanoparticles for Pancreatic Cancer Therapy. *J. Nanomed. Res.* **2014**, *1*, 00004. [CrossRef]
77. Wu, L.; Cai, X.; Nelson, K.; Xing, W.; Xia, J.; Zhang, R.; Stacy, A.J.; Luderer, M.; Lanza, G.M.; Wang, L.V. A Green Synthesis of Carbon Nanoparticles from Honey and Their Use in Real-Time Photoacoustic Imaging. *Nano Res.* **2013**, *6*, 312–325. [CrossRef] [PubMed]

78. Huang, H.; Xu, Y.; Tang, C.-J.; Chen, J.-R.; Wang, A.-J.; Feng, J.-J. Facile and green synthesis of photoluminescent carbon nanoparticles for cellular imaging. *New J. Chem.* **2014**, *38*, 784–789. [CrossRef]
79. Bhunia, S.K.; Saha, A.; Maity, A.; Ray, S.C.; Jana, N.R. Carbon Nanoparticle-based Fluorescent Bioimaging Probes. *Sci. Rep.* **2013**, *3*, 1473. [CrossRef]
80. Raja, S.; Ramesh, V.; Thivaharan, V. Green biosynthesis of silver nanoparticles using *Calliandra haematocephala* leaf extract, their antibacterial activity and hydrogen peroxide sensing capability. *Arab. J. Chem.* **2017**, *10*, 253–261. [CrossRef]
81. Zheng, Y.; Wang, Z.; Peng, F.; Fu, L. Application of biosynthesized ZnO nanoparticles on an electrochemical H_2O_2 biosensor. *Braz. J. Pharm. Sci.* **2016**, *52*, 781–786. [CrossRef]
82. Varun, S.; Daniel, S.K.; Gorthi, S.S. Rapid sensing of melamine in milk by interference green synthesis of silver nanoparticles. *Mater. Sci. Eng. C* **2017**, *74*, 253–258. [CrossRef]
83. Maiti, S.; Barman, G.; Laha, J.K. Detection of heavy metals (Cu^{+2}, Hg^{+2}) by biosynthesized silver nanoparticles. *Appl. Nanosci.* **2016**, *6*, 529–538. [CrossRef]
84. Hoyos, L.E.S.-D.; Sánchez-Mendieta, V.; Vilchis-Nestor, A.R.; Camacho-López, M.A. Biogenic Silver Nanoparticles as Sensors of Cu^{2+} and Pb^{2+} in Aqueous Solutions. *Univers. J. Mater. Sci.* **2017**, *5*, 29–37. [CrossRef]
85. Rojas-Perez, A.; Adorno, L.; Cordero, M.; Ruiz, A.; Mercado-Diaz, Z.; Rodriguez, A.; Betancourt, L.; Velez, C.; Feliciano, I.; Cabrea, C. Biosynthesis of Gold Nanoparticles Using Osmudaria Obtusiloba Extract and Their Potential Use in Optical Sensing Application. *Austin J. Biosens. Bioelectron.* **2015**, *1*, 1–9.
86. Praveena, V.D.; Kumar, K.V. A Thiocyanate Sensor Based on Ecofriendly Silver Nanoparticles Thin Film Composite. *Int. J. Innov. Sci. Eng. Technol.* **2015**, *2*, 741–752.
87. Tiquia-Arashiro, S.; Rodrigues, D. Thermophiles and Psychrophiles in Nanotechnology. In *Extremophiles: Applications in Nanotechnology*; Springer: Berlin/Heidelberg, Germany, 2016; pp. 89–127.
88. Saravanan, A.; Kumar, P.S.; Yashwanthraj, M. Sequestration of toxic Cr(VI) ions from industrial wastewater using waste biomass: A review. *Desalination Water Treat.* **2017**, *68*, 245–266. [CrossRef]
89. Jayaprakash, K.; Govarthanan, M.; Mythili, R.; Selvankumar, T.; Chang, Y.-C. Bioaugmentation and Biostimulation Remediation Technologies for Heavy Metal Lead Contaminant. *Microb. Biodegrad. Xenobiotic Compd.* **2019**, 24–36. [CrossRef]
90. Chidambaram, R. Isotherm Modelling, Kinetic Study and Optimization of Batch Parameters Using Response Surface Methodology for Effective Removal of Cr(VI) Using Fungal Biomass. *PLoS ONE* **2015**, *10*, e0116884. [CrossRef]
91. Samuel, M.S.; Chidambaram, R. Hexavalent chromium biosorption studies using *Penicillium griseofulvum* MSR1 a novel isolate from tannery effluent site: Box–Behnken optimization, equilibrium, kinetics and thermodynamic studies. *J. Taiwan Inst. Chem. Eng.* **2015**, *49*, 156–164. [CrossRef]
92. Bhandari, G. Environmental Nanotechnology: Applications of Nanoparticles for Bioremediation. In *Approaches in Bioremediation*; Springer: Berlin/Heidelberg, Germany, 2018; pp. 301–315.
93. El-Dafrawy, S.M.; Fawzy, S.; Hassan, S.M. Preparation of Modified Nanoparticles of Zinc Oxide for Removal of Organic and Inorganic Pollutant. *Trends Appl. Sci. Res.* **2016**, *12*, 1–9. [CrossRef]
94. Butnariu, I.C.; Stoian, O.; Voicu, Ş.; Iovu, H.; Paraschiv, G. Nanomaterials Used in Treatment of Wastewater: A Review. *Ann. Fac. Eng. Hunedoara* **2019**, *17*, 175–179.
95. Abigail, M.E.A.; Samuel, S.M.; Ramalingam, C. Addressing the environmental impacts of butachlor and the available remediation strategies: A systematic review. *Int. J. Environ. Sci. Technol.* **2015**, *12*, 4025–4036. [CrossRef]
96. Samuel, M.S.; Bhattacharya, J.; Parthiban, C.; Viswanathan, G.; Singh, N.P. Ultrasound-assisted synthesis of metal organic framework for the photocatalytic reduction of 4-nitrophenol under direct sunlight. *Ultrason. Sonochem.* **2018**, *49*, 215–221. [CrossRef] [PubMed]
97. Samuel, M.S.; Shah, S.S.; Subramaniyan, V.; Qureshi, T.; Bhattacharya, J.; Singh, N.P. Preparation of graphene oxide/chitosan/ferrite nanocomposite for Chromium(VI) removal from aqueous solution. *Int. J. Biol. Macromol.* **2018**, *119*, 540–547. [CrossRef] [PubMed]
98. Chidambaram, R. Application of rice husk nanosorbents containing 2,4-dichlorophenoxyacetic acid herbicide to control weeds and reduce leaching from soil. *J. Taiwan Inst. Chem. Eng.* **2016**, *63*, 318–326. [CrossRef]
99. Needhidasan, S.; Ramalingam, C. Stratagems employed for 2,4-dichlorophenoxyacetic acid removal from polluted water sources. *Clean Technol. Environ. Policy* **2017**, *19*, 1607–1620. [CrossRef]
100. Needhidasan, S.; Samuel, M.; Chidambaram, R. Electronic Waste—An Emerging Threat to the Environment of Urban India. *J. Environ. Health Sci. Eng.* **2014**, *12*, 36. [CrossRef]
101. Samuel, M.S.; Bhattacharya, J.; Raj, S.; Santhanam, N.; Singh, H.; Singh, N.P. Efficient removal of Chromium(VI) from aqueous solution using chitosan grafted graphene oxide (CS-GO) nanocomposite. *Int. J. Biol. Macromol.* **2019**, *121*, 285–292. [CrossRef]
102. Samuel, M.S.; Subramaniyan, V.; Bhattacharya, J.; Parthiban, C.; Chand, S.; Singh, N.P. A GO-CS@MOF [Zn(BDC)(DMF)] material for the adsorption of chromium(VI) ions from aqueous solution. *Compos. Part B Eng.* **2018**, *152*, 116–125. [CrossRef]
103. Hameed, A.; Fatima, G.R.; Malik, K.; Muqadas, A.; Fazalur-Rehman, M. Scope of Nanotechnology in Cosmetics: Dermatology and Skin Care Products. *J. Med. Chem. Sci.* **2019**, *2*, 9–16.
104. Khajehei, F.; Piatti, C.; Graeff-Hönninger, S. Novel Food Technologies and Their Acceptance. In *Food Tech Transitions*; Springer: Berlin/Heidelberg, Germany, 2019; pp. 3–22.

105. Ma, H.; Chen, Z.; Gao, X.; Liu, W.; Zhu, H. 3D hierarchically gold-nanoparticle-decorated porous carbon for high-performance supercapacitors. *Sci. Rep.* **2019**, *9*, 17065. [CrossRef]
106. Du, W.; Zhu, Z.; Wang, Y.; Liu, J.; Yang, W.; Qian, X.; Pang, H. One-step synthesis of $CoNi_2S_4$ nanoparticles for supercapacitor electrodes. *RSC Adv.* **2014**, *4*, 6998–7002. [CrossRef]

catalysts

Review

Biological Synthesis of Nanocatalysts and Their Applications

Arpita Roy [1,*], Amin Elzaki [2], Vineet Tirth [3,4], Samih Kajoak [2], Hamid Osman [2], Ali Algahtani [3,4], Saiful Islam [5], Nahla L. Faizo [2], Mayeen Uddin Khandaker [6], Mohammad Nazmul Islam [7], Talha Bin Emran [8] and Muhammad Bilal [9]

1 Department of Biotechnology, School of Engineering & Technology, Sharda University, Greater Noida 201310, India
2 Department of Radiological Sciences, College of Applied Medical Sciences, Taif University, Taif 21944, Makkah, Saudi Arabia; a.zaki@tu.edu.sa (A.E.); s.kajoak@tu.edu.sa (S.K.); ha.osman@tu.edu.sa (H.O.); drnfaizo@hotmail.com (N.L.F.)
3 Mechanical Engineering Department, College of Engineering, King Khalid University, Abha 61411, Asir, Saudi Arabia; vtirth@kku.edu.sa (V.T.); alialgahtani@kku.edu.sa (A.A.)
4 Research Center for Advanced Materials Science (RCAMS), King Khalid University Guraiger, Abha 61413, Asir, Saudi Arabia
5 Civil Engineering Department, College of Engineering, King Khalid University, Abha 61413, Asir, Saudi Arabia; sfakrul@kku.edu.sa
6 Centre for Applied Physics and Radiation Technologies, School of Engineering and Technology, Sunway University, Bandar Sunway, Petaling Jaya 47500, Selangor, Malaysia; mayeenk@sunway.edu.my
7 Department of Pharmacy, International Islamic University Chittagong, Chittagong 4318, Bangladesh; nazmul@iiuc.ac.bd
8 Department of Pharmacy, BGC Trust University Bangladesh, Chittagong 4381, Bangladesh; talhabmb@bgctub.ac.bd
9 Huaiyin Institute of Technology, School of Life Science and Food Engineering, Huai'an 223003, China; bilaluaf@hotmail.com
* Correspondence: arbt2014@gmail.com

Citation: Roy, A.; Elzaki, A.; Tirth, V.; Kajoak, S.; Osman, H.; Algahtani, A.; Islam, S.; Faizo, N.L.; Khandaker, M.U.; Islam, M.N.; et al. Biological Synthesis of Nanocatalysts and Their Applications. *Catalysts* **2021**, *11*, 1494. https://doi.org/10.3390/catal11121494

Academic Editor: Beom Soo Kim

Received: 7 November 2021
Accepted: 5 December 2021
Published: 8 December 2021

Abstract: Over the past few decades, the synthesis and potential applications of nanocatalysts have received great attention from the scientific community. Many well-established methods are extensively utilized for the synthesis of nanocatalysts. However, most conventional physical and chemical methods have some drawbacks, such as the toxicity of precursor materials, the requirement of high-temperature environments, and the high cost of synthesis, which ultimately hinder their fruitful applications in various fields. Bioinspired synthesis is eco-friendly, cost-effective, and requires a low energy/temperature ambient. Various microorganisms such as bacteria, fungi, and algae are used as nano-factories and can provide a novel method for the synthesis of different types of nanocatalysts. The synthesized nanocatalysts can be further utilized in various applications such as the removal of heavy metals, treatment of industrial effluents, fabrication of materials with unique properties, biomedical, and biosensors. This review focuses on the biogenic synthesis of nanocatalysts from various green sources that have been adopted in the past two decades, and their potential applications in different areas. This review is expected to provide a valuable guideline for the biogenic synthesis of nanocatalysts and their concomitant applications in various fields.

Keywords: green source; nanocatalysts; nanoparticles; bacteria; fungi; algae; applications

1. Introduction

Nanotechnology has evolved as a highly technical research arena with potential applications in all spheres of life. The term "nano" has been derived from the Greek for "dwarf." With a clear idea of the extremity of something in a nano, a nanoparticle can be defined as particles that have at least one dimension below 100 nm [1]. Several bulk materials show completely different properties when they are studied on the nanoscale. One known reason for this phenomenon is their higher surface-to-volume aspect ratio.

For different nanoparticles, this can result in a variety of characteristics. For example, the higher aspect ratio of silver nanoparticles allows them to have increased efficacy in antibacterial properties. Consequently, silver nanoparticles can have diverse applications in cosmetics, packaging, electronics, coatings, and biotechnology [2,3]. A unique property of nanoparticles is that they have the ability to combine or form a solid at lower temperatures without melting. This property helps to achieve improved coatings on capacitors and other electronic components. Nanoparticles are also transparent, which allows them to be utilized in packaging, coating, and cosmetics (e.g., scratchproof eyeglasses, crack-resistant paints). When metallic nanoparticles are attached to single-stranded DNA, they can travel through the bloodstream and confine any target organ. This characteristic can be exploited in medical diagnostics, therapeutics, and other biomedical applications. Due to their significant potentials, nanoparticles must be further studied to find unexplored uses in everyday life [4].

Nanocatalysts are usually differentiated based on their dimension, composition, morphology, material nature, agglomeration, and uniformity. The morphology and shape of nanoparticles have vital functions, such as their toxic effect on mankind or the environment [5]. On the basis of dimension, nanoparticles can be one-dimensional, two-dimensional, and three-dimensional. Thin-film coatings used in sensors and electronic devices come under 1D, whereas carbon nanotubes, wires, fibers, etc. belong to 2D nanoparticles. Three dimensional nanoparticles include quantum dots, hollow spheres, and dendrimers. On the basis of morphology, they can be spherical, flat, crystalline, cubic, etc. structures and present in either single or composite form.

Numerous physical and chemical approaches can be effectively utilized for nanocatalyst synthesis. These include aerosol technologies, microemulsion, microwave, laser ablation, lithography, photochemical reduction, ion sputtering, sol–gel, sonochemical, ultrasonic spark discharge, and template synthesis [6]. However, most approaches have some nonnegligible drawbacks as these processes use expensive and hazardous chemicals and consume a lot of energy. Chemical synthesis has proved to be useful and can be used for a long time, but they have certain demerits such as the aggregation of particles when allowed to react for a long time, instability of the final product, and improper control of crystal growth [7]. Moreover, this method is not environmentally friendly, as a lot of toxic wastes and pollutants are generated as by-products. In particular, both physical and chemical techniques produce harmful pollutants such as harmful capping and reducing agents and organic solvents. Therefore, the use of harmful chemicals and organic solvents involved in nanomaterial synthesis should be reduced [8]. Hence, both conventional methods of nanoparticle synthesis, i.e., physical and chemical methods, have evolved as costly and are not friendly to the ecosystem. The demerits of these methods lead to the development of novel methods for the synthesis of nanomaterials that should be environmentally friendly, cheap, nonhazardous, clean, and energy-efficient [9]. Recently, the focus has shifted to the utilization of biological agents for the synthesis of nanomaterials due to their various advantages as compared to chemical and physical ones. Biological methods of synthesis are generally utilized by biological entities such as algae, fungi, and bacteria [10].

There are different groups of nanoparticles available, which include carbon-based nanoparticles, ceramic nanoparticles, semiconductor nanoparticles, metal/metal oxide nanoparticles, and polymeric nanoparticles [11–17]. Metal/metal oxide nanoparticles have gained significant interest due to their wide range of applications such as the detection and imaging of biomolecules, antimicrobial activity, removal of environmental pollutants, and bioanalytical applications [11]. These nanoparticles are prepared from the metal/metal oxide precursors. Metal/metal oxide nanoparticles include silver, copper, gold, titanium oxide, iron oxide, and zinc oxide [11]. They can be synthesized by chemical, physical, electrochemical, or photochemical approaches. However, due to their negative impact, biological methods have been currently in demand. Therefore, in this review, an overview of different methods of synthesis, and the use of various biological agents such as algae, fungi, and bacteria, which are used for the synthesis of metal/metal oxide nanocatalysts, is discussed. Further discussion of the application of nanocatalysts in different sectors is conducted.

2. Different Methods of Nanocatalyst Synthesis

Generally, there are two techniques for nanocatalyst synthesis: top-down and bottom-up. In the first method, the bulk material is broken down into smaller nanosized particles [18,19]. Various metallic nanoparticles are composed of top-down methods such as etching, sputtering, and laser ablation. On the other hand, the bottom-up method involves joining a molecule by a molecule, atom by atom, and clusters by cluster. In this method, single molecules are explored to form a complex structure of nanoscale size [20]. Various methods that use bottom-up techniques include supercritical fluid synthesis, plasma or flame spraying synthesis, laser pyrolysis, molecular condensation, the sol–gel process, chemical reduction, and green synthesis (Figure 1). In this technique, physicochemical reactions occur that may affect the properties of nanoparticles, and the nanoparticles are collected from smaller units. Therefore, both techniques are controlled by kinetic processes, which regulate the size and shapes of the synthesized nanoparticles.

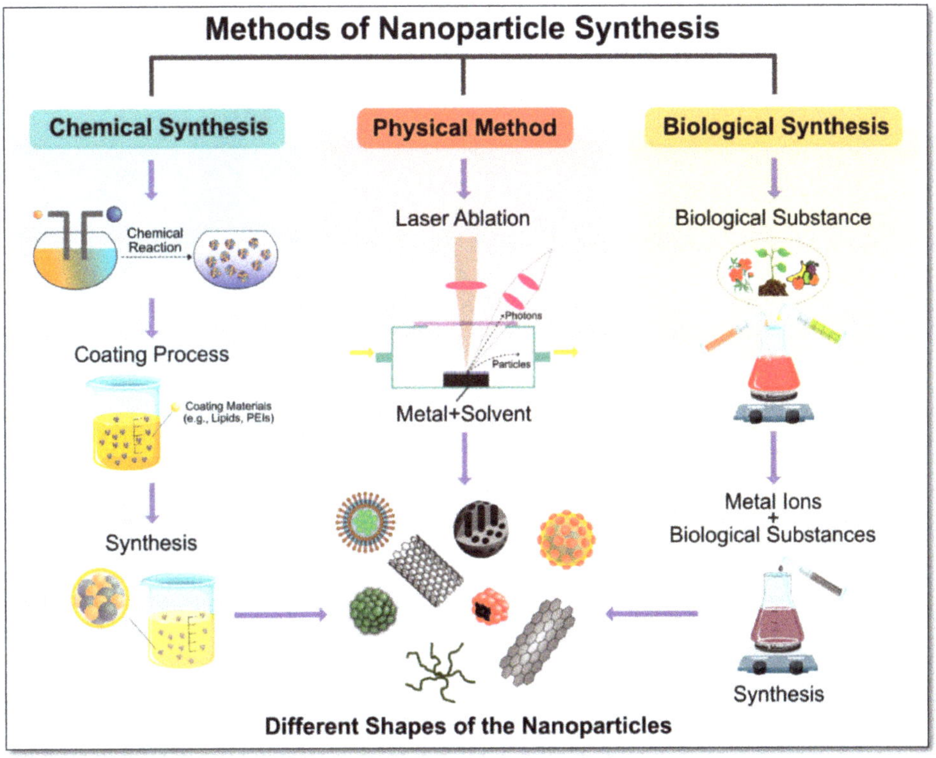

Figure 1. Method of nanoparticles synthesis [21].

3. Biological Approach for Nanocatalyst/Nanoparticle Synthesis

The biological method is preferred over the other two conventional methods (top-down and bottom-up) as it is a green method, environmentally friendly, and does not require a higher energy consumption [22]. Nanoparticles obtained through the biological approach have a greater specific surface area, increase the rate of catalysis, and have metal salt and improved enzymes [23]. Hence, the main objective in the synthesis of nanoparticles using a biological approach is to utilize cheap resources and facilitate a continuous production of nanoparticles. Biological sources that are used for nanoparticle synthesis provide a simple method and easy increase in biomass, ensuring a uniform particle size, as well as multiplication. The use of microbes is one of the most prominent methods among the biological approaches of nanoparticle synthesis. It utilizes different biological sources such as bacteria, fungi, and algae (Figure 2). Bacteria are the most commonly found organism in our biosphere. Under optimal conditions such as pH, temperature, and pressure, bacteria show the capability to synthesize various nanoparticles [24] (Figure 3). The ability of bacterial cells to survive and proliferate under extreme climatic conditions make them the most ideal organisms for nanoparticle synthesis. They can reproduce and multiply even under high metal concentrations, which may be due to particular resistance mechanisms.

Strains of bacteria that are not resistant to high metal concentrations can also be employed as appropriate microbes. The nanoparticles produced by microorganisms have important uses in the biological field such as bioleaching, biocorrosion, biomineralization, and bioremediation. In addition to bacteria, fungi and algae are two other green sources that are capable of synthesizing nanoparticles. Fungi have an outstanding ability for the synthesis of various bioactive compounds that have potential for numerous applications. They are widely used as reducing and stabilizing agents and can be easily grown on a large scale for the production of nanoparticles with controlled shape and size [25]. Similarly, algae have the ability to synthesize various bioactive compounds, pigments, and proteins, which help in the reduction of salts and act as capping agents in the synthesis mechanism [26].

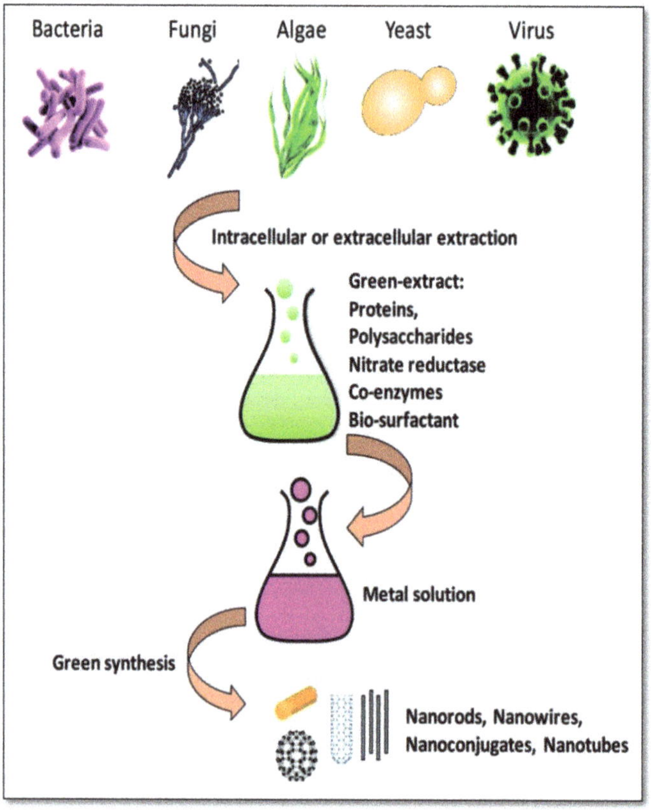

Figure 2. Biological approach to nanoparticle synthesis [26].

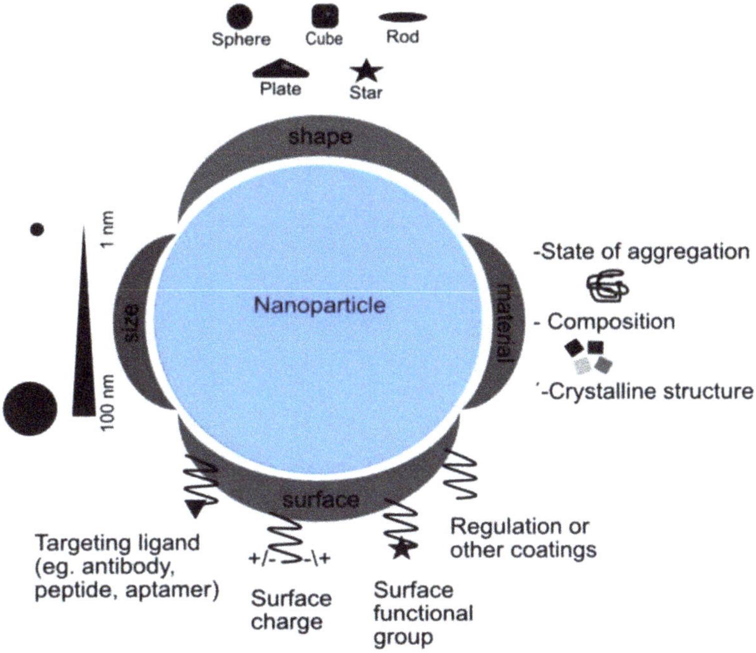

Figure 3. Different physiochemical properties of nanoparticles [24].

3.1. Bacterial Synthesis

In the vast field of biological resources, prokaryotic bacteria are the most researched for the synthesis of metallic nanoparticles. This is because they are relatively easy to manipulate compared to any other living organisms. Various researchers have shown the bacterial synthesis of nanoparticles that were responsible for the reduction of metal ions. The main benefit of bacterial synthesis is their higher reproduction with negligible uses of toxic chemicals. Nevertheless, there are some problems associated with the bacterial synthesis of nanoparticles such as the time-consuming culturing process, and difficulties in controlling the shape, size, and distribution. A study conducted with Lactobacillus strains that were extracted from common buttermilk showed a highly concentrated metallic ion. This process produced multiple, highly structured gold and silver nanoparticles. Lactobacillus was observed to synthesize nanoparticles within the plasma membrane and remained viable [27]. Ahmad et al. [28] reported the synthesis of gold nanoparticles using *Thermomonospora* sp.—an extremophilic actinomycetes strain and with another novel alkalotolerant actinomycete, i.e., *Rhodococcus* species. Another study reported the influence of the microbial synthesis of gold nanoparticles by changing the pH conditions of *Shewanella algae*. It was found that at pH 7, the nanoparticles synthesized ranged from 10 to 20 nm; however, when the pH was adjusted to 1, the size was modified to 50–500 nm [29]. In a study, *Rhodopseudomonas casulata* was used for the synthesis of gold nanoparticle synthesis. It was found that at pH 7, spherical-shaped nanoparticles were synthesized, but as the pH dropped to 4, the nanoparticles produced were plate-shaped [30]. Parikh et al. [31] reported the synthesis of silver nanoparticles using a bacterial strain, i.e., *Morganella* sp., which was isolated from an insect midgut. When exposed to silver nitrate, *Morganella* sp. synthesized crystalline silver nanoparticles extracellularly. Reddy et al. [32] reported the synthesis of silver and gold nanoparticles using Bacillus subtilis. They observed that silver nanoparticles were entirely synthesized extracellularly and formed after 7 days of the addition of silver ions, whereas gold nanoparticles were synthesized both extracellularly and intracellularly and formed after 1 day of addition of salt. Bruna et al. [33] reported the synthesis of CdS fluorescent nanoparticles using *Halobacillus* sp. They observed that the synthesized nanoparticles were hexagonal at 2–5 nm in size. Further details of nanoparticles synthesized using bacterial cultures are presented in Table 1.

Table 1. Nanoparticles synthesized from bacteria.

Source	Metal	Size (nm)	Shape	Location	References
Actinobacter spp.	M	10–40 (24 h) 50–150 (48 h)	Quasi-spherical cubic	E	[34]
Verticillium luteoalbum	G	100	Spherical, Triangular, Hexagonal	I and E	[35]
Bacillus selenitireducens	T	~10	Nanorods	E	[36]
Escherichia coli DH5a	G	25 ± 8	Spherical	I	[37]
Klebsiella pneumoniae	S	5–32	-	E	[38]
Bacillus licheniformis	S	50	-	E	[39]
Stenotrophomonas maltophilia	G	~40	-	I	[40]
Bacillus sp.	S	5–15	-	I	[41]
Bacillus megatherium D01	G	1.9 ± 0.8	Spherical	E	[42]
Shewanella algae	G	9.6	Spherical	E	[43]
Trichoderma viride	S	2–4	Spherical	E	[44]
Streptomyces sp.	S	10–100	Spherical	E	[45]
Bacillus cereus	S	10–30	Spherical	E	[46]
Pseudomonas aeruginosa	S	13	Spherical	E	[47]
Idiomarina sp. *PR58-8*	S	26	-	I	[48]
Pseudomonas fluorescens	G	50–70	Spherical	E	[49]
Vibrio alginolyticus	S	50–100	Spherical	I and E	[50]
Azospirillum brasilense	G	5–50	Nanospheres	E	[51]
Planomicrobium sp.	TO	8.89	Spherical	E	[52]
Salmonella typhirium	S	50–150	-	E	[53]
Geobacillus sp.	G	5–50	Quasi-hexagonal	I	[54]
Lactobacillus crispatus	T	70.98	-	E	[55]
Bacillus strain CS 11	S	42–94	Spherical	I and E	[56]
Pseudomonas fluorescens	C	49	Spherical, Hexagonal	E	[57]
Stereum hirsutum	C/CO	5–20	Spherical	E	[58]
Salmonella typhimurium	C	40–60	-	E	[59]
Bhargavaea indica	S	30–100	Pentagon, spherical, icosahedron, nanobar, hexagonal, truncated triangle, and triangular	E	[60]
Exiguobacterium mexicanum	S	5–40	-	I and E	[61]
Deinococcus radiodurans	G	43.75	Spherical, triangular, and irregular	I and E	[62]
Sporosarcina koreensis	G and S	-	Spherical	E	[63]
Bacillus brevis	S	41–68	Spherical	E	[64]
Alcaligenes sp.	S	30–50	Spherical	-	[65]
Marinobacter algicola	G	4–168	Spherical, triangular, pentagonal, and hexagonal	I	[66]
Paracoccus haeundaensis	G	20.93 ± 3.46	Spherical	E	[67]
Bacillus sp.	S	5–15	Spherical	-	[68]

S = silver, G = gold, C = copper, CO = copper oxide, T = titanium, TO = titanium oxide, TE = tellurium, M = magnetite, Z = zirconia, ZO = zinc oxide, IO = iron oxide, P = palladium, E = extracellular, I = intracellular.

3.2. Fungal Synthesis

The use of fungi as a biological agent to synthesize metal nanoparticles has become popular because they show some advantages over bacteria. The presence of mycelia that enhances the surface area of fungi, and the economic utility and simplicity of the scaleup and downstream processing of fungi offer significant merits in using fungi as an agent for the synthesis of nanoparticles [69]. Fungi are also able to produce various enzymes, which help in nanoparticles synthesis with different shapes and sizes. As a result of their larger biomass compared to that of bacteria, the production of nanoparticles is higher. Various fungal species such as *Fusarium oxysporum*, *Asperigillus oryzae*, *Verticillium luteoalbum*, *Alternata alternata*, and *Collitotrichum* sp. were utilized for nanoparticles synthesis. However, some of the drawbacks of using fungi include laborious and more costly downstream processes. Nanoparticle synthesis using fungal culture can be intracellular or extracellular. In intracellular synthesis, the metal precursor is added to the fungal culture and biomass internalization of the precursor. Nanoparticle extraction is performed by breaking cells using different methods such as chemical treatment, centrifugation, and filtration [70]. In the case of extracellular synthesis, aqueous filtrate that contains fungal bioactive compounds are mixed with the metal precursor; hence, the synthesis process occurs easily. Extracellular synthesis is the most commonly used technique [71]. Mukherjee et al. [72] reported the production of gold nanoparticles by *Verticillium* sp. where the intracellular gold nanoparticles were located on the mycelial surface. A study reported the use of *Fusarium oxysporum* for the synthesis of silver nanoparticles where pure silver nanoparticles were produced with size ranges from 5 to 15 nm, and their capping was performed in such a way that they could be stabilized via the fungal proteins [28]. Bhainsa and D'Souza [73] reported the use of *Aspergillus fumigatus* to produce extracellular silver nanoparticles of 5–25 nm in size. Riddin et al., [74] reported the synthesis of platinum nanoparticles by *F. oxysporum*. The intracellular and extracellular synthesis of the nanoparticles was observed, but the extracellular synthesis was more prominent, and the production of extracellular nanoparticles was found to be 5.66 mg/L. Rai et al. [75] used *Fusarium oxysporum* for the synthesis of zinc sulfide, sulfur, molybdenum sulfide, cadmium sulfide, and lead sulfide nanoparticles. Sanghi et al. [76] used *Coriolus versicolor* for the synthesis of intracellular silver nanoparticles. When the reaction conditions were changed, it was observed that both the extracellular and intracellular synthesis of nanoparticles could be performed by the fungus. Vahabi et al. [77] used *Trichoderma reesei* for the synthesis of extracellular silver nanoparticles and found size ranges of 5–50 nm. Castro-Longoria et al. [78] reported the production of platinum nanoparticles by *Neurospora crassa*. Intracellular platinum nanoparticles were in the size range of 4–35 nm and spherical. Arun et al. [79] reported the synthesis of silver nanoparticles using *Schizophyllum commune* and found that the synthesized nanoparticles were spherical with sizes ranging from 54 to 99 nm. Gudikandula et al. [71] used 55 strains of white rot fungi (basidomycetes) for the synthesis of silver nanoparticles. They found that the synthesized nanoparticles were 15–25 nm in size with a spherical to round shape. Molnár et al. [70] used 29 different thermophilic filamentous fungi for the synthesis of gold nanoparticles, and the mechanism was intracellular where the synthesized nanoparticles were 1–80 nm in size with a hexagonal and spherical shape. A study reported the synthesis of silver nanoparticles using *Botryosphaeria rhodiana*, and it found that the synthesized nanoparticles were spherical with sizes ranging from 2 to 50 nm [80]. More details of the nanoparticles synthesized using fungal cultures are presented in Table 2.

Table 2. Nanoparticles synthesized from fungus.

Source	Metal	Size (nm)	Shape	Location	Reference
Fusarium oxysporum	G	20–40	Spherical; triangular	E	[81]
Phoma sp. 3.2883	S	71.06–74.46	-	E	[82]
Fusarium oxysporum	Z	3–11	Regular	E	[83]
Trichothecium sp.	G	10–25	Hexagonal, triangular	E	[84]
Fusarium oxysporum	M	20–50	Quasi-spherical	E	[85]
Phaenerochaete chrysosporium	S	50–200	Pyramid	E	[86]
Fusarium oxysporum	S	1.6	Spherical	E	[87]
Trichoderma asperellum	S	13–18	-	E	[88]
Fusarium acuminatum	S	5–40	Spherical	E	[89]
Rhizopus oryzae	G	10	Spherical	-	[90]
Aspergillus clavatus	S	10–25	Spherical, hexagonal	E	[91]
Aspergillus clavatus	G	20–35	nanotriangle	I	[92]

Table 2. *Cont.*

Source	Metal	Size (nm)	Shape	Location	Reference
Rhizopus stolonier	S	5–50	Spherical	E	[93]
Aspergillus terreus	S	1–20	Spherical	E	[94]
Aspergillus fumigatus	ZO	1.2–6.8	Spherical	E	[95]
Macrophomina phaseolina	S	5–40	Spherical	E	[96]
Penicillium chrysogenum	S	19–60	Spherical	E	[97]
Penicillium nalgiovense	S	15.2 ± 2.6	Spherical	E	[98]
Aspergillus flavus	TO	12–15	-	E	[99]
Trichoderma viride	S	1–50	Globular	-	[100]
Phoma exigua	S	22	Spherical	E	[101]
Phenerochaete chrysosporium	S	34–90	Spherical, oval	E	[102]
Beauveria bassiana	S	10–50	Circular, triangular, hexagonal	E	[103]
Cladosporium cladosporioides	S	30–60	Spherical	E	[104]
Phomopsis helianthin	S	5–60	Spherical, hexagonal	E	[105]
Fusarium solani	G	40–45	Needle, flower-like	-	[106]
Aspergillus niger	IO	20–40	Flake	E	[107]

S = silver, G = gold, C = copper, CO = copper oxide, T = titanium, TO = titanium oxide, TE = tellurium, M = magnetite, Z = zirconia, ZO = zinc oxide, IO = iron oxide, P = palladium, E = extracellular, I = intracellular.

3.3. Algal Synthesis

Algae are an economically important group of organisms and are unicellular or multicellular. They are present in various environments such as marine water and freshwater. They are classified into macroalgae and microalgae and are used for various commercial purposes. They possess various advantages such as less toxicity, requiring low temperature for synthesis. The production of nanoparticles by algae involves three major steps, i.e., the algal extract is obtained by boiling or heating algae in an organic solvent or water for a fixed period. Then, molar solutions of ionic metal compounds are prepared. Lastly, the molar solutions of the ionic metal compounds and the algal extract solution are mixed and incubated under controlled culture conditions with continuous mixing or without mixing for a fixed period [108]. The production of metallic nanoparticles depends largely on the algal species being used and its concentration. The reduction of metal ions is carried out by various biomolecules, such as peptides, polysaccharides, and pigments. The aqueous solution of metal nanoparticles is capped and stabilized with the help of cysteine residues and amino groups present in different proteins or by polysaccharides having a sulfur group [109]. Production of nanoparticles using algae occurs at a faster rate as compared to the synthesis using other biological agents. Various seaweeds such as *Ulva faciata*, *Sargassum wightii*, and *Chaetomorpha linum* have been employed to produce silver nanoparticles with various sizes and shapes (Table 3). Marine algae species are rarely studied for the production of nanoparticles. *Chlorella vulgaris* can bind to tetrachloroaurate ions firmly to reduce the bonding of gold to Au (0). Gold bound to the algal species was converted into a metallic state in almost 90% of cases, and gold crystals were deposited inside and outside the cells and they had icosahedral, decahedral, and tetrahedral structures [110]. Mata et al. [111] reported a reduction of Au (III) to Au (0) using the biomass of the brown alga *Fucus vesiculosus*, and the synthesized nanoparticles were spherical. Shakibaie et al. [112] used a marine green microalgae *Tetraselmis suecica* for the synthesis of gold nanoparticles and found that the synthesized nanoparticles were spherical with a size range of 51–59 nm. A study on the in vitro and in vivo synthesis of silver nanoparticles using *Chlamydomonas reinhardtii* was reported in [113]. It was found that the in vitro synthesis was slower and produced round-shaped nanoparticles 5–15 nm in size, whereas, in the case of in vivo synthesis, it was moderately faster and produced rectangular-shaped nanoparticles 5–35 nm in size [113]. A study reported the synthesis of silver nanoparticles using *Spirulina plantesis*, where the average size of the nanoparticles was approximately 12 nm [114]. Senapati et al. [115] reported the synthesis of gold nanoparticles using *T. kochinensis* and it was found that the size of the synthesized nanoparticles was 18 nm. A study reported that *Spirogyra submaxima* was able to convert Au^{3+} to Au^0 ions, which means that they were able to synthesize gold nanoparticles that were spherical, hexagonal, and triangular-shaped with a size of 2–50 nm [116]. Suganya et al. [117] reported the synthesis of gold nanoparticles using *Spirulina platensis*, and they were uniform in shape with an average size of 5 nm. The *Amphiroa fragilissima* was used in a separate study for the synthesis of silver nanoparticles, and it was found that crystalline nanoparticles were produced. Arsiya et al. [118] used *Chlorella vulgaris* for palladium nanoparticle synthesis, and the synthesized nanoparticles were crystalline with a size range of 5–20 nm. Sanaeimeh et al. [119] used *Sargassum muticum* for the synthesis of zinc oxide nanoparticles and evaluated its potential anticancer activity against liver cancer cell lines. A study reported the synthesis of silver nanoparticles using *Portieria hornemannii*, where the nanoparticles were spherical in shape with a size between 70 and 75 nm [120]. There are different shapes of nanomaterials are produced by differently bacterial synthesis (Figure 4).

Table 3. Nanoparticles synthesized from algae.

Source	Metal Involved	Size	Shape	References
Sargassum spp.	G	-	Hexagonal, triangular	[121]
Sargassum wightii	G	8–12	Thin planar structures	[109]
Sargassum Wightii	S	8–27	Spherical	[122]
Gelidiella acerosa	S	22	Spherical	[123]
Stoechospermum marginatum	G	18.7–93.7	Spherical	[124]
Ulva fasciata	S	28–41	Spherical	[124]
Sargassum myriocystum	G	10–23	Triangular and spherical	[125]
Ulva reticulata	S	40–50	Spherical	[126]
Chaetomorpha linum	S	3–44	-	[127]
Gracilaria corticate	G	45–57	-	[128]
Bifurcaria bifurcate	CO	5–45	Spherical	[129]
Enteromorpha flexuosa	S	2–32	Circular	[130]
Prasiola crispa	G	5–25	Cubic	[131]
Sargassum Alga	P	5–10	Octahedral	[132]
Caulerpa racemose	S	5–25	Spherical	[133]
Acanthophora specifera	S	33–81	Cubic	[134]
Isochrysis sp.	S	98.1–193	Spherical	[135]
Laurencia papillosa	S	-	Cubic	[136]
Spirulina platensis	S	5–50	Spherical	[137]
Caulerpa serrulate	S	10	Spherical	[138]
Botryococcus braunii	S	40–100	Cubical, spherical, and truncated triangular	[139]
Padina pavonia	S	49.58–86.37	Spherical, triangular, rectangle, polyhedral, and hexagonal	[140]
Gelidium amansii	S	27–54	Spherical	[141]
Padina sp.	S	25–60	Spherical	[142]
Gelidium corneum	S	20–50	Spherical	[143]

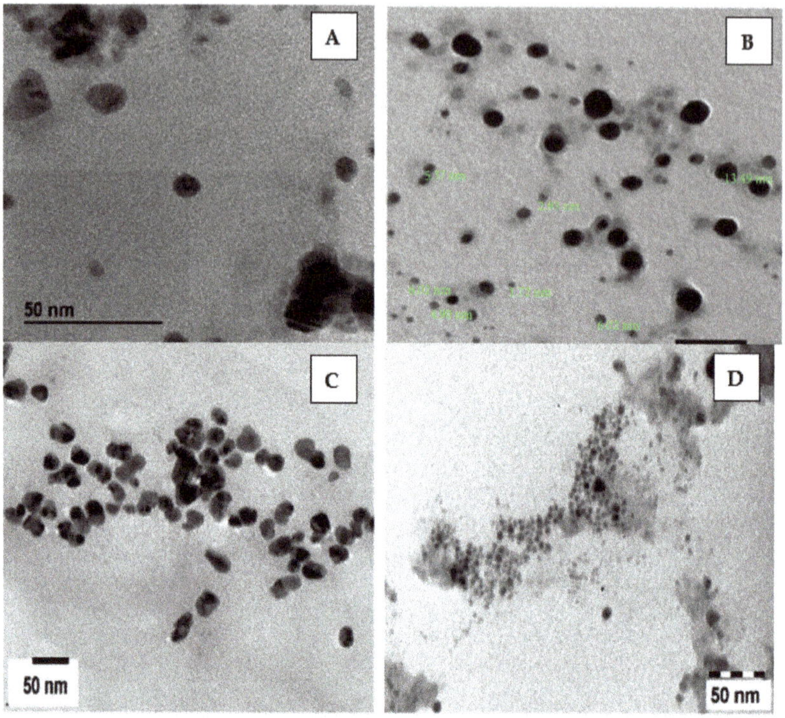

Figure 4. TEM images of different bacterial synthesized nanoparticles. (**A**) Silver nanoparticles (from *Pseudoduganella eburnea*); (**B**) copper oxide nanoparticles (from *Streptomyces*); (**C**) silver nanoparticles (from *Sargassum incisifolium*), and (**D**) gold nanoparticles (from *Sargassum incisifolium*) [144–146].

4. Applications of Nanomaterials/Nanocatalysts

4.1. Nanocatalysts in Biological Applications

Biologically synthesized nanoparticles have been extensively used in various applications (Figure 5). Silver and gold nanoparticles were also found to generally be used as antimicrobial agents against several microorganisms. They also possess anti-cancerous, anti-viral, antimalarial, and antifungal activities [1,3]. In addition to biomedical applications, they are also used in electronics, optics, cosmetics, coatings, sensing devices, therapeutics, environmental health, and chemical industries. [12]. They have appeared as a new drug delivery system for drug and gene transportation. A study reported that silver nanoparticles with a different shape can show varied antimicrobial activity due to their different surface area and active faces [147]. Mishra et al. [148] reported the synthesis of gold nanoparticles mediated by *Penicillium brevicompactum* and evaluated its potential role against mouse mayo blast cancer cells. Chauhan et al. [149] reported the synthesis of gold nanoparticles using *Candida albicans* and evaluated its anticancer potential against liver cancer cells. In another study, silver nanoparticles synthesized from *Stoechospermum marginatum* were used to evaluate the antibacterial activity against *Enterobacter faecalis*. It was found to have a higher antibacterial activity compared to tetracycline, whereas, in the case of *E. coli*, no positive effect was observed [124]. Soni and Prakash [150] used *Aspergillus niger* for the synthesis of gold nanoparticles and evaluated its anti-larval activity against *Anopheles stephensi*, *Aedes aegypti*, and *Culex quinquefasciatus*. It was observed that nanoparticles were more effective against *C. quinquefasciatus*. Sunkar and Nachiyar [151] reported the synthesis of silver nanoparticles from *Bacillus cereus*. It was found to have antibacterial activity against *S. aureus*, *K. pneumonia*, *E. coli*, *S. typhi*, and *P. aeruginosa* [151].

Abdeen et al. [152] reported the synthesis of silver and iron nanoparticles using *Fusarium oxysporum* and evaluated their antimicrobial properties against *E. coli, Bacillus, P. aeruginosa, K. pneumoniae, Proteus vulgaris*, and *Staphylococcus* sp. A study reported the synthesis of silver nanoparticles using *Ulva lactuca* and found that at low concentration, it was able to inhibit the growth of *Plasmodium falciparum* [153]. A study reported the synthesis of selenium nanorods using *Streptomyces bikiniensis* and evaluated their potential anticancer activity against human cancer cells [154]. Borse et al. [155] tested the anticancer activity of platinum nanoparticles against MCF-7 and A431 cell lines, which were synthesized from *Saccharomyces boulardii*. Mohamed et al. [156] synthesized iron nanoparticles using *Alternaria alternate* and tested their antibacterial properties against *E. coli, B. subtilis, S. aureus*, and *P. aeruginosa*. It was found that iron nanoparticles possess the maximum inhibition of *B. subtilis*. In another study, *Streptomyces cyaneus* was used to synthesize gold nanoparticles and to investigate their anticancer activity against liver and breast cells. It was found that gold nanoparticles stimulated mitochondrial apoptosis, DNA damage, and induced cytokinesis arrest [157]. A study reported the synthesis of gold and silver nanoparticles using *Streptomyces* sp. and evaluated its potential antibacterial activity against *Salmonella infantis, S. aureus, Bacillus subtilis, Proteus mirabilis, K. pneumoniae, P. aeruginosa*, and *E. coli* [158]. Arya et al. [139] used *Botryococcus braunii* for the synthesis of copper nanoparticles and evaluated their antimicrobial activity. It was found to show toxicity against *E. coli, K. pneumoniae, P. aeruginosa*, and *S. aureus*. Husain et al. [159] reported that the synthesis of silver nanoparticles using cyanobacteria showed potential photocatalytic activity against dye. Dananjaya et al. [160] reported the synthesis of *Spirulina maxima*-mediated gold nanoparticles and evaluated its cytotoxicity and anticandidal activity. It was found that nanoparticles do not possess any cytotoxicity against cell lines, because they act as potent anticandidal agents. In a study, *Escherichia* sp. was used to synthesize copper nanoparticles and was utilized for the degradation of azo dye and textile effluent treatment. It was found that 83.90% of the congo red dye was removed and, in textile effluents, there was a significant reduction in electrical conductivity, pH, turbidity, total dissolved solids, total suspended solids, hardness, chlorides, and sulfates compared to nontreated samples [161].

4.2. Nanocatalysts in Dye Degradation

Due to the increasing population and use of dye in textiles, food, and other industries, the release of untreated waste into the water bodies is contaminating the environment at a rapid rate, due to which there is a growing demand for newer and more efficient technologies for the removal of these substances from the environment. Nanosized materials can be used to detoxify harmful organic and inorganic chemicals from the environment due to their ultrafine size, high aspect ratio, and interaction-dominating characteristics. Nanoparticles have generated a lot of interest due to their numerous uses in disciplines such as catalysts, detection, and environmental cleanup, such as the adsorption and degradation of different pollutants from liquid medium (Figure 6). Various bacterial and fungal species have been used for the synthesis of nanoparticles, and these synthesized nanocatalysts can be further used for the degradation of dyes (Table 4).

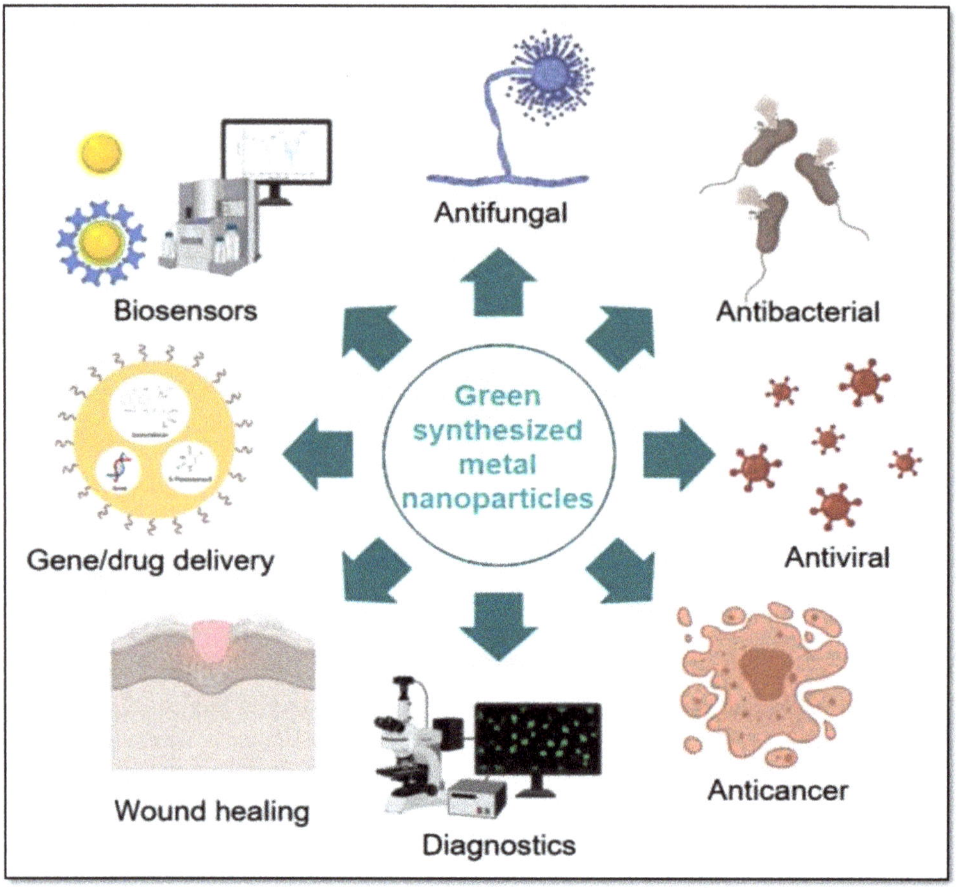

Figure 5. Biological applications of nanocatalyst [162].

Table 4. Degradation of dyes using bacterial-derived nanoparticles.

Dye	Nanoparticle Used	Synthesis of Nanoparticles Source	% Degradation	References
Malachite green	Silver	*Bacillus paralicheniformis*	90	[163]
Reactive black 5	Palladium	*Pseudomonas putida*	100	[164]
Methyl orange	Palladium	*Clostrodium* sp.	90	[165]
Amaranth	Iron	*Shewanella decolorotionis*	90.5	[166]
Methyl orange	Tin(iv) oxide	*Erwinia herbicola*	94	[167]
Congo red	Silver	*Pleurotus sajor caju*	78	[168]
Malachite green	Silver	*Acremonium kiliense*	95.4	[169]
Methylene Blue	Silver	*Saccharomyces cerevisiae*	80	[170]
Congo red	Silver	*Pestalotiopsis versicolor*	91.56	[171]
Direct blue 71	Palladium	*Saccharomyces cerevisiae*	98	[172]
Naphthol Green B	Iron–sulfur	*Pseudoalteromonas* sp. *CF10-13*	19.46	[173]
Bismarck brown	Zinc-oxide	*Aspergillus niger*	89	[174]
Methyl orange	Platinum	*Fusarium oxysporum*	-	[175]
Acid Brilliant Scarlet GR	Gold	*Trichoderma* sp.	94.7	[176]
Rhodamine B	Gold	*Cladosporium oxysporum AJP03*	-	[177]
Malachite green	Copper	*Escherichia* sp. *SINT7*	90.55	[178]
Rhodamine B	Gold	*Turbinaria conoides*	-	[179]
Malachite green	Silver	*Gracilaria corticata*	-	[180]
Methylene blue, rhodamine B, and methyl orange	Gold and silver	*Sargassum serratifolium*	-	[181]

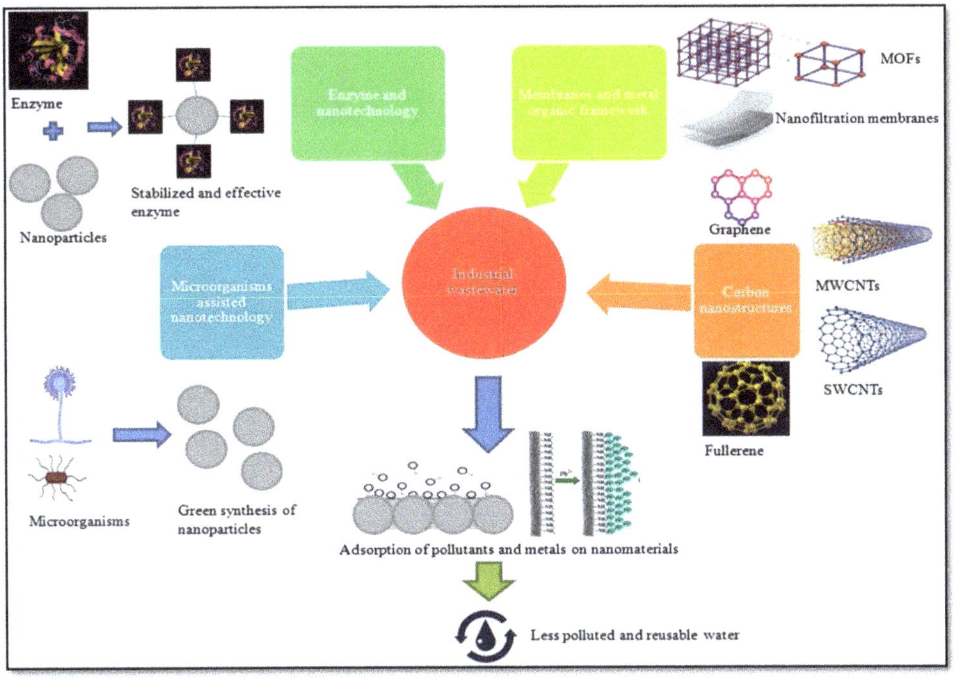

Figure 6. Remediation of industrial wastewater using a nanocatalyst [179].

4.3. Nanocatalysts in Heavy Metal Remediation

Heavy metals are detrimental contaminants that are toxic both in soluble and elemental forms. Diverse activities, including the development of industries, absurd waste management, defective landfill operations, and manufacturing and mining, lead to increased contamination of metals in soil and water [182]. The traditional method for heavy metal removal includes reverse osmosis, chemical precipitation, ion exchange filtration, evaporation, and membrane technology. However, the cost of these methods is occasionally higher; therefore, a cost-efficient and environmentally friendly method is of prime importance [183,184]. Various microbial-derived nanocatalysts have been reported to show remediated heavy metals (Figure 7). In a study, palladium nanoparticles were synthesized from *Enterococcus faecalis*, and these nanoparticles were used for the removal of hexavalent chromium from contaminated water. [185]. In another study, iron oxide nanoparticles derived from *Aspergillus tubingensis* were able to remove heavy metals such as copper 92.19%, nickel 96.45%, lead 98%, and zinc 93.99% from wastewater, and the reusability study showed that iron nanoparticles possess a high regeneration capacity [186].

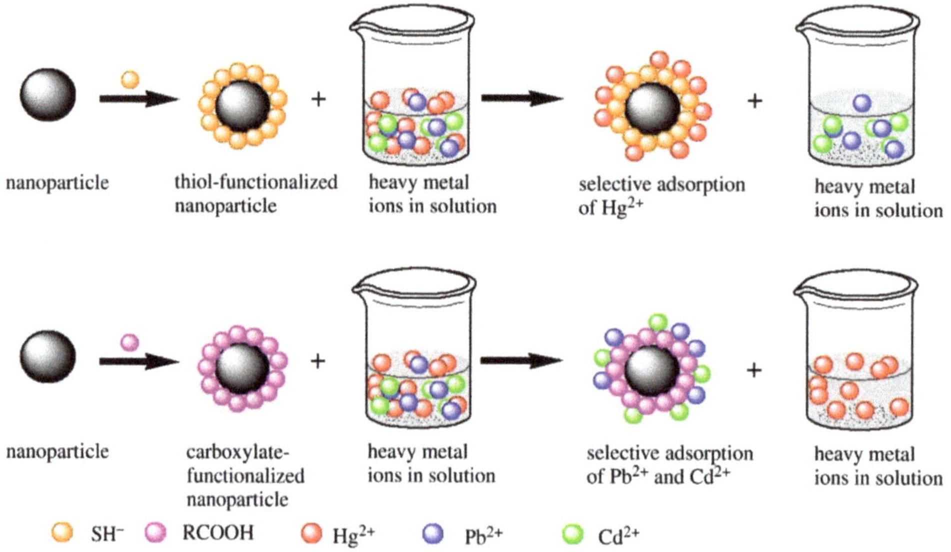

Figure 7. Nanocatalyst in heavy metal remediation [187].

5. Conclusions

The development of environmentally friendly and cost-effective techniques for producing nanomaterials and their concomitant applications in various fields is in great demand. Although a range of physical and chemical techniques have been found to be suitable for the synthesis of various nanomaterials, these methods show a nonnegligible concern because of the production of various toxic and nonbiodegradable by products. In this regard, the biogenic synthesis of nanomaterials offers an alternative solution to overcome the existing drawbacks that come from physical and chemical methods. The biological method offers a rigid control on the synthesized particles size and uniform shape, while the physical characteristics are retained at the same level as physical and chemical methods. Biologically synthesized nanomaterials are more prepared for biomedical application because of their lower toxicity. Nanomaterials provide various applications such as dye degradation, heavy metal remediation, and biological activity. However, in biological methods, many parameters affect the synthesis of nanoparticles, including pH, temperature, and whether they are manufactured internally or externally in the cell. These parameters should be studied in order to optimize the synthesis process. It is easy to manipulate bacteria genetically, whereas, in the case of fungi, the downstream process has been shown to be suitable for the large-scale production of nanomaterials. Regardless of the biological sources used, it is crucial to recognize the mechanism behind nanoparticle synthesis for maximum synthesis, which is important for commercialization purposes. Nonpathogenic sources are beneficial for the production of nanoparticles in an effective way. In addition, nanotoxicity should also be considered because it sometimes causes adverse effects on human health and animals. This can be tackled by the implementation of regulation and legislation, and researchers must conduct joint multidisciplinary studies in various fields of medical sciences, nanomedicine, nanotechnology, and biomedical engineering.

Author Contributions: Conceptualization, A.R.; writing—original draft preparation, A.R. and M.N.I.; writing—review and editing, A.R., A.E., S.K., V.T., H.O., A.A., S.I., N.L.F., M.U.K., M.N.I., T.B.E. and M.B. All authors have read and agreed to the published version of the manuscript.

Funding: This research received no external funding.

Institutional Review Board Statement: Not applicable.

Informed Consent Statement: Not applicable.

Data Availability Statement: Not applicable.

Acknowledgments: The authors gratefully acknowledge the Deanship of Scientific Research, King Khalid University (KKU), Abha-Asir, Kingdom of Saudi Arabia for funding this work under the grant number RGP.2/58/42. Arpita Roy is grateful to Sharda University (Seed fund-4 2001 (SUSF2001/12)). All the authors are grateful to their university for constant support.

Conflicts of Interest: The authors declare no conflict of interest.

References

1. Roy, A.; Bharadvaja, N. Silver nanoparticle synthesis from *Plumbago zeylanica* and its dye degradation activity. *Bioinspired Biomim. Nanobiomaterials* **2019**, *8*, 130–140. [CrossRef]
2. Ahmed, H.M.; Roy, A.; Wahab, M.; Ahmed, M.; Othman-Qadir, G.; Elesawy, B.H.; Khandaker, M.U.; Islam, M.N.; Emran, T.B. Applications of Nanomaterials in Agrifood and Pharmaceutical Industry. *J. Nanomater.* **2021**, *2021*, 1472096. [CrossRef]
3. Roy, A. Plant Derived Silver Nanoparticles and their Therapeutic Applications. *Curr. Pharm. Biotechnol.* **2021**, *22*, 1834–1847. [CrossRef] [PubMed]
4. Thakkar, K.N.; Mhatre, S.S.; Parikh, R.Y. Biological synthesis of metallic nanoparticles. *Nanomed. Nanotechnol. Biol. Med.* **2010**, *6*, 257–262. [CrossRef]
5. Jeevanandam, J.; Barhoum, A.; Chan, Y.S.; Dufresne, A.; Danquah, M.K. Review on nanoparticles and nanostructured materials: History, sources, toxicity and regulations. *Beilstein J. Nanotechnol.* **2018**, *9*, 1050–1074. [CrossRef]
6. Remya, V.R.; Abitha, V.K.; Rajput, P.S.; Rane, A.V.; Dutta, A. Silver nanoparticles green synthesis: A mini review. *Chem. Int.* **2017**, *3*, 165–171.
7. Iravani, S.; Korbekandi, H.; Mirmohammadi, S.V.; Zolfaghari, B. Synthesis of silver nanoparticles: Chemical, physical and biological methods. *Res. Pharm. Sci.* **2014**, *9*, 385.
8. Raina, S.; Roy, A.; Bharadvaja, N. Degradation of dyes using biologically synthesized silver and copper nanoparticles. *Environ. Nanotechnol. Monit. Manag.* **2020**, *13*, 100278. [CrossRef]
9. Nagore, P.; Ghotekar, S.; Mane, K.; Ghoti, A.; Bilal, M.; Roy, A. Structural Properties and Antimicrobial Activities of *Polyalthia longifolia* Leaf Extract-Mediated CuO Nanoparticles. *BioNanoScience* **2021**, *11*, 579–589. [CrossRef]
10. Shah, M.; Fawcett, D.; Sharma, S.; Tripathy, S.K.; Poinern, G.E.J. Green synthesis of metallic nanoparticles via biological entities. *Materials* **2015**, *8*, 7278–7308. [CrossRef] [PubMed]
11. Loomba, L.; Scarabelli, T. Metallic nanoparticles and their medicinal potential. Part II: Aluminosilicates, nanobiomagnets, quantum dots and cochleates. *Ther. Deliv.* **2013**, *4*, 1179–1196. [CrossRef] [PubMed]
12. Tajzad, I.; Ghasali, E. Production methods of CNT-reinforced Al matrix composites: A review. *J. Compos. Compd.* **2020**, *2*, 1–9.
13. Puente Santiago, A.R.; Fernandez-Delgado, O.; Gomez, A.; Ahsan, M.A.; Echegoyen, L. Fullerenes as Key Components for Low-Dimensional (Photo) Electrocatalytic Nanohybrid Materials. *Angew. Chem.* **2021**, *133*, 124–143. [CrossRef]
14. C Thomas, S.; Kumar Mishra, P.; Talegaonkar, S. Ceramic nanoparticles: Fabrication methods and applications in drug delivery. *Curr. Pharm. Des.* **2015**, *21*, 6165–6188. [CrossRef] [PubMed]
15. Chaudhary, R.G.; Bhusari, G.S.; Tiple, A.D.; Rai, A.R.; Somkuvar, S.R.; Potbhare, A.K.; Abdala, A.A. Metal/metal oxide nanoparticles: Toxicity, applications, and future prospects. *Curr. Pharm. Des.* **2019**, *25*, 4013–4029. [CrossRef] [PubMed]
16. Suresh, S. Semiconductor nanomaterials, methods and applications: A review. *Nanosci. Nanotechnol.* **2013**, *3*, 62–74.
17. Crucho, C.I.; Barros, M.T. Polymeric nanoparticles: A study on the preparation variables and characterization methods. *Mater. Sci. Eng. C* **2017**, *80*, 771–784. [CrossRef]
18. Khan, I.; Saeed, K.; Khan, I. Nanoparticles: Properties, applications and toxicities. *Arab. J. Chem.* **2019**, *12*, 908–931. [CrossRef]
19. Roy, A.; Bharadvaja, N. Qualitative analysis of phytocompounds and synthesis of silver nanoparticles from *Centella asiatica*. *Innov. Tech. Agric.* **2017**, *1*, 88–95.
20. Nath, D.; Banerjee, P. Green nanotechnology—A new hope for medical biology. *Environ. Toxicol. Pharmacol.* **2013**, *36*, 997–1014. [CrossRef]
21. Jeyaraj, M.; Gurunathan, S.; Qasim, M.; Kang, M.H.; Kim, J.H. A comprehensive review on the synthesis, characterization, and biomedical application of platinum nanoparticles. *Nanomaterials* **2019**, *9*, 1719. [CrossRef] [PubMed]
22. Roy, A.; Bharadvaja, N. Silver nanoparticles synthesis from a pharmaceutically important medicinal plant *Plumbago zeylanica*. *MOJ Bioequivalence Bioavailab.* **2017**, *3*, 00046.
23. Liang, X.J.; Kumar, A.; Shi, D.; Cui, D. Nanostructures for medicine and pharmaceuticals. *J. Nanomater.* **2012**, *2012*, 921897. [CrossRef]
24. Auría-Soro, C.; Nesma, T.; Juanes-Velasco, P.; Landeira-Viñuela, A.; Fidalgo-Gomez, H.; Acebes-Fernandez, V.; Gongora, R.; Almendral Parra, M.J.; Manzano-Roman, R.; Fuentes, M. Interactions of nanoparticles and biosystems: Microenvironment of nanoparticles and biomolecules in nanomedicine. *Nanomaterials* **2019**, *9*, 1365. [CrossRef] [PubMed]

25. Guilger-Casagrande, M.; de Lima, R. Synthesis of Silver Nanoparticles Mediated by Fungi: A Review. *Front. Bioeng. Biotechnol.* **2019**, *7*, 287. [CrossRef] [PubMed]
26. Huynh, K.H.; Pham, X.H.; Kim, J.; Lee, S.H.; Chang, H.; Rho, W.Y.; Jun, B.H. Synthesis, properties, and biological applications of metallic alloy nanoparticles. *Int. J. Mol. Sci.* **2020**, *21*, 5174. [CrossRef] [PubMed]
27. Nair, B.; Pradeep, T. Coalescence of nanoclusters and formation of submicron crystallites assisted by *Lactobacillus strains*. *Cryst. Growth Des.* **2002**, *2*, 293–298. [CrossRef]
28. Ahmad, A.; Senapati, S.; Khan, M.I.; Kumar, R.; Sastry, M. Extracellular biosynthesis of monodisperse gold nanoparticles by a novel extremophilic actinomycete, *Thermomonospora* sp. *Langmuir* **2003**, *19*, 3550–3553. [CrossRef]
29. Konishi, Y.; Ohno, K.; Saitoh, N.; Nomura, T.; Nagamine, S. Microbial synthesis of gold nanoparticles by metal reducing bacterium. *Trans. Mater. Res. Soc. Jpn.* **2004**, *29*, 2341–2343.
30. He, S.; Guo, Z.; Zhang, Y.; Zhang, S.; Wang, J.; Gu, N. Biosynthesis of gold nanoparticles using the bacteria *Rhodopseudomonas capsulata*. *Mater. Lett.* **2007**, *61*, 3984–3987. [CrossRef]
31. Parikh, R.Y.; Singh, S.; Prasad, B.L.V.; Patole, M.S.; Sastry, M.; Shouche, Y.S. Extracellular synthesis of crystalline silver nanoparticles and molecular evidence of silver resistance from *Morganella* sp.: Towards understanding biochemical synthesis mechanism. *ChemBioChem* **2008**, *9*, 1415–1422. [CrossRef] [PubMed]
32. Reddy, A.S.; Chen, C.Y.; Chen, C.C.; Jean, J.S.; Chen, H.R.; Tseng, M.J.; Fan, C.-W.; Wang, J.C. Biological synthesis of gold and silver nanoparticles mediated by the bacteria *Bacillus subtilis*. *J. Nanosci. Nanotechnol.* **2010**, *10*, 6567–6574. [CrossRef] [PubMed]
33. Bruna, N.; Collao, B.; Tello, A.; Caravantes, P.; Díaz-Silva, N.; Monrás, J.P.; Órdenes-Aenishanslins, N.; Flores, M.; Espinoza-Gonzalez, R.; Bravo, D.; et al. Synthesis of salt-stable fluorescent nanoparticles (quantum dots) by polyextremophile halophilic bacteria. *Sci. Rep.* **2019**, *9*, 1953. [CrossRef] [PubMed]
34. Bharde, A.; Wani, A.; Shouche, Y.; Joy, P.A.; Prasad, B.L.; Sastry, M. Bacterial aerobic synthesis of nanocrystalline magnetite. *J. Am. Chem. Soc.* **2005**, *127*, 9326–9327. [CrossRef]
35. Gericke, M.; Pinches, A. Microbial production of gold nanoparticles. *Gold Bull.* **2006**, *39*, 22–28. [CrossRef]
36. Baesman, S.M.; Bullen, T.D.; Dewald, J.; Zhang, D.; Curran, S.; Islam, F.S.; Beveridge, T.J.; Oremland, R.S. Formation of tellurium nanocrystals during anaerobic growth of bacteria that use Te oxyanions as respiratory electron acceptors. *Appl. Environ. Microbiol.* **2007**, *73*, 2135–2143. [CrossRef]
37. Du, L.; Jiang, H.; Liu, X.; Wang, E. Biosynthesis of gold nanoparticles assisted by *Escherichia coli* DH5α and its application on direct electrochemistry of hemoglobin. *Electrochem. Commun.* **2007**, *9*, 1165–1170. [CrossRef]
38. Shahverdi, A.R.; Fakhimi, A.; Shahverdi, H.R.; Minaian, S. Synthesis and effect of silver nanoparticles on the antibacterial activity of different antibiotics against *Staphylococcus aureus* and *Escherichia coli*. *Nanomed. Nanotechnol. Biol. Med.* **2007**, *3*, 168–171. [CrossRef]
39. Kalishwaralal, K.; Deepak, V.; Ramkumarpandian, S.; Nellaiah, H.; Sangiliyandi, G. Extracellular biosynthesis of silver nanoparticles by the culture supernatant of *Bacillus licheniformis*. *Mater. Lett.* **2008**, *62*, 4411–4413. [CrossRef]
40. Nangia, Y.; Wangoo, N.; Goyal, N.; Shekhawat, G.; Suri, C.R. A novel bacterial isolate *Stenotrophomonas maltophilia* as living factory for synthesis of gold nanoparticles. *Microb. Cell Factories* **2009**, *8*, 39. [CrossRef]
41. Pugazhenthiran, N.; Anandan, S.; Kathiravan, G.; Prakash, N.K.U.; Crawford, S.; Ashokkumar, M. Microbial synthesis of silver nanoparticles by *Bacillus* sp. *J. Nanoparticle Res.* **2009**, *11*, 1811. [CrossRef]
42. Wen, L.; Lin, Z.; Gu, P.; Zhou, J.; Yao, B.; Chen, G.; Fu, J. Extracellular biosynthesis of monodispersed gold nanoparticles by a SAM capping route. *J. Nanoparticle Res.* **2009**, *11*, 279–288. [CrossRef]
43. Ogi, T.; Saitoh, N.; Nomura, T.; Konishi, Y. Room-temperature synthesis of gold nanoparticles and nanoplates using *Shewanella algae* cell extract. *J. Nanoparticle Res.* **2010**, *12*, 2531–2539. [CrossRef]
44. Fayaz, M.; Tiwary, C.S.; Kalaichelvan, P.T.; Venkatesan, R. Blue orange light emission from biogenic synthesized silver nanoparticles using *Trichoderma viride*. *Colloids Surf. B Biointerfaces* **2010**, *75*, 175–178. [CrossRef]
45. Zonooz, N.F.; Salouti, M. Extracellular biosynthesis of silver nanoparticles using cell filtrate of *Streptomyces* sp. ERI-3. *Sci. Iran.* **2011**, *18*, 1631–1635. [CrossRef]
46. Prakash, A.; Sharma, S.; Ahmad, N.; Ghosh, A.; Sinha, P. Synthesis of AgNps By Bacillus cereus bacteria and their antimicrobial potential. *J. Biomater. Nanobiotechnol.* **2011**, *2*, 155. [CrossRef]
47. Kumar, C.G.; Mamidyala, S.K. Extracellular synthesis of silver nanoparticles using culture supernatant of *Pseudomonas aeruginosa*. *Colloids Surf. B Biointerfaces* **2011**, *84*, 462–466. [CrossRef] [PubMed]
48. Seshadri, S.; Prakash, A.; Kowshik, M. Biosynthesis of silver nanoparticles by marine bacterium, *Idiomarina* sp. PR58-8. *Bull. Mater. Sci.* **2012**, *35*, 1201–1205. [CrossRef]
49. Rajasree, S.R.; Suman, T.Y. Extracellular biosynthesis of gold nanoparticles using a gram negative bacterium *Pseudomonas fluorescens*. *Asian Pac. J. Trop. Dis.* **2012**, *2*, S796–S799. [CrossRef]
50. Rajeshkumar, S.; Malarkodi, C.; Paulkumar, K.; Vanaja, M.; Gnanajobitha, G.; Annadurai, G. Intracellular and extracellular biosynthesis of silver nanoparticles by using marine bacteria *Vibrio alginolyticus*. *J. Nanosci. Nanotechnol.* **2013**, *3*, 21–25.
51. Kupryashina, M.A.; Vetchinkina, E.P.; Burov, A.M.; Ponomareva, E.G.; Nikitina, V.E. Biosynthesis of gold nanoparticles by *Azospirillum brasilense*. *Microbiology* **2013**, *82*, 833–840. [CrossRef]
52. Malarkodi, C.; Chitra, K.; Rajeshkumar, S.; Gnanajobitha, G.; Paulkumar, K.; Vanaja, M.; Annadurai, G. Novel eco-friendly synthesis of titanium oxide nanoparticles by using *Planomicrobium* sp. and its antimicrobial evaluation. *Der Pharm. Sin.* **2013**, *4*, 59–66.

53. Ghorbani, H.R. Biosynthesis of silver nanoparticles using *Salmonella typhirium*. *J. Nanostructure Chem.* **2013**, *3*, 29. [CrossRef]
54. Correa-Llantén, D.N.; Muñoz-Ibacache, S.A.; Castro, M.E.; Muñoz, P.A.; Blamey, J.M. Gold nanoparticles synthesized by *Geobacillus* sp. strain ID17 a thermophilic bacterium isolated from Deception Island, Antarctica. *Microb. Cell Factories* **2013**, *12*, 75. [CrossRef] [PubMed]
55. Ibrahem, K.H.; Salman, J.A.S.; Ali, F.A. Effect of Titanium Nanoparticles Biosynthesis by *Lactobacillus crispatus* on Urease, Hemolysin & Biofilm Forming by Some Bacteria Causing Recurrent UTI in Iraqi Women. *Eur. Sci. J.* **2014**, *10*, 324–338.
56. Das, V.L.; Thomas, R.; Varghese, R.T.; Soniya, E.V.; Mathew, J.; Radhakrishnan, E.K. Extracellular synthesis of silver nanoparticles by the Bacillus strain CS 11 isolated from industrialized area. *3 Biotech* **2014**, *4*, 121–126. [CrossRef] [PubMed]
57. Shantkriti, S.; Rani, P. Biological synthesis of copper nanoparticles using *Pseudomonas fluorescens*. *Int. J. Curr. Microbiol. Appl. Sci.* **2014**, *3*, 374–383.
58. Cuevas, R.; Durán, N.; Diez, M.C.; Tortella, G.R.; Rubilar, O. Extracellular biosynthesis of copper and copper oxide nanoparticles by *Stereum hirsutum*, a native white-rot fungus from chilean forests. *J. Nanomater.* **2015**, *2015*, 57. [CrossRef]
59. Ghorbani, H.R.; Mehr, F.P.; Poor, A.K. Extracellular synthesis of copper nanoparticles using culture supernatants of *Salmonella typhimurium*. *Orient. J. Chem.* **2015**, *31*, 527–529. [CrossRef]
60. Singh, P.; Kim, Y.J.; Singh, H.; Mathiyalagan, R.; Wang, C.; Yang, D.C. Biosynthesis of anisotropic silver nanoparticles by *Bhargavaea indica* and their synergistic effect with antibiotics against pathogenic microorganisms. *J. Nanomater.* **2015**, *2015*, 4. [CrossRef]
61. Padman, A.J.; Henderson, J.; Hodgson, S.; Rahman, P.K. Biomediated synthesis of silver nanoparticles using *Exiguobacterium mexicanum*. *Biotechnol. Lett.* **2014**, *36*, 2079–2084. [CrossRef]
62. Li, J.; Li, Q.; Ma, X.; Tian, B.; Li, T.; Yu, J.; Dai, S.; Weng, Y.; Hua, Y. Biosynthesis of gold nanoparticles by the extreme bacterium *Deinococcus radiodurans* and an evaluation of their antibacterial properties. *Int. J. Nanomed.* **2016**, *11*, 5931. [CrossRef] [PubMed]
63. Singh, P.; Singh, H.; Kim, Y.J.; Mathiyalagan, R.; Wang, C.; Yang, D.C. Extracellular synthesis of silver and gold nanoparticles by *Sporosarcina koreensis* DC4 and their biological applications. *Enzym. Microb. Technol.* **2016**, *86*, 75–83. [CrossRef]
64. Saravanan, M.; Barik, S.K.; MubarakAli, D.; Prakash, P.; Pugazhendhi, A. Synthesis of silver nanoparticles from *Bacillus brevis* (NCIM 2533) and their antibacterial activity against pathogenic bacteria. *Microb. Pathog.* **2018**, *116*, 221–226. [CrossRef] [PubMed]
65. Divya, M.; Kiran, G.S.; Hassan, S.; Selvin, J. Biogenic synthesis and effect of silver nanoparticles (AgNPs) to combat catheter-related urinary tract infections. *Biocatal. Agric. Biotechnol.* **2019**, *18*, 101037. [CrossRef]
66. Gupta, R.; Padmanabhan, P. Biogenic synthesis and characterization of gold nanoparticles by a novel marine bacteria marinobacter algicola: Progression from nanospheres to various geometrical shapes. *J. Microbiol. Biotechnol. Food Sci.* **2019**, *2019*, 732–737. [CrossRef]
67. Patil, M.P.; Kang, M.J.; Niyonizigiye, I.; Singh, A.; Kim, J.O.; Seo, Y.B.; Kim, G.D. Extracellular synthesis of gold nanoparticles using the marine bacterium *Paracoccus haeundaensis* BC74171T and evaluation of their antioxidant activity and antiproliferative effect on normal and cancer cell lines. *Colloids Surf. B: Biointerfaces* **2019**, *183*, 110455. [CrossRef]
68. Almalki, M.A.; Khalifa, A.Y. Silver nanoparticles synthesis from *Bacillus* sp. KFU36 and its anticancer effect in breast cancer MCF-7 cells via induction of apoptotic mechanism. *J. Photochem. Photobiol. B Biol.* **2020**, *204*, 111786. [CrossRef]
69. Dorcheh, S.K.; Vahabi, K. Biosynthesis of nanoparticles by fungi: Large-scale production. In *Fungal Metabolites*; Springer International Publishing: Cham, Switzerland, 2016; pp. 1–20.
70. Molnár, Z.; Bódai, V.; Szakacs, G.; Erdélyi, B.; Fogarassy, Z.; Sáfrán, G.; Lagzi, I. Green synthesis of gold nanoparticles by thermophilic filamentous fungi. *Sci. Rep.* **2018**, *8*, 3943. [CrossRef] [PubMed]
71. Gudikandula, K.; Vadapally, P.; Charya, M.S. Biogenic synthesis of silver nanoparticles from white rot fungi: Their characterization and antibacterial studies. *OpenNano* **2017**, *2*, 64–78. [CrossRef]
72. Mukherjee, P.; Ahmad, A.; Mandal, D.; Senapati, S.; Sainkar, S.R.; Khan, M.I.; Parishcha, R.; Ajaykumar, P.V.; Alam, M.; Kumar, R.; et al. Fungus-mediated synthesis of silver nanoparticles and their immobilization in the mycelial matrix: A novel biological approach to nanoparticle synthesis. *Nano Lett.* **2001**, *1*, 515–519. [CrossRef]
73. Bhainsa, K.C.; D'souza, S.F. Extracellular biosynthesis of silver nanoparticles using the fungus *Aspergillus fumigatus*. *Colloids Surf. B Biointerfaces* **2006**, *47*, 160–164. [CrossRef] [PubMed]
74. Riddin, T.L.; Gericke, M.; Whiteley, C.G. Analysis of the inter-and extracellular formation of platinum nanoparticles by *Fusarium oxysporum* f. sp. lycopersici using response surface methodology. *Nanotechnology* **2006**, *17*, 3482. [CrossRef] [PubMed]
75. Rai, M.; Yadav, A.; Gade, A. Silver nanoparticles as a new generation of antimicrobials. *Biotechnol. Adv.* **2009**, *27*, 76–83. [CrossRef] [PubMed]
76. Sanghi, R.; Verma, P. A facile green extracellular biosynthesis of CdS nanoparticles by immobilized fungus. *Chem. Eng. J.* **2009**, *155*, 886–891. [CrossRef]
77. Vahabi, K.; Mansoori, G.A.; Karimi, S. Biosynthesis of silver nanoparticles by fungus *Trichoderma reesei* (a route for large-scale production of AgNPs. *Insciences J.* **2011**, *1*, 65–79. [CrossRef]
78. Castro-Longoria, E.; Vilchis-Nestor, A.R.; Avalos-Borja, M. Biosynthesis of silver, gold and bimetallic nanoparticles using the filamentous fungus *Neurospora crassa*. *Colloids Surf. B Biointerfaces* **2011**, *83*, 42–48. [CrossRef]
79. Arun, G.; Eyini, M.; Gunasekaran, P. Green synthesis of silver nanoparticles using the mushroom fungus *Schizophyllum commune* and its biomedical applications. *Biotechnol. Bioprocess Eng.* **2014**, *19*, 1083–1090. [CrossRef]

80. Akther, T.; Mathipi, V.; Kumar, N.S.; Davoodbasha, M.; Srinivasan, H. Fungal-mediated synthesis of pharmaceutically active silver nanoparticles and anticancer property against A549 cells through apoptosis. *Environ. Sci. Pollut. Res.* **2019**, *26*, 13649–13657. [CrossRef] [PubMed]
81. Mukherjee, P.; Senapati, S.; Mandal, D.; Ahmad, A.; Khan, M.I.; Kumar, R.; Sastry, M. Extracellular synthesis of gold nanoparticles by the fungus *Fusarium oxysporum*. *ChemBioChem* **2002**, *3*, 461–463. [CrossRef]
82. Chen, J.C.; Lin, Z.H.; Ma, X.X. Evidence of the production of silver nanoparticles via pretreatment of *Phoma* sp. 3.2883 with silver nitrate. *Lett. Appl. Microbiol.* **2003**, *37*, 105–108. [CrossRef] [PubMed]
83. Bansal, V.; Rautaray, D.; Ahmad, A.; Sastry, M. Biosynthesis of zirconia nanoparticles using the fungus *Fusarium oxysporum*. *J. Mater. Chem.* **2004**, *14*, 3303–3305. [CrossRef]
84. Ahmad, A.; Senapati, S.; Khan, M.I.; Kumar, R.; Sastry, M. Extra-/intracellular biosynthesis of gold nanoparticles by an alkalotolerant fungus, *Trichothecium* sp. *J. Biomed. Nanotechnol.* **2005**, *1*, 47–53. [CrossRef]
85. Bharde, A.; Rautaray, D.; Bansal, V.; Ahmad, A.; Sarkar, I.; Yusuf, S.M.; Sanyal, M.; Sastry, M. Extracellular biosynthesis of magnetite using fungi. *Small* **2006**, *2*, 135–141. [CrossRef]
86. Vigneshwaran, N.; Kathe, A.A.; Varadarajan, P.V.; Nachane, R.P.; Balasubramanya, R.H. Biomimetics of silver nanoparticles by white rot fungus, *Phaenerochaete chrysosporium*. *Colloids Surf. B Biointerfaces* **2006**, *53*, 55–59. [CrossRef]
87. Durán, N.; Marcato, P.D.; De Souza, G.I.; Alves, O.L.; Esposito, E. Antibacterial effect of silver nanoparticles produced by fungal process on textile fabrics and their effluent treatment. *J. Biomed. Nanotechnol.* **2007**, *3*, 203–208. [CrossRef]
88. Mukherjee, P.; Roy, M.; Mandal, B.P.; Dey, G.K.; Mukherjee, P.K.; Ghatak, J.; Tyagi, A.K.; Kale, S.P. Green synthesis of highly stabilized nanocrystalline silver particles by a non-pathogenic and agriculturally important fungus *T. asperellum*. *Nanotechnology* **2008**, *19*, 075103. [CrossRef]
89. Ingle, A.; Gade, A.; Pierrat, S.; Sonnichsen, C.; Rai, M. Mycosynthesis of silver nanoparticles using the fungus *Fusarium acuminatum* and its activity against some human pathogenic bacteria. *Curr. Nanosci.* **2008**, *4*, 141–144. [CrossRef]
90. Das, S.K.; Das, A.R.; Guha, A.K. Gold nanoparticles: Microbial synthesis and application in water hygiene management. *Langmuir* **2009**, *25*, 8192–8199. [CrossRef]
91. Verma, V.C.; Kharwar, R.N.; Gange, A.C. Biosynthesis of antimicrobial silver nanoparticles by the endophytic fungus *Aspergillus clavatus*. *Nanomedicine* **2010**, *5*, 33–40. [CrossRef]
92. Verma, V.C.; Singh, S.K.; Solanki, R.; Prakash, S. Biofabrication of anisotropic gold nanotriangles using extract of endophytic *Aspergillus clavatus* as a dual functional reductant and stabilizer. *Nanoscale Res. Lett.* **2011**, *6*, 16. [CrossRef] [PubMed]
93. Banu, A.; Rathod, V. Synthesis and characterization of silver nanoparticles by *Rhizopus stolonier*. *Int. J. Biomed. Adv. Res.* **2011**, *2*, 148–158. [CrossRef]
94. Li, G.; He, D.; Qian, Y.; Guan, B.; Gao, S.; Cui, Y.; Yokoyama, K.; Wang, L. Fungus-mediated green synthesis of silver nanoparticles using *Aspergillus terreus*. *Int. J. Mol. Sci.* **2012**, *13*, 466–476. [CrossRef] [PubMed]
95. Raliya, R.; Tarafdar, J.C. ZnO nanoparticle biosynthesis and its effect on phosphorous-mobilizing enzyme secretion and gum contents in Clusterbean (*Cyamopsis tetragonoloba* L.). *Agric. Res.* **2013**, *2*, 48–57. [CrossRef]
96. Chowdhury, S.; Basu, A.; Kundu, S. Green synthesis of protein capped silver nanoparticles from phytopathogenic fungus *Macrophomina phaseolina* (Tassi) Goid with antimicrobial properties against multidrug-resistant bacteria. *Nanoscale Res. Lett.* **2014**, *9*, 365. [CrossRef] [PubMed]
97. Pereira, L.; Dias, N.; Carvalho, J.; Fernandes, S.; Santos, C.; Lima, N. Synthesis, characterization and antifungal activity of chemically and fungal-produced silver nanoparticles against *Trichophyton rubrum*. *J. Appl. Microbiol.* **2014**, *117*, 1601–1613. [CrossRef]
98. Maliszewska, I.; Juraszek, A.; Bielska, K. Green synthesis and characterization of silver nanoparticles using ascomycota fungi *Penicillium nalgiovense* AJ12. *J. Clust. Sci.* **2014**, *25*, 989–1004. [CrossRef]
99. Raliya, R.; Biswas, P.; Tarafdar, J.C. TiO_2 nanoparticle biosynthesis and its physiological effect on mung bean (*Vigna radiata* L.). *Biotechnol. Rep.* **2015**, *5*, 22–26. [CrossRef] [PubMed]
100. Elgorban, A.M.; Al-Rahmah, A.N.; Sayed, S.R.; Hirad, A.; Mostafa, A.A.F.; Bahkali, A.H. Antimicrobial activity and green synthesis of silver nanoparticles using *Trichoderma viride*. *Biotechnol. Biotechnol. Equip.* **2016**, *30*, 299–304. [CrossRef]
101. Shende, S.; Gade, A.; Rai, M. Large-scale synthesis and antibacterial activity of fungal-derived silver nanoparticles. *Environ. Chem. Lett.* **2017**, *15*, 427–434. [CrossRef]
102. Saravanan, M.; Arokiyaraj, S.; Lakshmi, T.; Pugazhendhi, A. Synthesis of silver nanoparticles from *Phenerochaete chrysosporium* (MTCC-787) and their antibacterial activity against human pathogenic bacteria. *Microb. Pathog.* **2018**, *117*, 68–72. [CrossRef]
103. Tyagi, S.; Tyagi, P.K.; Gola, D.; Chauhan, N.; Bharti, R.K. Extracellular synthesis of silver nanoparticles using entomopathogenic fungus: Characterization and antibacterial potential. *SN Appl. Sci.* **2019**, *1*, 1545. [CrossRef]
104. Hulikere, M.M.; Joshi, C.G. Characterization, antioxidant and antimicrobial activity of silver nanoparticles synthesized using marine endophytic fungus-*Cladosporium cladosporioides*. *Process Biochem.* **2019**, *82*, 199–204. [CrossRef]
105. Gond, S.K.; Mishra, A.; Verma, S.K.; Sharma, V.K.; Kharwar, R.N. Synthesis and Characterization of Antimicrobial Silver Nanoparticles by an Endophytic Fungus Isolated from *Nyctanthes arbortristis*. *Proc. Natl. Acad. Sci. India Sect. B Biol. Sci.* **2019**, *90*, 641–645. [CrossRef]

106. Clarance, P.; Luvankar, B.; Sales, J.; Khusro, A.; Agastian, P.; Tack, J.C.; Al Khulaifi, M.M.; Al-Shwaiman, H.A.; Elgorban, A.M.; Syed, A.; et al. Green synthesis and characterization of gold nanoparticles using endophytic fungi *Fusarium solani* and its in-vitro anticancer and biomedical applications. *Saudi J. Biol. Sci.* **2020**, *27*, 706–712. [CrossRef] [PubMed]
107. Chatterjee, S.; Mahanty, S.; Das, P.; Chaudhuri, P.; Das, S. Biofabrication of iron oxide nanoparticles using manglicolous fungus *Aspergillus niger* BSC-1 and removal of Cr (VI) from aqueous solution. *Chem. Eng. J.* **2020**, *385*, 123790. [CrossRef]
108. LewisOscar, F.; Vismaya, S.; Arunkumar, M.; Thajuddin, N.; Dhanasekaran, D.; Nithya, C. Algal nanoparticles: Synthesis and biotechnological potentials. *Algae–Org. Imminent Biotechnol.* **2016**, *7*, 157–182.
109. Singaravelu, G.; Arockiamary, J.S.; Kumar, V.G.; Govindaraju, K. A novel extracellular synthesis of monodisperse gold nanoparticles using marine alga, *Sargassum wightii* Greville. *Colloids Surf. B Biointerfaces* **2007**, *57*, 97–101. [CrossRef]
110. Xie, J.; Lee, J.Y.; Wang, D.I.; Ting, Y.P. Identification of active biomolecules in the high-yield synthesis of single-crystalline gold nanoplates in algal solutions. *Small* **2007**, *3*, 672–682. [CrossRef] [PubMed]
111. Mata, Y.N.; Torres, E.; Blazquez, M.L.; Ballester, A.; González, F.M.J.A.; Munoz, J.A. Gold (III) biosorption and bioreduction with the brown alga *Fucus vesiculosus*. *J. Hazard. Mater.* **2009**, *166*, 612–618. [CrossRef]
112. Shakibaie, M.; Forootanfar, H.; Mollazadeh-Moghaddam, K.; Bagherzadeh, Z.; Nafissi-Varcheh, N.; Shahverdi, A.R.; Faramarzi, M.A. Green synthesis of gold nanoparticles by the marine microalga *Tetraselmis suecica*. *Biotechnol. Appl. Biochem.* **2010**, *57*, 71–75. [CrossRef] [PubMed]
113. Barwal, I.; Ranjan, P.; Kateriya, S.; Yadav, S.C. Cellular oxido-reductive proteins of *Chlamydomonas reinhardtii* control the biosynthesis of silver nanoparticles. *J. Nanobiotechnol.* **2011**, *9*, 56. [CrossRef] [PubMed]
114. Mahdieh, M.; Zolanvari, A.; Azimee, A.S. Green biosynthesis of silver nanoparticles by *Spirulina platensis*. *Sci. Iran.* **2012**, *19*, 926–929. [CrossRef]
115. Senapati, S.; Syed, A.; Moeez, S.; Kumar, A.; Ahmad, A. Intracellular synthesis of gold nanoparticles using alga *Tetraselmis kochinensis*. *Mater. Lett.* **2012**, *79*, 116–118. [CrossRef]
116. Roychoudhury, P.; Pal, R. Spirogyra submaxima—A green alga for nanogold production. *J. Algal Biomass Util.* **2014**, *5*, 15–19.
117. Suganya, K.U.; Govindaraju, K.; Kumar, V.G.; Dhas, T.S.; Karthick, V.; Singaravelu, G.; Elanchezhiyan, M. Blue green alga mediated synthesis of gold nanoparticles and its antibacterial efficacy against Gram positive organisms. *Mater. Sci. Eng. C* **2015**, *47*, 351–356. [CrossRef]
118. Arsiya, F.; Sayadi, M.H.; Sobhani, S. Green synthesis of palladium nanoparticles using *Chlorella vulgaris*. *Mater. Lett.* **2017**, *186*, 113–115. [CrossRef]
119. Sanaeimehr, Z.; Javadi, I.; Namvar, F. Antiangiogenic and antiapoptotic effects of green-synthesized zinc oxide nanoparticles using *Sargassum muticum* algae extraction. *Cancer Nanotechnol.* **2018**, *9*, 3. [CrossRef] [PubMed]
120. Fatima, R.; Priya, M.; Indurthi, L.; Radhakrishnan, V.; Sudhakaran, R. Biosynthesis of silver nanoparticles using red algae *Portieria hornemannii* and its antibacterial activity against fish pathogens. *Microb. Pathog.* **2020**, *138*, 103780. [CrossRef]
121. Liu, B.; Xie, J.; Lee, J.Y.; Ting, Y.P.; Chen, J.P. Optimization of high-yield biological synthesis of single-crystalline gold nanoplates. *J. Phys. Chem. B* **2005**, *109*, 15256–15263. [CrossRef]
122. Govindaraju, K.; Kiruthiga, V.; Kumar, V.G.; Singaravelu, G. Extracellular synthesis of silver nanoparticles by a marine alga, *Sargassum wightii* Grevilli and their antibacterial effects. *J. Nanosci. Nanotechnol.* **2009**, *9*, 5497–5501. [CrossRef] [PubMed]
123. Vivek, M.; Kumar, P.S.; Steffi, S.; Sudha, S. Biogenic silver nanoparticles by *Gelidiella acerosa* extract and their antifungal effects. *Avicenna J. Med. Biotechnol.* **2011**, *3*, 143. [PubMed]
124. Rajathi, F.A.A.; Parthiban, C.; Kumar, V.G.; Anantharaman, P. Biosynthesis of antibacterial gold nanoparticles using brown alga, *Stoechospermum marginatum* (kützing). *Spectrochim. Acta Part A Mol. Biomol. Spectrosc.* **2012**, *99*, 166–173. [CrossRef] [PubMed]
125. Dhas, T.S.; Kumar, V.G.; Abraham, L.S.; Karthick, V.; Govindaraju, K. *Sargassum myriocystum* mediated biosynthesis of gold nanoparticles. *Spectrochim. Acta Part A Mol. Biomol. Spectrosc.* **2012**, *99*, 97–101. [CrossRef] [PubMed]
126. Dhanalakshmi, P.K.; Azeez, R.; Rekha, R.; Poonkodi, S.; Nallamuthu, T. Synthesis of silver nanoparticles using green and brown seaweeds. *Phykos* **2012**, *42*, 39–45.
127. Kannan, R.R.R.; Arumugam, R.; Ramya, D.; Manivannan, K.; Anantharaman, P. Green synthesis of silver nanoparticles using marine macroalga *Chaetomorpha linum*. *Appl. Nanosci.* **2013**, *3*, 229–233. [CrossRef]
128. Naveena, B.E.; Prakash, S. Biological synthesis of gold nanoparticles using marine algae *Gracilaria corticata* and its application as a potent antimicrobial and antioxidant agent. *Asian J. Pharm. Clin. Res.* **2013**, *6*, 179–182.
129. Abboud, Y.; Saffaj, T.; Chagraoui, A.; El Bouari, A.; Brouzi, K.; Tanane, O.; Ihssane, B. Biosynthesis, characterization and antimicrobial activity of copper oxide nanoparticles (CONPs) produced using brown alga extract (*Bifurcaria bifurcate*). *Appl. Nanosci.* **2014**, *4*, 571–576. [CrossRef]
130. Yousefzadi, M.; Rahimi, Z.; Ghafori, V. The green synthesis, characterization and antimicrobial activities of silver nanoparticles synthesized from green alga *Enteromorpha flexuosa* (wulfen) J. Agardh. *Mater. Lett.* **2014**, *137*, 1–4. [CrossRef]
131. Sharma, B.; Purkayastha, D.D.; Hazra, S.; Gogoi, L.; Bhattacharjee, C.R.; Ghosh, N.N.; Rout, J. Biosynthesis of gold nanoparticles using a freshwater green alga, *Prasiola crispa*. *Mater. Lett.* **2014**, *116*, 94–97. [CrossRef]
132. Momeni, S.; Nabipour, I. A simple green synthesis of palladium nanoparticles with *Sargassum alga* and their electrocatalytic activities towards hydrogen peroxide. *Appl. Biochem. Biotechnol.* **2015**, *176*, 1937–1949. [CrossRef]
133. Kathiraven, T.; Sundaramanickam, A.; Shanmugam, N.; Balasubramanian, T. Green synthesis of silver nanoparticles using marine algae *Caulerpa racemosa* and their antibacterial activity against some human pathogens. *Appl. Nanosci.* **2015**, *5*, 499–504. [CrossRef]

134. Ibraheem, I.B.M.; Abd-Elaziz, B.E.E.; Saad, W.F.; Fathy, W.A. Green biosynthesis of silver nanoparticles using marine Red Algae *Acanthophora specifera* and its antimicrobial activity. *J. Nanomed. Nanotechnol.* **2016**, *7*, 1–4.
135. Tevan, R.; Jayakumara, S.; Sahimia, N.H.A.; Iqbala, N.F.A.; Zapria, I.; Govindana, N.; Ichwan, S.J.A.; Ichwanb, G.P.M. Biosynthesis of silver nanoparticles using marine microalgae *Isochrysis* sp. *J. Chem. Eng. Ind. Biotechnol.* **2017**, *1*, 12.
136. Omar, H.H.; Bahabri, F.S.; El-Gendy, A.M. Biopotential application of synthesis nanoparticles as antimicrobial agents by using *Laurencia papillosa*. *Int. J. Pharmacol.* **2017**, *13*, 303–312. [CrossRef]
137. Muthusamy, G.; Thangasamy, S.; Raja, M.; Chinnappan, S.; Kandasamy, S. Biosynthesis of silver nanoparticles from *Spirulina microalgae* and its antibacterial activity. *Environ. Sci. Pollut. Res.* **2017**, *24*, 19459–19464. [CrossRef]
138. Aboelfetoh, E.F.; El-Shenody, R.A.; Ghobara, M.M. Eco-friendly synthesis of silver nanoparticles using green algae (*Caulerpa serrulata*): Reaction optimization, catalytic and antibacterial activities. *Environ. Monit. Assess.* **2017**, *189*, 349. [CrossRef]
139. Arya, A.; Gupta, K.; Chundawat, T.S.; Vaya, D. Biogenic synthesis of copper and silver nanoparticles using green alga *Botryococcus braunii* and its antimicrobial activity. *Bioinorg. Chem. Appl.* **2018**, *2018*, 7879403. [CrossRef]
140. Abdel-Raouf, N.; Al-Enazi, N.M.; Ibraheem, I.B.M.; Alharbi, R.M.; Alkhulaifi, M.M. Biosynthesis of silver nanoparticles by using of the marine brown alga *Padina pavonia* and their characterization. *Saudi J. Biol. Sci.* **2019**, *26*, 1207–1215. [CrossRef]
141. Pugazhendhi, A.; Prabakar, D.; Jacob, J.M.; Karuppusamy, I.; Saratale, R.G. Synthesis and characterization of silver nanoparticles using *Gelidium amansii* and its antimicrobial property against various pathogenic bacteria. *Microb. Pathog.* **2018**, *114*, 41–45. [CrossRef] [PubMed]
142. Bhuyar, P.; Rahim, M.H.A.; Sundararaju, S.; Ramaraj, R.; Maniam, G.P.; Govindan, N. Synthesis of silver nanoparticles using marine macroalgae *Padina* sp. and its antibacterial activity towards pathogenic bacteria. *Beni-Suef Univ. J. Basic Appl. Sci.* **2020**, *9*, 1–15. [CrossRef]
143. Öztürk, B.Y.; Gürsu, B.Y.; Dağ, İ. Antibiofilm and antimicrobial activities of green synthesized silver nanoparticles using marine red algae *Gelidium corneum*. *Process Biochem.* **2020**, *89*, 208–219. [CrossRef]
144. Huq, M. Green synthesis of silver nanoparticles using *Pseudoduganella eburnea* MAHUQ-39 and their antimicrobial mechanisms investigation against drug resistant human pathogens. *Int. J. Mol. Sci.* **2020**, *21*, 1510. [CrossRef]
145. Bukhari, S.I.; Hamed, M.M.; Al-Agamy, M.H.; Gazwi, H.S.; Radwan, H.H.; Youssif, A.M. Biosynthesis of copper oxide nanoparticles using Streptomyces MHM38 and its biological applications. *J. Nanomater.* **2021**, *2021*, 6693302. [CrossRef]
146. Mmola, M.; Roes-Hill, M.L.; Durrell, K.; Bolton, J.J.; Sibuyi, N.; Meyer, M.E.; Beukes, D.R.; Antunes, E. Enhanced antimicrobial and anticancer activity of silver and gold nanoparticles synthesised using *Sargassum incisifolium* aqueous extracts. *Molecules* **2016**, *21*, 1633. [CrossRef]
147. Pal, S.; Tak, Y.K.; Song, J.M. Does the antibacterial activity of silver nanoparticles depend on the shape of the nanoparticle? A study of the gram-negative bacterium *Escherichia coli*. *Appl. Environ. Microbiol.* **2007**, *73*, 1712–1720. [CrossRef] [PubMed]
148. Mishra, A.; Tripathy, S.K.; Wahab, R.; Jeong, S.H.; Hwang, I.; Yang, Y.B.; Kim, Y.-S.; Shin, H.-S.; Yun, S.I. Microbial synthesis of gold nanoparticles using the fungus *Penicillium brevicompactum* and their cytotoxic effects against mouse mayo blast cancer C 2 C 12 cells. *Appl. Microbiol. Biotechnol.* **2011**, *92*, 617–630. [CrossRef]
149. Chauhan, A.; Zubair, S.; Tufail, S.; Sherwani, A.; Sajid, M.; Raman, S.C.; Azam, A.; Owais, M. Fungus-mediated biological synthesis of gold nanoparticles: Potential in detection of liver cancer. *Int. J. Nanomed.* **2011**, *6*, 2305.
150. Soni, N.; Prakash, S. Synthesis of gold nanoparticles by the fungus *Aspergillus niger* and its efficacy against mosquito larvae. *Rep. Parasitol.* **2012**, *2*, 1–7.
151. Sunkar, S.; Nachiyar, C.V. Biogenesis of antibacterial silver nanoparticles using the endophytic bacterium *Bacillus cereus* isolated from *Garcinia xanthochymus*. *Asian Pac. J. Trop. Biomed.* **2012**, *2*, 953–959. [CrossRef]
152. Abdeen, S.; Isaac, R.R.; Geo, S.; Sornalekshmi, S.; Rose, A.; Praseetha, P.K. Evaluation of Antimicrobial Activity of Biosynthesized Iron and Silver Nanoparticles Using the Fungi *Fusarium oxysporum* and *Actinomycetes* sp. on Human Pathogens. *Nano Biomed. Eng.* **2013**, *5*, 39–45. [CrossRef]
153. Murugan, K.; Samidoss, C.M.; Panneerselvam, C.; Higuchi, A.; Roni, M.; Suresh, U.; Chandramohan, B.; Subramaniam, J.; Madhiyazhagan, P.; Dinesh, D.; et al. Seaweed-synthesized silver nanoparticles: An eco-friendly tool in the fight against *Plasmodium falciparum* and its vector *Anopheles stephensi*? *Parasitol. Res.* **2015**, *114*, 4087–4097. [CrossRef] [PubMed]
154. Ahmad, M.S.; Yasser, M.M.; Sholkamy, E.N.; Ali, A.M.; Mehanni, M.M. Anticancer activity of biostabilized selenium nanorods synthesized by *Streptomyces bikiniensis* strain Ess_amA-1. *Int. J. Nanomed.* **2015**, *10*, 3389.
155. Borse, V.; Kaler, A.; Banerjee, U.C. Microbial synthesis of platinum nanoparticles and evaluation of their anticancer activity. *Int. Conf. Recent Trends Eng. Technol.* **2015**, *11*, 26–31.
156. Mohamed, Y.M.; Azzam, A.M.; Amin, B.H.; Safwat, N.A. Mycosynthesis of iron nanoparticles by *Alternaria alternata* and its antibacterial activity. *Afr. J. Biotechnol.* **2015**, *14*, 1234–1241. [CrossRef]
157. El-Batal, A.I.; Mona, S.; Al-Tamie, M. Biosynthesis of gold nanoparticles using marine *Streptomyces cyaneus* and their antimicrobial, antioxidant and antitumor (in vitro) activities. *J. Chem. Pharm. Res.* **2015**, *7*, 1020–1036.
158. Składanowski, M.; Wypij, M.; Laskowski, D.; Golińska, P.; Dahm, H.; Rai, M. Silver and gold nanoparticles synthesized from *Streptomyces* sp. isolated from acid forest soil with special reference to its antibacterial activity against pathogens. *J. Clust. Sci.* **2017**, *28*, 59–79. [CrossRef]

159. Husain, S.; Afreen, S.; Yasin, D.; Afzal, B.; Fatma, T. Cyanobacteria as a bioreactor for synthesis of silver nanoparticles—An effect of different reaction conditions on the size of nanoparticles and their dye decolorization ability. *J. Ofmmicrobiological Methods* **2019**, *162*, 77–82. [CrossRef]
160. Dananjaya, S.H.S.; Thao, N.T.; Wijerathna, H.M.S.M.; Lee, J.; Edussuriya, M.; Choi, D.; Kumar, R.S. In vitro and in vivo anticandidal efficacy of green synthesized gold nanoparticles using *Spirulina maxima* polysaccharide. *Process Biochem.* **2020**, *92*, 138–148. [CrossRef]
161. Noman, M.; Shahid, M.; Ahmed, T.; Niazi, M.B.K.; Hussain, S.; Song, F.; Manzoor, I. Use of biogenic copper nanoparticles synthesized from a native *Escherichia* sp. as photocatalysts for azo dye degradation and treatment of textile effluents. *Environ. Pollut.* **2020**, *257*, 113514. [CrossRef]
162. Rónavári, A.; Igaz, N.; Adamecz, D.I.; Szerencsés, B.; Molnar, C.; Kónya, Z.; Pfeiffer, I.; Kiricsi, M. Green silver and gold nanoparticles: Biological synthesis approaches and potentials for biomedical applications. *Molecules* **2021**, *26*, 844. [CrossRef]
163. Allam, N.G.; Ismail, G.A.; El-Gemizy, W.M.; Salem, M.A. Biosynthesis of silver nanoparticles by cell-free extracts from some bacteria species for dye removal from wastewater. *Biotechnol. Lett.* **2019**, *41*, 379–389. [CrossRef] [PubMed]
164. Batool, S.; Akib, S.; Ahmad, M.; Balkhair, K.S.; Ashraf, M.A. Study of modern nano enhanced techniques for removal of dyes and metals. *J. Nanomater.* **2014**, *2014*, 1–20. [CrossRef]
165. Johnson, A.; Merilis, G.; Hastings, J.; Palmer, M.E.; Fitts, J.; Chidambaram, D. Nanotechnology and microbial electrochemistry for environmental remediation. *ECS Trans.* **2011**, *33*, 103. [CrossRef]
166. Fang, Y.; Xu, M.; Wu, W.M.; Chen, X.; Sun, G.; Guo, J.; Liu, X. Characterization of the enhancement of zero valent iron on microbial azo reduction. *BMC Microbiol.* **2015**, *15*, 85. [CrossRef] [PubMed]
167. Srivastava, N.; Mukhopadhyay, M. Biosynthesis of SnO_2 nanoparticles using bacterium *Erwinia herbicola* and their photocatalytic activity for degradation of dyes. *Ind. Eng. Chem. Res.* **2014**, *53*, 13971–13979. [CrossRef]
168. Nithya, R.; Ragunathan, R. Decolorization of the dye congo red by pleurotus sajor caju silver nanoparticle. In Proceedings of the International Conference on Food Engineering and Biotechnology, Bangkok, Thailand, 7–9 May 2011; Volume 9, pp. 12–15.
169. Youssef, A.S.; El-Sherif, M.F.; El-Assar, S.A. Studies on the decolorization of malachite green by the local isolate *Acremonium kiliense*. *Biotechnology* **2008**, *7*, 213–223. [CrossRef]
170. Roy, K.; Sarkar, C.K.; Ghosh, C.K. Photocatalytic activity of biogenic silver nanoparticles synthesized using yeast (*Saccharomyces cerevisiae*) extract. *Appl. Nanosci.* **2015**, *5*, 953–959. [CrossRef]
171. Kavish, R.; Shruti, A.; Jyoti, S.; Agrawal, P.K. Mycosynthesis of silver nanoparticles using endophytic fungus *Pestalotiopsis versicolor* and investigation of its antibacterial and azo dye degradation efficacy. *Kavaka* **2018**, *49*, 65–71.
172. Sriramulu, M.; Sumathi, S. Biosynthesis of palladium nanoparticles using Saccharomyces cerevisiae extract and its photocatalytic degradation behaviour. *Adv. Nat. Sci. Nanosci. Nanotechnol.* **2018**, *9*, 025018. [CrossRef]
173. Cheng, S.; Li, N.; Jiang, L.; Li, Y.; Xu, B.; Zhou, W. Biodegradation of metal complex Naphthol Green B and formation of iron–sulfur nanoparticles by marine bacterium *Pseudoalteromonas* sp. CF10-13. *Bioresour. Technol.* **2019**, *273*, 49–55. [CrossRef] [PubMed]
174. Kalpana, V.N.; Kataru, B.A.S.; Sravani, N.; Vigneshwari, T.; Panneerselvam, A.; Rajeswari, V.D. Biosynthesis of zinc oxide nanoparticles using culture filtrates of *Aspergillus niger*: Antimicrobial textiles and dye degradation studies. *OpenNano* **2018**, *3*, 48–55. [CrossRef]
175. Gupta, K.; Chundawat, T.S. Bio-inspired synthesis of platinum nanoparticles from fungus *Fusarium oxysporum*: Its characteristics, potential antimicrobial, antioxidant and photocatalytic activities. *Mater. Res. Express* **2019**, *6*, 1050d6. [CrossRef]
176. Qu, Y.; Shen, W.; Pei, X.; Ma, F.; You, S.; Li, S.; Wang, J.; Zhou, J. Biosynthesis of gold nanoparticles by *Trichoderma* sp. WL-Go for azo dyes decolorization. *J. Environ. Sci.* **2017**, *56*, 79–86. [CrossRef]
177. Bhargava, A.; Jain, N.; Khan, M.A.; Pareek, V.; Dilip, R.V.; Panwar, J. Utilizing metal tolerance potential of soil fungus for efficient synthesis of gold nanoparticles with superior catalytic activity for degradation of rhodamine B. *J. Environ. Manag.* **2016**, *183*, 22–32. [CrossRef]
178. Mandeep, P.S. Microbial nanotechnology for bioremediation of industrial wastewater. *Front. Microbiol.* **2020**, *11*, 590631. [CrossRef] [PubMed]
179. Ramakrishna, M.; Babu, D.R.; Gengan, R.M.; Chandra, S.; Rao, G.N. Green synthesis of gold nanoparticles using marine algae and evaluation of their catalytic activity. *J. Nanostructure Chem.* **2016**, *6*, 1–13. [CrossRef]
180. Poornima, S.; Valivittan, K. Degradation of malachite green (dye) by using photo-catalytic biogenic silver nanoparticles synthesized using red algae (*Gracilaria corticata*) aqueous extract. *Int. J. Curr. Microbiol. Appl. Sci.* **2017**, *6*, 62–70. [CrossRef]
181. Kim, B.; Song, W.C.; Park, S.Y.; Park, G. Green Synthesis of Silver and Gold Nanoparticles via *Sargassum serratifolium* Extract for Catalytic Reduction of Organic Dyes. *Catalysts* **2021**, *11*, 347. [CrossRef]
182. Kaur, S.; Roy, A. Bioremediation of heavy metals from wastewater using nanomaterials. *Environ. Dev. Sustain.* **2021**, *23*, 9617–9640. [CrossRef]
183. Roy, A.; Bharadvaja, N. Removal of toxic pollutants using microbial fuel cells. In *Removal of Toxic Pollutants through Microbiological and Tertiary Treatment*; Elsevier: Amsterdam, The Netherlands, 2020; pp. 153–177.
184. Roy, A.; Bharadvaja, N. Efficient removal of heavy metals from artificial wastewater using biochar. *Environ. Nanotechnol. Monit. Manag.* **2021**, *16*, 100602. [CrossRef]

185. Ha, C.; Zhu, N.; Shang, R.; Shi, C.; Cui, J.; Sohoo, I.; Wu, P.; Cao, Y. Biorecovery of palladium as nanoparticles by *Enterococcus faecalis* and its catalysis for chromate reduction. *Chem. Eng. J.* **2016**, *288*, 246–254. [CrossRef]
186. Mahanty, S.; Chatterjee, S.; Ghosh, S.; Tudu, P.; Gaine, T.; Bakshi, M.; Das, S.; Das, P.; Bhattacharyya, S.; Bandyopadhyay, S.; et al. Synergistic approach towards the sustainable management of heavy metals in wastewater using mycosynthesized iron oxide nanoparticles: Biofabrication, adsorptive dynamics and chemometric modeling study. *J. Water Process Eng.* **2020**, *37*, 101426. [CrossRef]
187. Mensah, M.B.; Lewis, D.J.; Boadi, N.O.; Awudza, J.A. Heavy metal pollution and the role of inorganic nanomaterials in environmental remediation. *R. Soc. Open Sci.* **2021**, *8*, 201485. [CrossRef] [PubMed]

catalysts

MDPI

Article

The Catalytic Activity of Biosynthesized Magnesium Oxide Nanoparticles (MgO-NPs) for Inhibiting the Growth of Pathogenic Microbes, Tanning Effluent Treatment, and Chromium Ion Removal

Ebrahim Saied [1], Ahmed M. Eid [1], Saad El-Din Hassan [1,*], Salem S. Salem [1], Ahmed A. Radwan [1], Mahmoud Halawa [2], Fayez M. Saleh [3], Hosam A. Saad [4], Essa M. Saied [5,6] and Amr Fouda [1,*]

1 Department of Botany and Microbiology, Faculty of Science, Al-Azhar University, Nasr City, Cairo 11884, Egypt; hema_almassry2000@azhar.edu.eg (E.S.); aeidmicrobiology@azhar.edu.eg (A.M.E.); salemsalahsalem@azhar.edu.eg (S.S.S.); ahmedradwan@azhar.edu.eg (A.A.R.)

2 SAJA Pharmaceuticals, 6th of October, Giza 12573, Egypt; mahmoudwaleedh76@gmail.com

3 Department of Medical Microbiology, Faculty of Medicine, University of Tabuk, Tabuk 71491, Saudi Arabia; fsaleh@ut.edu.sa

4 Department of Chemistry, College of Science, Taif University, P.O. Box 11099, Taif 21944, Saudi Arabia; h.saad@tu.edu.sa

5 Chemistry Department, Faculty of Science, Suez Canal University, Ismailia 41522, Egypt; saiedess@hu-berlin.de

6 Institute for Chemistry, Humboldt Universität zu Berlin, Brook-Taylor-Str. 2, 12489 Berlin, Germany

* Correspondence: Saad.el-din.hassan@umontreal.ca (S.E.-D.H.); amr_fh83@azhar.edu.eg (A.F.)

Citation: Saied, E.; Eid, A.M.; Hassan, S.E.-D.; Salem, S.S.; Radwan, A.A.; Halawa, M.; Saleh, F.M.; Saad, H.A.; Saied, E.M.; Fouda, A. The Catalytic Activity of Biosynthesized Magnesium Oxide Nanoparticles (MgO-NPs) for Inhibiting the Growth of Pathogenic Microbes, Tanning Effluent Treatment, and Chromium Ion Removal. *Catalysts* **2021**, *11*, 821. https://doi.org/10.3390/catal11070821

Academic Editors: Beom Soo Kim and Pritam Kumar Dikshit

Received: 28 June 2021
Accepted: 5 July 2021
Published: 6 July 2021

Abstract: Magnesium oxide nanoparticles (MgO-NPs) were synthesized using the fungal strain *Aspergillus terreus* S1 to overcome the disadvantages of chemical and physical methods. The factors affecting the biosynthesis process were optimized as follows: concentration of $Mg(NO_3)_2{\cdot}6H_2O$ precursor (3 mM), contact time (36 min), pH (8), and incubation temperature (35 °C). The characterization of biosynthesized MgO-NPs was accomplished using UV-vis spectroscopy, Fourier transform infrared (FT-IR) spectroscopy, transmission electron microscopy (TEM), scanning electron microscopy—energy dispersive X-ray (SEM-EDX), X-ray diffraction (XRD), and dynamic light scattering (DLS). Data confirmed the successful formation of crystallographic, spherical, well-dispersed MgO-NPs with a size range of 8.0–38.0 nm at a maximum surface plasmon resonance of 280 nm. The biological activities of biosynthesized MgO-NPs including antimicrobial activity, biotreatment of tanning effluent, and chromium ion removal were investigated. The highest growth inhibition of pathogenic *Staphylococcus aureus*, *Bacillus subtilis*, *Pseudomonas aeruginosa*, *Escherichia coli*, and *Candida albicans* was achieved at 200 µg mL^{-1} of MgO-NPs. The biosynthesized MgO-NPs exhibited high efficacy to decolorize the tanning effluent (96.8 ± 1.7% after 150 min at 1.0 µg mL^{-1}) and greatly decrease chemical parameters including total suspended solids (TSS), total dissolved solids (TDS), biological oxygen demand (BOD), chemical oxygen demand (COD), and conductivity with percentages of 98.04, 98.3, 89.1, 97.2, and 97.7%, respectively. Further, the biosynthesized MgO-NPs showed a strong potential to remove chromium ions from the tanning effluent, from 835.3 mg L^{-1} to 21.0 mg L^{-1}, with a removal percentage of 97.5%.

Keywords: biogenic synthesis; *Aspergillus terreus*; tanning effluent; chromium ion; pathogenic microbes; nanoparticle characterization

1. Introduction

Environmental and water pollution dramatically increase due to industrial development, populational growth, and energy production [1]. These pollutants are considered the main factors for disease, illness, and death due to their toxicity, non-degradability, and tendency to accumulate in the food chain [2]. The different industrial activities are the main sources of contaminated wastewater because they produce highly toxic waste that has a long-term negative impact [3,4]. The tannery and leather sectors are considered the largest industries that use hazardous chemical compounds and require a high amount of water [5]. These sectors produce high amounts of effluents containing highly toxic compounds and are characterized by increase total dissolved solids (TDS), total suspended solids (TSS), conductivity, chemical oxygen demand (COD), and biochemical oxygen demand (BOD) [6]. Therefore, the tanning and leather effluents require extra treatment before discharge into the eco-system. Chromium is considered one of the most widely used heavy metals in the tanning and leather industry and is widely discharged as part of the effluent [7]. Further, chromium is incorporated into electroplating and paint manufacturing, resulting in large quantities being discharged into the environment [8]. Moreover, the development of new compounds to overcome the resistance properties of microbes to different antibiotic is the main goal for investigators [9,10].

Nanotechnology sciences provide a promising tool for the synthesis of new active compounds with superior properties to be breakthrough applications in various biomedical and biotechnological sectors [11,12]. Among these new nanoparticles, magnesium oxide nanoparticles (MgO-NPs) are characterized by excellent optical, thermal, mechanical, and chemical properties [13]. MgO-NPs have high reactivity due to the presence of highly reactive edges and a high surface area [14]. Therefore, MgO-NPs have a variety of applications in various fields such as catalyst supports, agricultural products, paints, superconductor products, antimicrobial materials, photonic devices, sensors, and adsorbents [15–17]. Notably, MgO-NPs have antibacterial properties against harmful microbes, for instance, *S. aureus* and *E. coli* [18,19]. Further, they can be utilized as adsorbent materials due to the high removal efficiency [20]. MgO-NPs act as excellent adsorbents for various chemical species and this property increases with a decrease in MgO size [21]. Recently, MgO nanoparticles have obtained popularity in environmental science due to their fascinating and intrinsic properties [22].

Several chemical and physical methods have been used to fabricate MgO-NPs such as chemical precipitation, thermal decomposition, sol-gel, combustion, and chemical vapor deposition [13,23]. These methods predominantly require several processing steps, controlled pH, high temperature and pressure, expensive equipment, and toxic chemicals. These techniques produce numerous by-products that may be toxic to ecosystems. Therefore, there is a need to develop a low-cost, eco-friendly method for nanoparticle synthesis [11]. Recently, the biogenic synthesis of NPs has a wide range of interest because of the reduction or elimination of toxic substances that are present in the environment from chemical and physical methods [24]. Microorganisms such as fungi, yeast, actinomycetes, and bacteria can reduce metal and their oxides to NPs. The biogenic synthesis of NPs using fungi (eukaryotic organisms) has numerous advantages over the prokaryotic organisms, e.g., easy to multiply, grow, handle, and downstream process for nano-biosynthesis [25].

In the present study, we have tried to explore a rapid, cost-effective, eco-friendly method for fabricating MgO-NPs using the fungal strain *Aspergillus terreus* S1. The optimized biosynthesis process was investigated by studying the effect of metal precursors, incubation temperature, pH, and contact time. The biogenically synthesized MgO-NPs were characterized using various techniques consisting of UV-vis spectroscopy, X-Ray diffraction (XRD), scanning and energy dispersive X-Ray spectroscopy (SEM-EDX), transmission electron microscopes (TEM), dynamic light scattering (DLS), and Fourier-transform infrared spectroscopy (FT–IR). The efficacy of biosynthesized MgO-NPs to inhibit the growth of different pathogenic bacterial and fungal strains was assessed. Moreover, utilizing MgO-NPs for decolorization and treatment of tanning effluent and removal of heavy metals are among the main goals.

2. Results and Discussion

2.1. Isolation and Identification of the Fungal Isolates

In the current study, five fungal isolates were obtained from cultivated soil samples; we selected fungal isolate S1 for MgO-NPs based on its best and rapid NP synthesis. The selected fungal isolate underwent primary identification using morphological and microscopic analysis. Original identification was done according to standard keys based on morphological and cultural characteristics. The fungal isolate appears brownish, with compact conidial heads, biseriate, and densely columnar. Conidiophores are smooth and hyaline. The conidia are small, about 2 μm in diameter, globose-shaped, and smooth-walled. According to morphological and cultural characterization, the fungal isolate S1 belongs to *Aspergillus* sp. [26,27]. The primary identification was confirmed by amplification and sequencing of the internal transcribed spacer (ITS) gene. The sequence analysis revealed that the fungal strain S1 is strongly related to *Aspergillus terreus* (accession number: MT558939) with a similarity percentage of 93%. The fungal strain obtained in this study was identified as *Aspergillus terreus* strain S1 (Figure 1). The sequence analysis acquired from the current study was deposit in GenBank under accession number MW774586.

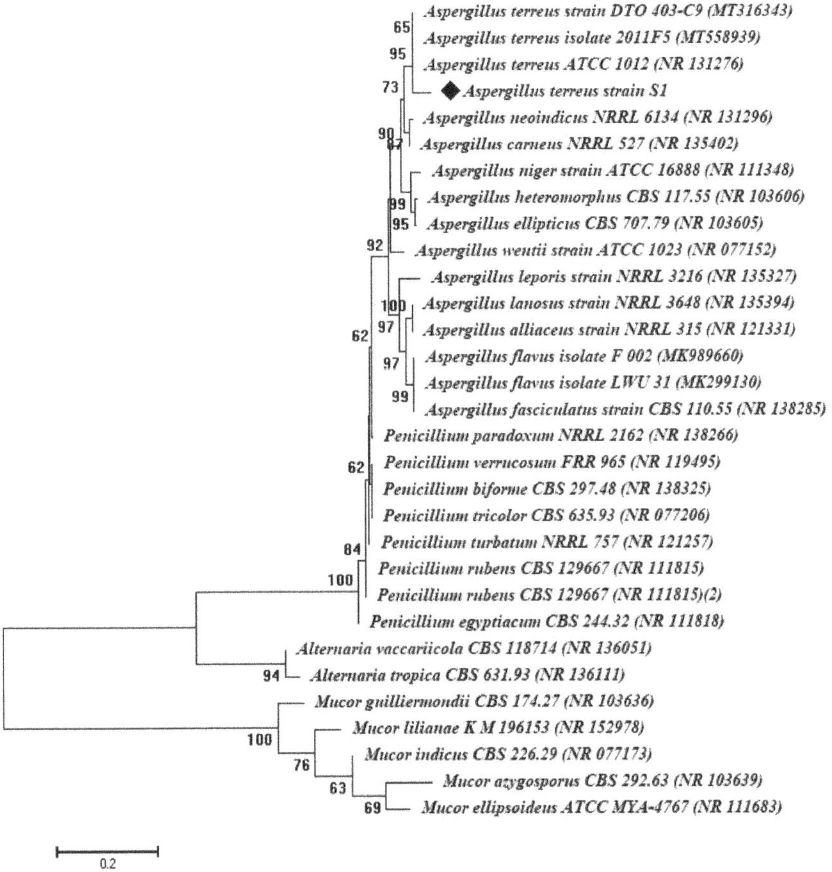

Figure 1. Phylogenetic tree of the fungal strain S1 with the sequences from NCBI. The symbol ♦ refers to ITS fragments retrieved from this study. The tree was constructed with MEGA 6.1 using the neighbor-joining method.

Aspergillus terreus has been characterized by its high secretion of various secondary metabolites such as chemicals (i.e., terrein, terreic acid, and terretonin), enzymes (Lipase, amylase, and reductase), fermentative compounds (i.e., polyketide compounds), and a wide range of by-products [28–30]. These various metabolites enable faster incorporation of *A. terreus* into various biotechnological applications. Therefore, we can benefit from these metabolites in the green synthesis of metal and metal oxide nanoparticles. Interestingly, *A. terreus* was utilized as a biocatalyst for the green synthesis of zinc, titanium, and magnesium nanoparticles [31].

2.2. Biogenic Synthesis of MgO-NPs

Metal and metal oxide nanoparticle synthesis by biological approaches can be used as an alternative to chemical and physical methods [32]. This phenomenon can be attributed to the advantages of biological synthesis such as the cost, the environmentally friendly nature, biocompatibility, scalability, and the avoidance of harsh synthesis conditions such as high temperature and pH [11]. Among biological entities, fungi can be identified as a promising tool for biogenic nanoparticle synthesis because of diverse metabolites and high metal tolerances [33]. Proteins and enzymes secreted by *A. terreus* strain S1 have an important role in the reduction of $Mg(NO_3)_2 \cdot 6H_2O$ to MgO-NPs and confer capping and stabilization features [31,34].

The initial observation for successful MgO-NP production is the color change from colorless to turbid white after stirring Mg $(NO_3)_2 \cdot 6H_2O$ with biomass filtrate. This change can be attributed to the role of *A. terreus*-secreted metabolites in the reduction of NO_3^- to NO_2 and then the reduction of Mg^{2+} to $Mg(OH)_2$ by liberated electrons. The as-formed $Mg(OH)_2$ was calcinated at 400 °C to form MgO-NPs [35].

The production of MgO-NPs was confirmed by measuring the maximum surface plasmon resonance (SPR) by UV-Vis spectroscopy. The morphological characteristics (size and shape), as well as distribution of biogenically synthesized NPs, are usually correlated with SPR [36]. In this respect, Nguyen et al. [37] reported that the size of biogenic MgO-NPs was smaller or larger according to $300 \leq SPR \geq 300$. In the current study, the maximum SPR value of biogenic MgO-NPs was detected at a wavelength of 280 nm (Figure 2), which confirms the formation of particles at the nanoscale. Further, the MgO-NPs synthesized by different extracts (flower, bark, leaf) of *Tecoma stans* (L.) showed maximum SPR peaks at 281 nm [37]. Moreover, the maximum absorption band of MgO-NPs synthesized by the floral extract of *Matricaria chamomilla* L. was observed at 230 nm [38].

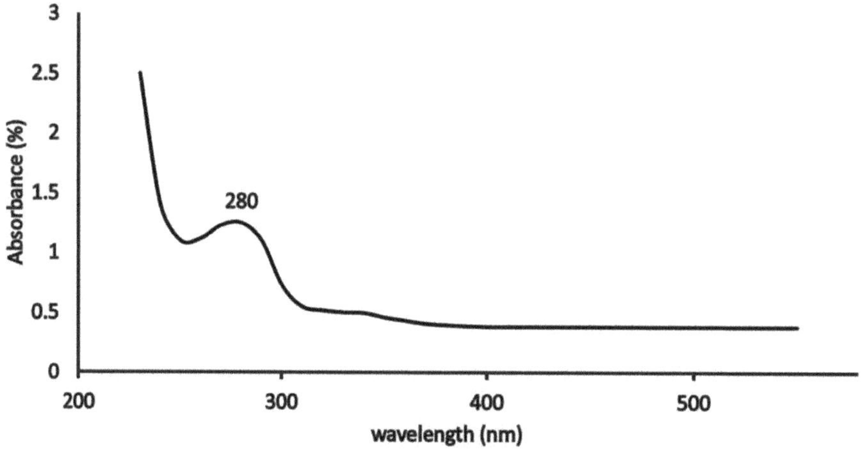

Figure 2. UV-Vis spectroscopy of myco-synthesized MgO-NPs showed maximum SPR at 280 nm.

2.3. Optimizing Biosynthesized MgO-NPs

The stability and biological activity of biogenic nanoparticles are usually influenced by environmental factors such as precursor concentration, contact time or incubation time, pH values, and incubation temperature. The investigated factors have various impacts on fungal-secreted metabolites such as enzymes, proteins, carbohydrates, and hence the reducing and stabilizing processes are affected [39]. Therefore, the optimization of these environmental factors will decrease the times required for biosynthesis, increase the NP stability, reduce the NP agglomeration, and finally support the productivity [40].

The activity of reducing agents differs according to the metal precursor concentration. In the current study, the absorbance band at λmax_{280} was increased by increasing the precursor concentration, and the maximum absorbance was achieved at 3 mM. By increasing the $Mg(NO_3)_2{\cdot}6H_2O$ concentration up to 3 mM, the absorbance was decreased (Figure 3A). According to obtained data, the fungal metabolites exhibited the optimum reduction of the metal precursor at 3 mM, whereas above and below this concentration, the biosynthesized MgO-NPs aggregated and hence decreased the absorbance band [41]. Muangban and Jaroenapiba [42] reported that tungsten oxide nanofibers tended to agglomerate by increasing the metal precursor concentration because of increasing nanoparticle size. Further, Jeevanandam et al. [43] study the effect of $Mg(NO_3)_2{\cdot}6H_2O$ concentration on the average particle size of MgO-NPs synthesized by *Aloe barbadensis* aqueous extract. It can be concluded that an increase in the concentration of the metal precursor leads to increased nanoparticle size and hence an increase in NP aggregation.

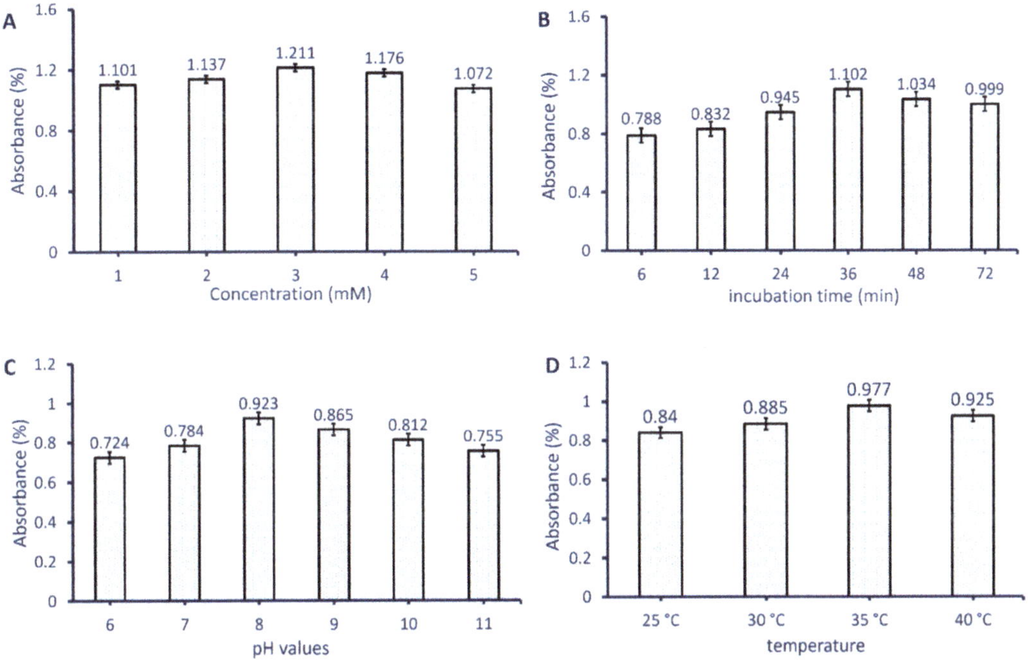

Figure 3. Optimizing factors for biogenic MgO-NPs using *A. terreus* strain S1. (**A**) denotes the different $Mg(NO_3)_2{\cdot}6H_2O$ concentrations; (**B**) denotes the contact time between biomass filtrate and optimum $Mg(NO_3)_2{\cdot}6H_2O$ concentration; (**C**) illustrates the effect of pH values, and (**D**) denotes the effect of incubation temperature on biogenic MgO-NPs.

The contact time or reaction time between the fungal biomass filtrate and optimum $Mg(NO_3)_2{\cdot}6H_2O$ concentration (3 mM) is considered a critical factor affecting the biogenic synthesis of MgO-NPs. The intensity of the color that formed was monitored by detecting the maximum absorbance band at λmax_{280}. Data illustrated in Figure 3B show that the optimum contact time between the metal precursor and fungal biomass filtrate was 36 min. At this time, the maximum color intensity was achieved because a large number of metal ions were reduced. On the other hand, the absorbance intensity was decreased by increasing contact time due to the aggregation of some MgO-NPs, and then the color intensity and particle size were reduced. Moreover, in the early contact time stages, the low number of metal ions was reduced and hence the SPR peak appeared broader [44,45]. Compatible with our study, the optimum contact time for the biogenic synthesis of MgO-NPs by *Aloe barbadensis* plant extract was 30 min [43]. Synthesis by chemical and physical methods such as microemulsion, sol-gel, co-precipitation, and solvothermal methods required a contact time of more than 48 min [46].

The effect of different pH values ranging between 6 and 11 on the biogenic synthesis of MgO-NPs by *Aspergillus terreus* strain S1 was shown (Figure 3C). Data analysis showed that the highest absorbance at λmax_{280} was accomplished at pH 8, which is evidence of maximum MgO-NP production. This behavior could be attributed to the stabilizing fungal metabolites including proteins and enzymes secreted by *Aspergillus terreus* strain S1 in an alkaline medium [47]. The metabolites present in the biomass filtrate of *A. carbonarious* D-1 were more active to reduce $FeCl_3 \cdot 6H_2O$ and $Mg(NO_3)_2.6H_2O$ to fabricate α-Fe_2O_3-NPs and MgO-NPs in an alkaline medium [48].

The activity of reducing agents involved in the biomass filtrate of *A. terreus* strain S1 is correlated with the incubation temperature. Therefore, it is important to detect the optimum temperature required for reducing, capping, and stabilizing MgO-NPs. In the current study, the effects of different incubation temperatures (25–40 °C) on the color intensity and hence biogenic MgO-NPs synthesis were investigated. Data showed that the metabolites involved in the *A. terreus* biomass filtrate were highly active as reducing agents at 35 °C (Figure 3D). The absorbance intensity was decreased at a temperature of more or less than 35 °C. This was attributed to the enzymes and proteins being more stable at this temperature. Hassan et al. [41] reported that the biomass filtrate of *Rhizopus oryaze* was more stable at an incubation temperature of 35 °C during MgO nanoparticle synthesis. Moreover, the size, shape, and stability of nanoparticles are correlated with the incubation temperature [43]. Interestingly, increasing the incubation temperature will lead to a higher diffusion coefficient, which decreases the reaction time needed to form stable particles and hence decreases the induction time [49].

2.4. Characterizations of Biogenically Synthesized MgO-NPs

2.4.1. Fourier Transform Infrared (FT-IR) Spectroscopy

FT-IR analysis is a powerful technique used for identifying the possible functional groups in the biomass filtrate of *A. terreus* strain S1 that are responsible for the reduction of metal precursors to form MgO-NPs [50]. FT-IR analysis scans at a wavenumber between 400 to 4000 cm^{-1} as shown in Figure 4. The result showed that several intense absorption peaks appeared at 3700, 3420, 2850, 2727, 2398, 1630, 1370, 1027, and 520 cm^{-1}. The peak observed at 3700 cm^{-1} signifies the –OH stretching band [51]. The broad peak at 3420 cm^{-1} corresponds to hydrogen bonds arising from NH_2 and OH groups in protein molecules [52]. The peaks observed at 2850, 2727, and 2398 cm^{-1} may correspond to the C-H stretching of methylene groups of proteins. The medium peaks observed at 1630 cm^{-1} correspond to the bending mode of primary amine (N—H) overlapped with either amide or carboxylate salt. The medium peak at 1370 cm^{-1} can be related to C–H bending vibrations of the aromatic tertiary amine group [53,54], whereas the peak at 1027 cm^{-1} matched the Mg–OH stretching [55] with the C-H out-of-plane bend. The peaks that appear between wavenumber 400–700 cm^{-1} confirm the presence of MgO at the nanoscale [51,53,56]. The peaks observed in FT-IR spectra reflect the capacity of metabolites present in the *A. terreus* strain S1 biomass filtrate to reduce, cap, and stabilize MgO-NPs.

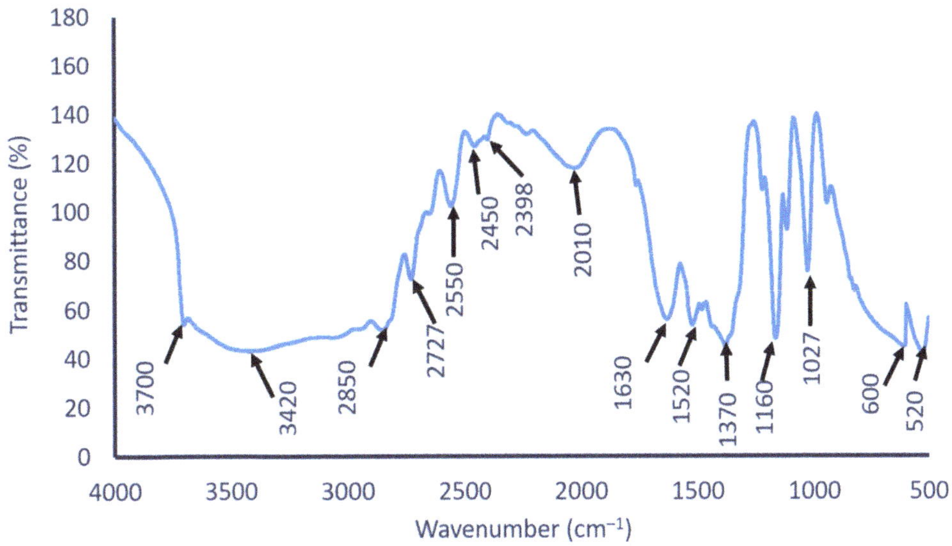

Figure 4. The FT-IR spectrum of myco-synthesized MgO-NPs fabricated by metabolites of *A. terreus* strain S1.

2.4.2. Transmission Electron Microscopy

TEM analysis was carried out to determine the approximate size and shape of the *A. terreus*-mediated MgO-NPs biosynthesized. Data illustrated in Figure 5A showed that the biogenic MgO-NPs synthesized by harnessing metabolites of *A. terreus* strain S1 had a spherical shape and well-dispersed narrow-sized particles surrounded with capping proteins and enzymes. TEM image measurement revealed that the sizes of biogenic MgO-NPs ranged between 8.0 and 38.0 nm with an average diameter of 19.91 $\pm$ 9.9 nm (Figure 5B). In our recent study, *Aspergillus carbonarious* D-1 mediated green synthesis of spherical MgO-NPs with an average size of 20–80 nm [48]. Compatible with our study, the particle size of spherical MgO-NPs synthesized by *Aspergillus terreus* TFR was 10 nm with a PDI value of 0.236 and 100% conversion of the precursor compound into nanoparticles [31]. Moreover, the plant extract of *Pisidium guvajava* and *Aloe vera* mediated biosynthesis of MgO-NPs with an average size range of 50 nm [57]. It is well known from previously published studies that the biological activity of NPs is increased by decreasing the average size [58,59]. The growth inhibition percentages of *Bacillus subtilis* after treatment with different sizes of biosynthesized MgO-NPs (35.9 nm, 47.3 nm, and micron size 2145.9 nm) were 96.1%, 94.5%, and 75.7%, respectively [60]. In this study, the size of fabricated MgO-NPs was small (8.0–39.0 nm), and we therefore predicted their integration in different biomedical and biotechnological applications.

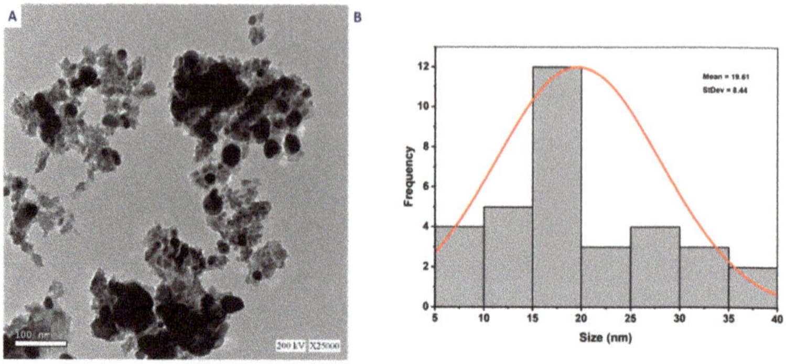

Figure 5. Characterization of MgO-NPs synthesized by *A. terreus* strain S1, (**A**) TEM image; (**B**) size distribution according to the TEM image.

2.4.3. Scanning Electron Microscopy—Energy Dispersive X-ray (SEM-EDX)

The SEM-EDX analysis is considered a useful technique to study the topographical structure of biosynthesized MgO-NPs, aggregation, and chemical compositions. As seen in Figure 6A the MgO-NPs synthesized by *A. terreus* strain S1 were well-dispersed and spherical. Moreover, the presence of Mg and O ions in the sample was confirmed by the EDX profile. Data showed that the weight percentages of Mg and O were 18.3% and 28.1%, respectively, whereas the atomic percentages were 10.9% and 25.8%, respectively (Figure 6B). Further, the successful fabrication of MgO-NPs was confirmed by the presence of a Mg peak at an energy of 0.5 to 1.5 KeV [53] as shown in Figure 6B.

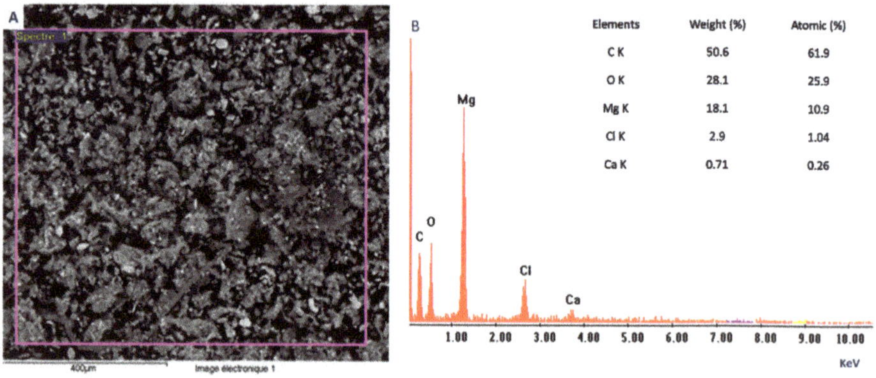

Figure 6. Characterization of MgO-NPs synthesized by *A. terreus* strain S1, (**A**) SEM image; (**B**) EDX profile.

On the other hand, other elements, C, Cl, and Ca, in the MgO-NP sample were detected by EDX profiles with weight percentages of 50.6%, 2.49%, and 0.71%, respectively. The presence of these additional peaks indicates the presence of some impurities in the sample, which was confirmed by XRD analysis. Some investigators attributed the presence of additional peaks in the EDX profile to the hydrolysis of enzymes, proteins, and other fungal metabolites that act as capping and stabilizing agents by X-ray [61].

2.4.4. X-ray Diffraction (XRD) Analysis

The crystallographic structure of optimized MgO-NPs was studied using XRD analysis. XRD spectra (Figure 7A) showed major five intense peaks at 2θ values of 36.94° (111), 42.68° (200), 62.4° (220), 74.28° (311), and 78.62° (222). The identified diffraction peaks matched well with the crystallographic structure according to JCPDS standard (JCPDS file No. 89-7746) [62]. The presence of fine additional peaks in XRD spectra confirms data obtained by EDX analysis, i.e., the sample contained some impurities. According to XRD spectra, oxides represented by $Mg(OH)_2$ and MgO existed in the biosynthesized sample. The observed peaks at 2θ° of 36.9° (111), 75.08° (311), and 78.64° (222) corresponded to $Mg(OH)_2$, whereas, the diffraction peaks at 2θ° of 42.16° (200), and 62.6° (220) signified cubic MgO-NPs [37]. The average crystallite size can be calculated according to XRD analysis using the Debye–Scherrer equation, which was found to be <20 nm.

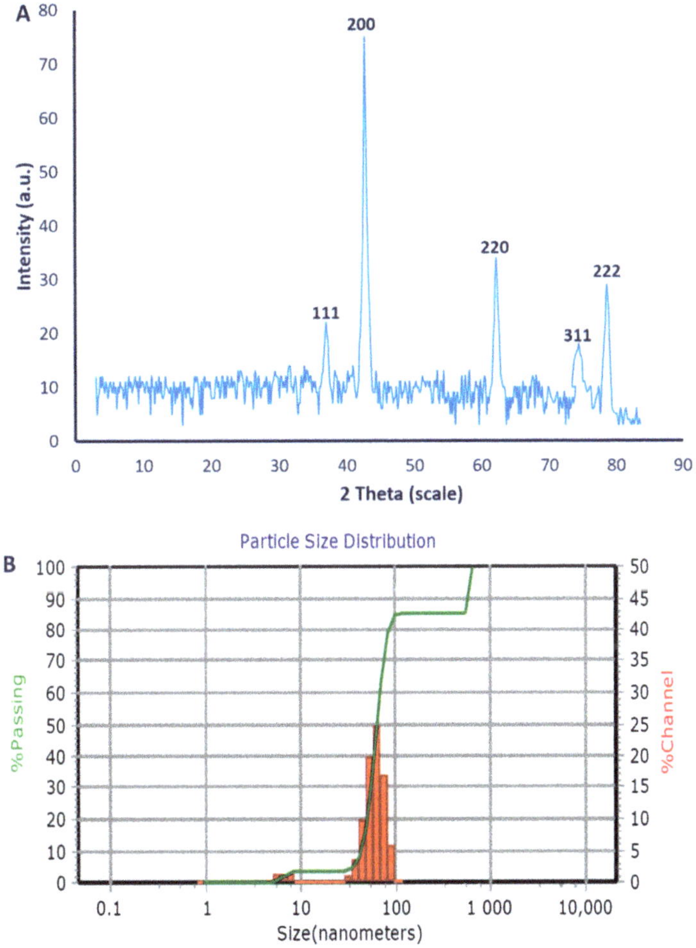

Figure 7. (**A**) The XRD analysis of the crystallographic structure; (**B**) the DLS analysis of biogenically synthesized MgO-NPs.

2.4.5. Dynamic Light Scattering (DLS)

The DLS technique is used to investigate the size and dispersion of MgO-NPs in the colloidal solution through a reaction of light beams with biogenically synthesized MgO-NPs [63]. In the current study, the average size of biogenic MgO-NPs was 40.6 nm, 60.1 nm, and 5.6 nm for volume intensities of 10%, 81.3%, and 8.7% of the colloidal solution (Figure 7B). As shown from the DLS analysis, the size of MgO-NPs was larger than that acquired from other techniques such as TEM and XRD. This is attributed to the coating agent that capped and stabilized the surface of NPs [64,65]. Furthermore, the larger size from DLS may be due to the non-homogenous NP distribution in colloidal solution [66]. Moreover, the Solvation spheres around the nanoparticles may be a factor for the size increase.

The polydispersity index (PDI) refers to the homogeneity percentages of NPs in the colloidal solution. The homogeneity percentages are increased or decreased when the PDI value is lower than or higher than 0.4, respectively. On the other hand, the NP colloidal solution is heterogenous when the PDI value ≥ 1. The obtained data demonstrated that the PDI value of MgO-NPs synthesized by the *A. terreus* strain S1 was 0.2, which indicates the high homogeneity of the colloidal solution.

2.5. Antimicrobial Activity

The activity of biogenically synthesized MgO-NPs to inhibit the growth of pathogenic Gram-positive bacteria represented by *Staphylococcus aureus, Bacillus subtilis*, Gram-negative bacteria including *Pseudomonas aeruginosa*, and *Escherichia coli*, and unicellular fungi of *Candida albicans* was studied by the agar well-diffusion method. Analysis of variance showed that the antimicrobial activity of MgO-NPs against selected pathogenic microbes was dependent on the concentration; the activity increased by increasing the NP concentration. The obtained data are compatible with published investigations about the relationship between the activity of NPs and their concentrations [57,59,67]. Results showed that the biogenic MgO-NPs synthesized by *A. terreus* strain S1 exhibited antimicrobial activity at 200 μg mL^{-1} against all tested pathogenic microbes as follows: *C. albicans* (12.8 ± 0.3 mm), *E. coli* (11.3 ± 0.6 mm), *P. aeruginosa* (14.7 ± 1.9 mm), *S. aureus* (11.3 ± 0.6 mm), and *B. subtilis* (13.3 ± 1.9 mm) (Figure 8). Recently, MgO-NPs synthesized by *Rhizopus* oryaze E3 showed antimicrobial activity with varied ZOIs, e.g., *B. subtilis* (11.5 ± 0.5 mm), *S. aureus* (10.6 ± 0.4 mm), *E. coli* (14.3 ± 0.7 mm), *P. aeruginosa* (13.7 ± 0.5 mm), and *C. albicans* (14.7 ± 0.6 mm) at a concentration of 200 μg mL^{-1} [41]. Moreover, a *Swertia chirayaita* plant extract mediated the green synthesis of MgO-NPs with antibacterial activity against *S. aureus, E. coli*, and *S. epidermidis*, with ZOIs of 14, 15, and 12 mm, respectively [18].

The minimum inhibitory concentration (MIC) is defined as the lowest concentration of an active substance that inhibits microbial growth. It is important to detect MIC values for active compounds against pathogenic microbes especially if these compounds are integrated into biomedical applications. To achieve this goal in the current study, the activity of different concentrations (150, 100, 50, and 25 μg mL^{-1}) of MgO-NPs was investigated. Data analysis showed that the MIC value for *P. aeruginosa* was 50 μg mL^{-1} with a ZOI of 8.3 ± 0.3 mm, whereas *E. coli, C. albicans, S. aureus*, and *B. subtilis* had an MIC value of 100 μg mL^{-1} with ZOIs of 8.0 ± 0.0, 8.7 ± 0.9, 8.0 ± 0.0, and 9.3 ± 0.6 mm, respectively (Figure 8).

The inhibitory effect of biogenic MgO-NPs can be attributed to different mechanisms such as (1) producing reactive oxygen species (ROS), (2) interaction between MgO-NPs and microbial cell walls that ultimately lead to cell death, (3) discharge of Mg^{2+} ions into the cell, and (4) alkaline effects of MgO on the microbial cell [68–70]. In the current study, Gram-negative *P. aeruginosa* was the most sensitive microorganism toward biosynthesized MgO-NPs, and this phenomenon can be attributed to differences in cell wall structures between Gram-positive and Gram-negative bacteria. The cell wall of Gram-positive bacteria is characterized by a thick layer of peptidoglycan in contrast to Gram-negative bacteria that have a thin layer of peptidoglycan plus lipopolysaccharides (LPS). The positive charge of NPs is strongly attracted to the LPS-negative charge, and hence it is the deposit on the bacterial cell membrane that ultimately disrupts selective permeability [71]. Moreover, MgO-NPs can stop the communication tools, which is quorum sensing between microbial strains, and hence the physiological functions and various microbial activities fail to continue [72,73].

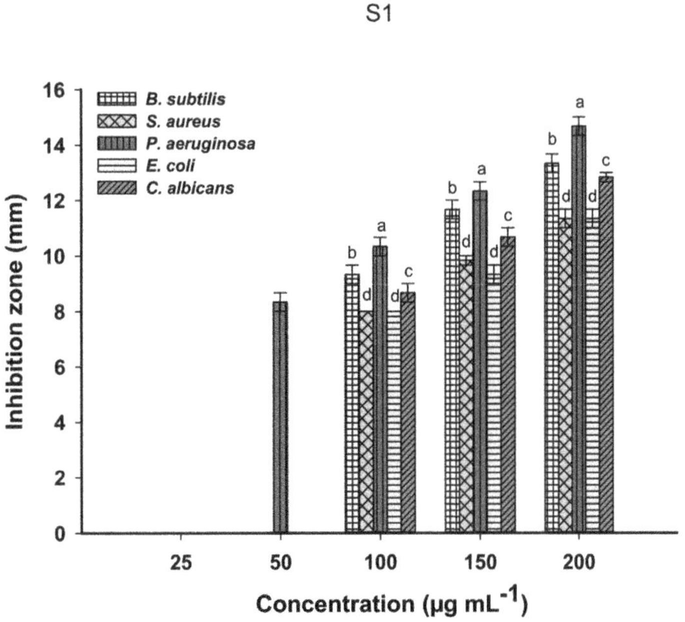

Figure 8. The antimicrobial activity of different concentrations of biogenically synthesized MgO-NPs against *Staphylococcus aureus*, *Bacillus subtilis*, *Pseudomonas aeruginosa*, and *Escherichia coli*, and *Candida albicans*. Different letters (a, b, c, and d) on bars at the same concertation refer that the mean values are significantly different ($p \leq 0.05$) (n = 3).

2.6. Biotreatment of Tanning Effluent

The main challenge facing different countries is discovering new active compounds that can be utilized in the treatment of different industrial effluents especially the countries that suffer from water scarcity. Nanotechnology especially that defined as green nanotechnology provides a new approach for producing new active compounds characterized as eco-friendly, with a large surface area, high stability, cost-effectiveness, and hence can be utilized to adsorb different contaminants [74,75]. Among highly contaminated industrial effluent is tanning wastewater, which appears a greenish-blue color because of the presence of chrome ions and other materials at high concentrations [76]. Therefore, the discharge of tanning effluent directly in the surrounding environment without treatment hinders sunlight penetration and hence decreases the pollutant oxidation process [77]. In the current study, the potential of different concentrations (0.25, 0.5, 0.75, and 1.0 µg mL^{-1}) of biogenically synthesized MgO-NPs for treatment of tanning effluent at different interval times (30, 60, 90, 120, 150, 180, and 240 min) was investigated. Data recorded in Table 1 showed that the efficiency of biogenic MgO-NPs to decolorize the tanning effluent was concentration and time-dependent, meaning the decolorization percentages were increased as concentrations and contact time increased. This phenomenon can be attributed to the increase in adsorption sites by increasing the concentration of adsorbents [78]. At the lowest MgO-NP concentration (0.25 µg mL^{-1}), the decolorization percentages ranged from 16.1 ± 1.6% after 30 min contact time to 46.6 ± 3.2% after 240 min as compared to the control (5.8 ± 0.4% after 240 min). At the highest concentration (1.0 µg mL^{-1}), the adsorption sites increased and hence the decolorization increased to reach 89.1 ± 1.6% after 120 min. At this high concentration, the decolorization percentages were not significant at times 150, 180, and 240, i.e., 96.8 ± 1.7%, 97.5 ± 1.6%, and 97.7 ± 1.7% (Table 1, Figure 9). The time and concentration are considered the main factors that should be taken into consideration on a large or industrial scale; therefore, 1.0 µg mL^{-1} of MgO-NPs and 150 min were chosen as the optimal conditions for decolorization of tanning wastewater and to study the physicochemical parameters that indicate successful treatment.

Table 1. Decolorization percentages (%) of tanning effluent using different concentrations (0.25, 0.5, 0.75, and 1.0 μg mL^{-1}) of myco-synthesized MgO-NPs at different contact times (30, 60, 90, 120, 150, 180, and 240 min).

MgO-NPs Concentration	Decolorization Percentages (%) after the Time (min)						
	30 min	60 min	90 min	120 min	150 min	180 min	240 min
Control	1.9 ± 0.2	2.2 ± 0.2	3.1 ± 0.3	3.8 ± 0.4	4.5 ± 0.3	5.2 ± 0.3	5.8 ± 0.4
0.25 μg mL^{-1}	16.1 ± 1.6	18.4 ± 1.7	21.3 ± 2.1	26.5 ± 2.2	31.3 ± 2.2	38.5 ± 2.05	46.6 ± 3.2
0.5 μg mL^{-1}	28.5 ± 2.7	38.3 ± 2.2	45.7 ± 3.1	50.7 ± 3.7	54.8 ± 2.01	59.4 ± 2.5	61.6 ± 1.8
0.75 μg mL^{-1}	37.8 ± 2.2	49.3 ± 3.3	58.9 ± 2.6	69.7 ± 1.9	78.4 ± 1.7	81.4 ± 0.3	82.2 ± 1.7
1.0 μg mL^{-1}	53.5 ± 3.6	67.4 ± 1.9	77.8 ± 1.3	89.1 ± 1.6	96.8 ± 1.7	97.5 ± 1.6	97.7 ± 1.7

Data are represented as the mean ± SD (n = 3).

Figure 9. The decolorization of tanning effluent using biogenically synthesized MgO-NPs by *A. terreus* strain S1.

The successful treatment process by MgO-NPs is monitored by measuring the main factors including pH, BOD, COD, TDS, TSS, and conductivity. These factors are high in tanning effluent due to hazardous chemicals, bicarbonates, calcium phosphates, chlorides, sulfates, nitrates, potassium, sodium, and various dissolved salts [79]. Moreover, the values of these factors are wide-ranging according to the chemicals used, tannery size, type of products, and water used [80]. The alkalinity of crude tanning wastewater is because of high amounts of carbonates and bicarbonates used during tanning steps [81]. Moreover, the high conductivity values of crude wastewater are because of the high content of salts and acids such as sodium and chrome salts. Further, high values of other factors such as TDS, TSS, and conductivity have adverse impacts on plants and aquatic eco-systems [6]. Data recorded in Table 2 showed a high level of measured factors in un-treated tanning effluents as follows: pH (10.5), TSS (8776.3 ± 5.8 mg L^{-1}), TDS (15,720 ± 4.1 mg L^{-1}), BOD (2345.7 ± 7.0 mg L^{-1}), COD (641.7 ± 4.7 mg L^{-1}), and conductivity (26,750.7 ± 6.0 S m^{-1}). MgO-NPs exhibited high efficacy to decrease the factors of tanning effluent as follows: pH (8), TSS (172.0 ± 4.8 mg L^{-1}), TDS (252.0 ± 4.1 mg L^{-1}), BOD (255.0 ± 5.1 mg L^{-1}), COD (18.0 ± 1.9 mg L^{-1}), and conductivity (628.0 ± 3.8 S m^{-1}). As shown, the MgO-NPs can remove BOD, COD, TSS, TDS, and conductivity with percentages of 89.1%, 97.2%, 98.04%, 98.3%, and 97.7%.

Table 2. Physicochemical characterization and chromium ion adsorption from tanning effluent by MgO-NPs.

Physicochemical Parameters	Control	After MgO-NPs Treatment	Removal Percentages (%)
pH	10.5	8	-
TSS (mg L^{-1})	8776.3 ± 5.8 ^{a}	172.0 ± 4.8 ^{b}	98.04
TDS (mg L^{-1})	15,720 ± 4.1 ^{a}	252.0 ± 4.1 ^{b}	98.3
BOD (mg L^{-1})	2345.7 ± 7.0 ^{a}	255.0 ± 5.1 ^{b}	89.1
COD (mg L^{-1})	641.7 ± 4.7 ^{a}	18.0 ± 1.9 ^{b}	97.2
Conductivity (S m^{-1})	26,750.7 ± 6.0 ^{a}	628.0 ± 3.8 ^{b}	97.7
Cr mg L^{-1}	835.3 ± 2.5 ^{a}	21.0 ± 0.7 ^{b}	97.5

Different letters in the same row are significantly different ($p \leq 0.05$) based on the Tukey LSD test. Data are presented as the mean ± SD (n = 3).

2.7. Chromium Ion Removal

Chromium is the main heavy metal released into the environment from different industries such as textiles, electroplating, mining, and fertilizer manufacturing. Leather tanning is considering the main source for discharge of chromium [82]. The toxicity of chromium ions can be attributed to their mutagenic and carcinogenic properties, causing cancer of the lung and digestive tract, nausea, vomiting, diarrhea, epigastric pain, and hemorrhaging [83]. Therefore, there is an urgent need to discover new, high efficacy and eco-friendly adsorption compounds to remove heavy metals. MgO-NPs are characterized by cost-effectiveness, nontoxicity, high adsorption efficacy, abundance, eco-friendly, and biocompatibility [13]. Data represented in Table 2 showed the high efficacy of biogenically synthesized MgO-NPs to decrease the chromium ion concentration from 835.3 ± 2.5 mg L^{-1} to 21.0 ± 0.7 mg L^{-1} with a removal percentage of 97.5%. The removal mechanism of heavy metals by MgO-NPs is dependent on precipitation and adsorption, whereas other nanomaterials such as nanotubes, NiO, ZrO_2, and TiO_2 are dependent on adsorption only [84]. The adsorption process for these nanomaterials is controlled by the size, shape, surface, and textural properties [85]. On the other hand, the dissociation of OH^- from the pre-synthesized $Mg(OH)_2$ and the synergistic effects between precipitation and adsorption can be a reason for high MgO-NP adsorption [86]. Another mechanism to explain Cr removal is electrostatic attraction/repulsion [87]. The main Cr form that exists in an aquatic medium at pH > 6.0 is CrO_4^{2-} [88]. The electrostatic attraction was achieved between the positive charge of NPs and the negative charge on the surface of chromate [89]. In our recent study, MgO-NPs synthesized by harnessing metabolites of *Aspergillus niger* F1 and *Rhizopus oryzae E3* showed removal of Cr from tanning wastewater with percentages of 94.2 ± 1.2% and 95.6 ± 1.6, respectively [41,76]. Therefore, this study provides a new, highly active nanomaterial that has the potential to adsorb various contaminants including heavy metals from tanning effluent and has antimicrobial activity.

3. Materials and Methods

3.1. Reagents and Materials

Chemicals used in the current study including magnesium nitrate hexahydrate (Mg $(NO_3)_2{\cdot}6H_2O$) and sodium hydroxide (NaOH) are analytical grade and were obtained from Sigma Aldrich, Cairo, Egypt. Malt Extract agar (MEA) media for fungal isolations and Muller Hinton agar media for antimicrobial activity were readymade (Oxoid, Thermo Fisher Scientific Inc. USA). The tannery wastewater was collected from Robbiki Leather City, 10th of Ramadan, Cairo, Egypt (GPS: N: 30° 17′ 898″, E: 31° 76′ 840″).

3.2. Isolation and Identification of the Fungal Strain

The fungal strain S1 used in the current study for the biosynthesis of MgO-NPs was isolated from a soil sample collected from El-Sharqia Governorate, Egypt (GPS: N: 30°41′588.38″, E: 31°56′211.84″). The isolation procedures were achieved according to Hashem, et al. [90] as follows: 100 µL of the fifth dilution of a diluted soil sample was inoculated onto MEA plates and incubated for 3–4 days at 28 ± 2 °C. All different shapes and colors of fungal colonies were picked and re-inoculated again onto new MEA plates for purification. Finally, the purified colony was preserved on an MEA slant for further work.

The identification was accomplished by routine work including morphological and microscopic characterization. The primary identification was confirmed using internal transcribed spacer (ITS) sequence analysis. The ITS rDNA region was amplified using primers for ITS1 f (5-CTTGGTCATTTAGAGGAAGTAA-3) and ITS4 (5-TCCTCCGCTTATTGATATGC-3) [91]. The PCR mixture contained 1X PCR buffer, 0.5 mM $MgCl_2$, 2.5 U Taq DNA polymerase (QIAGEN, Germantown, MD 20874, USA), 0.25 mM dNTP, 0.5 µL of each primer, and 1 µg of extracted genomic DNA. The PCR was performed in a DNA Engine Thermal Cycler (PTC-200, BIO-RAD, USA) with a program of 94 °C for 3 min, followed by 30 cycles of 94 °C for 30 s, 55 °C for 30 s, and 72 °C for 1 min, followed by a final extension performed at 72 °C for 10 min. The PCR product was checked for the expected sizes on 1% agarose gel and was sequenced by Sigma Company for scientific research, Egypt, with the two primers. The sequence was compared against the GenBank database using the NCBI BLAST tool. Multiple sequence alignment was done using the Clustal Omega software package (https://www.ebi.ac.uk/Tools/msa/clustalo (accessed on 28 June 2021)), and a phylogenetic tree was constructed using the neighbor-joining method with MEGA (Version 6.1) software, with confidence tested by bootstrap analysis (1000 repeats).

3.3. Biogenic Synthesis of MgO-NPs

The fungal biomass filtrate utilized as a biocatalyst for the biogenic synthesis of MgO-NPs was prepared through the inoculation of three disks (0.8 cm in diameter) of fungal strain S1 into 100 mL of malt extract broth (MAB) medium and incubated for five days at 28 ± 2 °C under shaking conditions (150 rpm). At the end of the incubation period, the inoculated MAB medium was centrifuged to collect the fungal biomass. The collected fungal biomass (10 g) was resuspended in 100 mL distilled water for 48 h. at 28 ± 2 °C under shaking conditions. The previous mixture was centrifuged at 15,000 rpm for three minutes; the upper layer (supernatant) was collected and used for biosynthesis of MgO-NPs as follows.

$Mg(NO_3)_2{\cdot}6H_2O$ (76.6 mg) was dissolved in 10 mL dis. H_2O and the volume made to 100 mL by adding 90 mL fungal biomass filtrate to a final concentration of 3 mM. The previous mixture was incubated for 24 h at room temperature. At first, $Mg(OH)_2$ appeared as a turbid white precipitate that was collected and washed with dis. H_2O before drying at 100 °C for one hour (Equation (1)). After that, $Mg(OH)_2$ was subjected to calcination at 400 °C for 3 h to form MgO-NPs (Equation (2)) [92].

$$Mg(NO_3)_2{\cdot}6H_2O + H_2O \underset{\text{Metabolites}}{\overset{\text{Fungal}}{\rightarrow}} Mg(OH)_2 \quad (1)$$

$$Mg(OH)_2 \overset{400^{\circ}C}{\rightarrow} MgO \quad (2)$$

Optimization factors of MgO-NPs biosynthesis.

The physical and chemical factors affecting the production as well as the distribution of MgO-NPs were optimized. Different factors such as temperature, contact times, different pH values, and Mg $(NO_3)_2{\cdot}6H_2O$ concentrations are investigated by detecting the maximum surface plasmon resonance by a UV-Vis spectrophotometer (Jenway 6305, Staffordshire, UK). The different contact times (6, 12, 24, 36, 48, and 72 min) between fungal biomass filtrate and Mg $(NO_3)_2{\cdot}6H_2O$ were assessed. Moreover, the incubation temperatures (25 °C, 30 °C, 35 °C, and 40 °C), different pH values (6, 7, 8, 9, 10, and 11), and different Mg $(NO_3)_2{\cdot}6H_2O$ concentrations (1–5 mM) were assessed. At the end of each experiment, 1.0 mL of the sample was withdrawn to measure the color intensity at maximum SPR at λ_{max} = 280 nm.

3.4. Characterization of Biosynthesized MgO-NPs

Fourier transform infrared (FT-IR) spectroscopy (Agilent system Cary 660 FT-IR model) was used to inspect the functional groups present in the fungal biomass filtrate and involved in the reduction and stabilization of MgO-NPs. The MgO-NP sample was mixed with KBr and scanned in the range of 400 to 4000 cm^{-1}.

The physicochemical characterizations of biogenically synthesized MgO-NPs were accomplished using Transmission Electron Microscopy (TEM) (JEOL 1010, Japan, acceleration voltage of 200 KV) to detect the MgO-NP sizes and shapes. A few drops of MgO-NPs suspension were added to the carbon-copper grid, which was subjected to vacuum desiccation before placing on a TEM-holder for analysis [93]. The elemental analysis of biogenically synthesized MgO-NPs was measured using Scanning Electron Microscopy connected to energy dispersive X-rays (SEM-EDX) ((JEOL, JSM-6360LA, Japan). Moreover, the crystallographic structure of MgO-NPs was investigated using X-ray diffraction (XRD) analysis by an X'Pert pro diffractometer (Philips, Eindhoven, Netherlands). The XRD analysis condition was achieved at 2θ values of 4° to 80°, Ni-filtered Cu Ka as an X-ray radiation source, and the operating voltage and current were 40 KV and 30 mA, respectively. Based on XRD analysis, the average size of MgO-NPs was measured using the Debye–Scherrer equation [94] as follows:

$$D = 0.9\lambda / \beta \mathrm{Cos}\theta \quad (3)$$

where D is the average particle size; 0.9 is the Scherrer's' constant; λ is the wavelength of X-ray radiation (0.154 nm); β and θ are the half of maximum intensity and Bragg's angle, respectively.

The size distribution of biogenic MgO-NPs in colloidal solution was detected by dynamic light scattering (DLS) analysis. The sample was subjected to measurement by Zeta sizer nano series (Nano ZS), Malvern, UK.

3.5. Antimicrobial Activity

The efficacy of biogenic MgO-NPs to inhibit the growth of pathogenic Gram-positive bacteria (*Bacillus subtilis* ATCC 6633 and *Staphylococcus aureus* ATCC 6538), Gram-negative bacteria (*Pseudomonas aeruginosa* ATCC 9022 and *Escherichia coli* ATCC 8739), and unicellular fungi represented by *Candida albicans* ATCC 10231 was investigated using the agar well-diffusion method. Under aseptic conditions, each bacterial strain was inoculated into Mueller–Hinton agar medium (Oxid, ready-prepared), whereas the unicellular fungi were inoculated into yeast extract peptone dextrose (YEPD) agar medium (containing g L^{-1}:glucose, 20; peptone, 20; yeast extract, 10; agar, 20; distilled water, 1000 mL). Three wells (0.7 cm diameter) were prepared in the inoculated plates and filled with 100 μL of biosynthesized MgO-NPs (200 μg mL^{-1}). Different concentrations of MgO-NPs (150, 100, 50, and 25 μg mL^{-1}) were prepared to detect the minimum inhibitory concentrations (MIC). The plates were kept in the refrigerator for 1.0 h before incubation at 35 ± 2 °C for 24 h. At the end of the incubation period, the diameters of the inhibition zone (ZOI) that appeared around each well were measured in mm [76]. The experiment was carried out in triplicate.

3.6. Tanning Effluent Treatment and Bio-Adsorption of Chromium Ions

The potential of MgO-NPs to decolorize the tanning effluent was investigated. Briefly, the tanning effluent was mixed with MgO-NPs (0.25, 0.5, 0.75, and 1.0 µg mL^{-1}) for different contact times (30, 60, 90, 120, 150, 180, and 240 min) in a 250 mL conical flask under shaking conditions (150 rpm). Each treatment was performed in triplicate. The formed mixture was stirred for 30 min before the experiment reached absorption/desorption equilibrium. The decolorization ratio was measured at the end of each different contact time as follows.

Approximately 1.0 mL of the mixture (tanning effluent with MgO-NPs) was withdrawn, centrifuged at 5000 rpm for 8 min, and used to detect the optical density of the treated tanning effluent at the maximum λ_{max} = 550 nm by a spectrophotometer (721 spectrophotometers, M-ETCAL). The decolorization ratio (%) of the tanning effluent was measured using the following equation [48].

$$D(\%) = \frac{C_0 - C_t}{C_0} \times 100 \quad (4)$$

where, D is the decolorization ratio %; C_0 is the absorbance at zero time; C_t is the absorbance after specific time t (min).

At the optimum contact time and suitable MgO-NP concentration, the chemical parameters including biological oxygen demand (BOD), chemical oxygen demand (COD), total dissolved solids (TDS), total suspended solids (TSS), and conductivity were calculated according to the standard recommended methods [95].

Chromium (Cr) ion was the main common heavy metal present in the tanning effluent. Therefore, the Cr ion concentration before and after treatment with optimum MgO-NP concentration and optimum contact time was measured using atomic adsorption spectroscopy (A PerkinElmer Analyst 800 atomic spectrometer).

3.7. Statistical Analysis

The means of three replications and standard error (±SE) were calculated for all the results obtained. Data were subjected to statistical analysis by the statistical package SPSS v17. The mean difference comparison between the treatments was analyzed by *t*-tests or analysis of variance (ANOVA) and subsequently by Tukey's HSD test at $p < 0.05$.

4. Conclusions

In the current study, MgO-NPs were fabricated through the reduction of Mg $(NO_3)_2 \cdot 6H_2O$ by metabolites secreted by *A. terreus* strain S1. The first indicator for successful MgO-NP synthesis was a color change from colorless to turbid white and detection of maximum surface plasmon resonance at 280 nm. The various parameters such as metal precursor concentration, contact time, temperature, and pH values that affect the production process were optimized. The physicochemical characterization was achieved by TEM, SEM-EDX, XRD, DLS, and FT-IR spectroscopy. The role of fungal metabolites in the reduction, capping, and the stabilizing process was detected by FT-IR. Moreover, the MgO-NP size (8.0 to 38.0 nm), crystallographic structure, qualitative and quantitative compositions, and dispersion of NPs in colloid solution were confirmed by TEM, XRD, SEM-EDX, and DLS analyses, respectively. The antimicrobial activity of biosynthesized MgO-NPs was assessed against pathogenic *S. aureus*, *B. subtilis*, *P. aeruginosa*, *E. coli*, and *C. albicans*. Data showed that the inhibitory action of MgO-NPs was concentration-dependent. Further, the MIC value was detected as 50 µg mL^{-1} for *P. aeruginosa* with ZOI of 8.3 ± 0.3 mm, whereas *E. coli*, *C. albicans*, *S. aureus*, and *B. subtilis* had an MIC value 100 µg mL^{-1} with ZOIs of 8.0 ± 0.0, 8.7 ± 0.9, 8.0 ± 0.0, and 9.3 ± 0.6 mm, respectively. Moreover, biogenically synthesized MgO-NPs exhibited the ability to decolorize the greenish-blue color of tanning effluent with a percentage of 96.8 ± 1.7% after 150 min. At these decolorization percentages, the physicochemical parameters of tanning effluent including TSS, TDS, BOD, COD, and conductivity were highly reduced. Finally, the MgO-NPs showed high removal of chromium ions with a percentage of 97.5%. This study provides a promising eco-friendly, cost-effective, and biocompatible nanomaterial for various applications.

Author Contributions: Conceptualization, S.E.-D.H. and A.F.; methodology, E.S.; A.M.E.; S.E.-D.H.; S.S.S.; A.A.R.; M.H.; A.F.; software, E.S.; S.E.-D.H.; A.F.; validation, E.S.; A.M.E.; S.E.-D.H.; S.S.S.; A.A.R.; A.F.; formal analysis, E.S.; A.M.E.; S.E.-D.H.; S.S.S.; A.A.R.; M.H.; A.F.; investigation, E.S.; A.M.E.; S.E.-D.H.; S.S.S.; A.A.R.; M.H.; A.F.; resources, E.S.; A.M.E.; S.E.-D.H.; S.S.S.; A.A.R.; A.F. data curation, E.S.; A.M.E.; S.E.-D.H.; S.S.S.; A.A.R.; M.H.; A.F. writing—original draft preparation, E.S.; A.M.E.; A.A.R.; A.F. writing—review and editing, E.S.; A.M.E.; S.E.-D.H.; S.S.S.; A.A.R.; M.H.; A.F.; H.A.S.; E.M.S.; F.M.S. visualization, E.S.; A.M.E.; S.E.-D.H.; A.F.; supervision, E.S., S.E.-D.H.; A.F.; project administration, E.S., S.E.-D.H.; A.F.; funding acquisition, F.M.S., E.M.S. and H.A.S. All authors have read and agreed to the published version of the manuscript.

Funding: This research received no external funding.

Data Availability Statement: The data presented in this study are available on request from the corresponding author.

Acknowledgments: We appreciate and thank Taif University for the financial support for Taif University Researchers Supporting Project (TURSP-2020/07), Taif University, Taif, Saudi Arabia.

Conflicts of Interest: The authors declare no conflict of interest.

References

1. Song, M.; Wang, R.; Zeng, X. Water resources utilization efficiency and influence factors under environmental restrictions. *J. Clean. Prod.* **2018**, *184*, 611–621. [CrossRef]
2. Jobby, R.; Jha, P.; Yadav, A.K.; Desai, N. Biosorption and biotransformation of hexavalent chromium [Cr(VI)]: A comprehensive review. *Chemosphere* **2018**, *207*, 255–266. [CrossRef] [PubMed]
3. Selim, M.T.; Salem, S.S.; Mohamed, A.A.; El-Gamal, M.S.; Awad, M.F.; Fouda, A. Biological Treatment of Real Textile Effluent Using *Aspergillus flavus* and *Fusarium oxysporium* and Their Consortium along with the Evaluation of Their Phytotoxicity. *J. Fungi* **2021**, *7*, 193. [CrossRef] [PubMed]
4. Hamza, M.F.; Hamad, N.A.; Hamad, D.M.; Khalafalla, M.S.; Abdel-Rahman, A.A.; Zeid, I.F.; Wei, Y.; Hessien, M.M.; Fouda, A.; Salem, W.M. Synthesis of Eco-Friendly Biopolymer, Alginate-Chitosan Composite to Adsorb the Heavy Metals, Cd(II) and Pb(II) from Contaminated Effluents. *Materials* **2021**, *14*, 2189. [CrossRef]
5. Salem, S.S.; Mohamed, A.; El-Gamal, M.; Talat, M.; Fouda, A. Biological Decolorization and Degradation of Azo Dyes from Textile Wastewater Effluent by *Aspergillus niger*. *Egypt. J. Chem.* **2019**, *62*, 1799–1813.
6. Ali, A.; Shaikh, I.A.; Abbasi, N.A.; Firdous, N.; Ashraf, M.N. Enhancing water efficiency and wastewater treatment using sustainable technologies: A laboratory and pilot study for adhesive and leather chemicals production. *J. Water Process Eng.* **2020**, *36*, 101308. [CrossRef]
7. Wang, D.; He, S.; Shan, C.; Ye, Y.; Ma, H.; Zhang, X.; Zhang, W.; Pan, B. Chromium speciation in tannery effluent after alkaline precipitation: Isolation and characterization. *J. Hazard. Mater.* **2016**, *316*, 169–177. [CrossRef]
8. Bhattacharya, A.; Gupta, A. Evaluation of Acinetobacter sp. B9 for Cr (VI) resistance and detoxification with potential application in bioremediation of heavy-metals-rich industrial wastewater. *Environ. Sci. Pollut. Res. Int.* **2013**, *20*, 6628–6637. [CrossRef]
9. Soliman, A.M.; Abdel-Latif, W.; Shehata, I.H.; Fouda, A.; Abdo, A.M.; Ahmed, Y.M. Green Approach to Overcome the Resistance Pattern of *Candida* spp. Using Biosynthesized Silver Nanoparticles Fabricated by *Penicillium chrysogenum* F9. *Biol. Trace Elem. Res.* **2021**, *199*, 800–811. [CrossRef]
10. Hamza, M.F.; Fouda, A.; Elwakeel, K.Z.; Wei, Y.; Guibal, E.; Hamad, N.A. Phosphorylation of Guar Gum/Magnetite/Chitosan Nanocomposites for Uranium (VI) Sorption and Antibacterial Applications. *Molecules* **2021**, *26*, 1920. [CrossRef]
11. Salem, S.; Fouda, A. Green Synthesis of Metallic Nanoparticles and Their Prospective Biotechnological Applications: An Overview. *Biol. Trace Elem. Res.* **2020**, *199*, 344–370. [CrossRef]
12. Badawy, A.A.; Abdelfattah, N.A.H.; Salem, S.S.; Awad, M.F.; Fouda, A. Efficacy Assessment of Biosynthesized Copper Oxide Nanoparticles (CuO-NPs) on Stored Grain Insects and Their Impacts on Morphological and Physiological Traits of Wheat (*Triticum aestivum* L.) Plant. *Biology* **2021**, *10*, 233. [CrossRef]
13. Abinaya, S.; Kavitha, H.P.; Prakash, M.; Muthukrishnaraj, A. Green synthesis of magnesium oxide nanoparticles and its applications: A review. *Sustain. Chem. Pharm.* **2021**, *19*, 100368. [CrossRef]
14. Vergheese, M.; Vishal, S.K. Green synthesis of magnesium oxide nanoparticles using *Trigonella foenum-graecum* leaf extract and its antibacterial activity. *J. Pharm. Phytochem.* **2018**, *7*, 1193–1200.
15. Abdulkhaleq, N.A.; Nayef, U.M.; Albarazanchi, A.K.H. MgO nanoparticles synthesis via laser ablation stationed on porous silicon for photoconversion application. *Optik* **2020**, *212*, 164793. [CrossRef]
16. Sofi, A.H.; Akhoon, S.A.; Mir, J.F.; Rather, M.U.D. Magnesium Oxide (MgO): A Viable Agent for Antimicrobial Activity. In *Applications of Nanomaterials in Agriculture, Food Science, and Medicine*; IGI Global: Hershey, PA, USA, 2021; pp. 98–105.

17. Fouda, A.; Awad, M.A.; Eid, A.M.; Saied, E.; Barghoth, M.G.; Hamza, M.F.; Awad, M.F.; Abdelbary, S.; Hassan, S.E. An Eco-Friendly Approach to the Control of Pathogenic Microbes and *Anopheles stephensi* Malarial Vector Using Magnesium Oxide Nanoparticles (Mg-NPs) Fabricated by *Penicillium chrysogenum*. *Int. J. Mol. Sci.* **2021**, *22*, 5096. [CrossRef]
18. Refat, M.S.; Ibrahim, H.K.; Sowellim, S.Z.A.; Soliman, M.H.; Saeed, E.M. Spectroscopic and Thermal Studies of Mn(II), Fe(III), Cr(III) and Zn(II) Complexes Derived from the Ligand Resulted by the Reaction Between 4-Acetyl Pyridine and Thiosemicarbazide. *J. Inorg. Organomet. Polym.* **2009**, *19*, 521. [CrossRef]
19. Jagadeeshan, S.; Parsanathan, R. Nano-metal oxides for antibacterial activity. In *Advanced Nanostructured Materials for Environmental Remediation*; Springer: Cham, Switzerland, 2019; pp. 59–90.
20. Srivastava, V.; Sharma, Y.C.; Sillanpää, M. Green synthesis of magnesium oxide nanoflower and its application for the removal of divalent metallic species from synthetic wastewater. *Ceram. Int.* **2015**, *41*, 6702–6709. [CrossRef]
21. Parham, S.; Wicaksono, D.H.B.; Bagherbaigi, S.; Lee, S.L.; Nur, H. Antimicrobial Treatment of Different Metal Oxide Nanoparticles: A Critical Review. *J. Chin. Chem. Soc.* **2016**, *63*, 385–393. [CrossRef]
22. Saravanathamizhan, R.; Perarasu, V.T. Chapter 6—Improvement of Biodegradability Index of Industrial Wastewater Using Different Pretreatment Techniques. In *Wastewater Treatment*; Shah, M.P., Sarkar, A., Mandal, S., Eds.; Elsevier: Amsterdam, The Netherlands, 2021; pp. 103–136.
23. Rani, P.; Kaur, G.; Rao, K.V.; Singh, J.; Rawat, M. Impact of Green Synthesized Metal Oxide Nanoparticles on Seed Germination and Seedling Growth of *Vigna radiata* (Mung Bean) and *Cajanus cajan* (Red Gram). *J. Inorg. Organomet. Polym. Mater.* **2020**, *30*, 4053–4062. [CrossRef]
24. Eid, A.M.; Fouda, A.; Niedbała, G.; Hassan, S.E.-D.; Salem, S.S.; Abdo, A.M.; Hetta, H.F.; Shaheen, T.I. Endophytic *Streptomyces laurentii* Mediated Green Synthesis of Ag-NPs with Antibacterial and Anticancer Properties for Developing Functional Textile Fabric Properties. *Antibiotics* **2020**, *9*, 641. [CrossRef] [PubMed]
25. Samak, D.H.; El-Sayed, Y.S.; Shaheen, H.M.; El-Far, A.H.; Abd El-Hack, M.E.; Noreldin, A.E.; El-Naggar, K.; Abdelnour, S.A.; Saied, E.M.; El-Seedi, H.R.; et al. Developmental Toxicity of Carbon Nanoparticles during Embryogenesis in Chicken. *Environ Sci Pollut Res* **2020**, *27*, 19058–19072. [CrossRef] [PubMed]
26. McClenny, N. Laboratory detection and identification of *Aspergillus* species by microscopic observation and culture: The traditional approach. *Med. Mycol.* **2005**, *43* (Suppl. 1), S125–S128. [CrossRef] [PubMed]
27. Diba, K.; Kordbacheh, P.; Mirhendi, S.; Rezaie, S.; Mahmoudi, M. Identification of *Aspergillus* species using morphological characteristics. *Pak. J. Med Sci.* **2007**, *23*, 867.
28. Risslegger, B.; Zoran, T.; Lackner, M.; Aigner, M.; Sánchez-Reus, F.; Rezusta, A.; Chowdhary, A.; Taj-Aldeen, S.J.; Arendrup, M.C.; Oliveri, S.; et al. A prospective international *Aspergillus terreus* survey: An EFISG, ISHAM and ECMM joint study. *Clin. Microbiol. Infect. Off. Publ. Eur. Soc. Clin. Microbiol. Infect. Dis.* **2017**, *23*, 776.e1–776.e5. [CrossRef]
29. Subhan, M.; Faryal, R.; Macreadie, I. Exploitation of *Aspergillus terreus* for the Production of Natural Statins. *J. Fungi* **2016**, *2*, 13. [CrossRef]
30. Sethi, B.K.; Nanda, P.K.; Sahoo, S. Characterization of biotechnologically relevant extracellular lipase produced by *Aspergillus terreus* NCFT 4269.10. *Braz. J. Microbiol.* **2016**, *47*, 143–149. [CrossRef]
31. Raliya, R.; Tarafdar, J.C. Biosynthesis and characterization of zinc, magnesium and titanium nanoparticles: An eco-friendly approach. *Int. Nano Lett.* **2014**, *4*, 93. [CrossRef]
32. Lashin, I.; Fouda, A.; Gobouri, A.A.; Azab, E.; Mohammedsaleh, Z.M.; Makharita, R.R. Antimicrobial and In Vitro Cytotoxic Efficacy of Biogenic Silver Nanoparticles (Ag-NPs) Fabricated by Callus Extract of *Solanum incanum* L. *Biomolecules* **2021**, *11*, 341. [CrossRef]
33. Grijseels, S.; Nielsen, J.C.; Nielsen, J.; Larsen, T.O.; Frisvad, J.C.; Nielsen, K.F.; Frandsen, R.J.N.; Workman, M. Physiological characterization of secondary metabolite producing *Penicillium* cell factories. *Fungal Biol. Biotechnol.* **2017**, *4*, 8. [CrossRef]
34. Shaheen, T.I.; Fouda, A.; Salem, S.S. Integration of Cotton Fabrics with Biosynthesized CuO Nanoparticles for Bactericidal Activity in the Terms of Their Cytotoxicity Assessment. *Ind. Eng. Chem. Res.* **2021**, *60*, 1553–1563. [CrossRef]
35. Hassan, S.E.-D.; Fouda, A.; Radwan, A.A.; Salem, S.S.; Barghoth, M.G.; Awad, M.A.; Abdo, A.M.; El-Gamal, M.S. Endophytic actinomycetes *Streptomyces* spp mediated biosynthesis of copper oxide nanoparticles as a promising tool for biotechnological applications. *JBIC J. Biol. Inorg. Chem.* **2019**, *24*, 377–393. [CrossRef]
36. Loo, Y.Y.; Rukayadi, Y.; Nor-Khaizura, M.-A.-R.; Kuan, C.H.; Chieng, B.W.; Nishibuchi, M.; Radu, S. In Vitro Antimicrobial Activity of Green Synthesized Silver Nanoparticles Against Selected Gram-negative Foodborne Pathogens. *Front. Microbiol.* **2018**, *9*, 1555. [CrossRef]
37. Nguyen, D.T.C.; Dang, H.H.; Vo, D.-V.N.; Bach, L.G.; Nguyen, T.D.; Tran, T.V. Biogenic synthesis of MgO nanoparticles from different extracts (flower, bark, leaf) of *Tecoma stans* (L.) and their utilization in selected organic dyes treatment. *J. Hazard. Mater.* **2021**, *404*, 124146. [CrossRef]
38. Ogunyemi, S.O.; Zhang, F.; Abdallah, Y.; Zhang, M.; Wang, Y.; Sun, G.; Qiu, W.; Li, B. Biosynthesis and characterization of magnesium oxide and manganese dioxide nanoparticles using *Matricaria chamomilla* L. extract and its inhibitory effect on *Acidovorax oryzae* strain RS-2. *Artif. Cells Nanomed. Biotechnol.* **2019**, *47*, 2230–2239. [CrossRef]
39. Hamida, R.S.; Ali, M.A.; Abdelmeguid, N.E.; Al-Zaban, M.I.; Baz, L.; Bin-Meferij, M.M. Lichens—A Potential Source for Nanoparticles Fabrication: A Review on Nanoparticles Biosynthesis and Their Prospective Applications. *J. Fungi* **2021**, *7*, 291. [CrossRef]

40. Mohd Yusof, H.; Mohamad, R.; Zaidan, U.H.; Abdul Rahman, N.A. Microbial synthesis of zinc oxide nanoparticles and their potential application as an antimicrobial agent and a feed supplement in animal industry: A review. *J. Anim. Sci. Biotechnol.* **2019**, *10*, 57. [CrossRef]
41. Hassan, S.E.; Fouda, A.; Saied, E.; Farag, M.M.S.; Eid, A.M.; Barghoth, M.G.; Awad, M.A.; Hamza, M.F.; Awad, M.F. *Rhizopus oryzae*-Mediated Green Synthesis of Magnesium Oxide Nanoparticles (MgO-NPs): A Promising Tool for Antimicrobial, Mosquitocidal Action, and Tanning Effluent Treatment. *J. Fungi* **2021**, *7*, 372. [CrossRef]
42. Muangban, J.; Jaroenapibal, P. Effects of precursor concentration on crystalline morphologies and particle sizes of electrospun WO_3 nanofibers. *Ceram. Int.* **2014**, *40*, 6759–6764. [CrossRef]
43. Jeevanandam, J.; San Chan, Y.; Jing Wong, Y.; Siang Hii, Y. Biogenic synthesis of magnesium oxide nanoparticles using *Aloe barbadensis* leaf latex extract. *IOP Conf. Ser. Mater. Sci. Eng.* **2020**, *943*, 012030. [CrossRef]
44. Naseem, K.; Zia Ur Rehman, M.; Ahmad, A.; Dubal, D.; AlGarni, T.S. Plant Extract Induced Biogenic Preparation of Silver Nanoparticles and Their Potential as Catalyst for Degradation of Toxic Dyes. *Coatings* **2020**, *10*, 1235. [CrossRef]
45. Irfan, M.; Moniruzzaman, M.; Ahmad, T.; Samsudin, M.F.R.; Bashir, F.; Butt, M.T.; Ashraf, H. Identifying the role of process conditions for synthesis of stable gold nanoparticles and insight detail of reaction mechanism. *Inorg. Nano Met. Chem.* **2021**, 1–14. [CrossRef]
46. Sahu, P.K.; Sahu, P.K.; Agarwal, D.D. Role of basicity and the catalytic activity of KOH loaded MgO and hydrotalcite as catalysts for the efficient synthesis of 1-[(2-benzothiazolylamino)arylmethyl]-2-naphthalenols. *RSC Adv.* **2015**, *5*, 69143–69151. [CrossRef]
47. Vijayanandan, A.S.; Balakrishnan, R.M. Biosynthesis of cobalt oxide nanoparticles using endophytic fungus *Aspergillus nidulans*. *J. Environ. Manag.* **2018**, *218*, 442–450. [CrossRef]
48. Fouda, A.; Hassan, S.E.-D.; Abdel-Rahman, M.A.; Farag, M.M.S.; Shehal-deen, A.; Mohamed, A.A.; Alsharif, S.M.; Saied, E.; Moghanim, S.A.; Azab, M.S. Catalytic degradation of wastewater from the textile and tannery industries by green synthesized hematite (α-Fe_2O_3) and magnesium oxide (MgO) nanoparticles. *Curr. Res. Biotechnol.* **2021**, *3*, 29–41. [CrossRef]
49. Ahmadi, M.; Ghasemi, M.R.; Rafsanjani, H.H. Study of different parameters in TiO_2 nanoparticles formation. *J. Mater. Sci. Eng.* **2011**, *5*, 87.
50. Fouda, A.; Abdel-Maksoud, G.; Saad, H.A.; Gobouri, A.A.; Mohammedsaleh, Z.M.; El-Sadany, M.A. The efficacy of silver nitrate ($AgNO_3$) as a coating agentto protect paper against high deteriorating microbes. *Catalysts* **2021**, *11*, 310. [CrossRef]
51. Amina, M.; Al Musayeib, N.M.; Alarfaj, N.A.; El-Tohamy, M.F.; Oraby, H.F.; Al Hamoud, G.A.; Bukhari, S.I.; Moubayed, N.M.S. Biogenic green synthesis of MgO nanoparticles using *Saussurea costus* biomasses for a comprehensive detection of their antimicrobial, cytotoxicity against MCF-7 breast cancer cells and photocatalysis potentials. *PLoS ONE* **2020**, *15*, e0237567. [CrossRef] [PubMed]
52. Asami, H.; Tokugawa, M.; Masaki, Y.; Ishiuchi, S.-i.; Gloaguen, E.; Seio, K.; Saigusa, H.; Fujii, M.; Sekine, M.; Mons, M. Effective Strategy for Conformer-Selective Detection of Short-Lived Excited State Species: Application to the IR Spectroscopy of the N1H Keto Tautomer of Guanine. *J. Phys. Chem. A* **2016**, *120*, 2179–2184. [CrossRef] [PubMed]
53. Dobrucka, R. Synthesis of MgO Nanoparticles Using *Artemisia abrotanum* Herba Extract and Their Antioxidant and Photocatalytic Properties. *Iran. J. Sci. Technol. Trans. A Sci.* **2018**, *42*, 547–555. [CrossRef]
54. Brotton, S.J.; Lucas, M.; Jensen, T.N.; Anderson, S.L.; Kaiser, R.I. Spectroscopic Study on the Intermediates and Reaction Rates in the Oxidation of Levitated Droplets of Energetic Ionic Liquids by Nitrogen Dioxide. *J. Phys. Chem. A* **2018**, *122*, 7351–7377. [CrossRef]
55. Karimi, B.; Khorasani, M.; Vali, H.; Vargas, C.; Luque, R. Palladium Nanoparticles Supported in the Nanospaces of Imidazolium-Based Bifunctional PMOs: The Role of Plugs in Selectivity Changeover in Aerobic Oxidation of Alcohols. *ACS Catal.* **2015**, *5*, 4189–4200. [CrossRef]
56. Suresh, J.; Pradheesh, G.; Alexramani, V.; Sundrarajan, M.; Hong, S.I. Green synthesis and characterization of hexagonal shaped MgO nanoparticles using insulin plant (*Costus pictus* D. Don) leave extract and its antimicrobial as well as anticancer activity. *Adv. Powder Technol.* **2018**, *29*, 1685–1694. [CrossRef]
57. Umaralikhan, L.; Jamal Mohamed Jaffar, M. Green Synthesis of MgO Nanoparticles and it Antibacterial Activity. *Iran. J. Sci. Technol. Trans. A Sci.* **2018**, *42*, 477–485. [CrossRef]
58. Fouda, A.; Abdel-Maksoud, G.; Abdel-Rahman, M.A.; Eid, A.M.; Barghoth, M.G.; El-Sadany, M.A.-H. Monitoring the effect of biosynthesized nanoparticles against biodeterioration of cellulose-based materials by *Aspergillus niger*. *Cellulose* **2019**, *26*, 6583–6597. [CrossRef]
59. Aref, M.S.; Salem, S.S. Bio-callus synthesis of silver nanoparticles, characterization, and antibacterial activities via *Cinnamomum camphora* callus culture. *Biocatal. Agric. Biotechnol.* **2020**, *27*, 101689. [CrossRef]
60. Huang, L.; Li, D.; Lin, Y.; Evans, D.G.; Duan, X. Influence of nano-MgO particle size on bactericidal action againstBacillus subtilis var. niger. *Chin. Sci. Bull.* **2005**, *50*, 514–519. [CrossRef]
61. Jian, W.; Ma, Y.; Wu, H.; Zhu, X.; Wang, J.; Xiong, H.; Lin, L.; Wu, L. Fabrication of highly stable silver nanoparticles using polysaccharide-protein complexes from abalone viscera and antibacterial activity evaluation. *Int. J. Biol. Macromol.* **2019**, *128*, 839–847. [CrossRef]
62. Abdallah, Y.; Ogunyemi, S.O.; Abdelazez, A.; Zhang, M.; Hong, X.; Ibrahim, E.; Hossain, A.; Fouad, H.; Li, B.; Chen, J. The Green Synthesis of MgO Nano-Flowers Using *Rosmarinus officinalis* L. (Rosemary) and the Antibacterial Activities against *Xanthomonas oryzae* pv. *oryzae*. *BioMed Res. Int.* **2019**, *2019*, 5620989. [CrossRef]

63. Jhansi, K.; Jayarambabu, N.; Reddy, K.P.; Reddy, N.M.; Suvarna, R.P.; Rao, K.V.; Kumar, V.R.; Rajendar, V. Biosynthesis of MgO nanoparticles using mushroom extract: Effect on peanut (*Arachis hypogaea* L.) seed germination. *3 Biotech* **2017**, *7*, 263. [CrossRef]
64. Alsharif, S.M.; Salem, S.S.; Abdel-Rahman, M.A.; Fouda, A.; Eid, A.M.; El-Din Hassan, S.; Awad, M.A.; Mohamed, A.A. Multifunctional properties of spherical silver nanoparticles fabricated by different microbial taxa. *Heliyon* **2020**, *6*, e03943. [CrossRef]
65. Salem, S.S.; El-Belely, E.F.; Niedbała, G.; Alnoman, M.M.; Hassan, S.E.; Eid, A.M.; Shaheen, T.I.; Elkelish, A.; Fouda, A. Bactericidal and In-Vitro Cytotoxic Efficacy of Silver Nanoparticles (Ag-NPs) Fabricated by Endophytic Actinomycetes and Their Use as Coating for the Textile Fabrics. *Nanomaterials* **2020**, *10*, 2082. [CrossRef]
66. Fouda, A.; Hassan, S.E.; Abdo, A.M.; El-Gamal, M.S. Antimicrobial, Antioxidant and Larvicidal Activities of Spherical Silver Nanoparticles Synthesized by Endophytic *Streptomyces* spp. *Biol. Trace Elem. Res.* **2020**, *195*, 707–724. [CrossRef]
67. Fouda, A.; Hassan, S.E.-D.; Saied, E.; Azab, M.S. An eco-friendly approach to textile and tannery wastewater treatment using maghemite nanoparticles (γ-Fe_2O_3-NPs) fabricated by *Penicillium expansum* strain (K-w). *J. Environ. Chem. Eng.* **2021**, *9*, 104693. [CrossRef]
68. Al-Hazmi, F.; Alnowaiser, F.; Al-Ghamdi, A.A.; Al-Ghamdi, A.A.; Aly, M.M.; Al-Tuwirqi, R.M.; El-Tantawy, F. A new large-scale synthesis of magnesium oxide nanowires: Structural and antibacterial properties. *Superlattices Microstruct.* **2012**, *52*, 200–209. [CrossRef]
69. Karthik, K.; Dhanuskodi, S.; Gobinath, C.; Prabukumar, S.; Sivaramakrishnan, S. Fabrication of MgO nanostructures and its efficient photocatalytic, antibacterial and anticancer performance. *J. Photochem. Photobiol. B Biol.* **2019**, *190*, 8–20. [CrossRef] [PubMed]
70. Wang, L.; Hu, C.; Shao, L. The antimicrobial activity of nanoparticles: Present situation and prospects for the future. *Int. J. Nanomed.* **2017**, *12*, 1227–1249. [CrossRef]
71. Shaikh, S.; Nazam, N.; Rizvi, S.M.D.; Ahmad, K.; Baig, M.H.; Lee, E.J.; Choi, I. Mechanistic Insights into the Antimicrobial Actions of Metallic Nanoparticles and Their Implications for Multidrug Resistance. *Int. J. Mol. Sci.* **2019**, *20*, 2468. [CrossRef]
72. Chiddarwar, S. Novel Approaches of Magnesium Oxide Nanoparticles in MIC, MBC, Antibiofilm and Antimicrobial Activities against Bacteria, Yeast and Biofilms. In Proceedings of the MBC, Antibiofilm and Antimicrobial Activities against Bacteria, Yeast and Biofilms, Hyderabad, India, 19 February 2020.
73. Wong, C.W.; Chan, Y.S.; Jeevanandam, J.; Pal, K.; Bechelany, M.; Abd Elkodous, M.; El-Sayyad, G.S. Response Surface Methodology Optimization of Mono-dispersed MgO Nanoparticles Fabricated by Ultrasonic-Assisted Sol–Gel Method for Outstanding Antimicrobial and Antibiofilm Activities. *J. Clust. Sci.* **2020**, *31*, 367–389. [CrossRef]
74. Fouda, A.; Salem, S.S.; Wassel, A.R.; Hamza, M.F.; Shaheen, T.I. Optimization of green biosynthesized visible light active CuO/ZnO nano-photocatalysts for the degradation of organic methylene blue dye. *Heliyon* **2020**, *6*, e04896. [CrossRef]
75. Ye, W.; Liu, H.; Lin, F.; Lin, J.; Zhao, S.; Yang, S.; Hou, J.; Zhou, S.; van der Bruggen, B. High-flux nanofiltration membranes tailored by bio-inspired co-deposition of hydrophilic g-C_3N_4 nanosheets for enhanced selectivity towards organics and salts. *Environ. Sci. Nano* **2019**, *6*, 2958–2967. [CrossRef]
76. Fouda, A.; Hassan, S.E.-D.; Saied, E.; Hamza, M.F. Photocatalytic degradation of real textile and tannery effluent using biosynthesized magnesium oxide nanoparticles (MgO-NPs), heavy metal adsorption, phytotoxicity, and antimicrobial activity. *J. Environ. Chem. Eng.* **2021**, *9*, 105346. [CrossRef]
77. Kishor, R.; Purchase, D.; Saratale, G.D.; Saratale, R.G.; Ferreira, L.F.R.; Bilal, M.; Chandra, R.; Bharagava, R.N. Ecotoxicological and health concerns of persistent coloring pollutants of textile industry wastewater and treatment approaches for environmental safety. *J. Environ. Chem. Eng.* **2021**, *9*, 105012. [CrossRef]
78. Li, S. Combustion synthesis of porous MgO and its adsorption properties. *Int. J. Ind. Chem.* **2019**, *10*, 89–96. [CrossRef]
79. Ilyas, M.; Ahmad, W.; Khan, H.; Yousaf, S.; Yasir, M.; Khan, A. Environmental and health impacts of industrial wastewater effluents in Pakistan: A review. *Rev. Environ. Health* **2019**, *34*, 171–186. [CrossRef]
80. Laxmi, V.; Kaushik, G. Toxicity of hexavalent chromium in environment, health threats, and its bioremediation and detoxification from tannery wastewater for environmental safety. In *Bioremediation of Industrial Waste for Environmental Safety*; Springer: Singapore, 2020; pp. 223–243.
81. Sawalha, H.; Alsharabaty, R.; Sarsour, S.; Al-Jabari, M. Wastewater from leather tanning and processing in Palestine: Characterization and management aspects. *J. Environ. Manag.* **2019**, *251*, 109596. [CrossRef]
82. Barnhart, J. Occurrences, uses, and properties of chromium. *Regul. Toxicol. Pharmacol. RTP* **1997**, *26*, S3–S7. [CrossRef]
83. Mohanty, K.; Jha, M.; Meikap, B.C.; Biswas, M.N. Removal of chromium (VI) from dilute aqueous solutions by activated carbon developed from *Terminalia arjuna* nuts activated with zinc chloride. *Chem. Eng. Sci.* **2005**, *60*, 3049–3059. [CrossRef]
84. Mahdavi, S.; Jalali, M.; Afkhami, A. Heavy metals removal from aqueous solutions using TiO_2, MgO, and Al_2O_3 nanoparticles. *Chem. Eng. Commun.* **2013**, *200*, 448–470. [CrossRef]
85. Gusain, R.; Gupta, K.; Joshi, P.; Khatri, O.P. Adsorptive removal and photocatalytic degradation of organic pollutants using metal oxides and their composites: A comprehensive review. *Adv. Colloid Interface Sci.* **2019**, *272*, 102009. [CrossRef]
86. Yang, J.; Hou, B.; Wang, J.; Tian, B.; Bi, J.; Wang, N.; Li, X.; Huang, X. Nanomaterials for the Removal of Heavy Metals from Wastewater. *Nanomaterials* **2019**, *9*, 424. [CrossRef]
87. Jain, M.; Garg, V.K.; Kadirvelu, K. Equilibrium and kinetic studies for sequestration of Cr(VI) from simulated wastewater using sunflower waste biomass. *J. Hazard. Mater.* **2009**, *171*, 328–334. [CrossRef]

88. Mohan, D.; Pittman, C.U., Jr. Activated carbons and low cost adsorbents for remediation of tri- and hexavalent chromium from water. *J. Hazard. Mater.* **2006**, *137*, 762–811. [CrossRef]
89. Jain, M.; Yadav, M.; Kohout, T.; Lahtinen, M.; Garg, V.K.; Sillanpää, M. Development of iron oxide/activated carbon nanoparticle composite for the removal of Cr(VI), Cu(II) and Cd(II) ions from aqueous solution. *Water Resour. Ind.* **2018**, *20*, 54–74. [CrossRef]
90. Hashem, A.H.; Saied, E.; Hasanin, M.S. Green and ecofriendly bio-removal of methylene blue dye from aqueous solution using biologically activated banana peel waste. *Sustain. Chem. Pharm.* **2020**, *18*, 100333. [CrossRef]
91. Fouda, A.; Abdel-Maksoud, G.; Abdel-Rahman, M.A.; Salem, S.S.; Hassan, S.E.-D.; El-Sadany, M.A.-H. Eco-friendly approach utilizing green synthesized nanoparticles for paper conservation against microbes involved in biodeterioration of archaeological manuscript. *Int. Biodeterior. Biodegrad.* **2019**, *142*, 160–169. [CrossRef]
92. Essien, E.R.; Atasie, V.N.; Okeafor, A.O.; Nwude, D.O. Biogenic synthesis of magnesium oxide nanoparticles using Manihot esculenta (Crantz) leaf extract. *Int. Nano Lett.* **2020**, *10*, 43–48. [CrossRef]
93. Fouda, A.; El-Din Hassan, S.; Salem, S.S.; Shaheen, T.I. In-Vitro cytotoxicity, antibacterial, and UV protection properties of the biosynthesized Zinc oxide nanoparticles for medical textile applications. *Microb. Pathog.* **2018**, *125*, 252–261. [CrossRef]
94. El-Belely, E.F.; Farag, M.M.S.; Said, H.A.; Amin, A.S.; Azab, E.; Gobouri, A.A.; Fouda, A. Green Synthesis of Zinc Oxide Nanoparticles (ZnO-NPs) Using *Arthrospira platensis* (Class: Cyanophyceae) and Evaluation of their Biomedical Activities. *Nanomaterials* **2021**, *11*, 95. [CrossRef]
95. Oladipo, A.A.; Adeleye, O.J.; Oladipo, A.S.; Aleshinloye, A.O. Bio-derived MgO nanopowders for BOD and COD reduction from tannery wastewater. *J. Water Process Eng.* **2017**, *16*, 142–148. [CrossRef]

catalysts

Article

Bio-Catalytic Activity of Novel *Mentha arvensis* Intervened Biocompatible Magnesium Oxide Nanomaterials

Shah Faisal [1,*], Abdullah [2], Hasnain Jan [1,3], Sajjad Ali Shah [1], Sumaira Shah [4], Muhammad Rizwan [5], Nasib Zaman [5], Zahid Hussain [5], Muhammad Nazir Uddin [5], Nadia Bibi [6], Aishma Khattak [7], Wajid Khan [5], Arshad Iqbal [5], Muhammad Idrees [8] and Rehana Masood [9]

1 Institute of Biotechnology and Microbiology, Bacha Khan University, Charsadda 24460, KPK, Pakistan; rhasnain849@gmail.com (H.J.); sajjadbiotec@gmail.com (S.A.S.)
2 Department of Microbiology, Abdul Wali Khan University, Mardan 23200, KPK, Pakistan; abdul.9353chd@gmail.com
3 Department of Biotechnology, Quaid-i-Azam University, Islamabad 45320, Pakistan
4 Department of Botany, Bacha Khan University, Charsadda 24460, KPK, Pakistan; sunehra23@gmail.com
5 Centre for Biotechnology and Microbiology University of Swat, Mingora 19200, KPK, Pakistan; muhammad.rizwan@uswat.edu.pk (M.R.); nasibzaman@uswat.edu.pk (N.Z.); zahid@uswat.edu.pk (Z.H.); nazir@uswat.edu.pk (M.N.U.); sherafghan.shah@gmail.com (W.K.); arshad.iqbal@uswat.edu.pk (A.I.)
6 Department of Microbiology, Shaheed Benazir Bhutto Women University, Peshawar 25000, KPK, Pakistan; nadiabibi@sbbwu.edu.pk
7 Department of Bioinformatics, Shaheed Benazir Bhutto Women University, Peshawar 25000, KPK, Pakistan; aishma.khattak@yahoo.com
8 Department of Biotechnology, University of Swabi, Swabi 23430, KPK, Pakistan; midrees@uoswabi.edu.pk
9 Department of Biochemistry, Shaheed Benazir Bhutto Women University, Peshawar 25000, KPK, Pakistan; rehana.masood@sbbwu.edu.pk
* Correspondence: shahfaisal_std@bkuc.edu.pk; Tel.: +92-3159353867

Abstract: In the present study *Mentha arvensis* medaited Magnesium oxide nanoparticles were synthesized by novel green route followed by advanced characterization via XRD, FTIR, UV, SEM, TEM, DLS and TGA. The mean grain size of 32.4 nm and crystallite fcc morphology were confirmed by X-ray diffractive analysis. Scanning and Transmission electron microscopy analysis revealed the spherical and elliptical morphologies of the biosynthesized nanoparticles. Particle surface charge of −16.1 mV were determined by zeta potential and zeta size of 30–120 nm via dynamic light scattering method. Fourier transform spectroscopic analysis revealed the possible involvement of functional groups in the plant extract in reduction of Mg^{2+} ions to Mg^0. Furthermore, the antioxidant, anti-Alzheimer, anti-cancer, and anti-*H. pylori* activities were performed. The results revealed that MgO-NPs has significant anti-*H. pyloric* potential by giving ZOI of 17.19 ± 0.83 mm against *Helicobacter felis* followed by *Helicobacter suis*. MgO-NPs inhibited protein kinase enzyme up to 12.44 ± 0.72% at 5 mg/mL and thus showed eminent anticancer activity. Significant free radicals scavenging and hemocompatability was also shown by MgO-NPs. MgO-NPs also displayed good inhibition potential against Hela cell lines with maximum inhibition of 49.49 ± 1.18 at 400 µg/mL. Owing to ecofriendly synthesis, non-toxic and biocompatible nature, *Mentha arvensis* synthesized MgO-NPs can be used as potent antimicrobial agent in therapeutic applications.

Keywords: *Mentha arvensis*; green synthesis; magnesium oxide; antibacterial; bio-compatible; anti-cancer

check for updates

Citation: Faisal, S.; Abdullah; Jan, H.; Shah, S.A.; Shah, S.; Rizwan, M.; Zaman, N.; Hussain, Z.; Uddin, M.N.; Bibi, N.; et al. Bio-Catalytic Activity of Novel *Mentha arvensis* Intervened Biocompatible Magnesium Oxide Nanomaterials. *Catalysts* **2021**, *11*, 780. https://doi.org/10.3390/catal11070780

Academic Editor: Beom Soo Kim

Received: 24 May 2021
Accepted: 21 June 2021
Published: 27 June 2021

1. Introduction

Nanotechnology, with various branches rooted in industrial sector such as biomedical, nanomedicine, cosmetics, pharmaceutical, and food manufacturing, is now seen as an established, state of the art technology [1]. Many nano-scaled structures have been produced using various techniques. Yet the synthesis of green nanoparticles is a method of choice that is easy to plan and engineer [2]. Traditional approaches to nanoparticle development have a number of drawbacks, including long-term processing, high prices, time-consuming methods, and the use of toxic compounds in particular. Because of these drawbacks, much of the relevant research has focused on developing environment friendly nanoparticles synthetic approach [3,4].

Plant mediated Synthetic methods that are less harmful to the environment are being created. In recent years, material scientists have prioritised the production of environmentally sustainable methods for synthesising nanoscale materials. Green NP synthesis, especially using various plant extracts, is an emerging field in trend in environment friendly chemistry and thought to be simple, inexpensive, and non-toxic [5–7]. Nanoscience has improved the human life by addressing a wide range of issues, and play an important role in treating many diseases [8,9]. Magnesium oxide nanoparticles has ionic properties with good crystal structure and has a vital role in novel applications such as adsorption, electronics, catalysis, and in extraction of petrochemicals [10–12]. It can be synthesized by different approaches but the green route is more advantageous. The physico-morphic properties i.e., size, shape and crystallanity of MgO nanoparticles depending on the reaction time and condition [13–15]. Material scientists are now using the enormous potentiality of medicinal plants as a reliable basis for the development of (MgO) nanoparticles as an alternative to traditional approaches, according to the aforementioned disadvantages of synthetic methods other than green.

Mentha arvensis is a 40-cm tall herbaceous plant with a pleasant, calming odour. Hortel-vick is the common name for it. It's used to treat skin infections and is prescribed as a gastrointestinal, carminative, and nasal decongestant in conventional medicine [16,17]. This species' essential oil is rich in menthol (70%) and is used in nasal inhalants, perfumes, tobacco, and the pharmaceutical industry. Mentha arvensis was found to contain menthol, -terpineol, p-menthone, menthol acetate, and other compounds [17]. UV spectroscopy, FTIR, XRD, SEM, TEM, DLS, EDX, and TGA were used to classify the bio-inspired synthesised MgO-NPs. Furthermore, inhibition of protein kinase enzyme, antidiabetic, anti-Alzheimer's, antimicrobial, antioxidant, and cytotoxic applications of the synthesised nanoparticles were assessed. Biocompatibility of MgO-NPs against human enterocytes were also tested to ensure the engineered nanoparticles are bio safe.

2. Results

2.1. Biosynthesis of MgO-NPs

Mentha arvensis, also known as "wild mint" and is used in drug production, it has a variety of pharmacological practises in different countries. They also contain therapeutic compounds that are well known for their medicinal properties, such as anti-inflammatory, antioxidant, and gastro protective properties [18]. HPLC studies have identified the most active compounds in *M. arvensis* extracts, including hesperidin, ferulic acid, rosmarinic acid, diosmin, didymin, buddleoside, acacetin, and linarin, and have reported a variety of biological effects. Rosmarinic acid, which is present in plants, has been studied for its antioxidant and anti-inflammatory properties [19,20]. As a reducing, capping, and stabilising agent, aqueous leaf extract of *M. arvensis* was used in the biosynthesis of MgO-NPs. The constituents contained in *M. arvensis* extracts are thought to have played a significant role in the production of environmentally sustainable and biomedically essential MgO-NPs. The colour of the mixture changes from light brown to darkish brown when the reaction between ($Mg(NO_3)_2{\cdot}6H_2O$ + *M. arvensis*) is carried out, confirming the development of MgO-NPs [20]. Figure 1. After Washing, drying, grinding, and calcination is followed by the development of a dark brown powder of MgO-NPs. The fine powder was extracted and deposited at room temperature in an airtight glass vial labelled MgO-NPs for physicochemical, morphological, and biological applications. The results of the literature review showed that the physicochemical and morphological properties of metallic nanoparticles depends upon the type of plant and reaction conditions [21].

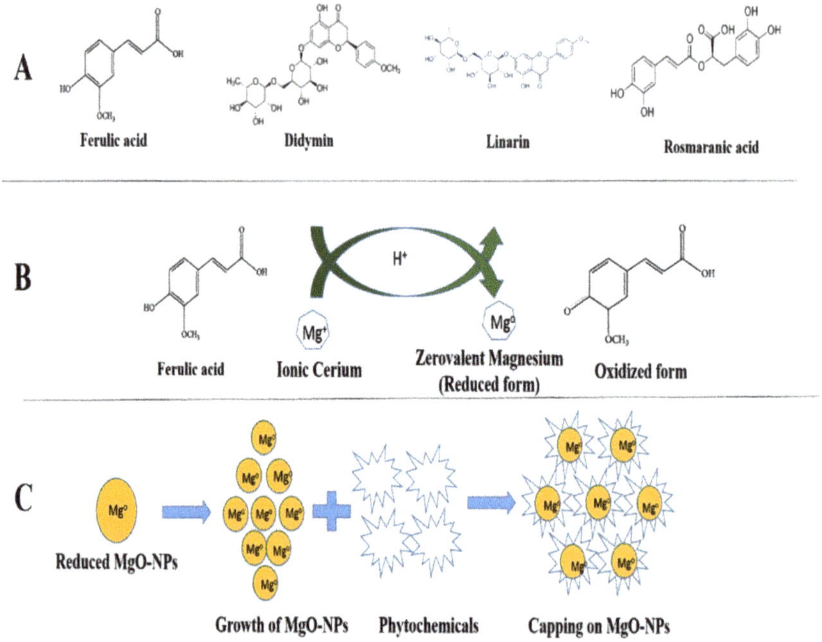

Figure 1. Mechanistic approach for synthesis of MgO-NPs using aqueous extract of Mentha arvensis. (**A**) The structural and chemical formulas of Ferulic acid, Didymin, Linarin, Rosmaranic acid these all compounds are present in Mentha arvensis extract. (**B**) Ferulic acid interacts with magnesium ion and reduced it to Mg^{+1} to Mg^0. (**C**) Reduced form of magnesium nanoparticles and phytochemicals that are taking part in in capping o magnesium nanoparticles.

2.2. Optical Band Gap

Figure 2a shows how the optical band distance of MgO nanoparticles was calculated using the simple relationship between absorbance and incident photon energy (hv). The TauC plot was used to find the direct band gap for the NPs that is 3.3 eV. The smaller band leads to increased photo degradation behaviour, as electrons are more quickly excited from the valence to conduction band and leads to degradation of dye [22]. The presence of MgO-NPs in the reaction media, which are formed by reducing $Mg(NO_3)_2$, is indicated by the sharp peak at 280 nm. Our findings are consistent with previous research [23]. Band gap is influenced by grain size, lattice structure and surface physique [24].

2.3. Powder X-Ray Diffraction

The crystalline structure of the particles was verified using X-ray diffraction. Figure 2c shows the XRD patterns of MgO-NPs synthesised by *M. arvensis* leaf extract after total reduction of Mg^{2+} to Mg^0. The biosynthesized MgO-NPs showed strong peaks at 2 (degree) 12.57°, 20.8°, 25.68°, and 32.56°, which correspond to miller's indices (111), (002), (202), and (113), respectively, conforming to the polycrystalline cubic structure. In XRD patterns, no other impurities were found. The average crystallite size was estimated to be 32.4 nm using the Debye Scherer equation. As mentioned in the process portion, the MgO-NPs were shaped, centrifuged, and redistributed in sterile distilled water prior to XRD examination, removing the existence of any substance that could trigger irregular effects. The appearance of structural peaks in the XRD models explicitly demonstrates that the MgO-NPs synthesised using our green approach are nano-crystalline.

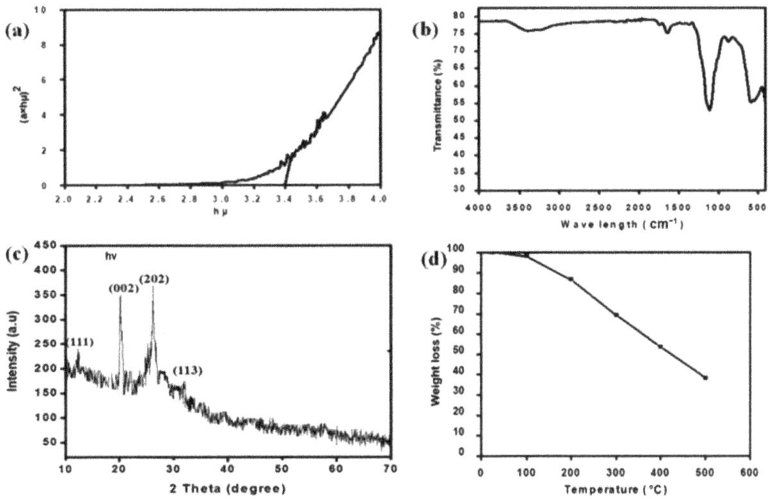

Figure 2. (**a**) Optical Band gap, (**b**) FTIR spectrum, (**c**) XRD representation and (**d**) TGA analysis of Biosynthesized MgO-NPs.

2.4. Fourier Transformed Infrared Spectroscopy

The different functional groups of potential bio molecules present in the plant extract, which serve as synthesised MgO-NPs reduction and capping agents, were identified using FT-IR spectroscopy [25]. Figure 2b shows how FTIR analysis is used to analyse the surface chemistry of biosynthesized NPs. MgO-NPs had notable peaks in their FTIR spectra at 3461 cm^{-1}, 1737 cm^{-1}, 1649 cm^{-1}, 1151 cm^{-1}, 858 cm^{-1}, and 604 cm^{-1}. The O-H stretching vibration in hydroxyl pairs, C=O stretching vibrations, and C–O–C stretching vibrations are responsible for the absorption peaks at 3461 cm^{-1}, 1737 cm^{-1}, and 1649 cm^{-1} [26,27]. Peaks at 1148 cm^{-1} corresponded to O-H folding and C-O stretching of main alcohol functional groups. The broad absorption peak on 607 cm^{-1} clearly indicated Mg-O bond stretching, indicating that MgO-NPs were successfully synthesized [28]. The FTIR spectra, which also parallels previous studies [25–27], supports the active capping of plant metabolites on the surface of MgO-NPs.

2.5. Thermo Galvanometric Analysis

TGA is a thermal analysis tool that measures changes in material physical and chemical properties as a result of increasing temperature with constant heating. It's used to figure out the properties of materials loss or gain mass due to decomposition, oxidation, or the loss of volatiles [29]. TGA analysis was carried out in the temperature range of 25 °C to 600 °C, as seen in Figure 2d. MgO-NPs lost a total of 61.9 percent of their weight. Dehydration and lack of moisture content from the samples was due to the original weight loss up to 150 °C [30]. Nanoparticles having more plant molecules are susceptible to more weight loss with the increase of temperature. This is due to the breaking of bonds between plant molecules and magnesium ions.

2.6. SEM, TEM and EDX Analysis of MgO-NPs

SEM was used to investigate the size, distribution, and morphology of MgO-NPs, as seen in Figure 3a. The majority of the particles are spherical and can be classified as nanoparticles with an overall size of 29.72–40 nm and varying levels of aggregation [25]. The dosage, temperature, and pH of extracts, on the other hand, affect parameters such as nanoparticle size, shape, and agglomeration. Figure 3b shows TEM micrographs of synthesised NPs with circular or elliptical morphology and a mean size of 29.72–36 nm.

Using Image J programme, the measurements of about 50 particles were determined for each sample. Similar morphologies have previously been found in the literature [31]. The biomolecules capped the plant sample because the shapes of the particles were brighter than the centres. The MgO-NPs are spherical, as seen by scanning electron micrography, and this is due to interactions and Vander Waals forces. These results are agree with those of previous studies [32]. Figure 3c EDX shows that the MgO-NPs had a high percentage of magnesium as well as a high percentage of oxygen, indicating that MgO-NPs were formed. The extra peaks indicates the elements present in biomolecules of plant [33].

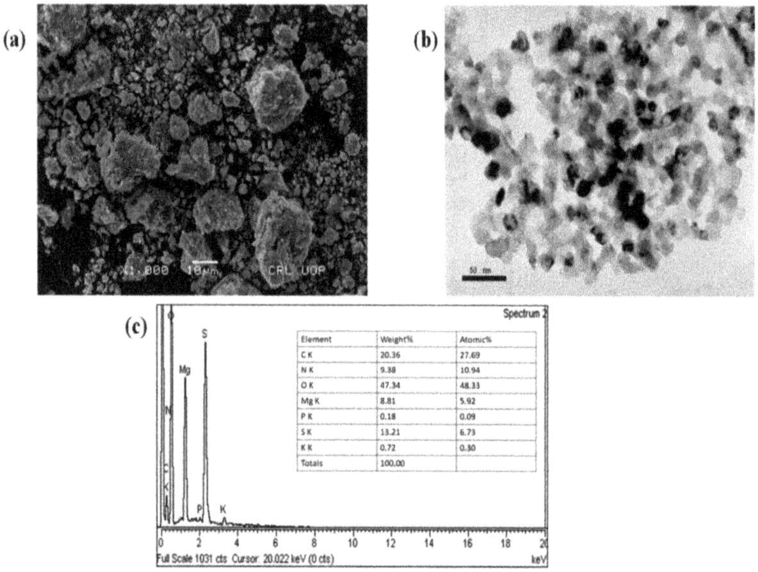

Figure 3. (**a**) SEM micrograph, (**b**) TEM micrograph and (**c**) EDX analysis of MgO-NPs.

2.7. Zeta Size and Zeta Potential of MgO-NPs

The DLS technique is used to investigate the size distribution and zeta potential (ζ) of biosynthesized MgO-NPs. The zeta potential (ζ) is a common calculation of a particle's surface charge which determines colloidal stability. Stable colloids are described as suspensions with a voltage of 15 mV [34]. The zeta potential of MgO-NPs in distilled water was found to be -16.1 mV in the sample, indicating that the colloidal solution is stable. The dispersion power of the greenly synthesized MgO-NPs is thus checked and supported by the zeta potential measurements. Biomolecules in plant extract and there possible coating and capping during synthesis of nanoparticles leads to negative surface charge [35]. The Zeta hydrodynamic size were 30 to 120 ± 2 nm observed via DLS as shown in Figure 4. There were found polydispersity index of 0.56 ± 0.04 in particle size in the size distribution graph obtained via DLS. The tendency of the technique against the estimation of larger particles (or even aggregates) explains the increased scale of the MgO-NPs determined by DLS [34]. The zeta potential of NPs may be affected by distinct functional groups present in plant extract that adsorbed on their surface. Same were also observed by [36].

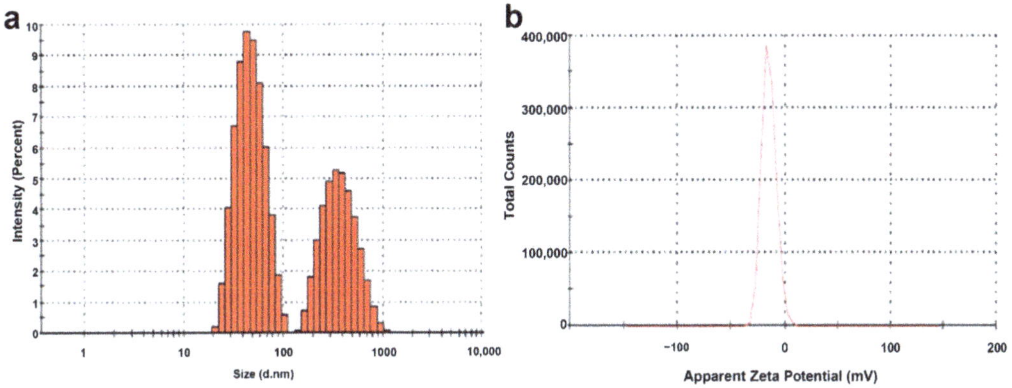

Figure 4. (**a**) Zeta size measurement and (**b**) Zeta potential measurement of MgO-NPs.

2.8. Antibacterial Assay against H. pylori Bacterial Isolates

Antimicrobial resistance challenging the word health care system and multidrug-resistant diseases have had a major impact on current antibacterial therapy. Plant-mediated nanomaterials, have a wide therapeutic applications and thus proved itself a new source of antimicrobials [37–39]. In the current situation, environmentally friendly approaches for producing nanomaterials have been a significant technology. Plant mediated nanomaterials gain popularity due to dual function of capping and reduction [40]. In the current research, we synthesised MgO-NPs from a traditional medicinal plant and measured their antibacterial effectiveness against *H. pylori* bacterial strains [41]. The antimicrobial potential of MgO-NPs against test species were shown in Table 1. The study found that various doses of NPs had varying degrees of antimicrobial activity against all microorganisms studied (5 mg/mL, 4 mg/mL, 2 mg/mL and 1 mg/mL). MgO-NPs solution (5 mg/mL) showed the greatest zone of inhibition against *Helicobacter felis* (17.19 ± 0.83 mm), followed by *Helicobacter suis* (16.49 ± 0.64 mm) and *Helicobacter salomonis* (16.09 ± 0.66 mm) in the present report. Our findings are consistent with previous research evaluating the killing ability of MgO-NPs and other metallic NPs against urinary tract infections and *H. pylori* bacterial isolates [42,43].

Table 1. Antibacterial potential of MgO-NPs against *H. pylori* isolates.

H. pylori Strains	MgO-NPs			
	5 mg/mL	**4 mg/mL**	**2 mg/mL**	**1 mg/mL**
Helicobacter felis	17.19 ± 0.83 *	13.32 ± 0.53 *	9.55 ± 0.56 *	6.23 ± 0.31 *
Helicobacter suis	16.49 ± 0.64 **	13.62 ± 0.51 *	9.29 ± 0.53 *	5.48 ± 0.37 **
Helicobacter salomonis	16.09 ± 0.66 **	9.92 ± 0.42 ***	7.70 ± 0.49 ***	5.71 ± 0.23 **
Helicobacter bizzozeronii	14.19 ± 0.51 **	10.23 ± 0.59 **	8.43 ± 0.41 **	4.79 ± 0.26 ***
Positive control (Kanamycin)	21.82 ± 0.74	16.71 ± 0.82	12.67 ± 0.58	12.14 ± 0.44

Star *–*** represent; * highly significant, ** slightly significant and *** non-significant difference from control at $p < 0.05$ by one-way ANOVA in the column Values are mean ± SD of triplicate.

2.9. Protein Kinase Inhibition Assay

Protein kinase inhibitors are a well-established class of clinically effective medications that has a major role in the treatment of cancer. Since achieving inhibitor selectivity remains a major challenge for researchers, synthesising alternative compounds for chemical biology study or new small molecules as drugs remains a viable option [44,45]. These enzymes phosphorylate serine-threonine and tyrosine amino acid residues, which play important roles in cell replication, differentiation, and apoptosis. Protein kinase deregulation can contribute to tumor development, so institutions that can suppress these enzymes are crucial in anticancer research [46]. The protein kinase inhibition ability of *M. arvensis* synthesised NPs was tested using the Streptomyces 85E strain. Figure 5a depicts the results. At 0.5 mg/mL, MgO-NPs solution (5 mg/mL) showed the highest zone of inhibition (12.44 ± 0.72) and (5.66 ± 0.44), respectively. Streptomyces strain was inhibited by MgO-NPs in a concentration-dependent manner (Figure 5a). Overall, the findings revealed that all research samples accumulate essential metabolites with anti-cancer properties in *M. arvensis*. The findings are backed up by a previous study that looked at hyphae formation inhibition in Streptomyces 85E and found that isolated compounds displayed a remarkable zone of inhibition at 80 g/disk, leading to the hypothesis that the compounds prevent the formation of hyphae in Streptomyces 85E, which may inhibit cancer proliferation [45,47].

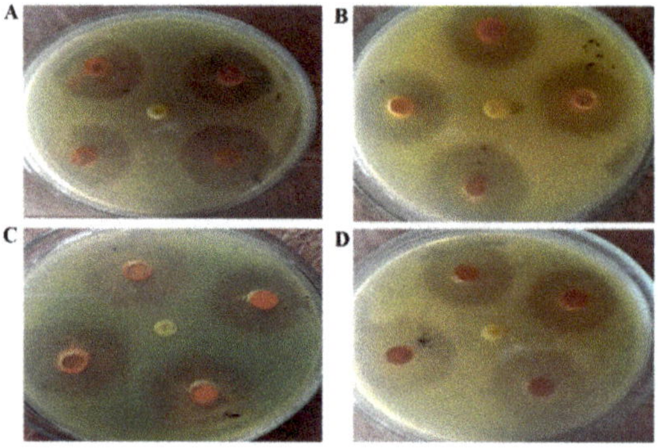

Figure 5. Assay picture of NPs against various strains (**A**) *Helicobacter felis,* (**B**) *Helicobacter suis,* (**C**) *Helicobacter salomonis* and (**D**) *Helicobacter bizzozeronii.*

2.10. In Vitro AChE and BChE Inhibition Assays

Alzheimer's is a neurodegenerative disease and worldwide accounts for sixty to 80 percent cases of dementia. This disease occurrence is alarming, with one person developing Alzheimer's disease every 65 s in the United States alone [48]. Cholinesterase inhibitors are currently available for patients at any stage of Alzheimer's disease. The successful inhibition of cholinesterase enzymes has been identified using a variety of synthetic and natural substances. Hydrolysis of acetyl choline to choline and acetic acid in synapsis in tissues and neuromuscular junctions is catalyzed by these enzymes. Reduced acetyl choline levels contribute to the development of Alzheimer's disease. acetylcholinesterase (AChE) and butrylcholineterase (BChE) inhibition was investigated using different concentrations of plant extracts [49]. The inhibition of esterases by MgO-NPs was dosage dependent. MgO-NPs were most potent at 400 μg/mL, inhibiting AChE by 73.82 ± 2.19 percent and BChE by 69.50 ± 1.82 percent. At 25 μg/mL, AChE had a 19.63 ± 0.47% inhibition response and BChE had a 23.53 ± 0.51% inhibition response. Overall, as seen by the values in Figure 5b, NPs is found to be highly active against both enzymes. Our findings are consistent with previous research [50,51].

2.11. In Vitro Cytotoxic Potential Against Hela Cell Lines

Only a few studies assessed the apoptotic potential of biosynthesized NPs. The effect of MgO-NPs on HeLa cell proliferation was determined using the MTT assay [52]. This is the first research to test the cytotoxicity of MgO-NPs extracted from *M. arvensis* HeLa cells. The cytotoxic effect against HeLa cells are due to biosynthesized NPs. At the applied concentration, cell death was estimated to be 49.49 ± 1.18 at 400 μg/mL.

Figures 6c and 7a. Doxorubicin was used as a supportive regulation which resulted in cell death of 97.11 ± 3.97 percent. Sriram et al. [53] and Safaepour et al. [54] addressed a similar cytotoxicity analysis. Metallic NPs are toxic to mammalian cells, according to several in vitro tests. Metallic NPs have been shown in some experiments to have the ability to interfere with genes involved in cell cycle development, as well as cause damage to nucleotides of DNA also induce programmed cell death of ancerous cells. Our findings are in correspondence with [55].

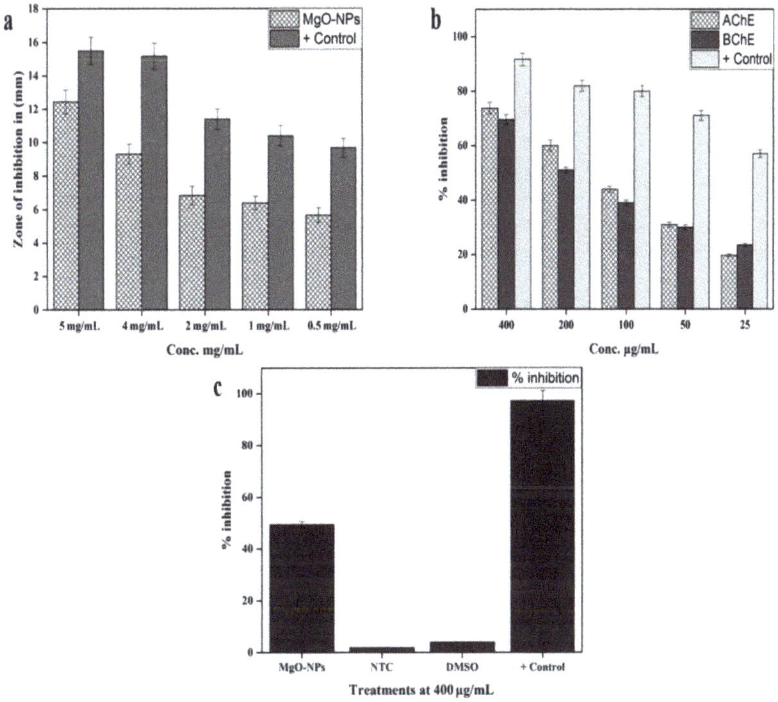

Figure 6. (**a**) Protein kinase inhibition, (**b**) in vitro AChE and BChE inhibition, (**c**) Anti-cancer potential of MgO-NPs against Hela Cell Lines.

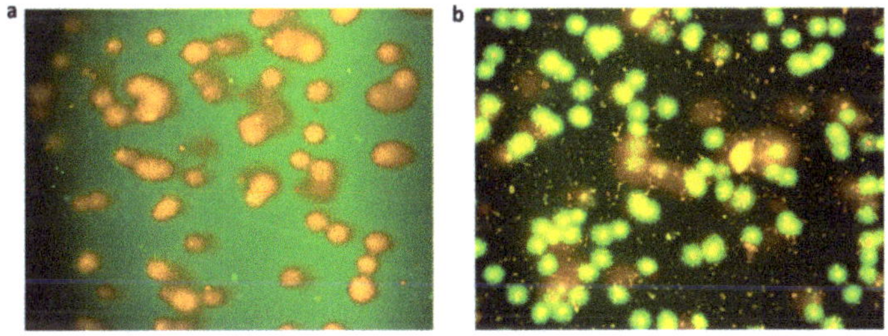

Figure 7. (**a**) Doxorubicin as positive control, (**b**) MgO-NPs against cell lines.

2.12. In Vitro Antioxidant Potential of MgO-NPs

Reactive oxygen species is responsible for the degradation of membrane lipids in cell membrane of plant, degradation of nucleic acids and aminoacids, which results in a shift in plant metabolic pathways [56]. Oxidative stress reaction nd plant biomolecules results in capping and reduction of nanoparticles [57,58]. Total reduction potential, DPPH free radicle scavenging, ABTS, and Total antioxidant capability of biosynthesized NPs were assessed. In Table 2, we summarise the antioxidant potential of biosynthesised nanoparticle. Phosphomolbdenum were used to investigate the antioxidant potential of synthesized NPs. The principle of this approach is the reduction of Mo (VI) to Mo (V) in the presence of antioxidant agent, the reduction of Mo form green phosphate molybdenum [59]. The antioxidant potential of biogenic MgO-NPs was at 61.1 ± 0.73 gAAE/mg at 400 µg/mL. With the total reducing power estimate (TRP) assay, total antioxidant potential (TAC) was amplified. The Fe^{3+} ion would be converted to Fe^{2+} ion if the measured sample has redox potential [60]. The largest TRP, like TAC, was 43.41 ± 0.23 at the highest concentration. ABTS and DPPH free radical scavenging assays were also performed to confirm the TAC and TRP results. The formation of the yellowish diphenyl picrylhydrazine molecule reduces DPPH, which is a stable free radical that is reduced by taking hydrogen or electron from a donor [61]. The antioxidant potential of the sample was confirmed by the quenching of DPPH and ABTS free radicals. MgO-NPs showed maximum scavenging of the DPPH and ABTS free radical at 400 µg/mL, which is 56.3 ± 0.38 and 77.12 ± 0.18 TEAC, respectively. All the assays repeated three times and the average were taken as final reading. Same results were also observed by [41,62].

Table 2. Antioxidants potentialities of biosynthesized MgO-NPs.

Conc. (µg/mL)	TAC (µg AAE/mg)	TRP (µg AAE/mg)	ABTS (TEAC)	DPPH (%FRSA)
400	61.1 ± 0.73	43.41 ± 0.23	77.12 ± 0.18	56.3 ± 0.38
200	55.37 ± 0.17	39.51 ± 0.47	63.63 ± 0.29	41.1 ± 0.61
100	33.86 ± 0.62	22.23 ± 0.16	44.64 ± 0.46	27.69 ± 0.42
50	25.29 ± 0.56	16.76 ± 0.28	25.47 ± 0.16	18.45 ± 0.88
25	19.16 ± 0.15	10.41 ± 0.86	16.39 ± 0.25	10.19 ± 0.38

2.13. Bio-Compatible Nature of MgO-NPs against Human (RBCs)

Human red blood cells were used in a biocompatibility experiment to demonstrate the biocompatibility of the green synthesised NPs. The haemolysis of erythrocytes against varying Conc. Of NPs (25 µg/mL to 400 µg/mL) is observed in this bioassay. A spectrophotometer is used to calculate RBCS haemolysis at 405 nm. Only if the sample has the potential to burst the cell will the RBCs hemolysis be detected. Table 3 shows the biocompatibility effects of our research. The American Society for testing materials has released several recommendations for biocompatibility of compounds, according to which substances with >2% haemolysis are called non-haemolytic, 2–5% mildly haemolytic, and >5% haemolysis are considered haemolytic [63]. As can be seen in Table 3, there are a number of factors to consider. Also at high concentrations, all of our stock solutions of synthesised nanoparticles exhibit fewer haemolysis, demonstrating their high biocompatibility. Also at high concentrations of 400 µg/mL, our biogenic MgO-NPs are hem compatible, and no haemolytic activity is observed at this concentration. As a result of our study's biocompatibility findings, we should conclude that mediated NPs are biosafe and that MgO-NPs can be used as a therapeutic agent.

Table 3. % Hemolysis of green synthesized MgO-NPs.

S. No	Conc. µg/mL	% Hemolysis
1	400	2.11 ± 0.13
2	200	2.06 ± 0.11
3	100	1.19 ± 0.09
4	50	0.83 ± 0.05

3. Discussion

Generally magnesium nanoparticles are synthesized by chemical and physical methods such as chemical vapor deposition, thermal evaporation, sol-gel, sonichemical and spray pyrolysis among others [64]. However, such procedures are high energy demanding, expensive, time consuming and are not eco-friendly [65]. Moreover, chemical methods may result in adsorption of toxic chemicals on the surface of nanoparticles that may lead to adverse effects in biomedical applications [66]. Biological methods that exploit living organism (microbes, plants) or living systems (enzymes) for the synthesis of nanoparticles is one possible alternative for eco-friendly and inexpensive synthesis of MgO-NPs. In the current study, an aqueous extract of *Mentha arvensis* was utilized as reducing and stabilizing agent for the synthesis of multifunctional silver nanoparticles (Ag-NPs). To the best of our knowledge it is the first ever study on *Mentha arvensis* mediated synthesis of magnesium nanoparticles. the main active compounds of *M. arvensis* extracts such as hesperidin, ferulic acid, rosmarinic acid, diosmin, didymin, buddleoside, acacetin and linarin, and have documented numerous biological effects [67].

The species are rich in medicinally important biochemicals including proteins and carbohydrates that played a major role in biosynthesis of MgO-NPs. The initial formation MgO-NPs was indicated by the color change and later on by UV- visible spectroscopy The bang gap of MgO-NPs were calculated as (3.3 eV) respectively, affirming successful synthesis of MgO-NPs [68]. After synthesis, stable MgO-NPs were well characterized by UV- visible spectroscopy, X-ray diffraction (XRD), Fourier transform infrared spectroscopy (FTIR), Dynamic Light Scattering (DLS), Scanning electron microscopy (SEM), Transmission electron microscopy (TEM) and Thermo galvanometric analysis (TGA). Moreover, aqueous dispersion stability of the nanoparticles with varied pH and polarity was also probed as function of storage time. Fourier transformed infrared spectroscopy revealed major functional groups on the synthesized nanoparticles. Major absorption peaks in the FTIR spectra were observed at 3461 cm^{-1}, 1737 cm^{-1}, 1649 cm^{-1}, 1151 cm^{-1}, 858 cm^{-1} and 604 cm^{-1}. The peaks at 1148 cm^{-1} corresponded to O-H bending and C-O stretching of the functional groups in primary alcohols. While the broad absorption peak on 607 cm^{-1} clearly indicated Mg-O bond stretching affirming the successful synthesis of MgO-NPs [28]. The FTIR spectra, thus confirms the successful capping of plant metabolites on the surface of MgO-NPs also resembles with previous reports [69].

X-ray diffraction analysis confirmed high purity of MgO-NPs with no extra peaks and indicated a face centered cubic (FCC) structure of biosynthesized MgO-NPs affirming the highly crystalline nature of the particles. Similar XRD patterns were also reported in previous studies. Moreover, the mean crystallite size as calculated using Scherer equation was found to be 32.4 nm Furthermore, Thermal gravimetric analysis (TGA) shown that total weight loss of MgO-NPs resulted to be 61.9%. The initial weight loss up to 150 °C is attributed to the dehydration and loss of moisture content from the samples [70].

Morphology, particle size and elemental composition of MgO-NPs were determined using SEM, TEM and EDX, respectively. SEM micrograph revealed spherical shaped morphology with little degree of aggregation. Similar morphological attributes were also observed in previous reports [71]. Moreover, average particle size of the particles as calculated using ImageJ software from TEM micrographs was found to be 29.72 ± 11.36. Furthermore, a strong peak at 3 KeV in EDX spectrograph confirmed purity NPs and augment the XRD results.

Dispersion stability plays a vital role in determining functional activities of the engineered nanoparticles in any biological system [72]. Several parameters affect the dispersion capacity of the nanoparticles such as presence of charged or uncharged molecules that are adsorbed on the particle surface and the ionic strength (pH) of the solvent [73]. Our study shown that MgO-NPs form highly stable dispersion in H_2O even after 24 h of sonication. Dynamic Light Scattering studies confirmed the stable Zeta potential (ζ) of -16.1 mV. Zeta potential (ζ) defines the colloidal stability and is a typical measurement of the surface charge on the particles. Suspensions that exhibits $|\zeta| \geq 15$ mV are generalized as stable colloids [73]. The Zeta potential measurement thus verifies and augments to the dispersion capacity of green synthesized MgO-NPs. The −ve surface charge is due to the binding affinity of the extract compounds on NPs conferring MgO nanoparticles stability and alleviates aggregation potential of the particles [35]. The size distribution graph shows that the particle size is polydispersed and is larger as compared to the TEM observations. The increased size of MgO-NPs measured via DLS is due to the biasness of the technique towards measurement of larger particles (even aggregates) [73]. After physicochemical and morphological characterization the biosynthesized nanoparticles were investigated for multifaceted invitro biomedical applications including Anti *H. pylori*, protein kinase inhibition, anti-Alzheimer's, antioxidant and anticancer potential and biocompatibility against isolated human red blood cells (hRBCs).

Pathogenic microbes cause wide range of diseases in both humans and animals. Novel approaches and sophisticated scientific research has become inevitable for the establishing novel therapeutic strategies to overcome antimicrobial resistance (AMR) and unsystemic use of antibiotics. In general, all the tested bacterial strains showed dose dependent sensitivity against MgO-NPs. Among the bacterial strains, *Helicobacter felis*, and *Helicobacter suis* exhibited high susceptible, displaying considerable zone of inhibition (ZOI) i.e., 17.19 ± 0.83 mm and 16.49 ± 0.64 mm, respectively. Factors such as size, shape, oxidation state and surface chemistry are considered as the most influential factors dictating antibacterial properties of MgO-NPs. In general recently metal oxide nanoparticles such as zinc, silver, titanium and magnesium based treatments have gained much attention owing to their significant cytotoxicity against *H. pylori* [74]. Ag-NPs has the potential to be used for targeted delivery of potential candidate drugs for Alzheimer's disease. We found interestingly, the inhibition response for both esterases was dose dependent. MgO-NPs were most active at 400 μg/mL resulted in 73.82 ± 2.19% inhibition of AChE and 69.50 ± 1.82% for BChE. Our findings, are in agreement with some of the previous studies reported [74]. For preliminary screening for anticancer activity, the biosynthesis nanoparticles were investigated for growth inhibition potential against Protein Kinase. It was observed that MgO-NPs solution (5 mg/mL) displayed maximum zone of inhibition (12.44 ± 0.72) and (5.66 ± 0.44) at 0.5 mg/mL respectively. Due to excellent inhibition activity against Protein Kinase enzyme the particles were further exploited against HeLa cells to verify and augment the anticancer potential.

In our study, MgO-NPs showed potential inhibition of 49.49 ± 1.18 at 400 μg/mL towards fresh HeLa cells line.There is dearth of data regarding detailed anti-cancerous mechanism of MgO-NPs however certain studies suggest that Ag-NPs results in DNA damage, lysosomal damage, mitochondrial chain and complex disruption and pro apoptotic activity that in turn effect proliferative system and cell cycle of cancer cells ultimately lead to complete inhibition of proliferation. Moreover, the ROS production and associated damages resulting from oxidative stress are highly size dependent with smaller size of MgO-NPs leads to enhanced overproduction of ROS. Smaller particles of MgO-NPs have the capability of increased interaction with cellular compartments and enhanced penetration to release Mg^{+} ions [75]. Our findings thus augment and support previously reported studies. Our study thus suggest that biosynthesized Ag-NPs could be used as novel therapeutic agent against HeLa cells. However, detailed in vivo study can be designed to augment the in vitro results and to investigate the detail mechanism involved in anti-cancerous effects.

Furthermore, the biosynthesized nanoparticles were also investigated for antioxidant potential via DPPH (2,2-diphenyl-1-picrylhydrazyl) and ABTS (2,2′-azino-bis(3-ethylbenzothiazoline-6- sulfonic acid) free radical scavenging activity (FRSA), total antioxidant activity (TAC) and total reducing power (TRP) at various concentrations. The particles exhibited dose dependent antioxidant capacity and displayed 61.1 ± 0.73 μgAAE/mg and 43.41 ± 0.23 μgAAE/mg TAC and TRP activity at the highest concentration of 400 μg/mL. While moderate DPPH and ABTS free radical scavenging activity (FRSA) of 56.3 ± 0.38 (IC50; 358 μg/mL) and 77.12 ± 0.18 TEAC, respectively was noted. From the results summarized, it can be suggested that some of the antioxidant compounds may be involved in the reduction and stabilization of the NPs during synthesis process and may be responsible for imparting overall moderate antioxidant potential to MgO-NPs. The biocompatibility of the engineered nanomaterial (ENMs) is an essential requirement for biomedical applications. The particles exhibited excellent hemocompatibility as even at the highest concentration of 400 μg/mL 2.11 ± 0.13 haemolysis activity was observed. Our finding thus, endorses the bio safe nature of the particles and thus pave the way for *Mentha arvensis* synthesized MgO-NPs be subjected for therapeutic applications.

4. Materials and Methods

4.1. Collection and Processing of the Plant Material

The herb used in this analysis was collected in the district of Charsadda in the Pakistani province of Khyber Pakhtunkhwa. The plant was confirmed as *Mentha* arvensis by the professors in Department of Botany, Bacha Khan University in Charsadda, Pakistan. Leaves were taken from the plant and dried in shade followed by grinding in to powder form and stored at 25 °C for extraction process. 30 g of plant powder were mixed with 200 mL of deionized water and properly shaked for 10 min, followed by incubation at 200 rpm in sun scientific orbital shaking incubator model number ES 20. The obtain extract was filtered thrice with nylon cloth and thrice with Whattman filter paper to remove any remaining residues. The extract were stored for future use in the experiment.

4.2. Biosynthesis of MgO Nanoparticles

With minor modifications, MgO-NPs were synthesised according to a published protocol [12]. In a nutshell, 100 mL plant extract was mixed with 6.0 g of $Mg(NO_3)_2.6H_2O$ and leftover at 60 °C for 2 h on a magnetic stirrer. The mixture were centrifuged at 1000 rpm just after the completion of the reaction. The pellets were washed three times with distilled water and then kept in oven to be dried and after that the particles were calcinated for 2 h at 500 °C. The nanoparticles were grinded in to fine powder and stored in a vial for physiochemical characterization.

4.3. Characterizations of Biosynthesized MgO-NPs

Physicochemical properties of *M. arvensis* synthesised MgO-NPs were investigated using various characterization techniques. 200 to 700 nm standard wavelenth was used in UV analysis of MgO-NPs [76]. The X-ray diffraction method was used to detect the crystallite nature of green synthesised MgO-NPs. XRD spectra were obtained by PANalyticaX'pert diffractometer (Company, City, State abbr, Country). The Scherer's standard equation were evaluated to determine the crystal size [77].

$$D = k\,\lambda/\beta\ \mathrm{Cos}\theta$$

D represents half-peak-height of an XRD line due to a specific crystalline plane K denotes shape factor (0.94), λ depicts X-rays wavelength of 1.5421Å while β and θ refers to FWHM in radians and Bragg's angle, respectively". Fourier transforming infrared spectroscopic analysis were performed in 400 cm^{-1} to 4000 cm^{-1} spectral range to detect the functional group responsible for nanoparticle formulation by using Jasco FT/IR-6000 FTIR spectrometer [78]. SEM (JSM-7600F, Japan) and TEM (JEM-2100F, Japan) were used to analyse morphology and physical measurements, while EDX with TEM (JEM2100) INCA100/Oxford instruments, U.K. were used to know about the elemental composition of MgO-NPs [79]. The electrostatic Zeta potential arises at a particle's shear plane and affects all surface charges and the particle's local medium. The Zeta Potential Analyzer is used to examine the Zeta potential. For logging all of the measurements, phase analysis Light Scattering Mastersizer 3000 were used calculate Zeta Potential were calculated via Smoluchowski equation.

$$V = (\varepsilon E/\eta)\xi$$

where v = electrophoretic velocity, η = viscosity, ε = electrical permittivity of the electrolytic solution and E = electric field [80]. Thermo Gravimetric Analysis was used to investigate thermal equilibrium using a Q500 thermo gravimetric analyser Pyris Diamond Series TG/DTA under flowing nitrogen gas at a. at a temperature of 30 °C to 600 °C.

4.4. Antibacterial Assay of MgO-NPs against H. pylori Bacterial Strains

The antibacterial activity of test samples was assessed using the agar well diffusion process, as previously described [81]. *Helicobacter bizzozeronii, Helicobacter felis, Helicobacter salomonis,* and *Helicobacter suis* were among the bacteria included in the report. Using the MacFarland specifications. Following that, 50 L of fresh culture is spilled onto nutrient agar plates and evenly scattered with cotton swabs. 5 mm wells were made with a sterile borer, and 10 L of the examined samples were applied, with the plates labelled accordingly. The concentrations ranged from 1 mg/mL to 5 mg/mL. Kanamycin and DMSO were used as positive and negative controls, respectively, in the assay. After that, the bacterial culture plates were incubated at 37 °C for overnight. Zones of inhibition were measured by using Vernier Calliper.

4.5. Protein Kinase Inhibition Assay

This assay is used to test the anticancer function of biosynthesized MgO-NPs. This is a bioassay for confirming the synthesised NPs' capacity to suppress protein kinases [47]. Adopted a protocol that was subtly different from ours. Streptomyces 85E was used as a research strain. 100 mL of Streptomyces 85E culture were poured into ISP4 medium plates. Each well (5 mm) was filled with around 5 L of MgO-NPs and labelled accordingly. Surfactin was used as a +ve, while DMSO as a −ve control. After that plates were incubated for two days at 28 degrees Celsius. Clear and bald areas around wells were observed, which indicate that phosphorylation, mycelia, and spore formation have been inhibited. The cytotoxic potential of MgO-NPs were observed by strong inhibition zones.

4.6. Anti-Alzheimer's Activity

Inhibition of the enzymes Acetylcholinesterase (AChE) and Butyrylcholinesterase (BChE) may be a target in Alzheimer's treatment. The inhibition power of MgO-NPs by AChE (Sigma "101292679") (St. Louis, MI, USA) and BChE (Sigma "101303874") was investigated using Elman's protocol [82], which was slightly modified. The concentration level for the reference sample was 12.5 µg/mL to 400 µg/mL. Phosphate buffer saline (PBS) solution was used to diffuse the NPs. In AChE, the final enzyme concentration was 0.03U/mL, while in BChE, it was 0.01 U/mL. The reaction mixture, which included DTNB (0.00022 M), BTchI (0.0005 M), and ATchI (0.0005 M), was prepared in purified water and stored at 4 °C. The positive control was Methanol-mediated Galanthamine hydrobromide (Sigma; GI660), while the negative control was the reaction combination without the reference sample. The anticholinesterase assay works by hydrolyzing ATchI and BTchI into AChE and BChE, respectively, resulting in the formation of 5-thio-2-nitrobenzoate anion. The latter forms further complexes with DTNB, resulting in a yellow colour. Absorbance was measured using a UV-VIS spectrophotometer set to 412 nm. With a decrease in absorption rate over time, galantamine and MgO-NPs can be used to calculate percent enzyme inhibition and percent enzyme activity.

$$V = \Delta Abs/\Delta t$$

$$\text{Enzyme activity } (\%) = V/Vmax \times 100$$

$$\text{Enzyme inhibition } (\%) = 100 - \text{Enzyme activity } (\%)$$

4.7. Anti-Cancer Potential of MgO-NPs against Hela Cell Lines

HeLa cells were obtained from the University of Peshawar in Pakistan. HeLa cells were cultured in MEM & McCoys 5a media, the media were supplemented with 10%

calf serum and incubated with 5% CO_2. HeLa cells were seeded in to 96 well plates 1×10^4 cells per plate and incubated for two days at 37 °C. HeLa cells were treated with biosynthesized MgO-NPs and controls at a fix concentration of 400 µg/mL. Simultaneously, HeLa cells were also treated with Doxorubicin (a well-known anticancer agent having Conc. of 100 mM). The paltes were incubated for 2–3 days to examine the cytotoxic potential of MgO-NPs. The MTT assay was then performed on these plates. MTT solution is prepared at a concentration of 5 mg/mL. Every well received 100 mL of MTT, which was incubated for 4 h. Since forming purple-colored formazone crystals, di-methyl sulphoxide was used to dissolve them (DMSO). At 620 nm In ELISA plate reader the samples were observed. The foolowing formula were used to calculate the anticancer potential of NPs.

Percentage of viability = OD value of experimental sample (MgO-NPs)/OD value of experimental control (untreated) × 100.

4.8. Estimation of Antioxidant Activity

4.8.1. DPPH Antioxidant Assay

The antioxidant function of DPPH (2,2-diphenyl-1-picrylhydrazyl) was calculated using the previously published protocol [83] with slight modifications. Sample extract (20 L) was combined with DPPH (3.2 mg/100 mL methanol) 180 L, and the mixture was incubated at 25 °C for 1 h before adding dH_2O (160 L). The absorbance at 517 nm was measured using an absorbance microplate reader. The methanolic extract and 0.5 mL of DPPH solution were used as standards to map the calibration curve (R2 = 0.989). The percent radical scaving potential of MgO-NPs were calculated by following formula.

$$Freeradicalscavengingactivity(\%) = 100 \times \left(1 - \frac{Ac}{As}\right)$$

4.8.2. Total Antioxidant Capacity Determination (TAC)

Protocol stated by [84] were used to determine overall antioxidant ability. Using a micropipette, 1 mL of sample was filled into Eppendorf tubing. Then we fill Eppendorf tubes with 0.9 mL of TAC reagent Followed by incubation at 90 °C for 48 h in a water bath. The sample absorbance were measured at 630 nm in microplate reader and total antioxidant capacity of MgO-NPs were calculated in ascorbic acid equivalent/mg units.

4.8.3. Total Reducing Power Determination (TRP)

The same technique as described by [85] were employed to determine TRP. The research sample was still in an Eppendorf tube, so 400 µL of 0.2 M phosphate buffer with pH 6.6 and potassium ferric cyanide (1 percent *w*/*v*) added to it and followed by incubation at 55 °C for half an hour in water bath and each Eppendorf tube was filled with 400 µL of trichloroacetic acid (10% *w*/*v*) after incubation which is followed centrifugation at 3000 rpm for 10 min. The supernatant (140 µL) obtained after centrifugation were poured into 96-well plate that already contained 60 µL of ferric cyanide solution (0.1 percent *w*/*v*). The absorbance of the sample in microplate reader was set to 630 nm for reading.

4.8.4. Antioxidant ABTS Assay

Previous methods [84] were used in this experiment. In a nutshell, the ABTS solution were prepared by combining 7 mM ABTS salt with 2.45 mM potassium persulphate in an equal proportion and storing the mixture in the dark for 16 h. Until combining with extracts, the solvent absorbance was measured at 734 nm and calibrated to 0.7. The mixture was left in the dark for another 15 min at 25 °C. The absorbance was measured at 734 nm.

4.9. Biocompatibility Studies

New human red blood cells (hRBCs) were used to explain biogenic MgO-NPs biocompatibility [85]. After the individual's consent, 1 mL of blood samples were collected in EDTA tubes from healthy individuals followed by centrifugation to isolate RBCs. Pellets were rinsed with PBS. Following centrifugation to isolate RBCs. 200 mL of RBCs and PBS (9.8 mL) at pH: 7.2 were gently mixed for preparation of suspension. MgO-NPs were reacted with erythrocyte suspension at different Conc. followed by incubation for 1 h at 35 °C. The reaction mixture were centrifuged and transfer to well plate to detect haemoglobin release at standart absorption peak of 450 nm. The formula for calculating percent haemolysis was:

$$(\%)\ \text{Haemolysis} = \left(\frac{sample\ Ab - negative\ control\ Ab}{Positive\ control\ Ab - Negative\ control\ Ab} \right) \times 100$$

5. Conclusions

This study focuses on an environmental friendly synthesis of MgO-NPs from *Mentha arvensis* XRD spectra confirms the crystalline structure of MgO-NPs. The presence of phytochemicals involved in the transition of ions was confirmed using Fourier transforming infrared spectroscopy (FTIR). SEM and TEM analysis were used to decide morphology and vibrational modes, while DLS was used to determine surface charge and stability, and TGA was used to determine stability. MgO-NPs that have been synthesised have proven to be effective antioxidants and antibacterial strains. Bioengineered MgO-NPs have a high inhibitory potential against AChE and BChE enzymes. Biogenic MgO-NPs were discovered to be highly effective against Hela cell lines. Human red blood cells have been shown to be biocompatible with synthesised MgO-NPs. Our study concluded that the biogenic MgO-NPs described above can be used in a variety of diseases, cosmetics, and cancer studies. More research into the use of magnesium oxide nanoparticles in biomedicine, both in vitro and in vivo, is required.

Author Contributions: Conceptualization, S.F.; methodology, S.F.; software, A., M.R., H.J., S.A.S. and S.S.; validation, M.I., N.B. and R.M.; formal analysis, M.R., Z.H., W.K. and A.I.; investigation, M.N.U. and N.Z.; data curation, M.R. and A.K.; A writing—original draft preparation, S.F.; A writing—review and editing, H.J.; visualization, S.S.; supervision, S.A.S. All authors have read and agreed to the published version of the manuscript.

Funding: This research received no external funding.

Data Availability Statement: All required data is present in this file.

Acknowledgments: We are thankful to Institute of Biotechnology and Microbiology, Bacha Khan University Charsadda, KPK, Pakistan.for their support and facilities.

Conflicts of Interest: The authors declare no conflict of interest.

References

1. Ramsden, J. *Nanotechnology: An Introduction*; William Andrew: Norwich, NY, USA, 2016.
2. Albrecht, M.A.; Evans, C.W.; Raston, C.L. Green chemistry and the health implications of nanoparticles. *Green Chem.* **2006**, *8*, 417–432. [CrossRef]
3. Herlekar, M.; Barve, S.; Kumar, R. Plant-mediated green synthesis of iron nanoparticles. *J. Nanoparticles* **2014**, *2014*. [CrossRef]
4. Simonis, F.; Schilthuizen, S. *Nanotechnology*; Innovation Opportunities for Tomorrow's Defence, Report; TNO Science & Industry Future Technology Center: Delft, The Netherlands, 2006.
5. Iravani, S. Green synthesis of metal nanoparticles using plants. *Green Chem.* **2011**, *13*, 2638–2650. [CrossRef]
6. Duan, H.; Wang, D.; Li, Y. Green chemistry for nanoparticle synthesis. *Chem. Soc. Rev.* **2015**, *44*, 5778–5792. [CrossRef]
7. Bala, N.; Saha, S.; Chakraborty, M.; Maiti, M.; Das, S.; Basu, R.; Nandy, P. Green synthesis of zinc oxide nanoparticles using Hibiscus subdariffa leaf extract: Effect of temperature on synthesis, anti-bacterial activity and anti-diabetic activity. *RSC Adv.* **2015**, *5*, 4993–5003. [CrossRef]
8. Hasan, S. A review on nanoparticles: Their synthesis and types. *Res. J. Recent Sci.* **2015**, *2277*, 2502.
9. Barzinjy, A.A.; Hamad, S.M.; Aydın, S.; Ahmed, M.H.; Hussain, F.H. Green and eco-friendly synthesis of Nickel oxide nanoparticles and its photocatalytic activity for methyl orange degradation. *J. Mater. Sci. Mater. Electron.* **2020**, *31*, 11303–11316. [CrossRef]
10. Tang, Z.-X.; Lv, B.-F. MgO nanoparticles as antibacterial agent: Preparation and activity. *Braz. J. Chem. Eng.* **2014**, *31*, 591–601. [CrossRef]

11. Fernández-García, M.; Rodriguez, J.A. Metal oxide nanoparticles. In *Encyclopedia of Inorganic and Bioinorganic Chemistry*; John Wiley & Sons, Ltd.: Hoboken, NJ, USA, 2011.
12. Vergheese, M.; Vishal, S.K. Green synthesis of magnesium oxide nanoparticles using Trigonella foenum-graecum leaf extract and its antibacterial activity. *J. Pharm. Phytochem.* **2018**, *7*, 1193–1200.
13. Bindhu, M.; Umadevi, M.; Micheal, M.K.; Arasu, M.V.; Al-Dhabi, N.A. Structural, morphological and optical properties of MgO nanoparticles for antibacterial applications. *Mater. Lett.* **2016**, *166*, 19–22. [CrossRef]
14. El-Moslamy, S.H. Bioprocessing strategies for cost-effective large-scale biogenic synthesis of nano-MgO from endophytic Streptomyces coelicolor strain E72 as an anti-multidrug-resistant pathogens agent. *Sci. Rep.* **2018**, *8*, 1–22. [CrossRef]
15. Jeevanandam, J.; San Chan, Y.; Danquah, M.K. Biosynthesis and characterization of MgO nanoparticles from plant extracts via induced molecular nucleation. *New J. Chem.* **2017**, *41*, 2800–2814. [CrossRef]
16. Thawkar, B.S. Phytochemical and pharmacological review of Mentha arvensis. *Int. J. Green Pharm. (IJGP)* **2016**, *10*, 2.
17. do Nascimento, E.M.M.; Rodrigues, F.F.G.; Campos, A.R.; da Costa, J.G.M. Phytochemical prospection, toxicity and antimicrobial activity of *Mentha arvensis* (Labiatae) from northeast of Brazil. *J. Young Pharm.* **2009**, *1*, 210.
18. Chan, K. *Mentha Spicata-A Potential Cover Crop for Tropical Conservation Agriculture*; University of Hawaii at Manoa: Honolulu, HI, USA, 2016.
19. Zhao, B.T.; Kim, T.I.; Kim, Y.H.; Kang, J.S.; Min, B.S.; Son, J.K.; Woo, M.H. A comparative study of *Mentha arvensis* L. and Mentha haplocalyx Briq. by HPLC. *Nat. Prod. Res.* **2018**, *32*, 239–242. [CrossRef]
20. Adomako-Bonsu, A.G.; Chan, S.L.; Pratten, M.; Fry, J.R. Antioxidant activity of rosmarinic acid and its principal metabolites in chemical and cellular systems: Importance of physico-chemical characteristics. *Toxicol. In Vitro* **2017**, *40*, 248–255. [CrossRef]
21. Guilger-Casagrande, M.; de Lima, R. Synthesis of Silver Nanoparticles Mediated by Fungi: A Review. *Front. Bioeng. Biotechnol.* **2019**, *7*. [CrossRef]
22. Balakrishnan, G.; Velavan, R.; Batoo, K.M.; Raslan, E.H. Microstructure, optical and photocatalytic properties of MgO nanoparticles. *Results Phys.* **2020**, *16*, 103013. [CrossRef]
23. Bhargava, R.; Khan, S. Superior dielectric properties and bandgap modulation in hydrothermally grown Gr/MgO nanocomposite. *Phys. Lett. A* **2019**, *383*, 1671–1676. [CrossRef]
24. Wang, F.-H.; Chang, C.-L. Effect of substrate temperature on transparent conducting Al and F co-doped ZnO thin films prepared by rf magnetron sputtering. *Appl. Surf. Sci.* **2016**, *370*, 83–91. [CrossRef]
25. Sushma, N.J.; Prathyusha, D.; Swathi, G.; Madhavi, T.; Raju, B.D.P.; Mallikarjuna, K.; Kim, H.-S. Facile approach to synthesize magnesium oxide nanoparticles by using *Clitoria ternatea*—Characterization and in vitro antioxidant studies. *Appl. Nanosci.* **2016**, *6*, 437–444. [CrossRef]
26. Jayaseelan, C.; Ramkumar, R.; Rahuman, A.A.; Perumal, P. Green synthesis of gold nanoparticles using seed aqueous extract of *Abelmoschus esculentus* and its antifungal activity. *Ind. Crops Prod.* **2013**, *45*, 423–429. [CrossRef]
27. Nguyen, D.T.C.; Dang, H.H.; Vo, D.-V.N.; Bach, L.G.; Nguyen, T.D.; Van Tran, T. Biogenic synthesis of MgO nanoparticles from different extracts (flower, bark, leaf) of *Tecoma stans* (L.) and their utilization in selected organic dyes treatment. *J. Hazard. Mater.* **2021**, *404*, 124146. [CrossRef]
28. Nandgaonkar, A.G. Bacterial Cellulose (BC) as a Functional Nanocomposite Biomaterial. Ph.D. Thesis, North Carolina State University, Raleigh, NC, USA, 2014.
29. Mašek, O.; Budarin, V.; Gronnow, M.; Crombie, K.; Brownsort, P.; Fitzpatrick, E.; Hurst, P. Microwave and slow pyrolysis biochar—Comparison of physical and functional properties. *J. Anal. Appl. Pyrolysis* **2013**, *100*, 41–48. [CrossRef]
30. Jhansi, K.; Jayarambabu, N.; Reddy, K.P.; Reddy, N.M.; Suvarna, R.P.; Rao, K.V.; Kumar, V.R.; Rajendar, V. Biosynthesis of MgO nanoparticles using mushroom extract: Effect on peanut (*Arachis hypogaea* L.) seed germination. *3 Biotech* **2017**, *7*, 1–11. [CrossRef]
31. Suresh, J.; Pradheesh, G.; Alexramani, V.; Sundrarajan, M.; Hong, S.I. Green synthesis and characterization of hexagonal shaped MgO nanoparticles using insulin plant (*Costus pictus* D. Don) leave extract and its antimicrobial as well as anticancer activity. *Adv. Powder Technol.* **2018**, *29*, 1685–1694. [CrossRef]
32. Rahmani-Nezhad, S.; Dianat, S.; Saeedi, M.; Hadjiakhoondi, A. Synthesis, characterization and catalytic activity of plant-mediated MgO nanoparticles using *Mucuna pruriens* L. seed extract and their biological evaluation. *J. Nanoanal.* **2017**, *4*, 290–298.
33. Wong, C.W.; San Chan, Y.; Jeevanandam, J.; Pal, K.; Bechelany, M.; Abd Elkodous, M.; El-Sayyad, G.S. Response surface methodology optimization of mono-dispersed MgO nanoparticles fabricated by ultrasonic-assisted sol–gel method for outstanding antimicrobial and antibiofilm activities. *J. Clust. Sci.* **2020**, *31*, 367–389. [CrossRef]
34. Mourdikoudis, S.; Pallares, R.M.; Thanh, N.T. Characterization techniques for nanoparticles: Comparison and complementarity upon studying nanoparticle properties. *Nanoscale* **2018**, *10*, 12871–12934. [CrossRef]
35. Vimala, K.; Sundarraj, S.; Paulpandi, M.; Vengatesan, S.; Kannan, S. Green synthesized doxorubicin loaded zinc oxide nanoparticles regulates the Bax and Bcl-2 expression in breast and colon carcinoma. *Process Biochem.* **2014**, *49*, 160–172. [CrossRef]
36. Lynch, I.; Dawson, K.A. Protein-nanoparticle interactions. *Nano Today* **2008**, *3*, 40–47. [CrossRef]
37. Romero, C.D.; Chopin, S.F.; Buck, G.; Martinez, E.; Garcia, M.; Bixby, L. Antibacterial properties of common herbal remedies of the southwest. *J. Ethnopharmacol.* **2005**, *99*, 253–257. [CrossRef]
38. Boucher, H.W.; Talbot, G.H.; Bradley, J.S.; Edwards, J.E.; Gilbert, D.; Rice, L.B.; Scheld, M.; Spellberg, B.; Bartlett, J. Bad bugs, no drugs: No ESKAPE! An update from the Infectious Diseases Society of America. *Clin. Infect. Dis.* **2009**, *48*, 1–12. [CrossRef] [PubMed]

39. Talbot, G.H.; Bradley, J.; Edwards, J.E., Jr.; Gilbert, D.; Scheld, M.; Bartlett, J.G. Bad bugs need drugs: An update on the development pipeline from the Antimicrobial Availability Task Force of the Infectious Diseases Society of America. *Clin. Infect. Dis.* **2006**, *42*, 657–668. [CrossRef]
40. Saif, S.; Tahir, A.; Chen, Y. Green synthesis of iron nanoparticles and their environmental applications and implications. *Nanomaterials* **2016**, *6*, 209. [CrossRef]
41. Rajeshkumar, S.; Menon, S.; Kumar, S.V.; Tambuwala, M.M.; Bakshi, H.A.; Mehta, M.; Satija, S.; Gupta, G.; Chellappan, D.K.; Thangavelu, L. Antibacterial and antioxidant potential of biosynthesized copper nanoparticles mediated through Cissus arnotiana plant extract. *J. Photochem. Photobiol. B Biol.* **2019**, *197*, 111531. [CrossRef]
42. Gurunathan, S.; Jeong, J.-K.; Han, J.W.; Zhang, X.-F.; Park, J.H.; Kim, J.-H. Multidimensional effects of biologically synthesized silver nanoparticles in Helicobacter pylori, Helicobacter felis, and human lung (L132) and lung carcinoma A549 cells. *Nanoscale Res. Lett.* **2015**, *10*, 1–17. [CrossRef] [PubMed]
43. Safarov, T.; Kiran, B.; Bagirova, M.; Allahverdiyev, A.M.; Abamor, E.S. An overview of nanotechnology-based treatment approaches against Helicobacter Pylori. *Expert Rev. Anti Infect. Ther.* **2019**, *17*, 829–840. [CrossRef] [PubMed]
44. Ferguson, F.M.; Gray, N.S. Kinase inhibitors: The road ahead. *Nat. Rev. Drug Discov.* **2018**, *17*, 353. [CrossRef] [PubMed]
45. Jan, H.; Shah, M.; Usman, H.; Khan, A.; Muhammad, Z.; Hano, C.; Abbasi, B.H. Biogenic Synthesis and Characterization of Antimicrobial and Anti-parasitic Zinc Oxide (ZnO) Nanoparticles using Aqueous Extracts of the Himalayan Columbine (*Aquilegia pubiflora*). *Front. Mater.* **2020**, *7*, 249. [CrossRef]
46. Bain, J.; Plater, L.; Elliott, M.; Shpiro, N.; Hastie, C.J.; Mclauchlan, H.; Klevernic, I.; Arthur, J.S.C.; Alessi, D.R.; Cohen, P. The selectivity of protein kinase inhibitors: A further update. *Biochem. J.* **2007**, *408*, 297–315. [CrossRef] [PubMed]
47. Jan, H.; Khan, M.A.; Usman, H.; Shah, M.; Ansir, R.; Faisal, S.; Ullah, N.; Rahman, L. The Aquilegia pubiflora (*Himalayan columbine*) mediated synthesis of nanoceria for diverse biomedical applications. *RSC Adv.* **2020**, *10*, 19219–19231. [CrossRef]
48. Weller, J.; Budson, A. Current understanding of Alzheimer's disease diagnosis and treatment. *F1000Research* **2018**, 7. [CrossRef] [PubMed]
49. Khalil, A.T.; Ayaz, M.; Ovais, M.; Wadood, A.; Ali, M.; Shinwari, Z.K.; Maaza, M. In vitro cholinesterase enzymes inhibitory potential and in silico molecular docking studies of biogenic metal oxides nanoparticles. *Inorg. Nano Met. Chem.* **2018**, *48*, 441–448. [CrossRef]
50. Hassan, D.; Khalil, A.T.; Saleem, J.; Diallo, A.; Khamlich, S.; Shinwari, Z.K.; Maaza, M. Biosynthesis of pure hematite phase magnetic iron oxide nanoparticles using floral extracts of *Callistemon viminalis* (bottlebrush): Their physical properties and novel biological applications. *Artif. Cells Nanomed. Biotechnol.* **2018**, *46 (Supp. 1)*, 693–707. [CrossRef]
51. Šinko, G.; Vrček, I.V.; Goessler, W.; Leitinger, G.; Dijanošić, A.; Miljanić, S. Alteration of cholinesterase activity as possible mechanism of silver nanoparticle toxicity. *Environ. Sci. Pollut. Res.* **2014**, *21*, 1391–1400. [CrossRef]
52. Sukirtha, R.; Priyanka, K.M.; Antony, J.J.; Kamalakkannan, S.; Thangam, R.; Gunasekaran, P.; Krishnan, M.; Achiraman, S. Cytotoxic effect of Green synthesized silver nanoparticles using Melia azedarach against in vitro HeLa cell lines and lymphoma mice model. *Process Biochem.* **2012**, *47*, 273–279. [CrossRef]
53. Sriram, M.I.; Kanth, S.B.M.; Kalishwaralal, K.; Gurunathan, S. Antitumor activity of silver nanoparticles in Dalton's lymphoma ascites tumor model. *Int. J. Nanomed.* **2010**, *5*, 753.
54. Safaepour, M.; Shahverdi, A.R.; Shahverdi, H.R.; Khorramizadeh, M.R.; Gohari, A.R. Green synthesis of small silver nanoparticles using geraniol and its cytotoxicity against fibrosarcoma-wehi 164. *Avicenna J. Med Biotechnol.* **2009**, *1*, 111. [PubMed]
55. Sanpui, P.; Chattopadhyay, A.; Ghosh, S.S. Induction of apoptosis in cancer cells at low silver nanoparticle concentrations using chitosan nanocarrier. *ACS Appl. Mater. Interfaces* **2011**, *3*, 218–228. [CrossRef]
56. Sergiev, I.; Todorova, D.; Shopova, E.; Jankauskiene, J.; Jankovska-Bortkevič, E.; Jurkonienė, S. Exogenous auxin type compounds amend PEG-induced physiological responses of pea plants. *Sci. Hortic.* **2019**, *248*, 200–205. [CrossRef]
57. Mohamed, H.I.; Akladious, S.A. Changes in antioxidants potential, secondary metabolites and plant hormones induced by different fungicides treatment in cotton plants. *Pestic. Biochem. Physiol.* **2017**, *142*, 117–122. [CrossRef]
58. Rehman, M.; Ullah, S.; Bao, Y.; Wang, B.; Peng, D.; Liu, L. Light-emitting diodes: Whether an efficient source of light for indoor plants? *Environ. Sci. Pollut. Res.* **2017**, *24*, 24743–24752. [CrossRef]
59. Prieto, M.; Curran, T.P.; Gowen, A.; Vázquez, J.A. An efficient methodology for quantification of synergy and antagonism in single electron transfer antioxidant assays. *Food Res. Int.* **2015**, *67*, 284–298. [CrossRef]
60. Skonieczna, M.; Hudy, D. Biological activity of Silver Nanoparticles and their applications in Anticancer Therapy. In *Biological Activity of Silver Nanoparticles and Their Applications in Anticancer Therapy*; IntechOpen: London, UK, 2018; p. 131.
61. Ul-Haq, I.; Ullah, N.; Bibi, G.; Kanwal, S.; Ahmad, M.S.; Mirza, B. Antioxidant and cytotoxic activities and phytochemical analysis of Euphorbia wallichii root extract and its fractions. *Iran. J. Pharm. Res. IJPR* **2012**, *11*, 241. [PubMed]
62. Khalil, A.T.; Ovais, M.; Ullah, I.; Ali, M.; Shinwari, Z.K.; Hassan, D.; Maaza, M. Sageretia thea (Osbeck.) modulated biosynthesis of NiO nanoparticles and their in vitro pharmacognostic, antioxidant and cytotoxic potential. *Artif. Cells Nanomed. Biotechnol.* **2018**, *46*, 838–852. [CrossRef] [PubMed]
63. Nasar, M.Q.; Khalil, A.T.; Ali, M.; Shah, M.; Ayaz, M.; Shinwari, Z.K. Phytochemical analysis, Ephedra Procera CA Mey. Mediated green synthesis of silver nanoparticles, their cytotoxic and antimicrobial potentials. *Medicina* **2019**, *55*, 369. [CrossRef]
64. Iravani, S.; Korbekandi, H.; Mirmohammadi, S.V.; Zolfaghari, B. Synthesis of silver nanoparticles: Chemical, physical and biological methods. *Res. Pharm. Sci.* **2014**, *9*, 385. [PubMed]

65. Vijayan, S.R.; Santhiyagu, P.; Ramasamy, R.; Arivalagan, P.; Kumar, G.; Ethiraj, K.; Ramaswamy, B.R. Seaweeds: A resource for marine bionanotechnology. *Enzym. Microb. Technol.* **2016**, *95*, 45–57. [CrossRef] [PubMed]
66. Jain, D.; Daima, H.K.; Kachhwaha, S.; Kothari, S. Synthesis of plant-mediated silver nanoparticles using papaya fruit extract and evaluation of their anti microbial activities. *Dig. J. Nanomater. Biostruct.* **2009**, *4*, 557–563.
67. Zhang, T.; Ye, J.; Xue, C.; Wang, Y.; Liao, W.; Mao, L.; Yuan, M.; Lian, S. Structural characteristics and bioactive properties of a novel polysaccharide from *Flammulina velutipes*. *Carbohydr. Polym.* **2018**, *197*, 147–156. [CrossRef] [PubMed]
68. Xiu, Z.M.; Zhang, Q.B.; Puppala, H.L.; Colvin, V.L.; Alvarez, P.J. Negligible particle-specific antibacterial activity of silver nanoparticles. *Nano Lett.* **2012**, *12*, 4271–4275. [CrossRef] [PubMed]
69. Ullah, R.; Shah, S.; Muhammad, Z.; Shah, S.A.; Faisal, S.; Khattak, U.; ul Haq, T.; Akbar, M.T. In vitro and in vivo applications of Euphorbia wallichii shoot extract-mediated gold nanospheres. *Green Process. Synth.* **2021**, *10*, 101–111. [CrossRef]
70. Faisal, S.; Khan, M.A.; Jan, H.; Shah, S.A.; Shah, S.; Rizwan, M.; Ullah, W.; Akbar, M.T. Edible mushroom (*Flammulina velutipes*) as biosource for silver nanoparticles: From synthesis to diverse biomedical and environmental applications. *Nanotechnology* **2020**, *32*, 065101. [CrossRef] [PubMed]
71. Nasr, H.; Nassar, O.; El-Sayed, M.; Kobisi, A. Characterization and antimicrobial activity of lemon peel mediated green synthesis of silver nanoparticles. *Int. J. Biol. Chem.* **2020**, *12*, 56–63. [CrossRef]
72. Bihari, P.; Vippola, M.; Schultes, S.; Praetner, M.; Khandoga, A.G.; Reichel, C.A.; Coester, C.; Tuomi, T.; Rehberg, M.; Krombach, F. Optimized dispersion of nanoparticles for biological in vitro and in vivo studies. *Part. Fibre Toxicol.* **2008**, *5*, 1–14. [CrossRef]
73. Modena, M.M.; Rühle, B.; Burg, T.P.; Wuttke, S. Nanoparticle characterization: What to measure? *Adv. Mater.* **2019**, *31*, 1901556. [CrossRef]
74. Jebali, A.; Kazemi, B. Nano-based antileishmanial agents: A toxicological study on nanoparticles for future treatment of cutaneous leishmaniasis. *Toxicol. In Vitro* **2013**, *27*, 1896–1904. [CrossRef]
75. Ratan, Z.A.; Haidere, M.F.; Nurunnabi, M.; Shahriar, S.M.; Ahammad, A.; Shim, Y.Y.; Reaney, M.J.; Cho, J.Y. Green chemistry synthesis of silver nanoparticles and their potential anticancer effects. *Cancers* **2020**, *12*, 855. [CrossRef]
76. Shah, S.; Shah, S.A.; Faisal, S.; Khan, A.; Ullah, R.; Ali, N.; Bilal, M. Engineering novel gold nanoparticles using Sageretia thea leaf extract and evaluation of their biological activities. *J. Nanostruct. Chem.* **2021**, 1–12. [CrossRef]
77. Suresh, J.; Yuvakkumar, R.; Sundrarajan, M.; Hong, S.I. Green Synthesis of Magnesium Oxide Nanoparticles. In *Advanced Materials Research*; Trans Tech Publications Ltd.: Freienbach, Switzerland, 2014; pp. 141–144.
78. Shah, R.; Shah, S.A.; Shah, S.; Faisal, S.; Ullah, F. Green Synthesis and Antibacterial Activity of Gold Nanoparticles of *Digera muricata*. *Indian J. Pharm. Sci.* **2020**, *82*, 374–378. [CrossRef]
79. Barzinjy, A.; Mustafa, S.; Ismael, H. Characterization of ZnO NPs prepared from green synthesis using Euphorbia Petiolata leaves. *EAJSE* **2019**, *4*, 74–83.
80. Morais, M.; Namouni, F. Asteroids in retrograde resonance with Jupiter and Saturn. *Mon. Not. R. Astron. Soc. Lett.* **2013**, *436*, L30–L34. [CrossRef]
81. Shah, M.; Nawaz, S.; Jan, H.; Uddin, N.; Ali, A.; Anjum, S.; Giglioli-Guivarc'h, N.; Hano, C.; Abbasi, B.H. Synthesis of bio-mediated silver nanoparticles from *Silybum marianum* and their biological and clinical activities. *Mater. Sci. Eng. C* **2020**, *112*, 110889. [CrossRef] [PubMed]
82. Imran, M.; Jan, H.; Faisal, S.; Shah, S.A.; Shah, S.; Khan, M.N.; Akbar, M.T.; Rizwan, M.; Jan, F.; Syed, S. In vitro Examination of Anti-parasitic, Anti-Alzheimer, Insecticidal and Cytotoxic Potential of Ajuga Bracteosa Wallich Leaves Extracts. *Saudi J. Biol. Sci.* **2021**, *28*, 3031–3036. [CrossRef]
83. Faisal, S.; Faisal, S.; Jan, H.; Shah, S.A.; Shah, S.; Khan, A.; Akbar, M.T.; Rizwan, M.; Jan, F.; Akhtar, N.; et al. (2021 Green Synthesis of Zinc Oxide (ZnO) Nanoparticles Using Aqueous2 Fruit Extracts of Myristica fragrans: Their Characterizations and Biological and Environmental Applications. *ACS Omega* **2021**, *14*, 9709–9722. [CrossRef] [PubMed]
84. Kainat; Khan, M.A.; Ali, F.; Faisal, S.; Rizwan, M.; Hussain, Z.; Zaman, N.; Afsheen, Z.; Uddin, M.N.; Bibi, N. Exploring the therapeutic potential of Hibiscus rosa sinensis synthesized cobalt oxide (Co3O4-NPs) and magnesium oxide nanoparticles (MgO-NPs). *Saudi J. Biol. Sci.* **2021**. [CrossRef]
85. Faisal, S.; Shah, S.A.; Shah, S.; Akbar, M.T.; Jan, F.; Haq, I.; Baber, M.E.; Aman, K.; Zahir, F.; Bibi, F.; et al. In Vitro Biomedical and Photo-Catalytic Application of Bio-Inspired *Zingiber officinale* Mediated Silver Nanoparticles. *J. Biomed. Nanotechnol.* **2020**, *16*, 492–504. [CrossRef]

MDPI
St. Alban-Anlage 66
4052 Basel
Switzerland
www.mdpi.com

Catalysts Editorial Office
E-mail: catalysts@mdpi.com
www.mdpi.com/journal/catalysts

www.ingramcontent.com/pod-product-compliance
Lightning Source LLC
LaVergne TN
LVHW070450100526
838202LV00014B/1700